Methods in Molecular Biology

Series Editor
John M. Walker
School of Life and Medical Sciences
University of Hertfordshire
Hatfield, Hertfordshire, AL10 9AB, UK

For further volumes:
http://www.springer.com/series/7651

Metabolic Network Reconstruction and Modeling

Methods and Protocols

Edited by

Marco Fondi

Department of Biology, University of Florence, Sesto Fiorentino, Florence, Italy

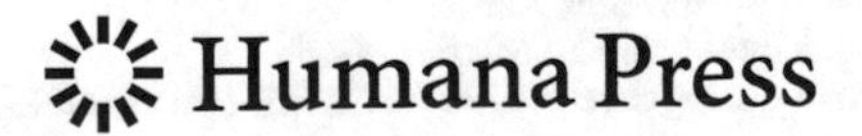

Editor
Marco Fondi
Department of Biology
University of Florence
Sesto Fiorentino, Florence, Italy

ISSN 1064-3745 ISSN 1940-6029 (electronic)
Methods in Molecular Biology
ISBN 978-1-4939-8511-1 ISBN 978-1-4939-7528-0 (eBook)
https://doi.org/10.1007/978-1-4939-7528-0

Printed on acid-free paper

This Humana Press imprint is published by Springer Nature
The registered company is Springer Science+Business Media, LLC
The registered company address is: 233 Spring Street, New York, NY 10013, U.S.A.

Preface

Thanks to the large diffusion of bacterial genome analysis, determining the genome sequence of (almost any) organisms has become a routine task. One of the most important drawbacks associated with the booming of genomics resides in the possibility to quickly derive a comprehensive metabolic reconstruction of a cell and to use computational simulations to predict its metabolic landscape.

Metabolic modeling confers to microbiologists the opportunity to further expand the knowledge offered by genome sequence alone. Indeed, despite the analysis of genome sequence *per se* that can provide interesting and fundamental hints on microbial life (e.g., genome structure, presence/absence patterns, and phylogenetic features), many questions may remain unresolved. This may limit the study of microbial genomics to the assembly and the interpretation of gene lists. To provide a system-level picture of cellular life and gain some insights on, for example, the effect of gene deletions and/or nutritional fluctuations, more sophisticated analytic tools have to be exploited. Constraint-based metabolic modeling (CBMM) represents one of these methodologies, combining good predictive abilities with a relatively simple conceptual and practical framework. This latter point probably best explains the success and the recent spreading of this approach as, ultimately, CBMM allows both the generation of testable hypotheses and the reduction of the amount of wet-lab experiments to be performed, saving time, efforts, and costs.

This book is intended to provide the most recent methodologies about the study of cellular metabolism using *in silico* approaches. The volume is ideally divided into three distinct parts. In the first part tools and methodologies for metabolic reconstructions and basic CBMM are presented (Chapters 1–10). The second part of the book (Chapters 11, 12, and 13) contains protocols for the generation of experimental data that can guide metabolic reconstruction and modeling, namely transcriptomics, proteomics, and mutants generation. The final part of the book (Chapters 14–18) covers more advanced methodologies for quantitative modeling of cellular metabolism, including dynamic Flux Balance Analysis, host-pathogen metabolic interactions, and multiobjective optimization. In each of these parts the most up-to-date protocols and procedures for metabolic reconstruction and CBMM will be provided. The aim of the present book is then to serve as a "field guide" both for qualified investigators on cellular metabolism (including computational biologists, microbiologists, physiologists, biochemists) who want to update their technical knowledge and for less-experienced researchers who want to start working with CBMM.

Sesto Fiorentino, Florence, Italy *Marco Fondi*

Contents

Contributors

BENJAMIN H. ALLEN • *Biosciences Division, Oak Ridge National Laboratory, Oak Ridge, TN, USA*
CLAUDIO ANGIONE • *Department of Computer Science and Information Systems, Teesside University, Middlesbrough, Tees Valley, UK*
SHANKAR BALACHANDRAN • *Department of Computer Science and Engineering, Indian Institute of Technology Madras, Chennai, India; Intel Research Labs, Bangalore, India*
PAUL I. BARTON • *Process Systems Engineering Laboratory, Cambridge, MA, USA*
EMANUELE BOSI • *Department of Biology, University of Florence, Sesto Fiorentino, Italy*
XIULAI CHEN • *State Key Laboratory of Food Science and Technology, Jiangnan University, Wuxi, Jiangsu, China*
LEONID CHINDELEVITCH • *School of Computing Sciences, Simon Fraser University, Burnaby, BC, Canada*
MAX CONWAY • *Computer Laboratory, University of Cambridge, Cambridge, UK*
OSCAR DIAS • *Centre of Biological Engineering, University of Minho, Braga, Portugal*
GEORGE C. DICENZO • *Department of Biology, McMaster University, Hamilton, ON, Canada*
KEITH DUFAULT-THOMPSON • *Department of Cell and Molecular Biology, College of the Environment and Life Sciences, University of Rhode Island, Kingston, RI, USA*
JANAKA N. EDIRISINGHE • *Computation Institute, University of Chicago, Chicago, IL, USA; Mathematics and Computer Science Division, Argonne National Laboratory, Argonne, IL, USA*
JOSÉ P. FARIA • *Computation Institute, University of Chicago, Chicago, IL, USA; Mathematics and Computer Science Division, Argonne National Laboratory, Argonne, IL, USA*
EUGÉNIO CAMPOS FERREIRA • *Centre of Biological Engineering, University of Minho, Braga, Portugal*
TURLOUGH M. FINAN • *Department of Biology, McMaster University, Hamilton, ON, Canada*
MARCO FONDI • *Department of Biology, University of Florence, Sesto Fiorentino, Florence, Italy*
GIUSEPPE GALLO • *Laboratory of Molecular Microbiology and Biotechnology, STEBICEF Department, University of Palermo, Palermo, Italy*
MATTHIAS P. GERSTL • *Austrian Centre of Industrial Biotechnology, Vienna, Austria; Department of Biotechnology, University of Natural Resources and Life Sciences, Vienna, Austria*
HUGO GIESTEIRA • *SilicoLife Lda, Braga, Portugal*
JOSE ALBERTO GOMEZ • *Process Systems Engineering Laboratory, Cambridge, MA, USA*
MICHAEL HANSCHO • *Austrian Centre of Industrial Biotechnology, Vienna, Austria; Department of Biotechnology, University of Natural Resources and Life Sciences, Vienna, Austria*
NOMI L. HARRIS • *Environmental Genomics and Systems Biology Division, E. O. Lawrence Berkeley National Laboratory, Berkeley, CA, USA*

CHRISTOPHER S. HENRY • *Computation Institute, University of Chicago, Chicago, IL, USA; Mathematics and Computer Science Division, Argonne National Laboratory, Argonne, IL, USA*

NEEMA JAMSHIDI • *Institute of Engineering in Medicine, University of California-San Diego, La Jolla, CA, USA; Department of Radiological Sciences, University of California, Los Angeles, CA, USA*

PAUL A. JENSEN • *Department of Bioengineering and Carl R. Woese Institute for Genomic Biology, University of Illinois at Urbana-Champaign, Urbana, IL, USA*

CHRISTIAN JUNGREUTHMAYER • *TGM - Technologisches Gewerbemuseum, HTBLuVA Wien XX, Vienna, Austria; Austrian Centre of Industrial Biotechnology, Vienna, Austria*

CHRISTOPHER LE • *School of Computing Sciences, Simon Fraser University, Burnaby, BC, Canada*

PIETRO LIÓ • *Computer Laboratory, University of Cambridge, Cambridge, UK*

LIMING LIU • *State Key Laboratory of Food Science and Technology, Jiangnan University, Wuxi, Jiangsu, China*

PAULO MAIA • *SilicoLife Lda, Braga, Portugal*

ALESSIO MENGONI • *Department of Biology, University of Florence, Sesto Fiorentino, Florence, Italy*

OMKAR MOHITE • *Department of Biotechnology, Bhupat and Jyoti Mehta School of Biosciences, Indian Institute of Technology Madras, Chennai, India*

JONATHAN MONK • *Department of Bioengineering, University of California, La Jolla, CA, USA*

DAVID A. PEÑA NAVARRO • *Austrian Centre of Industrial Biotechnology, Vienna, Austria; Department of Biotechnology, University of Natural Resources and Life Sciences, Vienna, Austria*

MARIA PIRES PACHECO • *Life Sciences Research Unit, University of Luxembourg, Belvaux, Luxembourg*

CLELIA PEANO • *Institute of Genetic and Biomedical Research, UoS Milan, National Research Council, Humanitas Clinical and Research Center, Milan, Italy*

EVA PINATEL • *Institute for Biomedical Technologies (ITB), National Research Council (CNR), Segrate, Milan, Italy*

ADITYA PRATAPA • *Department of Biotechnology, Bhupat and Jyoti Mehta School of Biosciences, Indian Institute of Technology Madras, Chennai, India; Department of Computer Science, Virginia Tech, Blacksburg, VA, USA*

ANU RAGHUNATHAN • *Chemical Engineering Division, National Chemical Laboratory, Pune, India*

KARTHIK RAMAN • *Department of Biotechnology, Bhupat and Jyoti Mehta School of Biosciences, Indian Institute of Technology Madras, Chennai, India; Robert Bosch Centre for Data Science and Artificial Intelligence (RBC-DSAI), Indian Institute of Technology Madras, Chennai, India; Initiative for Biological Systems Engineering (IBSE), Indian Institute of Technology Madras, Chennai, India*

ISABEL ROCHA • *Centre of Biological Engineering, University of Minho, Braga, Portugal; SilicoLife Lda, Braga, Portugal*

MIGUEL ROCHA • *Centre of Biological Engineering, University of Minho, Braga, Portugal; SilicoLife Lda, Braga, Portugal*

DAVID E. RUCKERBAUER • *Austrian Centre of Industrial Biotechnology, Vienna, Austria; Department of Biotechnology, University of Natural Resources and Life Sciences, Vienna, Austria*

THOMAS SAUTER • *Life Sciences Research Unit, University of Luxembourg, Belvaux, Luxembourg*
ANDREA SCALONI • *Proteomic and Mass Spectrometry Laboratory, ISPAAM, National Research Council, Naples, Italy*
JON LUND STEFFENSEN • *Department of Cell and Molecular Biology, College of the Environment and Life Sciences, University of Rhode Island, Kingston, RI, USA*
TIZIANO VIGNOLINI • *LENS, European Laboratory for Non-linear Spectroscopy, University of Florence, Sesto Fiorentino, Florence, Italy*
SUPREETA VIJAYAKUMAR • *Department of Computer Science and Information Systems, Teesside University, Middlesbrough, Tees Valley, UK*
PAULO VILAÇA • *SilicoLife Lda, Braga, Portugal*
NAN XU • *State Key Laboratory of Food Science and Technology, Jiangnan University, Wuxi, Jiangsu, China*
CHAO YE • *State Key Laboratory of Food Science and Technology, Jiangnan University, Wuxi, Jiangsu, China*
JÜRGEN ZANGHELLINI • *Austrian Centre of Industrial Biotechnology, Vienna, Austria; Department of Biotechnology, University of Natural Resources and Life Sciences, Vienna, Austria*
YING ZHANG • *Department of Cell and Molecular Biology, College of the Environment and Life Sciences, University of Rhode Island, Kingston, RI, USA*

Chapter 1

Reconstructing High-Quality Large-Scale Metabolic Models with *merlin*

Oscar Dias, Miguel Rocha, Eugénio Campos Ferreira, and Isabel Rocha

Abstract

Here, the basic principles of reconstructing genome-scale metabolic models with *merlin* are described. This tool covers the basic stages of this process, providing several tools that allow assembling models, using the sequenced genome as a starting point.

merlin has two main modules, separating the process of annotating (enzymes, transporters, and compartments) on the genome from the process of model assembly, though information from the former is integrated in the latter after curation. Moreover, *merlin* provides several tools to curate the model, including tools for generating reactions' gene rules and placeholder entities for biomass precursors, such as proteins (e-protein) or nucleotides (e-DNA and e-RNA) among others.

This tutorial covers each feature of *merlin* in detail, including the assessment of experimental data for the validation of the model.

Key words *merlin*, Genome-scale metabolic models, Genome functional annotation, Transport proteins annotation

1 Introduction

Before the dawn of the genomic era, strain evolution was underpinned by random mutagenesis followed by screening and selection of mutants or, later, by manipulation of genes directly associated with the product of interest. The latter strategy was supported by components biology and was often unsuccessful, while the former, though presenting interesting results, could not unveil the mechanisms justifying the outcomes [1].

Instead of exclusively studying each organism's components individually, systems biology studies also the interactions between them, to comprehend and predict the phenotypical behavior on conditions other than ones already characterized [2, 3].

Electronic supplementary material: The online version of this article (https://doi.org/10.1007/978-1-4939-7528-0_1) contains supplementary material, which is available to authorized users.

Marco Fondi (ed.), *Metabolic Network Reconstruction and Modeling: Methods and Protocols*, Methods in Molecular Biology, vol. 1716, https://doi.org/10.1007/978-1-4939-7528-0_1, © Springer Science+Business Media, LLC 2018

According to some definitions, a *gismo* is an advanced technological device which performs a particular task, usually in a new and efficient way [4, 5]. Currently, metabolic systems biology is becoming a standard field of study with its own *g*enome-w*i*de *s*cale *m*etabolic m*o*dels (GiSMos) on the forefront.

The reconstruction of a GiSMo involves four main stages [3] and has been detailed in a protocol with over 100 steps [6]. Those stages are genome annotation, assembling the genome-wide metabolic network, conversion of the network to a stoichiometric model and metabolic model validation.

These models are reconstructed based on the parts of the genome that encode metabolic functions. Hence, a robust and reliable genome functional annotation is of paramount importance. The genome structural annotation process is usually coupled to a draft functional annotation of the coding sequences (CDSs), which may be incorrect or at least incomplete. Thus, several frameworks (e.g., *merlin* [7], Model Seed [8]) provide tools to perform the genome re-annotation and the curation of the outcome.

When integrated with metabolic data, genome annotations are converted into *gen*ome-wide scale metabolic *net*works (GenNets). These are sets of reactions interconnected by metabolites, produced and consumed through reactions promoted by enzymes encoded in the genome. The process of creating a GenNet encompasses data collection from a variety of databases providing such metabolic information, namely the reactions and the enzymes that catalyze them. Curation of the network, such as gap finding and balance validation, should be performed at this stage.

Nevertheless, curation must continue in the third stage as the conversion of the GenNet to a GiSMo involves adding a biomass objective reaction and non-growth ATP requirements to the reaction set. This allows building a stoichiometric matrix that, together with the assumption that the rates of consumption are equal to the rates of production for all metabolites, originates the system of linear equations represented by $S.\ v = 0$, in which v is the flux vector and S is the stoichiometric matrix where the columns represent reactions and the rows the metabolites.

The GiSMo should then be saved in a standard format, i.e., the Systems Biology Markup Language (SBML) [9], to allow importing the models into tools specially developed for operating GiSMos, such as OptFlux [10] (More information on using OptFlux on Chapter XX) or COBRA [11]. Notwithstanding the use of SBML, MIRIAM [12] annotations should also be used for annotating GiSMos to enable, for instance, the comparison of distinct reconstructions of the same organism enhancing the overall comprehension of its metabolism.

merlin is a tool developed for accompanying the reconstruction process throughout all four stages. It is currently composed of two main modules, where the first is developed specifically to help on the genome annotation stage with dedicated tools and graphical user interfaces (GUIs), while the second module is oriented to the remaining stages, being that the last stage requires also a simulation platform. Moreover, the latter module offers a myriad of operations for curating the network and converting it to a GiSMo.

2 Materials

In this section, a brief description of the resources required for reconstructing GiSMos with *merlin* is provided.

2.1 *merlin*

merlin is an open source software tool fully implemented in Java™, distributed under the GNU General Public License, available for download in a single multi-platform (tested in Linux, Mac OSX, and Microsoft Windows) version at www.merlin-sysbio.org. The installation process is very straightforward as it only involves decompressing the downloaded .ZIP file into a folder. Whereas *Microsoft Windows* (*MS Windows*) users should use the *merlin.bat* file to run *merlin*, Linux/Mac OSX users must start the *run.sh* file.

In its core, *merlin* is supported by the AIBench [13] software development framework. The latter follows the model-view-controller (MVC) software architecture pattern, allowing the combination of new and existing software components.

It has a bimodular architecture with several subcomponents, as shown in Fig. 1. The three subcomponents of the genome annotation module allow decoding most metabolic capabilities available in the genome. Two of these components identify and select metabolic proteins' functions, whereas the last component predicts the location wherein these proteins operate. The second module is

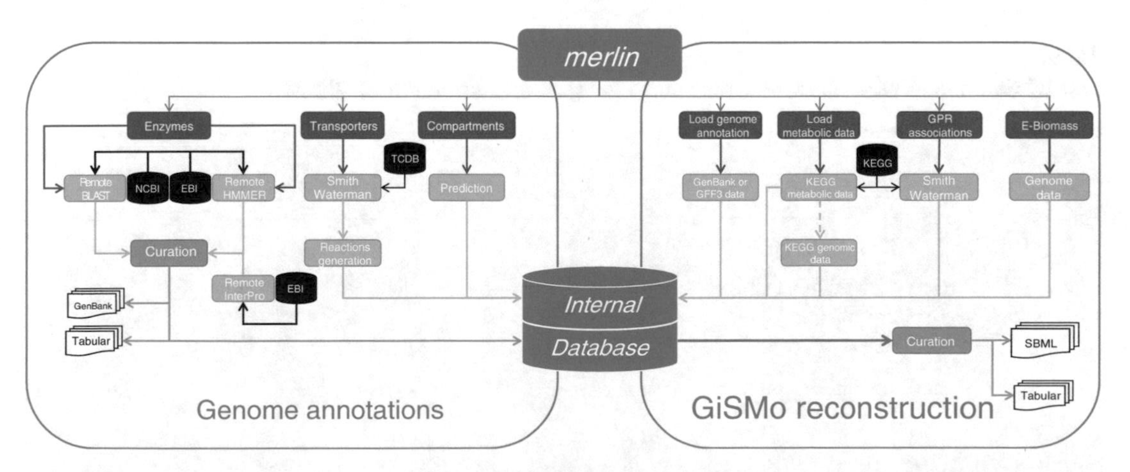

Fig. 1 *merlin*'s architecture. *merlin* is a bimodular framework with several submodules. The first module (genome annotations) allows decoding the metabolic capabilities of the genome while the latter module (GiSMo reconstruction) provides several operations to assemble the GiSMo

where the GiSMo is assembled. Its subcomponents allow loading existing annotations, loading metabolic data, determining *g*ene-*p*rotein-*r*eaction (GPR) rules, and adding an e-biomass reaction to the core of the GiSMo. All such operations will be detailed in the following sections.

merlin keeps all information in internal databases which are shared between both modules to ease the process of integrating genome annotations with the metabolic data. The information kept in the databases is accessed through *merlin*'s interface (*see* **Note 1** for a detailed description of the interface).

2.2 Online Resources

The reconstruction of metabolic models involves analyzing not only the genome of the organism under study, but also collecting information on proteins' functions, reactions, biomass composition, and other physiological data. Hence, besides organism-specific publications (books and journals' manuscripts), several databases must be accessed throughout the whole process (Table 1).

The Universal Protein Resource (UniProt) [14] is the combination of three databases (UniProt Knowledgebase, UniProt Reference Clusters, and UniProt Archive) and its curated version (UniProt-SwissProt) [15] is one of the best sources of curated protein annotations available to date. The National Center for Biotechnology Information (NCBI) maintains a repository of several biological databases, providing tools for the analysis and retrieval of this information [16]. BRENDA contains information curated by experts, thus being one of the most reliable databases for enzymatic information [17]. The Kyoto Encyclopedia of Genes and Genomes (KEGG) is a knowledge-base that maintains an extensive collection of information on genes, metabolites, reactions, and

Table 1
List of databases with relevant information for the reconstruction of GiSMOs

Name	Contents	Programmatic API	URL	Reference
UniProt	Proteins functions	Available	www.uniprot.org	[14]
NCBI	Genes, protein functions Taxonomic data Genome sequences	Available	https://www.ncbi.nlm.nih.gov/	[16]
KEGG	Genes, protein functions Metabolic data	Available	http://www.kegg.jp/	[18]
BRENDA	Proteins functions	Available	http://brenda-enzymes.org/	[17]
MetaCyc	Genes, protein functions Metabolic data	Not available	https://metacyc.org/	[48]

pathways [18]. All metabolic information used by *merlin* to assemble GiSMos is retrieved from KEGG. Moreover, when available, *merlin* can also retrieve KEGG's genome annotation.

3 Methods

A detailed description of all the steps required for reconstructing GiSMos with *merlin* is provided below. Though most databases and online tools are accessed automatically by *merlin*, some stages may require uploading results to *merlin*'s internal database.

3.1 Reconstruction Project

Starting a project in *merlin* requires the organisms' genome in the FASTA format [19] and its NCBI's taxonomic identifier. Whereas the former is optional, the latter is mandatory.

3.1.1 Download of Genome Sequence

Regarding the genome sequence, users are encouraged to use NCBI's Assembly database (www.ncbi.nlm.nih.gov/assembly) [21] to download the genome sequence, since *merlin* has parsers to process the headers in these files. As shown in Fig. S1 of the supplemental material, the NCBI assembly webpage provides links (green arrows) for downloading the genome in the GenBank and the RefSeq formats (*see* **Note 2** for a detailed explanation of the differences between these formats). These links redirect the user to a File Transfer Protocol (FTP) webpage where the user is presented with a list of several files. All the files are compressed in the gzip (GNU ZIP) format and must be decompressed to access their information. Whereas there are several tools available for *MS Windows* users to decompress these files, such as 7zip and WinRAR, Linux/Mac OSX users can access the compressed files contents using simple commands in the terminal. Files ending with *_protein.faa.gz* (protein products annotated on the genome assembly) and *_cds_from_genomic.fna.gz* (nucleotide sequences corresponding to all coding sequences annotated on the assembly) can be uploaded to *merlin*, after decompression. Other file types, like *_genomic.gbff.gz* (GenBank flat file format of the genomic sequences in the assembly), can also be used by *merlin* (after decompression) in other operations as shown next.

3.1.2 Taxonomic Identifier

The International Nucleotide Sequence Database Collaboration uses NCBI's Taxonomy database (www.ncbi.nlm.nih.gov/taxonomy) [20] as repository for standard nomenclature and classification. Hence, *merlin* requires the identifier provided by this database to univocally identify the organism under study throughout the reconstruction process. The link inside the red ellipse in Fig. S1 of the supplemental material contains a direct link to the taxonomic identifier (blue circle) of the case study organism. Nevertheless, for cases in which the genome sequence is not retrieved from NCBI, this database can be directly accessed to retrieve the taxonomic identifier.

3.1.3 Create Project

After retrieving the taxonomic identifier, and optionally the genome files, a new project can be created in *merlin*. When *merlin*'s interface is opened, users should choose "*project>create*" to start a new project. The minimum requirements to create a new project are setting the project name, the taxonomic identifier, selecting the type (H2 or MySQL, discussed below) and naming the database in which the data will be kept. Setting the FASTA files with genome sequences is optional at this phase.

3.2 Database

merlin supports two different relational database management system (RDBMS) for storing its data, H2 or MySQL. For users assembling GiSMos in personal computers, the H2 RDBMS is recommended, as *merlin* is ready for use with this system. For using MySQL, an external installation is required, which depends on the platform being used to host the database server.

3.2.1 MySQL

MySQL (www.mysql.com) is an open source RDBMS, written in C and C++, which supports the use of Structured Query Language (SQL) to access the information stored in its databases. It runs in practically all platforms, including *MS Windows, Linux and Mac OSX*. MySQL is a free, fast, stable multi-user, and multi-threaded database server [22], which makes it the most popular RDBMS in the world [23]. Hence, *merlin* embedded MySQL in its first versions, providing users with a reliable server for keeping data. Though this RDBMS is not currently included in its releases, *merlin* still supports connections to this server.

3.2.2 H2

H2 Database Engine (www.h2database.com) is an open-source RDBMS written purely in Java, which provides an alternative to traditional RDBMS by offering extremely fast performance and very small disk footprint. Like MySQL, H2 fully supports SQL and Java database connector (JDBC) application programming interface (API). In contrast, unlike MySQL all data are encrypted and each database uses just a couple of files for storage. Other Java-based databases, like Derby (http://db.apache.org/derby) and HSQLDB (http://hsqldb.org), do not support full text search or the Open Database Connectivity (ODBC) driver. Finally, the H2 database engine can be embedded in any Java applications or run in the client-server mode. As *merlin* is a cross-platform tool, which can be stored in an external hard drive and executed in computers with different operating systems, the features offered by this database integrate perfectly with *merlin*'s philosophy.

3.2.3 New Database

merlin is deployed ready to use with an internal database with the default name of "*my_dabatase*," which can be used to start a project (i.e., the reconstruction of a GiSMO). Nevertheless, users can create new databases, with different names, to create other projects for other organisms or just because they want to change the default name, by accessing the "*database>new*" menu.

3.2.4 Clean Database

merlin allows users to clean a project's database without affecting other projects. Users can clean the entire database ("*all information*"), or just specific parts of it, namely the "*model*", the "*enzymes annotation*", the "*transport proteins*", the "*transport annotations*" (TRIAGE), the "*compartments annotation*" or the "*InterPro annotation.*" While the former operation is inherently understandable by name and cleans the whole model, the latter operations clean specific annotations. These operations are available at "*database>clean*" and should be handled with care as their execution is permanent.

3.2.5 Delete Project

merlin also allows deleting a project, which involves deleting the entire database from the system. This option is available at "*project>delete*" and (as the previous operation) should be handled with care as the operation is irreversible.

3.3 Genome Annotation

The annotation module is where *merlin* processes genome-wide annotation data. This module is composed of three main procedures, which are discussed below.

3.3.1 Enzymes Annotation

Though not a mandatory step when reconstructing models, it is highly recommended that users perform their own annotation to have maximum confidence on the enzymatic potential of the genome. This unique feature is one of the finest resources offered by *merlin*, easing the process of assigning functions to genes that potentially encode enzymes. The enzymes annotation will be integrated in the model with the following assumption: genes encoding enzymatic proteins should have Enzyme Commission (EC) number assignments, as EC numbers can be mapped to reactions to build the metabolic network.

Similarity Searches

The enzymes functional annotation process in *merlin* relies on similarity searches to remote databases to find homologous protein sequences. The rationale for this approach is that gene or protein sequences with excess similarity (i.e., more similarity than expected by chance) should have arisen from a common ancestral lineage [24].

In *merlin*, users can use the Basic Local Alignment Search Tool (BLAST) [25] web-services from European Bioinformatics Institute (EBI) or the National Center for Biotechnology Information (NCBI), or HMMER remote similarity searches [26] to perform genome-wide sequence alignments. These tools allow configuring several parameters, well known by researchers that commonly use them, including the expected value threshold, the maximum number of hits and the remote database. The first and the second parameters are usually used as thresholds for the similarity search (*see* **Note 3a** for more information on these parameters).

Scoring Algorithm

merlin has a specific algorithm that calculates a score, between 0 and 1, for every possible annotation of each gene and automatically selects the annotation with the highest one. The score assigned to each annotation comprises two factors, being those the frequency and the taxonomy of the homologous protein annotations. As shown in Eq. 1, the first score is obtained by determining the number of times each EC number is found within the homologous gene records annotation (frequency), whereas the latter is associated with the taxonomy of the organisms to which these records belong. More information about this scoring algorithm is provided in [7].

$$\text{Score} = \alpha \times \text{Score}_{\text{f}} + (1 - \alpha) \times \text{Score}_{\text{t}}. \quad (1)$$

The score shown above reflects the degree of confidence of the annotation assigned by *merlin* to a given gene.

Nevertheless, several factors may influence the scoring results, like the quality of the genome assembly (and gene calling), the database selected for the remote similarity search and the taxonomy of the organism. The first factor regards the gene sequence sent to the alignment, as cases in which the sequence was incorrectly assembled or the gene calling is incorrect will affect the number of homologous genes found. In extreme cases, in which contigs have less than 50 nucleotides and these correspond to proteins' conserved domains, the frequency score may be positively affected because of having similarities with many sequences and consequently the calculated scores being very high. It should be noticed that these cases might happen when *merlin* is allowed to adjust the expected value for the alignment of smaller sequences. Regarding the remote database selected for alignment, databases such as "*swissprot*" are very biased, as annotations are propagated through clusters of homologous genes [27]. Thus, the frequency score will be remarkably high. Lastly, the number of bacterial genomes available in databases is much higher than the number of Eukaryotic ones. Moreover, there are quite a few organisms with numerous strains of the same species, which will also bias the results in terms of both frequency and taxonomy. Hence, when gene annotations, such as hypothetical or uncharacterized proteins, are propagated for different strains of the same organism, the similarity search results will also be biased and *merlin* will use these to calculate annotations, which do not provide useful information.

Therefore, *merlin* users should adopt the annotation strategy that best suits their project. For instance, organisms with several sequenced strains should be aligned against curated databases, to increase the confidence of the frequency scores provided by *merlin*. Genomes with many scaffolds and contigs should also be aligned to curated databases initially. Then, for CDSs without hits, a second alignment should be performed against a broader database (like

EBI's TrEMBL or NCBI's *nonredundant database*) to identify CDSs with similarities to known sequences and therefore find other potential enzymes.

Users can automatically accept all annotations with scores above a given threshold and reject the remaining, after an empirical analysis of the best alpha value. Instead, a (complete or partial) curation of the annotations may be performed. Regarding the latter option, selecting the best α value, and an upper and a lower threshold for the EC numbers' scores, allows decreasing the number of records to be manually curated. These parameters can be set after a rational analysis of the results. Usually, this analysis involves performing the manual curation of the annotations of a group of randomly selected genes, following a curation workflow, such as the one described in Subheading 3.3.1.4. The curated annotations are then assessed to the annotations automatically provided by *merlin*, for different α values. This strategy allows selecting the α value that provides the most correct annotations, as well as setting the upper and lower thresholds. All records with scores above the upper threshold should be automatically accepted. Likewise, all entries with scores below the lower threshold should be rejected. Annotations with scores in-between the thresholds should be revised according to the same curation workflow.

Curation Panel

The annotations proposed by *merlin* are based in the similarity searches and are presented in the "*enzymes*" panel of the "*annotation*" module (Fig. 2), which can be accessed from *merlin*'s

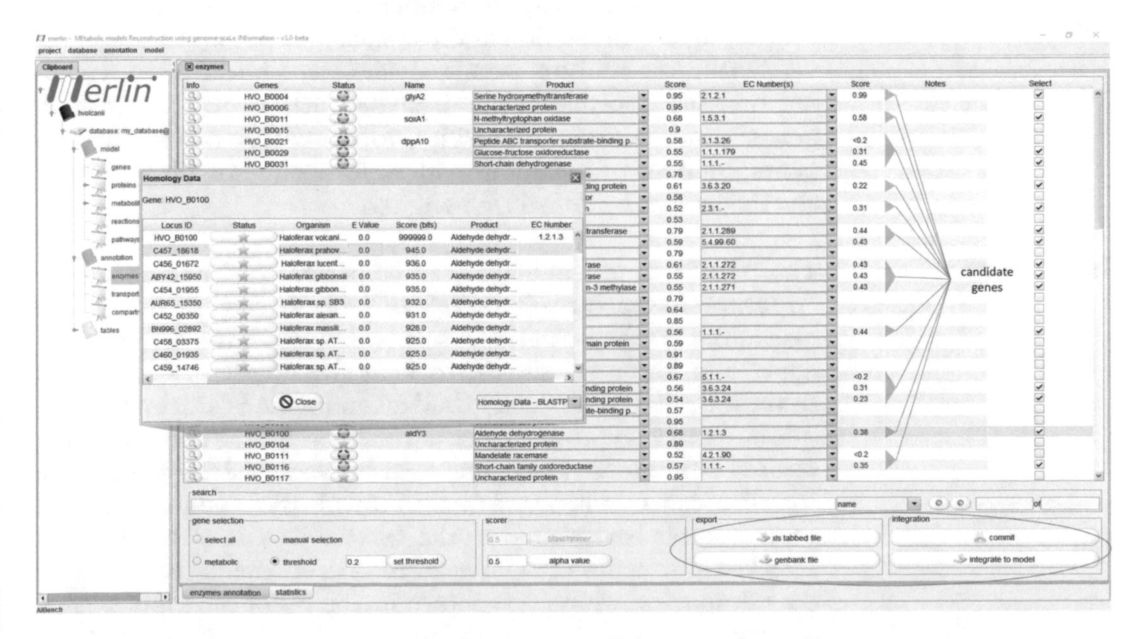

Fig. 2 *merlin*'s enzymes annotation panel. All genes with EC numbers are considered candidate genes. *merlin* allows exporting the annotation or integrating it in the model

clipboard. This panel contains a table with 10 columns and a line per gene. The first column contains buttons with "*magnifying glasses*" which allow accessing the similarity search information, namely the identifiers of the homologous genes, their UniProt status (if available), the similarity search tools' scores (expected value and bit scores), and the functions of the homologous genes. The second column displays the gene identifier (locus tag if available). The third column is very informative, as its buttons provide, when available, the annotation status of each studied gene in UniProt (reviewed entries in UniProt show a gold star and unreviewed a silver star), and the agreement of such annotation with *merlin*'s annotation (a green background reveals concordance and a red one shows divergence; a light green background establishes that *merlin* has identified all annotations available in UniProt plus additional EC numbers not proposed by UniProt, while an orange background represents the opposite, that is, UniProt's annotation provides EC numbers not available in *merlin*'s current assignments). The next column is the common gene name. The fifth and seventh columns (product name and EC number, respectively) are followed by column scores (sixth and eighth columns). These scores are calculated according to the scoring algorithm described below. The last two are: a column in which users can keep notes regarding each gene's annotation, and a column with a check box to select genes to be integrated in the model or exported. This panel also allows searching for text on the table, selecting a specific group of genes, configuring *merlin*'s scorer parameters, exporting the model, and finally saving the annotation data to the database and integrating the annotation in the model.

Finally, the statistics panel provides some information regarding the data available in the database, such as: total number of genes, genes without similarities, total number of homologous genes, number of organisms with at least one homologous gene and its division by domain, among others.

InterPro Annotation

The InterProScan annotation can be performed by accessing the "*annotation>enzymes>InterProScan*" menu, though this is not mandatory. As this operation is somewhat slow, it requires entering upper and lower thresholds, to limit the number of annotated records. Unlike BLAST and HMMER annotations, this operation's sole purpose is to provide more information for the curation of the records not automatically annotated by *merlin*. After the remote InterProScan search, *merlin*'s entries with these annotations will show a purple background on the "*magnifying glass*" button. The InterPro information can be accessed on a submenu (red circle) of the information window, as shown in Fig. S2 of the supplemental material.

Curation Workflow

The first step to perform the curation of the annotation is designing a workflow for reviewing the annotations of genes that potentially encode enzymes, i.e., genes with EC number assignments (Fig. 2), the so-called candidate genes (CG). Usually, this involves selecting a well-studied closely related organism to compare annotations. For instance, yeast reconstructions should use *Saccharomyces cerevisiae* as a model organism. The selection of this organism is quite important, as its annotation will be a pseudo-template for the GiSMo.

A typical workflow is presented in Fig. 3. Initially, the user should verify the status of the CG in UniProt and the agreement with *merlin*'s annotation. In case of a gold star and a green background on *merlin*'s status column then the gene's annotation should be accepted and, in the notes' field, the user can add a reference to the level of confidence of this annotation. As this annotation was confirmed with a reviewed entry, the level of confidence can be set to the maximum, i.e., confidence level (CL)-A. The level of confidence will assist in the process of curating the model. If the star is silver, and the background still green, the annotation is also accepted though with a classification of CL-B. When *merlin*'s assignment does not match with UniProt's, the annotation should be confirmed to determine the reason for the discrepancy. If the UniProt's entry is reviewed, *merlin*'s annotation is changed and classified with CL-B. Otherwise, or in the case that the CG does not have a UniProt annotation (no star on the status column), the similarity search results of the CG are analyzed, by seeking a record of the model organism. When such entry exists,

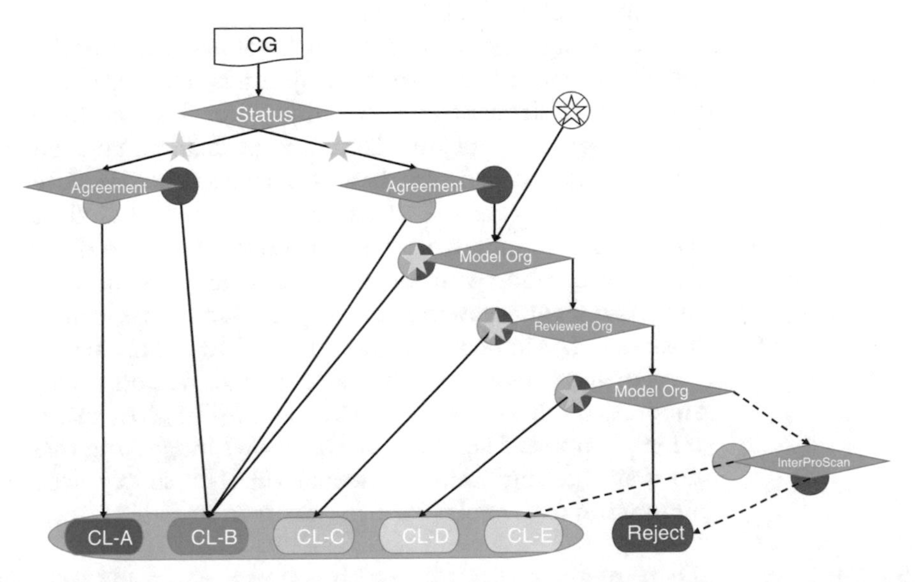

Fig. 3 Typical annotation workflow. CG—candidate gene; CL—confidence level classification. The CG's annotation provided by *merlin* is compared to its own annotation in UniProt, to a reference organism's annotation and to other reviewed entries in the homolgy search results. When one of the conditions is met, the CG's annotation is accepted and the classification can be added to the CG in *merlin*

and is reviewed in UniProt, the gene is annotated and classified with CL-B. In the case that the model organism's entry does not exist or exists, but is not reviewed, the CG's annotation is compared with the annotation of the homologous gene with lower expected value and higher score that has its annotation reviewed. In such a case, the classification should be decreased to CL-C. If the similarity search results do not encompass reviewed records but the model organism's annotation is available, this annotation should be considered with the classification of CL-D.

When available, the InterPro annotation results can be used as a final test to find evidences of enzymatic activity on CGs. If the InterPro annotation provides evidences that corroborate *merlin*'s annotation, it should be considered with the classification of CL-E. Alternatively, InterPro results can be used for other purposes, e.g., increasing the confidence level of annotations inferred from unreviewed records.

All other cases not encompassed in this workflow should be rejected, i.e., not considered enzyme-encoding genes, at this stage.

Moreover, all EC numbers verified using this workflow will be confirmed in BRENDA to determine if the function corresponds to the EC code, to prevent cases in which the EC number has been updated (*see* **Note 3b** for the examples of these cases).

Outputs

The enzymes annotation is the backbone of the metabolic model. Thus, a robust, reliable, and traceable genome annotation is mandatory. Annotations performed according to the set of steps described before fulfil these rules.

All changes performed in the "*enzymes*" panel of the "*annotations*" module are saved to the project file, unless the user clicks the "*commit*" to database button in this panel. This means that, before *committing*, none of the changes is permanent and new projects started with the same database will access raw data. After *committing*, new projects will have access to the curated annotation. Though not required, the curated annotations should be *committed* before integration with the model. Furthermore, the integration of the enzymes annotation must be preceded by the loading of metabolic data (*see* Subheading 3.4.1 below) in the model's database.

Finally, though usually the objective of performing genome annotations in *merlin* is reconstructing GiSMos, these can serve other purposes. Therefore, *merlin* allows integrating these annotations in existing GenBank format files or simply exporting the annotation in a tabular format, as shown in Fig. 2.

Previous Genome Annotations

There might be cases in which researchers already have access to good quality annotations performed within other contexts and might want to use them within *merlin*. For these cases, *merlin* provides operations to integrate annotations from GenBank or GFF3 files directly with previously loaded metabolic data. The only requirement is that these files have fields with *e*nzyme *c*ommission (EC) numbers associated with gene entries. These files are not required for creating a project and should be imported later.

3.3.2 Transporters Annotation

The transporters annotation is performed with the tool developed specifically for this purpose, TRIAGE [28]. This tool uses increasingly stringent conditions to predict transport protein encoding genes. The first assumption is that all transporter proteins are in membranes; hence, the initial step is filtering out all genes without transmembrane helices, thus identifying transporter candidate genes (TCGs). The second premise is that transporters should have similarities with proteins available in the Transporters Classification Database (TCDB); thus, TCGs are compared to the whole TCDB database. Lastly, when dealing with compartmentalized models, the first assumption is reinforced by excluding all TCGs not predicted to be in membranes by compartmentation software, such as PSortb 3.0 [29] or LocTree [30].

These proteins are annotated according to the frequency and taxonomy of the homologous proteins annotations in TCDB, in a similar manner to the enzymes annotation. In this case, the score assesses the TC family, metabolites, and transport mechanism of each TCG. A brief description of this process is presented below; for a more detailed explanation please refer to [28].

Transmembrane Helices

Transport proteins are usually located in membranes [31]. Thus, TRIAGE's first step is identifying these proteins. There are several bioinformatics tools such as TMHMM [32], Phobius [33], and several others, which predict the number of transmembrane helices from the amino acid sequence. TMHMM has been considered the best performing tool for this function [34], yet it does not provide an API for remotely accessing the server. Hence, users must perform the search on the website, or install the software locally, and then load the results (format: extensive, no graphics) into *merlin*. Phobius provides a programmatic API for performing remote transmembrane helices predictions, and thus it was possible to embed this tool in *merlin*. The output of either tools is the number of helices per gene, which can be used to filter genes for the next step.

Transporters Classification Database

This database is the most extensive and comprehensive resource of transport proteins available. TCDB implements a classification system analogous to the EC format, though including phylogenetic information that assigns TC numbers to proteins. These identifiers are formed by five components separated by a dot: #.*.#.#.#, in which # represent numbers and * a letter. TCDB records provide specific information regarding the transport activities of its entries, namely: TC number and generic description, accession number (UniProt), protein name, length, molecular weight, species (Organism), number of TM's, and location/topology/orientation.

Unfortunately, to date, TCDB does not provide in its records a specific field for reporting transported metabolites nor transport mechanisms. Thus, this information must be manually retrieved from the generic description or, often, the family description. Though containing valuable information, TCDB does not provide data in a readily accessible way. Hence, TRIAGE has an internal Transporters' Annotations Database (TAD), which keeps information, inferred from TCDB's annotations, useful for the reconstruction of GiSMos. Most information, such as UniProt's accession number, protein name, and species (organism), are retrieved from TCDB and stored as are. The remaining data (transport directions, transported metabolites, reversibility, reacting metabolites, and equation) must be extracted from TCDB (*see* **Note 4** for a detailed description of this process). TCDB contains (as of 17/04/2017) 15,021 records, 7383 of which are already curated and available in TRIAGE's TAD. The remaining entries should be curated as needed, following the transporters annotation workflow (*see* **Note 4** for a detailed description of this workflow).

After determining the number of helices in each protein sequence of the genome of the case study, TRIAGE compares each sequence with at least h (a number defined by the user, default = 1) helices to the whole set of proteins available in TCDB to find homologous proteins. This comparison is performed with the Smith-Waterman [35] algorithm, which guaranties optimality and high sensitivity. Overall, the alignment similarity threshold for considering homology between sequences is 10% which can be decreased in special cases [28]. The score for these alignments is calculated by the following equation:

$$\text{Similarity} = \frac{\text{score}_{\text{alignment}} - \text{score}_{\text{minimum}}}{\text{score}_{\text{maximum}} - \text{score}_{\text{minimum}}} \tag{2}$$

In Eq. 2 the minimum score is 0, the maximum score is the maximum between the maximum score of the TCDB protein sequence and the maximum score of the query sequence—and finally, the alignment score is the score of the alignment region between both the sequences. TCGs with at least one homologous protein in TCDB will be stored in TRIAGE's alignments database.

Afterward, transport reactions will be created to annotate the TCGs. This process is performed by retrieving TAD's annotations of each TCG's similarity hits. TAD's annotations are then used to calculate which metabolites and transport mechanisms should be assigned to each TCG. As stated before, TAD contains 7383 TCDB entries, yet TCDB is always adding new records. Hence, the similarity alignments may always find entries unavailable in TAD. The absent records should be annotated and included in TAD. Every user is invited to contribute to this effort, which already encompasses hundreds of person-hours, by following the transporters annotation workflow (*see* **Note 4** for a detailed description of this process).

After performing the similarity alignments, TRIAGE allows generating transport reactions associated to the genes with similarities to TCDB.

Membrane Locations

LocTree3 can be used to assign subcellular localizations both for eukaryotes and prokaryotes, whereas PSortb 3.0 can only be used for the latter. The integration of these predictions will be discussed in Subheading 3.3.3. Nevertheless, these locations will be used to filter and compartmentalize the transport reactions generated in the previous step.

Curation Panel

The annotation of transport proteins is initiated by running the transport proteins identification operation available at "*annotation>TRIAGE>transport proteins identification*." The transporters' annotations are presented in the "transporters" panel of the "annotation" module.

The panel shown in Fig. S3 of the supplemental material provides information on the transported metabolites and the type of transport associated with each gene. This panel allows tracing back the reason for associating a gene to the transport of a given metabolite and contains a table with seven columns and a line per gene. The first column contains buttons with "*magnifying glasses*," which when clicked open a smaller information window that provides information on the similarities, metabolites, and reactions associated with selected gene. The second column displays the gene identifier. The third column shows the number of transmembrane domains predicted to be associated with each gene. The fourth and fifth columns are only filled in after clicking the "*add TRIAGE data*" button (red ellipse in Fig. S3 of the supplemental material) and show the calculated TC family number and the number of metabolites associated with each gene. The sixth column shows the number of transport reactions generated after pressing the "*create transport reactions*" button (blue ellipse in Fig. S3 of the supplemental material). The last column allows selecting which genes and corresponding transport reactions will be integrated in the model. Lastly, the main table can be exported into an excel file and the transport reactions with gene associations integrated into the model (green ellipse).

The reactions will be created considering the parameters set in the panel. The metabolites associated with each gene will be classified with a score between 0 and 1. The α will weigh the frequency and the taxonomy of the homologous proteins, whereas the threshold will determine if the metabolite is selected or not. Compounds labeled as currency symport metabolites will not be classified in such reactions.

As shown in Fig. S3 of the supplemental material, the information window presents three types of information. The first table shows which TCDB entries are similar (and the similarity score) to the selected gene. The second table, which may be selected in the drop-down box (black circle), allows visualizing which metabolites are associated with each gene, their KEGG identifiers, their score, the direction and reversibility associated with each metabolite, and the transport types and corresponding scores. The information in the metabolites table is available after the integration with TRIAGE's database. Finally, the last table in these windows shows the generated transport reactions, and whether these originated from primary annotations or from ChEBI's ontologies.

3.3.3 Compartments

LocTree3 [30] and PSORTb 3.0 [29] are used to predict subcellular localizations of all proteins encoded in the genome. The former was selected because its previous version (LocTree2) behaved equal or better than all state-of-the-art location prediction tools [30]. The current version is an improvement of LocTree2 by adding homology-based inference. PSORTb 3.0 is the most widely employed localization prediction software for bacteria [36] and was already supported by *merlin*, thus it can still be used for predicting protein locations in bacteria and Archaea. Although none of these tools provides a web-service, both offer whole genome/proteome predictions, which can be loaded into *merlin*. If the case study organism is not available in the database, users may use email mode for submitting the genomes/proteomes and receive the results by emails. Examples of reports from these tools in formats supported by *merlin* are available at *merlin*'s homepage.

Compartmentalization is also related to the energetic requirements of the cells. The ATP synthase harnesses the energy of an unfavorable proton gradient, via allowing hydrogen cations to cross the membrane and using this drive for coupling inorganic phosphate to ADP. Transport reactions, associated with the oxidative phosphorylation and proton pumps, are often generated, based in the proteome similarities, by TRIAGE and the compartmentalization should be verified to determine the direction of the reaction.

Curation Panel

As shown in Fig. S4 of the supplemental material, this panel allows visualizing the compartment predictions. Though LocTree3 reports assign genes to a single location, PSORTb 3.0 often identifies several locations, with different scores, for each gene. Hence, *merlin* allows setting a percentage difference limit for considering alternative predictions. For instance, if the main compartment has a score of 0.4 and the secondary compartment a score of 0.35, the difference percentage between both compartments is 5%. If the difference percentage threshold for considering secondary compartments is set to 10%, the second compartment would be accepted. Similarly, as shown in Fig. S4 of the supplemental material, gene "*ADE01272.1*" the main location is *cytoplasmic* with a score of 0.75. The secondary compartment for this gene is *cytoplasmic membrane* with a score of 0.1. Hence, the acceptable difference between the main compartment prediction would have to be set to 66%, to accept the *cytoplasmic membrane*, which would be imprudent. Hence, the *cytoplasmic membrane* would not be considered when assigning gene products to compartments.

Clicking a given gene's "*magnifying glass*" provides information on the scores obtained with the compartments prediction tool.

3.4 Assembling the GiSMo

A curated genome annotation provides a good basis for reconstructing a GiSMo. However, metabolic information is also required for developing these models. The reconstruction of a GiSMo begins with the assembly of the genome-scale metabolic network (GenNet). This network represents the set of biochemical reactions encoded in the case study's genome. Though metabolic reactions are available in several databases, *merlin* retrieves them from KEGG as it provides a simple web-accessible API. Nevertheless, other metabolic data sources may also be included in upcoming versions of *merlin*.

3.4.1 Load Metabolic Information

Retrieving and loading metabolic data to *merlin*'s internal database is very simple, as it only requires accessing the "*model>load>load metabolic data*" menu. This operation retrieves all metabolic information from KEGG, which includes: compounds, glycans, drugs, reactions, enzymes, and pathways. Moreover, KEGG's genome annotation of the case study organism (if available) can also be retrieved, though *merlin* provides the enzymes annotation framework and supports loading annotations from different sources. This operation will load KEGG's information into *merlin*'s internal model database, setting all spontaneous reactions as integral to the model, thus launching the GenNet. The network will be further enlarged, when annotations are integrated to the GiSMo database, in the next step.

3.4.2 Integrate Annotations

The annotations performed earlier increase the reliability of the reconstructed GiSMo. However, if users have a former curated annotation, *merlin* provides, as mentioned earlier, operations for loading annotations from other sources in specific formats.

Enzymes Annotation

The enzymes annotation will define the topology and connectivity of the GenNet. The enzymatic activities (EC numbers) encoded in the genome will activate reactions according to the associations provided by the KEGG mappings. For this, *merlin* will include all reactions associated by KEGG to such EC numbers, provided that the reactions and enzymes are included in the same KEGG pathway or that an EC number is associated with a single reaction (more information on this principle on [7]). The resulting GenNet should then be curated and converted to a GiSMo.

Load External Annotations

merlin supports loading annotations from GFF (version 3) and GenBank file formats. Whereas the former does not natively provide a field for EC numbers, though these can be annotated in the notes' field, the latter has a specific field for this purpose. These operations can be performed by accessing the "*model>load*" menu and can only be performed if an annotation was not previously loaded in the model.

merlin's Annotation

The "*enzymes*" panel in the "*annotation*" module (Fig. 2) has a button ("*integrate to model*") for performing the integration of the annotation in the model database. Though it is possible to set a threshold and integrate the results directly in the model, users are encouraged to perform, at least, a partial curation of the annotations. This button opens a window, which offers different options for the integration process. The first option regards the genes' names, determining whether the given gene's name in the annotation should precede, be merged as a synonym, or rejected if the gene is already labeled in the model. The second is the most important as it decides if *merlin*'s EC numbers annotation should be favored, merged, or rejected concerning EC numbers already available in the model. Finally, the latter integration option is not mandatory, as every EC number is associated with a recommended name for the respective enzyme. Yet, *merlin* allows the integration of the enzymes' names in the same conditions as the previous options, by being preferred, merged into synonyms or rejected, regarding the existing annotations.

Transporters Annotation

The transporters annotation will set the GenNet's borders by restricting the number of metabolites that can be exchanged between compartments or with the exterior. The "*transporters*" panel in the "*annotation*" module (Fig. S3 of the supplemental material) provides a button ("*integrate to model*") for performing the integration of the annotation in the model database. This operation will integrate the transport reactions, gene annotations, and TRIAGE's transport proteins identifiers in the model.

Compartments Annotation

The compartments annotation will define the number of compartments and further restrict the connectivity of the GenNet. The same metabolite in different compartments is considered a distinct metabolite species in each compartment. Hence, the connectivity of the network may be impaired if reactions in the same pathway are allocated to different compartments.

The "*compartments*" panel in the "*annotation*" module (Fig. S4 of the supplemental material) provides a button ("*integrate to model*") for performing the integration of the compartments in the model database. This operation will reject the compartments set to be ignored in the main panel. Genes encoding proteins predicted to go to these compartments will be assigned to the internal compartment instead, cytosol and cytoplasm for prokaryotes and eukaryotes, respectively. There is also the option to compartmentalize the biochemical and/or the transport reactions.

This operation will create a new compartmentalized model. Reactions connected to distinct genes, which encode the same protein, assigned to different locations and genes assigned to multiple compartments will be replicated and versions of the same reaction in different locations will be included in the new model. This change in the internal model is transparent for the user and only the final model will be presented in the "*reactions*" panel of the "*model*" module.

3.4.3 Model Curation

The GenNet curation can start before the integration of the transporters and compartments. Indeed, the transport reactions can be added manually, and the compartments may be set to *inside* (pseudo-compartment representing the *interior* of the cell) and *outside* (representing the exterior of the cell). However, when added manually, transport reactions are seldom associated with genes. Moreover, a gene encoding a critical enzymatic activity inside the cell may have a paralogous gene encoding a similar protein in a different compartment. Thus, the final model will be impaired, as essentiality predictions will be affected. Nevertheless, the model must be iteratively curated to be able to mimic the organism phenotypical behavior. The curation phase comprises several steps, including its conversion to a GiSMo, which are described next.

Correct Reactions Reversibility and Direction

All reactions in KEGG are in the canonical format, i.e., all reactions are theoretically reversible. However, the reversibility and direction of the reactions can be determined by calculating the Gibbs free energy. Cases in which the free energy of the reactants is much greater than that of the products, the reaction takes place as written. Conversely, if the free energy of the products exceeds a lot that of the reactants, the reaction will tend to proceed in the reverse direction. Finally, if the free energy of the reactants is similar to the one of the products, the reaction may be reversible.

There are online tools to calculate the Gibbs free energy, such as the eQuilibrator (http://equilibrator.weizmann.ac.il). These tools allow establishing a procedure to determine the reactions direction and reversibility, through the analysis of the reactions' Gibbs energy, at 1 mM concentration. Additionally, databases like the ones developed by Ma and Zengh [37] or Stelzer et al. [38] provide information on the reversibility of reactions available in KEGG. Hence, these curated databases were used for correcting *merlin*'s internal metabolic database reactions. Nevertheless, KEGG is frequently updated and new reactions are frequently added, thus requiring the curation of the reactions not available in the aforementioned studies.

Unbalanced Reactions

A GiSMo must be balanced to yield reliable predictions. The number of atoms of each element in the input environmental conditions must be the same as in the output (including biomass). A model unbalanced at the stoichiometric level may sink elements or excrete compounds because of having unbalanced reactions, thus impairing simulations' results. Hence, *merlin* provides a feature designed for highlighting these reactions in bold and italics. After executing the operation available at "*model>unbalanced reactions>find*" a new tab becomes available in the reaction's properties panel ("*magnifying glass*"), which shows the sum of all elements in the reactants, in the products and finally the net result that should be 0. This way it is possible to pinpoint the possible errors in the reaction and correct them. There are several types of stoichiometric errors, including:

(a) Metabolites without formulae;

(b) Missing protons or water;

(c) Reactions for synthesis/hydrolysis of macromolecules.

The first case can be verified by determining the missing metabolites formulae. The second by adding/removing protons/water to the reaction, after confirmation in other data sources, whereas the third is usually fixed by eliminating the polymer from the reactants or products according to the reaction stoichiometry (e.g., reaction R01762> Arabinan + H_2O ⇔ Arabinan + Arabinose becomes Arabinan + H_2O ⇔ Arabinose).

Either way, there are no strict rules for dealing with these cases. Each situation should be assessed and other databases/tools like MetaCyc, BRENDA, and even eQuilibrator should be checked to determine the correct stoichiometry. Reactions whose stoichiometry cannot be verified should be removed from the model. It is worth re-emphasizing that this step is critical and every reaction marked as unbalanced by *merlin* must be corrected or removed.

Unconnected Reactions

There are cases of reactions in the GenNet that convert metabolites, not available in the remaining network, into metabolites not consumed by other reactions in the GenNet. Though the inclusion of these reactions does not usually impair model predictions, they can result from errors in the annotation, compartmentalization or simply be unnecessary to the GiSMo (although being indeed encoded in the genome). Hence, as before, *merlin* offers an operation (“*model>unconnected reactions>find*”) aimed at identifying these reactions, by coloring the reactions’ font in red. The analysis of these reactions may help in filling gaps in the GenNet or correcting mistakes in the annotation, improving the overall model performance. Alternatively, for cases in which there are too many of these reactions, they may be automatically removed (“*model>unconnected reactions>remove*”).

GPR Associations

The gene-protein-reaction (GPR) associations are one of the most important aspects of a GiSMo. These rules make direct associations between genes and reactions in the model, and are important to correctly predict mutant phenotypes. These rules are usually added manually to models, yet *merlin* provides a unique operation that accesses information available in the KEGG BRITE [39] database, including their subunits and stoichiometry, to automatically create these rules.

KEGG BRITE is a collection of manually created hierarchical entries, which include structural complex modules. These modules are collections of manually defined functional units, including protein structural complexes. Each module is composed by an identifier, name, class, a set of reactions associated with the module, sets of KEGG orthologous (KO) genes, and a definition, which describes the relationship between the KOs, i.e., the GP rule. The rules can be very complex with several alternative subunits and connected with AND/OR operators. Thus, *merlin* includes a mathematical parser for the rules, which calculates all possible KO combinations. Each KO comprises a set of genes with similar roles belonging to distinct organisms and horizontally conserved across several species.

The modus operandi for finding GPRs in *merlin* is briefly described next. For each EC number annotated in the case study, *merlin* scans KEGG BRITE searching for complex modules. The KO of the taxonomically closest organism is then aligned using SW, initially only to genes annotated with such an EC number, and if no hits are found, with the whole genome. Whenever *merlin* finds all KO comprised in the module definition it creates the gene rule and associates the rule to the reactions present in the KEGG BRITE module. Other reactions associated with genes identified as part of a rule, yet not available in the KEGG BRITE module, may be automatically removed from the model. Nevertheless, the user can insert/edit/remove any reaction to/from the model.

Adding GPRs to the GiSMo involves running the corresponding operation (“*model>gene-protein-reaction rules*”) and setting a few parameters. This operation requires setting distinct threshold for ortholog and paralog genes, as the operation will select KEGG's closest orthologues to compare against the case study genome. The default values are the result of empirical tests, yet users may prefer being either more or less restrictive. In addition, the operation can automatically integrate the rules after generating or just save them to a file on a report file to allow a previous inspection of the rule. Other options include whether to remove reactions not available in KEGG BRITE associated with enzymes assigned with GPR rules, or removing reactions with notes or manually inserted.

After determining the GPR rules, a new menu called "*GPRs*" becomes available in the dropdown box of the pop-up window shown when the "*magnifying glass*" of a reaction assigned with rules is clicked. These reactions are labeled on the notes field with the message "*merlin GPR*." Each line presented in the GPR table contains one or more gene subunits (AND rules) required for creating a protein complex that promotes the reaction of interest. Different lines offer alternative combinations of genes (OR rules).

The rules automatically generated by *merlin* can be edited or removed by using the edit reaction option on the "*reactions*" panel of the "*model*" module. Likewise, GPR rules can be manually added/edited/removed to/from the reaction using the same option.

Add Biomass Formation Reaction

This is the first step in converting the GenNet to a GiSMo. The equation of the biomass formation reaction denotes the necessary amount of each biomolecule (e.g., amino acids, lipids, etc.) required for cell replication [3]. This equation should also include growth-associated energy requirements, characterized by the number of ATP molecules required for synthesizing a gram of biomass. Though previous studies [40] propose that using the biomass equation of a related organism does not introduce significant errors in GiSMos, a recent study demonstrates that such an approach may be incorrect [41]. Moreover, according to [42], estimation of the average protein and (deoxy)ribonucleotides contents from the genome sequence (gene translated amino acid sequences for the former; mRNA, tRNA, and rRNA for ribonucleotides; gene nucleotide sequences for deoxyribonucleotides) provides better results than using the biomass equation from related species.

Hence, *merlin* provides a tool that allows adding an e-biomass equation to the GiSMo using the latter approach. Besides estimating the average protein and (deoxy)ribonucleotides contents *merlin*'s "*model> e-biomass equation*" operation adds several reactions to the model, representing the assembly of pseudo-compound macromolecules for the biomass, namely: e-Protein, e-Nucleotides, e-Lipids, e-Carbohydrates, and e-Cofactors. To perform this

Table 2
Relative contents of several macromolecules in different organisms

Organisms	Source	RNA % (g/g_{RNA}) mRNA	tRNA	rRNA	RNA % (g/g_{DW})	Protein % (g/g_{DW})	DNA % (g/g_{DW})	Lipids % (g/g_{DW})	Carbohydrates % (g/g_{DW})	Cofactors % (g/g_{DW})
Yeast	exp	0.05	0.15	0.8	0.049	0.401	0.007	0.045	0.497	n/a
	literature	0.05	0.15	0.8	0.067	0.425	0.026	0.055	0.353	0.074
Archaea	exp									n/a
	literature	n/a	n/a	n/a	n/a	n/a	n/a	n/a	n/a	n/a
Gram positive	exp	0.05	0.2	0.75	0.077	0.539	0.02	0.125	0.239	n/a
	literature	0.05	0.2	0.75	0.075	0.531	0.023	0.162	0.148	0.062
Gram negative	exp	0.05	0.15	0.8	0.154	0.582	0.021	0.096	0.147	n/a
	literature	0.05	0.15	0.8	0.152	0.591	0.025	0.094	0.084	0.054
Total Bacteria	exp	0.05	0.15	0.8	0.115	0.561	0.02	0.111	0.193	n/a
	literature	0.05	0.15	0.8	0.118	0.565	0.024	0.124	0.112	0.057

operation the relative amounts of proteins and nucleotides must be identified. Table 2 shows the relative amounts of several macromolecules, retrieved from the literature and experimentally determined in our group according to the methodology defined in [42], in different types of organisms. These values were determined by averaging the experimentally determined quantities of different organisms. Moreover, gene expression data can be used to adjust the protein contents to the ones expressed de facto. Besides the gene expression data, files required by this operation are the genome protein fasta file (.faa) for the amino acids, the whole genome sequence fasta file (.fna) for the DNA, genomic coding sequences fasta file (.fna) for the mRNA, and the RNA sequences fasta file (.fna) for rRNA and tRNA. The latter file will have to be preprocessed to extract the rRNA sequence into a file and the tRNA sequences into a separated file. All other RNA sequences are discarded. These files can be obtained on the same assembly FTP links shown in Fig. S1 of the supplemental material.

e-Protein

The e-Protein pseudo-compound is calculated by determining the frequency of each amino acid in the translated coding sequences. In this approach, the frequency of each amino acid, normalized by the total number of residues, will be equivalent to the mass fraction of each amino acid. This mass fraction is then converted to mol_{aa} per $g_{protein}$. Gene expression data can be used to correct this estimate for proteins effectively expressed in specific conditions. The only condition is that this data used the same gene identifiers as the genome protein fasta file (.faa). This calculation considers the amount of H_2O formed during protein polymerization, which is included in the e-Protein equation.

e-Nucleotides

The rationale for calculating the RNA and DNA composition is the same as before. The whole genome sequence is used to estimate the amount of each deoxyribonucleotide and the mRNA, rRNA, and tRNA are used to estimate the moles of total RNA in the cell. The contribution of each different type of RNA, for different types organisms, is shown in Table 2. As before, the amount of diphosphate formed for the polymerization of the nucleotides is considered and included in the e-DNA and e-RNA pseudo-compounds equations.

Cofactors

Cofactors are important in models as, though their biosynthesis is not a burden to the cell since these are recycled, the apparatus for their production must be available in the GiSMo. The inclusion of these compounds in the biomass forces the GiSMo to produce them. Hence, the cofactors pseudo-compound macromolecular composition includes trace amounts of all metabolites identified as essential in [43] and is included in the biomass. Nevertheless, the reactions' stoichiometries can be changed and elements added/removed to/from the reaction.

Polysaccharides, Lipids, and Fatty Acid Formulation

The polysaccharides and lipids pseudo-compounds composition must be determined experimentally. Alternatively, since it cannot be inferred from the genome sequencing data, the composition should be set to the composition of a closely related organism. The fatty acid pseudo-compound, though not usually directly included in the biomass reaction, is a precursor of the lipids. This average fatty acid composition should also be determined experimentally or alternatively from a closely related organism. A new reaction should be inserted in the model indicating the contribution (stoichiometry) of each fatty acid (e.g., Octanoic acid—C8, Decanoic acid—C10, Octadecenoic acid—C18, etc.).

Phosphorus to Oxygen Ratio

The phosphorus to oxygen (P/O) ratio is an indicator of the relationship between ATP and oxygen, which specifies the number of orthophosphate molecules used for ATP biosynthesis per atom of oxygen reduced through the electron transport chain in aerobic organisms. This ratio is directly proportional to the number of hydrogen atoms transported outward against the electrochemical gradient and inward through the ATP synthase. These reactions often involve the oxidation of ferrocytochrome to ferricytochrome with the reduction of oxygen and the reduction ubiquinol to ubiquinone to recycle the ferricytochrome. As before, the absence of specific studies to characterize the P/O ratio in the case studies should be overcome by assuming the same P/O ratio as of closely related organisms.

In *merlin*, this ratio is translated by curating a set of reactions that consume oxygen to transport hydrogen outward and a reaction that transports hydrogen inside, producing ATP. TRIAGE's analyses often detect genes associated with the oxidative phosphorylation pathway and create reactions involved in this process, as shown in the example in Table S1 of the supplemental material. Nevertheless, when assembling aerobic organisms' GiSMos, users should curate the stoichiometries of these reactions to guarantee the correct P/O ratio.

Growth and Maintenance ATP

The GiSMo should also include an equation that represents ATP depletion for cellular conservation and preservation, the maintenance ATP. Likewise, the ATP spent in cellular replication efforts should also be included in the model. The growth ATP flux includes all cellular processes involved in cell replication, such as proteins and nucleotide synthesis. There are some enzymes involved in the translation process, by spending energy to ligate amino acids to tRNA molecules. These reactions are now often included in GiSMos and should be considered when performing the calculations of the growth ATP flux.

The maintenance and growth ATP fluxes should be determined experimentally by plotting ATP metabolomics analyses against growth data from of chemostat growth experiments, as shown in Fig. S5 of the supplemental material. The slope of the linear regression of the ATP flux measurements vs. growth rate indicates the amount of ATP spent for cellular growth, whereas the *y*-intercept represents the amount of ATP consumed for cellular maintenance functions. Alternatively, the ATP fluxes can be determined by comparing the simulation results, with different maintenance and/or ATP flux values, to in vivo data. Simulations should be performed with the same environmental conditions as the experiments, namely the limiting carbon source flux. The linear regression slopes and *y*-intercept values of the simulated vs. experimental growth rates are then evaluated [44]. The better ATP flux values are provided by the simulation that yields the *y*-intercept value closest to zero and the slope value closest to the unit. Lastly, when experimental data is not an option, the ATP flux values of closely related organisms can be used.

In *merlin*, the maintenance ATP can be included in the GiSMo by setting the lower and upper thresholds of a clone of KEGG's reaction R00086. The original reaction is promoted by several enzymes identified with different EC numbers, but in such cases the reaction may serve specific purposes and a gene knockout may silence it. Thus, *merlin*'s duplicate function can be used to clone the reaction and the same, above zero, flux value should be set in both thresholds. The purpose of fixing the thresholds is guaranteeing that every simulation with the GiSMo produces enough ATP for cellular maintenance, besides growth.

The growth ATP requirements are added to the GiSMo by including the hydrolysis of ATP in the biomass equation with predetermined stoichiometric values, proportional to the requirements of the macromolecules biosynthesis.

Curation per Pathway

The reconstruction of a functional GiSMo must guarantee that routes from known carbon, nitrogen, sulfur, and phosphorous sources to the biomass' constituents and other identified products are present and gapless. The curation of the draft model can be performed while performing several of the steps described above. The first step of the curation should be analyzing the main pathways, to ensure the connectivity of the GiSMo. The main pathways may vary with the case study organism, depending on carbon sources and auxotrophies. Still, the following pathways should be examined, when available, as they are associated with the production of important metabolites: glycolysis, pentose phosphate, citrate cycle, (every) amino acid biosynthesis, pantothenate and CoA biosynthesis, fatty acid biosynthesis, glycerolipid metabolism, glycerophospholipid metabolism, purine metabolism, pyrimidine metabolism, folate biosynthesis and ubiquinone, and other terpenoid-quinone biosynthesis. Other pathways that may be of interest are: biotin metabolism, riboflavin metabolism, lipopolysaccharide biosynthesis, and steroid biosynthesis. As every GiSMo is distinct, pathways not listed here will also have to be curated.

The curation of the pathways should focus on finding a metabolic path from a starting substrate to one or more final products, within the several alternate *viae* in a metabolic pathway. For instance, in glycolysis the most common substrate is glucose, and a set of reactions converting this metabolite into pyruvate should be mapped. The path from pyruvate onward will highly depend on the type of organism. Though a part of the flux will in most cases be redirected to the citrate cycle, possible routes for the remainder include conversion to and excretion of lactate, ethanol, or acetate. Other curation actions include confirmation of the reversibility and direction of each reaction in the selected *viae*.

merlin allows selecting a KEGG pathway and using the "*draw in browser*" button to open a browser window where the pathway is opened and the enzymes and/or reactions available for such pathways in the GiSMo are colored. If the EC number, in the browser window, is written in green then the reaction is included in the GiSMo and the enzyme promoting the reaction in such a pathway is the same as the one highlighted. If it is written in dark blue then, though the reaction is present in the model, it is being promoted by an enzyme other than the one available for this via. Moreover, if the "*find unconnected reactions*" tool was used, then dead-end reactions and metabolites, in the browser window, will be colored (as in the reactions' panel) in red. Likewise, EC numbers associated to reaction connected to dead-end reactions will be colored in light blue.

These tools allow easily detecting gaps in the GiSMo which should be examined. A gap in the most obvious route of the pathway is, in most cases, probably related to annotation errors. Hence, other sources (such as MetaCyc or KEGG's own annotation) should be verified and annotations proposed by these analyzed to identify potential gap-filling options. Literature should also be considered as metabolites identified as dead-ends may be auxotrophic needs of the organisms and should be added to environmental conditions.

Users are encouraged to remove "*Metapathways*", such as "*Metabolic Pathways*", "*Biosynthesis of secondary metabolites*" and similar, from the GiSMo as the contribution of these to comprehension of the organisms' metabolism and the curation efforts is reduced. Other pathways that should be removed to keep the GiSMo as simple as possible include all other *viae* not available in the case study, like the "*Carbon fixation in photosynthetic organisms*" pathway for prokaryotes. This operation is easily performed in *merlin* by clicking the "*remove*" button in the "pathway" tab available in the "*reactions*" panel.

Drains

The last step of converting a GenNet to a GiSMo is adding drain boundaries to the model. Drains are exchange constraints set to mimic the environment conditions in which organisms live and grow. Usually, these constraints are set for external metabolites, thus allowing the GiSMo to control the uptake and excretion of metabolites. The constraints are pairs of lower and upper restrictions that limit the input and output fluxes of selected metabolites. To mimic the environmental conditions in which the organisms are grown, the drains of the metabolites that compose the growth media should be set to the fluxes measured during the experiment, yet with negative signs to label metabolites as consumed.

merlin provides an operation to automatically add drains to every external metabolite, through the creation of a reaction with the compound as a reactant, no products and setting the lower boundary to 0 and the upper boundary to 999999. Hence, drains are clustered in a pseudo-pathway to ease the process of finding drains for metabolites that compose the growth media.

3.4.4 Export

In *merlin*, GiSMos are kept in a database and should be exported to be validated and utilized in other platforms. Reactions can be exported in tabular format by pressing the "*xls tabbed file*" button in the "*reactions*" panel, but this operation will only export the panel contents. To export a fully formatted model in the SBML format with identifiers for reactions and metabolites, GPR rules, and MIRIAM annotations, users should go to "*model>export to SBML*" menu. *merlin* allows setting various options for exporting the model, such as whether generating or not a file with the metabolites formulae, validating the model online (may delay the operation), the biomass reaction name, the level and version of the SBML file, and export folder and file name.

3.5 Validation

The SBML file containing the GiSMo can be imported into several platforms like OptFlux and COBRA [11] to perform simulations and validate the GiSMo, by assessing the simulation results to experimental data. If the results of the growth rate and byproducts fluxes do not match, these steps should be repeated iteratively.

Validation tests that should be performed include, but should not be limited to, the following:

1. Spontaneous growth;
2. Auxotrophies;
3. Gene essentiality;
4. Alternative basic elements sources;
5. Assess growth rate.

These tests can be classified as qualitative and quantitative. The first four tests belong to the qualitative tests category, while the last one is quantitative and evaluates the GiSMo's ability to predict flux rates.

The first test is assessed by performing simulations with null environmental conditions that is by setting the lower boundary of all drains to zero. This test will evaluate if the GiSMo is able to grow without in silico growth media. The second test is performed by comparing simulations, in which the first is performed defining a minimal medium in the environment conditions and the second using the same medium except for the removal of a metabolite for which the case study is known to be auxotrophic. The third test involves mimicking growth for conditions in which genes are essential for growth, namely media that do not provide a certain metabolite, which must be biosynthesized. The fourth test depends on the availability of studies or tests such as Biolog phenotype microarrays that test for different conditions, providing data on which metabolites can be used as sources of different elements, namely carbon, nitrogen, sulfur, and phosphorous. Finally, the last test involves comparing experimental to in silico fluxes. This evaluates the quality of the model predictions, for the tested experimental conditions and is dependent on the availability of experimental data.

3.5.1 Experimental Data

These data should be, preferably, obtained from chemostat and fluxes normalized to $\frac{\text{mmol}}{h \times g_{\text{DW}}}$. Fluxes should be determined for the carbon, nitrogen, sulfur, and phosphorous sources (e.g., glucose, ammonium, oxygen, etc.) and for all fermentation byproducts (or at least the main ones such as carbon dioxide, ethanol, acetate, formate, etc.). Finally, the biomass growth rate is also required for the validation of the model.

Though the normalization process is straightforward for chemostat experiments, these data can also be obtained from batch experiments as shown in Sauer et al. [45] (*see* **Note 5** for a summarized description of this process).

4 Notes

1. *merlin*'s interface
   merlin's interface has three main components, as shown in Fig. S6 of the supplemental material. The first one (blue square—Fig. S6 of the supplemental material) is the *merlin*'s operations bar, where most procedures are called and which is divided into four tabs. The "*project*" operations' tab is where a project can be created, loaded, saved and deleted from the database. The "*database*" operations tab allows creating new internal databases or clean existing ones. The "*annotation*" tab includes three main sub-tabs. The first one, "*enzymes*", offers several options for identifying and annotating enzymes on a genome. Likewise, "*TRIAGE*" does the same but for transport proteins and transport reactions. The third sub-tab, "*compartments*", allows loading and processing compartments' prediction reports from different tools. Finally, the "*model*" operations tab offers several tools for developing and curating a GiSMo.

   merlin's clipboard (green square—Fig. S6 of the supplemental material) offers an intuitive schema for accessing the internal database data. It clusters information per project and within each project it further groups it into "*model*" and "*annotation*" data. Whereas the latter group provides instances for curating enzymes, transporters, and compartments annotations, the former is used to assemble the GiSMo by editing its subcomponents, namely genes, proteins, reactions, metabolites, and pathways.

   Lastly, the visualization area (red square—Fig. S6 of the supplemental material) is where the components selected in the clipboard are shown and in most cases, operations are provided for editing data in the database for curation of the annotation or the model itself. Most "*views*" allow inserting, editing, and removing information into/from the database. Furthermore, annotations' "*views*" also have operations for integrating these data with the model database.

2. GenBank versus RefSeq annotations
   The main difference between these is the identification of the genes products, in the FASTA format amino acids (.faa) files, which in RefSeq may contain records identified by WP_ + 9 digits + version number (e.g., WP_000000001.1). These records represent single, nonredundant, protein

sequences annotated on several different RefSeq genomes from different or the same (different strains) species, as shown in Fig. S7 of the supplemental material. Hence, the identifiers in these CDSs are generic and cannot be mapped to a specific genome.

3. Enzymes annotation
   (a) *Similarities Searches*
   Though gene or protein sequences with high similarity may have arisen from a common ancestor, the opposite is not always true, that is, homologous sequences do not always exhibit significant sequence similarity. Hence, it is possible to determine if two sequences are homologous by performing similarity searches, but it is not possible to sustain that two sequences are not homologous based on similarity searches to databases alone [24].

   When performing BLAST searches, each data source (NCBI or EBI) has its own databases with different accession numbers. Thus, when databases from the same data source are used in different tools, *merlin* may allow combining the results of both tools. For instance, if the database used in the NCBI BLAST is "*swissprot*", *merlin* allows performing a HMMER search to the same database and combine the results for annotating the genome. Likewise, any database selected for performing the EBI BLAST would be able to combine with the HMMER search to "*swissprot*", as the accession numbers are compatible. A list of compatibilities can be found in Table S2, of the supplemental material. Another important parameter (for BLAST) is the substitution-scoring matrix (matrix containing values relative to the probability that a given amino acid mutates into another amino acid for all pairs of amino acid). The matrix used in the similarity search may affect the results as the selection of the best matrix depends on the size of each sequence.
   (b) *EC Numbers Update*
   Genes initially annotated with partial EC numbers (e.g., 1.1.-.-) may eventually be assigned with complete EC numbers. These annotations may happen due to known enzymatic activities not yet assigned with EC codes or sequences showing evidence of generic enzymatic activities. In the former case, the function of the gene may be already known, but a complete EC number, at the time of the annotation, is not available. In the latter case, although the specific function (for instance a specific substrate) is not known, the protein sequence shows signs indicating that the gene encodes an enzyme that should belong to a given EC family.

Nonetheless, complete EC numbers can later be assigned with new codes. These are cases in which an EC number is discontinued and the enzymatic activity(ies) moved into a (sometimes more than one) new code, e.g., EC 1.1.1.128—created 1972, modified 1976, deleted 2012, L-idonate 2-dehydrogenase. The reaction described is now covered by EC 1.1.1.264.

4. *TAD Annotations*

   Whenever TRIAGE's alignments are integrated with TAD, *merlin* creates a report at the "*temp*" folder, within *merlin*'s main folder, with the extension "*.out*". The first two records of the report are examples of annotated entries. The following rows are TCDB records unavailable in TAD, which should be curated for posterior inclusion in the internal database.

   (a) *Report Structure*

   This report is organized in 18 columns (UniProt ID, TCDB ID, TCDB family, TCDB description, affinity, Transport type, TCDB location, YTPDB gene, YTPDB description, YTPDB type, YTPDB metabolites, YTPDB location, **TC #, direction, metabolite, reversibility, reacting metabolites and equation**). Information in columns 1–7 (UniProt and TCDB identifiers, protein family and description, transport affinity, type, and location) is retrieved from TCDB. Columns 8–12 contain information retrieved from the YTPDB [46] (http://ytpdb.biopark-it.be/ytpdb/index.php/Main_Page), a database specialized in the classification and annotation of yeast transporters. This resource is very useful because, unlike TCDB, it has a dedicated field for the substrate being transported. Hence, organisms with similarities to *Saccharomyces cerevisiae* will have another resource for inferring transport annotations. TRIAGE retrieves information on the yeast gene, protein, transport type, metabolites, and location from this database. Whereas the purpose of these columns (1–12) is to provide information, the bold columns (columns 13–18) must be filled out for adding new entries in TRIAGE's TAD.

   The *TC #* column is automatically populated by TRIAGE. The next column (direction) has controlled vocabulary, that is, there are a finite number of options to fill this column depending on the type of transport, being those: *in* or *out* (for uniport), *in:in* or *out:out* (for symport), and *in // out* (for antiport). If a protein has different behavior depending on the substrate, these can be combined by adding two vertical bars (||) between them (e.g., in || in:in).

   Column 15 holds the metabolites names. In this case, the vocabulary is not controlled, though using KEGG or

ChEBI nomenclature is encouraged, as TRIAGE uses an internal algorithm to assign KEGG and ChEBI identifiers to the metabolites. This algorithm uses ChEBI hierarchy to identify metabolites' second-generation elements. For instance, α-D-glucose (CHEBI:17925) and β-D-glucose (CHEBI:15903) are both second-generation elements of D-glucose (CHEBI:4167). For a detailed description of the algorithm, *see* [28]. Each metabolite transported by a given protein is separated by a semicolon (;) and (in cases where it applies) by the same symbol of the transport type. For instance, the symport of glycine with hydrogen and the symport of alanine with glycine with hydrogen by the 2.A.25.1.1 "*Alanine (or glycine):Na+ symporter*" (P30144) is denoted as "*alanine; glycine : Na+*". In these cases, every element on the left of the colon will be co-transported with every element on the right. Likewise, the antiport of oxalate with formate by the 2.A.1.11 oxalate formate antiporter (Q51330) is set as "oxalate // formate". Since the direction was set as "*in // out*", the oxalate will go in and the formate outward. However, as the next column identifies this transport as reversible, the metabolites' directions can be in fact reversed. Moreover, due to format parsing limitations (metabolite names often contain numbers and other characters) stoichiometry is represented by symport of the same molecule. For instance, the transport of two protons has a direction of *in* and the metabolites are *proton:proton*. Additionally, cases in which transporters may function with different mechanisms (||) should also reflect this separation in the metabolites column. For instance, 2.A.29.10.5 (P40556) may "*import NAD$^+$ into mitochondria by unidirectional transport or by exchange with intramitochondrially generated (d)AMP and (d)GMP*". In TAD, the direction for this transport will be in || in // out and the metabolites entry will be NAD || NAD// AMP; GMP. The subsequent column determines the reversibility of the transport reaction. It should be considered true by default, except in some cases discussed later, unless it is determined otherwise by the transporters annotation pipeline.

Some transport reactions may require energy provided by the chemical reactions, such as ATP hydrolysis. This information is added to TAD in column 17, using a specific notation. A exergonic reaction, such as ATP + water ⇒ ADP + P_i is represented as "*1:ATP; 1:water || 1:ADP; 1:orthophosphate*". In this case, the limited number of metabolites allows a straightforward representation of the stoichiometry and compound names. These reactions are often irreversible and the vertical bars separate the reactants from products. Moreover, some transport reactions involve modifications of substrates like, for instance, the phosphorylation of L-ascorbate to L-ascorbate-6-phosphate. In these cases, the transported metabolites field should be set to two hyphens ("--").

The last column is the family (or subfamily) equation. This field can be very informative, as in some cases (e.g., Quinol + 1/2 O_2 + $\underline{4H^+}$ [in] ⇒ Quinone + H_2O + $\underline{4H^+}$ [out]) it determines the direction, reversibility, and metabolites transported by a specific family of transporters (e.g., irreversible export of hydrogen promoted by the oxidation of a quinol to quinone).

(b) *Workflow*

A detailed explanation of the flowchart developed for annotating the TAD report and presented in Fig. S8 of the supplemental material is provided next. The first step is analyzing the TCDB entry description (Fig. S8 of the supplemental material—Level A) to identify the transported metabolite(s), transport type, and reversibility. Usually, this description is copied from publications, hence other information like reacting metabolites and equation is not available in this field. Information retrieved from this field is specific for each protein and should be favored regarding evidence collected from the next steps. The next step is checking the UniProt entry (Fig. S8 of the supplemental material—Level B) to find information on the reacting metabolites and fields unsuccessfully annotated in the previous step (metabolite(s), transport type, and reversibility). Likewise, after analyzing the UniProt entry, the TCDB sub-family (or family) (Fig. S8 of the supplemental material—Level C) should be checked to retrieve the transport equation. If this information cannot be retrieved from the sub-family (or family [47]), it can be searched in the level above (Fig. S8 of the supplemental material—Level D)—the TC family (or superfamily or class, according to the hierarchy of the TC number [47]). The default values for these fields, which should be set whenever data cannot be found, are direction -> in, metabolite -> unknown, reversibility -> true, reaction metabolites -> -- and equation -> --.

5. Determination of physiological parameters from batch experiments

Estimating physiological parameters from batch data involves identifying the exponential growth phase. This step can be performed by performing the log-linear regression of the biomass concentration vs. time, in which the growth rate (μ) is the regression coefficient. The regression should only include the phase in which growth is exponential, which is the phase wherein the plot produces a straight line as shown in Fig. S9 of the supplemental material (a).

The specific consumption/production rates (q_S/q_P) are the differential variation of the substrate/product with time normalized to the biomass concentration. Hence, these are calculated by performing the linear regression of the substrate/product concentration vs. the biomass concentration divided by μ $\left(\frac{X}{\mu}\right)$ in which the rate q_S/q_P is the regression coefficient, as shown in Fig. S9 of the supplemental material (b). Substrates will have negative values whereas products will have positive rates.

The consumption rates should be set in the environmental conditions before simulations (to override the default values) and the simulation output assessed for the production rates to evaluate the GiSMo.

References

1. Otero JM, Nielsen J (2010) Industrial systems biology. Biotechnol Bioeng 105:439–460. https://doi.org/10.1002/bit.22592
2. Kitano H (2002) Systems biology: a brief overview. Science 295:1662–1664. https://doi.org/10.1126/science.1069492
3. Dias O, Rocha I (2015) Systems biology in fungi. In: Paterson R (ed) Mol. Biol. Food water borne mycotoxigenic mycotic fungi. CRC Press, Boca Raton, FL, pp 69–92
4. gismo Meaning in the Cambridge English Dictionary. http://dictionary.cambridge.org/dictionary/english/gismo#translations. Accessed 13 Apr 2017
5. Gizmo definition and meaning | Collins English Dictionary. https://www.collinsdictionary.com/dictionary/english/gizmo. Accessed 13 Apr 2017
6. Thiele I, Palsson BØ (2010) A protocol for generating a high-quality genome-scale metabolic reconstruction. Nat Protoc 5:93–121. https://doi.org/10.1038/nprot.2009.203
7. Dias O, Rocha M, Ferreira EC, Rocha I (2015) Reconstructing genome-scale metabolic models with merlin. Nucleic Acids Res 43:3899–3910. https://doi.org/10.1093/nar/gkv294
8. Henry CS, DeJongh M, Best AA, Frybarger PM, Linsay B, Stevens RL (2010) High-throughput generation, optimization and analysis of genome-scale metabolic models. Nat Biotechnol 28:977–982. https://doi.org/10.1038/nbt.1672
9. Hucka M, Finney A, Sauro HM, Bolouri H, Doyle JC, Kitano H, Arkin AP, Bornstein BJ, Bray D, Cornish-Bowden A, Cuellar AA, Dronov S, Gilles ED, Ginkel M, Gor V, Goryanin II, Hedley WJ, Hodgman TC, Hofmeyr J-H, Hunter PJ, Juty NS, Kasberger JL, Kremling A, Kummer U, Le Novère N, Loew LM, Lucio D, Mendes P, Minch E, Mjolsness ED, Nakayama Y, Nelson MR, Nielsen PF, Sakurada T, Schaff JC, Shapiro BE, Shimizu TS, Spence HD, Stelling J, Takahashi K, Tomita M, Wagner J, Wang J (2003) The systems biology markup language (SBML): a medium for representation and exchange of biochemical network models. Bioinformatics 19:524–531. https://doi.org/10.1093/bioinformatics/btg015
10. Rocha I, Maia P, Evangelista P, Vilaça P, Soares S, Pinto JP, Nielsen J, Patil KR, Ferreira EC, Rocha M (2010) OptFlux: an open-source software platform for in silico metabolic engineering. BMC Syst Biol 4:45. https://doi.org/10.1186/1752-0509-4-45
11. Schellenberger J, Que R, Fleming RMT, Thiele I, Orth JD, Feist AM, Zielinski DC, Bordbar A, Lewis NE, Rahmanian S, Kang J, Hyduke DR, Palsson BØ (2011) Quantitative prediction of cellular metabolism with constraint-based models: the COBRA Toolbox v2.0. Nat Protoc 6:1290–1307. https://doi.org/10.1038/nprot.2011.308
12. Le Novère N, Finney A, Hucka M, Bhalla US, Campagne F, Collado-Vides J, Crampin EJ, Halstead M, Klipp E, Mendes P, Nielsen P, Sauro H, Shapiro B, Snoep JL, Spence HD, Wanner BL (2005) Minimum information requested in the annotation of biochemical

models (MIRIAM). Nat Biotechnol 23:1509–1515. https://doi.org/10.1038/nbt1156

13. Glez-Peña D, Reboiro-Jato M, Maia P, Rocha M, Díaz F, Fdez-Riverola F (2010) AIBench: a rapid application development framework for translational research in biomedicine. Comput Methods Programs Biomed 98:191–203. https://doi.org/10.1016/j.cmpb.2009.12.003
14. UniProt Consortium (2015) UniProt: a hub for protein information. Nucleic Acids Res 43: D204–D212. https://doi.org/10.1093/nar/gku989
15. Boutet E, Lieberherr D, Tognolli M, Schneider M, Bansal P, Bridge AJ, Poux S, Bougueleret L, Xenarios I (2016) UniProtKB/Swiss-Prot, the Manually Annotated Section of the UniProt KnowledgeBase: how to use the entry view. Methods Mol Biol 1374:23–54. https://doi.org/10.1007/978-1-4939-3167-5_2
16. Sayers EW, Barrett T, Benson DA, Bryant SH, Canese K, Chetvernin V, Church DM, DiCuccio M, Edgar R, Federhen S, Feolo M, Geer LY, Helmberg W, Kapustin Y, Landsman D, Lipman DJ, Madden TL, Maglott DR, Miller V, Mizrachi I, Ostell J, Pruitt KD, Schuler GD, Sequeira E, Sherry ST, Shumway M, Sirotkin K, Souvorov A, Starchenko G, Tatusova TA, Wagner L, Yaschenko E, Ye J (2009) Database resources of the National Center for Biotechnology Information. Nucleic Acids Res 37:D5–15. https://doi.org/10.1093/nar/gkn741
17. Schomburg I, Chang A, Schomburg D (2002) BRENDA, enzyme data and metabolic information. Nucleic Acids Res 30:47–49
18. Ogata H, Goto S, Sato K, Fujibuchi W, Bono H, Kanehisa M (1999) KEGG: Kyoto encyclopedia of genes and genomes. Nucleic Acids Res 27:29–34. https://doi.org/10.1093/nar/27.1.29
19. Lipman DJ, Pearson WRW (1985) Rapid and sensitive protein similarity searches. Science 227:1435–1441. PMID: 2983426
20. Federhen S (2012) The NCBI Taxonomy database. Nucleic Acids Res 40:D136–D143. https://doi.org/10.1093/nar/gkr1178
21. Kitts PA, Church DM, Thibaud-Nissen F, Choi J, Hem V, Sapojnikov V, Smith RG, Tatusova T, Xiang C, Zherikov A, DiCuccio M, Murphy TD, Pruitt KD, Kimchi A (2016) Assembly: a resource for assembled genomes at NCBI. Nucleic Acids Res 44: D73–D80. https://doi.org/10.1093/nar/gkv1226
22. mysql-server - Linux Mint Community. https://community.linuxmint.com/software/view/mysql-server. Accessed 13 Apr 2017
23. MySQL :: About MySQL. https://www.mysql.com/about/. Accessed 13 Apr 2017
24. Pearson WR (2013) An introduction to sequence similarity ("Homology") searching. In: Curr. Protoc. Bioinforma. John Wiley & Sons, Inc., Hoboken, NJ, pp 3.1.1–3.1.8
25. Altschul SF, Gish W, Miller W, Myers EW, Lipman DJ (1990) Basic local alignment search tool. J Mol Biol 215:403–410. https://doi.org/10.1016/S0022-2836(05)80360-2
26. Finn RD, Clements J, Eddy SR (2011) HMMER web server: interactive sequence similarity searching. Nucleic Acids Res 39: W29–W37. https://doi.org/10.1093/nar/gkr367
27. Magrane M, Consortium UP (2011) UniProt Knowledgebase: a hub of integrated protein data. Database. https://doi.org/10.1093/database/bar009
28. Dias O, Gomes D, Vilaca P, Cardoso J, Rocha M, Ferreira E, Rocha I (2017) Genome-wide semi-automated annotation of transporter systems. IEEE/ACM Trans Comput Biol Bioinforma 14:443. https://doi.org/10.1109/TCBB.2016.2527647
29. Yu NY, Wagner JR, Laird MR, Melli G, Rey S, Lo R, Dao P, Sahinalp SC, Ester M, Foster LJ, Brinkman FSL (2010) PSORTb 3.0: improved protein subcellular localization prediction with refined localization subcategories and predictive capabilities for all prokaryotes. Bioinformatics 26:1608–1615. https://doi.org/10.1093/bioinformatics/btq249
30. Goldberg T, Hecht M, Hamp T, Karl T, Yachdav G, Ahmed N, Altermann U, Angerer P, Ansorge S, Balasz K, Bernhofer M, Betz A, Cizmadija L, Do KT, Gerke J, Greil R, Joerdens V, Hastreiter M, Hembach K, Herzog M, Kalemanov M, Kluge M, Meier A, Nasir H, Neumaier U, Prade V, Reeb J, Sorokoumov A, Troshani I, Vorberg S, Waldraff S, Zierer J, Nielsen H, Rost B (2014) LocTree3 prediction of localization. Nucleic Acids Res 42:W350–W355. https://doi.org/10.1093/nar/gku396
31. Saier MH (2000) A functional-phylogenetic classification system for transmembrane solute transporters. Microbiol Mol Biol Rev 64:354–411
32. Sonnhammer EL, von Heijne G, Krogh A (1998) A hidden Markov model for predicting transmembrane helices in protein sequences. Proc Int Conf Intell Syst Mol Biol 6:175–182

33. Käll L, Krogh A, Sonnhammer ELL (2004) A combined transmembrane topology and signal peptide prediction method. J Mol Biol 338:1027–1036. https://doi.org/10.1016/j.jmb.2004.03.016
34. Moller S, Croning MDR, Apweiler R, Möller S (2001) Evaluation of methods for the prediction of membrane spanning regions. Bioinformatics 17:646–653. https://doi.org/10.1093/bioinformatics/17.7.646
35. Smith TF, Waterman MS (1981) Identification of common molecular subsequences. J Mol Biol 147:195–197. https://doi.org/10.1016/0022-2836(81)90087-5
36. Gardy JL, Brinkman FSL (2006) Methods for predicting bacterial protein subcellular localization. Nat Rev Microbiol 4:741–751. https://doi.org/10.1038/nrmicro1494
37. Ma H, Zeng A-P (2003) Reconstruction of metabolic networks from genome data and analysis of their global structure for various organisms. Bioinformatics 19:270–277. https://doi.org/10.1093/bioinformatics/19.2.270
38. Stelzer M, Sun J, Kamphans T, Fekete SP, Zeng A-P (2011) An extended bioreaction database that significantly improves reconstruction and analysis of genome-scale metabolic networks. Integr Biol (Camb) 3:1071–1086. https://doi.org/10.1039/c1ib00008j
39. Tanabe M, Kanehisa M (2012) Using the KEGG database resource. Curr Protoc Bioinformatics Chapter 1:Unit1.12. doi: https://doi.org/10.1002/0471250953.bi0112s38
40. Varma A, Palsson BO (1993) Metabolic capabilities of Escherichia coli II. Optimal growth patterns. J Theor Biol 165:503–522. https://doi.org/10.1006/jtbi.1993.1203
41. Santos ST (2013) Development of computational methods for the determination of biomass composition and evaluation of its impact in genome-scale models predictions. Universidade do Minho
42. Santos S, Rocha I (2016) Estimation of biomass composition from genomic and transcriptomic information. J Integr Bioinform. https://doi.org/10.2390/biecoll-jib-2016-285
43. Xavier JC, Patil KR, Rocha I (2017) Integration of biomass formulations of genome-scale metabolic models with experimental data reveals universally essential cofactors in prokaryotes. Metab Eng 39:200. https://doi.org/10.1016/j.ymben.2016.12.002
44. Dias O, Pereira R, Gombert AK, Ferreira EC, Rocha I (2014) iOD907, the first genome-scale metabolic model for the milk yeast Kluyveromyces lactis. Biotechnol J 9:776–790. https://doi.org/10.1002/biot.201300242
45. Sauer U, Lasko DR, Fiaux J, Hochuli M, Glaser R, Szyperski T, Wuthrich K, Bailey JE (1999) Metabolic flux ratio analysis of genetic and environmental modulations of escherichia coli central carbon metabolism. J Bacteriol 181:6679–6688
46. Brohée S, Barriot R, Moreau Y, André B (2010) YTPdb: a wiki database of yeast membrane transporters. Biochim Biophys Acta 1798:1908–1912. https://doi.org/10.1016/j.bbamem.2010.06.008
47. Saier MH, Reddy VS, Tamang DG, Västermark A (2014) The transporter classification database. Nucleic Acids Res 42:D251–D258. https://doi.org/10.1093/nar/gkt1097
48. Caspi R, Altman T, Dreher K, Fulcher CA, Subhraveti P, Keseler IM, Kothari A, Krummenacker M, Latendresse M, Mueller LA, Ong Q, Paley S, Pujar A, Shearer AG, Travers M, Weerasinghe D, Zhang P, Karp PD (2012) The MetaCyc database of metabolic pathways and enzymes and the BioCyc collection of pathway/genome databases. Nucleic Acids Res 40:D742–D753. https://doi.org/10.1093/nar/gkr1014

Chapter 2

Analyzing and Designing Cell Factories with OptFlux

Paulo Vilaça, Paulo Maia, Hugo Giesteira, Isabel Rocha, and Miguel Rocha

Abstract

OptFlux was launched in 2010 as the first open-source and user-friendly platform containing all the major methods for performing metabolic engineering tasks in silico. Main features included the possibility of performing microbial strain simulations with widely used methods such as Flux Balance Analysis and strain design using Evolutionary Algorithms. Since then, OptFlux suffered a major re-factoring to improve its efficiency and reliability, while many features were added in the form of novel plug-ins, such as the BioVisualizer and the over/under expression plug-ins. The current chapter described the main mathematical formulations of the major methods implemented within OptFlux, also providing a detailed guide on the usage of those functionalities.

Key words Computational biology and bioinformatics, Systems biology, Metabolic engineering, Flux-balance analysis

1 Introduction

In industrial biotechnology processes, microorganisms are usually used as microbial cell factories to produce chemical compounds of economic interest, from chemical building blocks, to food ingredients and bioplastics. In order to increase the yield, productivity, and specificity of these industrial processes, strain improvement methods were developed mostly based on random mutagenesis and on screening strains with naturally enhanced production levels [1, 2]. These techniques were labor intensive, but they were the best alternative for strain improvement before advanced genetic engineering procedures became available [3].

The buildup of knowledge on microbial metabolism, combined with the development of more sophisticated genome engineering techniques, gave rise to the rational modifications of microorganisms aiming to improve their phenotypical properties toward a certain goal. This discipline is usually referred to as Metabolic

Isabel Rocha and Miguel Rocha contributed equally to this work.

Marco Fondi (ed.), *Metabolic Network Reconstruction and Modeling: Methods and Protocols*, Methods in Molecular Biology, vol. 1716, https://doi.org/10.1007/978-1-4939-7528-0_2, © Springer Science+Business Media, LLC 2018

Engineering (ME) [3–6]. ME has been applied of a vast number of tasks over the past 20 years, due to the extraordinary growth in adoption of industrial biotechnological processes for the production of bulk chemicals, pharmaceuticals, food ingredients, enzymes, among other products [7, 8].

Many different approaches have been used to aid in ME efforts, from which Systems Biology (SB) deserves to be highlighted [9, 10]. SB addresses the computational and mathematical modeling of complex biological systems with the objective of examining the structure and dynamics of cells or organisms and understanding the properties of these systems. In other words, SB is focused on building and validating in silico models of biological systems that can be applied to generate novel, testable, and often quantitative predictions of cellular behavior, thus being able to support the rational development of optimized cell factories.

More recently, with the remarkable advances on genome sequencing technologies, culminated by the surge of the next-generation sequencing technologies [11], as well as semi-automated annotation techniques, an increasingly large number of fully annotated microbial genomes are being made available.

The advent of complete genomic sequences allowed the reconstruction of genome-scale networks that can be employed to generate models of diverse cellular processes, such as signaling transduction [12, 13], transcriptional regulation [14], and metabolism [15]. A reconstructed network is defined as a list of biochemical reactions occurring in a particular cellular system (such as metabolism) and the associations between these reactions and relevant proteins, transcripts and genes. A reconstruction can be converted to a mathematical model by including the assumptions necessary for the computational simulations, such as maximum reaction rates and nutrient uptake and production rates. An extensive collection of methods for analyzing metabolic genome-scale models (GSM) have been developed and applied to study a growing number of biological questions [16, 17].

The collection of the stoichiometry and reversibility of all the metabolic reactions from an annotated genome is the starting point for the construction of a GSM. Furthermore, many additional curation steps are required until models are complete. The full process of GSM reconstruction has been described in detail in several publications [18–20] and software tools that can help in the reconstruction process are also available (*see* Chapter 1 for more details) [21–24]. These software tools can be a great help in the reduction of the total time required to reconstruct a GSM and have proven extremely valuable for annotating genomes of less studied organisms.

Besides the stoichiometry and reversibility of all chemical reactions that can occur in a certain organism, GSMs can include additional details, such as the kinetic parameters for each enzymatic reaction. However, the availability of kinetic information is very scarce, which makes it very challenging to gather these data at the genome scale [25, 26]. Since the biochemical information included in most GSMs is limited to the stoichiometry and reversibility of all reactions, the application of these models is restricted to the steady-state modeling of intracellular fluxes [27, 28].

Since the advent of GSMs in 1999, when the *Haemophilus influenza* GSM was first published [29], the number of available GSMs grew to more than 100 models [30]. Some repositories facilitate the access to some of these models in standard formats, such as www.optflux.org/models or http://systemsbiology.ucsd.edu/InSilicoOrganisms/OtherOrganisms.

Many different approaches have been used to identify bottlenecks or targets for genetic engineering that take into account models together with mathematical tools and/or experimental data. Some of these techniques, like Metabolic Control Analysis (MCA), use dynamical representations of the metabolism, while others like Metabolic Flux Analysis (MFA) or Flux Balance Analysis (FBA) consider only the stoichiometry of the system to build a Constraint-Based Model, allowing the study of the phenotype of microorganisms, under different environmental and genetic conditions.

While the need for mathematical and computational tools to aid in ME efforts was already identified by James Bailey in 2001 [31], very few user-friendly software tools were available then and in the following years. *OptFlux* was introduced in 2010 [32] as a proposal to tackle this problem. OptFlux is a user-friendly software tool that aims to be a reference platform for the ME community, and it was developed with the objective of collecting several tools and algorithms that use GSMs in an integrated, extensible, and easy-to-use platform. Some of the main features of this tool are the following:

- Open-source—it allows all users to use the tool freely and invites the contribution of other researchers;
- User-friendly—facilitates its use by users with no/little background in modeling/informatics;
- Modular—facilitates the addition of specific features by computer scientists, given its plug-in based architecture;
- Compatible with standards—compatibility with the System Biology Markup Language (SBML) and layout standards as SBGN.

Optflux accommodates several tools and algorithms that have been developed to help in the analysis of GSMs. In the next sections, it will be explained how to apply these resources using Optflux.

2 Materials

OptFlux version 3 can be downloaded in www.optflux.org/downloads. The software is fully implemented in the Java Language. However, to perform the optimization of constraint-based model (CBM) problems, external software is required. Optflux has embedded two different open source external applications to solve these problems, CLP[1] and GLPK[2]. By default CLP is configured to solve Linear Programming (LP) and Quadratic Programming (QP) problems, while GLPK is configured to solve Mixed-Integer Linear Programming (MILP) problems. An interface for the commercial solver CPLEX[3] was implemented. CPLEX provides native support for several classes of constraint-based programming problems, such as LP, QP, and MILP.

Optflux was implemented in such a way that new features are easily plugged in. It is entirely implemented on the top of AIBench [33], a development framework that enforces the Model-View-Controller (MVC) design pattern, incorporating three types of well-defined artifacts: operations (controller), data types (model), and views. This leads to units of work with high coherence that can be easily combined and reused.

Furthermore, it is plug-in based: applications are developed by incorporating components, called plug-ins, each containing a set of functionalities, allowing the reuse and integration of functionalities from past and future developments. The management of OptFlux's plug-ins is easily achieved by a repository manager, which provides an intuitive graphical user interface (GUI), where users can select which plug-ins are installed or updated at each time.

All the OptFlux's source code is available in git repositories located in SourceForge[4], where it is also possible to submit support requests and bug tickets, keeping the developers aware of the problems reported by the users and helping in the planning of code development timelines.

Figure 1 depicts how OptFlux looks like and highlights the global layout of the platform. Most of OptFlux's main features and operations are accessible to the user either through the *Menu* or the *Toolbar*. Users can also access many of the operations by right-clicking in the *Clipboard* area. All the data types, i.e., projects, metabolic models, environmental conditions, simulation/optimization results, layouts for visualization, etc., are always placed in the *Clipboard* area, data type names can be changed by right clicking the object and selecting the "rename" option. The *Visualization*

[1] https://projects.coin-or.org/Clp

[2] https://www.gnu.org/software/glpk/

[3] https://www-01.ibm.com/software/commerce/optimization/cplex-optimizer/

[4] https://sourceforge.net/projects/optflux/

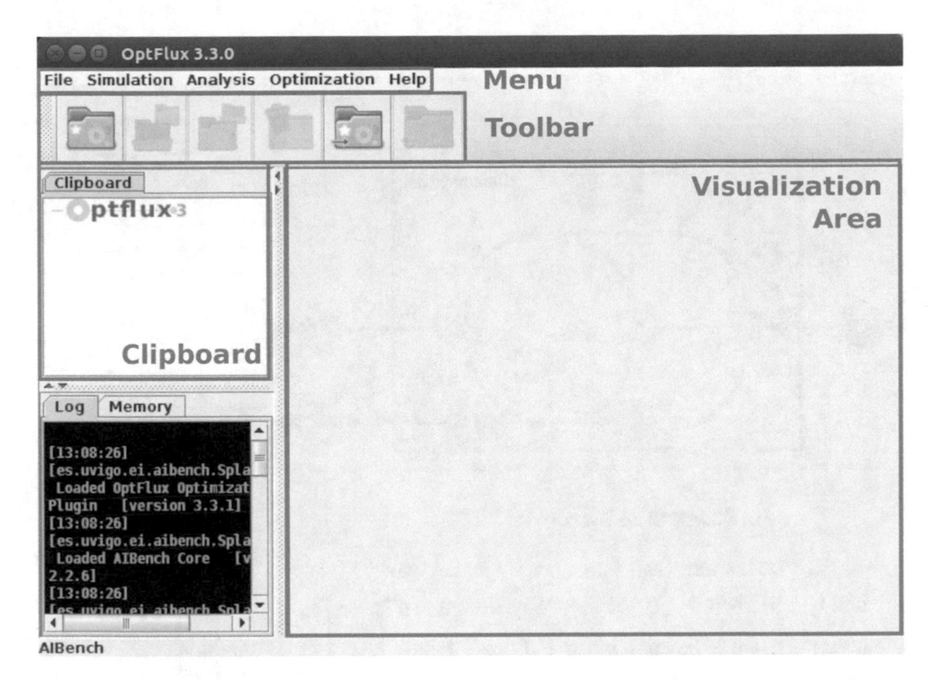

Fig. 1 The OptFlux application interface is split into four main parts identified in the figure: (a) Menu—where the user has access to all the operations existent in OptFlux application; (b) Toolbar—where the user has access to some important short-cuts for operations related to projects such as create, open, close, delete, import, and export project. (c) Clipboard—where the data types will be placed after being created by the operations. (d) Visualization area—where the user will be able to see the data types' contents when clicked on the top of those

Area is where the user can examine those data types in greater detail. When the user clicks in a data type, different views for that object are displayed in this area.

3 Methods

3.1 Constraint-Based Modeling

The formulation of Constraint-based models requires two levels of metabolic information (Fig. 2a). First, metabolic stoichiometry is required to write down all chemical reactions that take place in a metabolic network of interest. The second item is the information of the demands that are placed on the metabolic system. These demands include non-growth-associated maintenance requirements, biomass synthesis, and nutrients inputs. The non-growth-associated maintenance requirements can be obtained from strain-specific experiments [34]. In terms of the biomass synthesis, it is necessary the inclusion of an artificial reaction in the network that represents the cellular growth. Such a reaction is built taking into account the contribution, in millimoles per gram of cell dry weight, of all macromolecules or building blocks and essential cofactors, based on the biomass composition. The next Equation represents

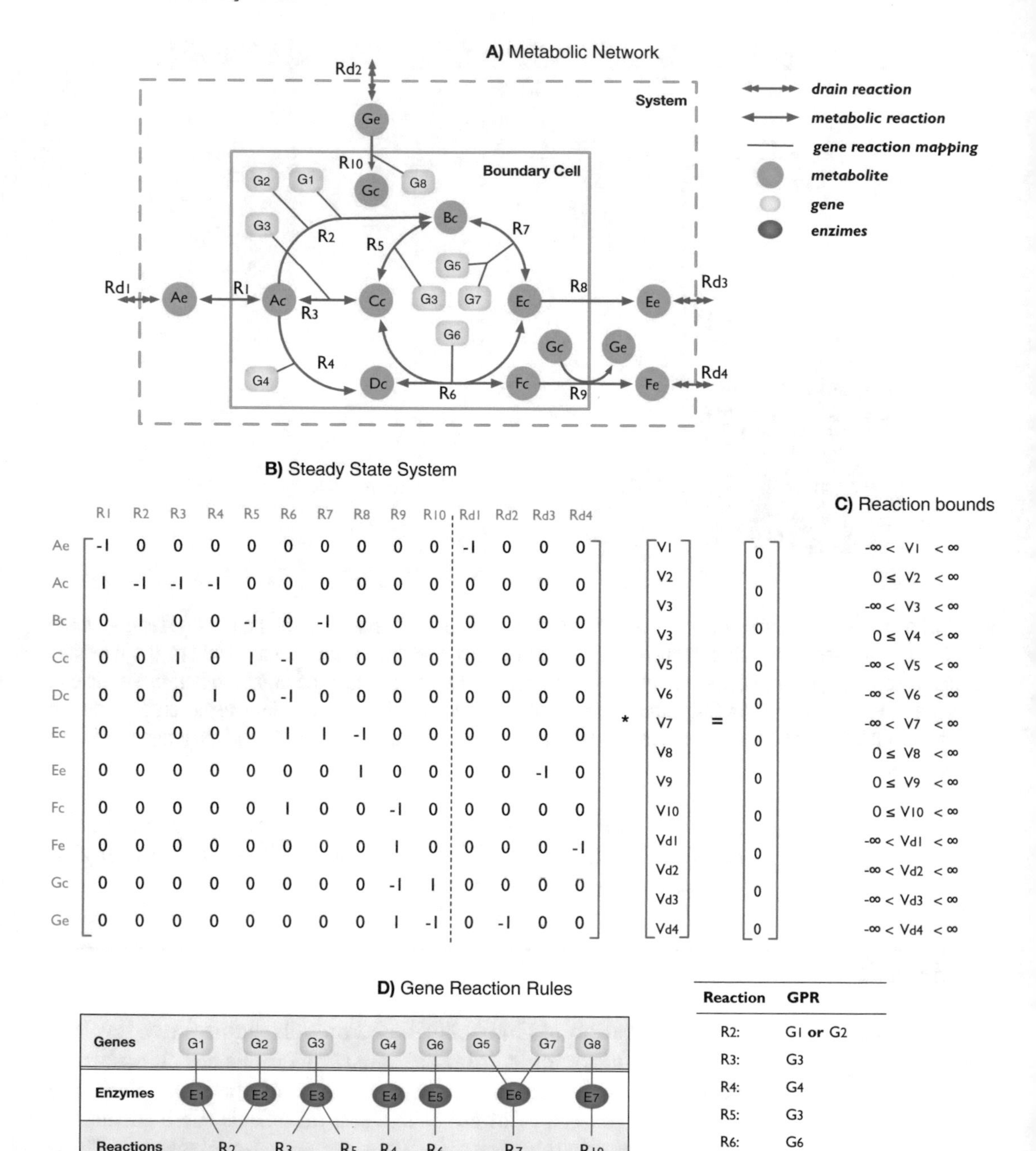

	R1	R2	R3	R4	R5	R6	R7	R8	R9	R10	Rd1	Rd2	Rd3	Rd4
Ae	-1	0	0	0	0	0	0	0	0	0	-1	0	0	0
Ac	1	-1	-1	-1	0	0	0	0	0	0	0	0	0	0
Bc	0	1	0	0	-1	0	-1	0	0	0	0	0	0	0
Cc	0	0	1	0	1	-1	0	0	0	0	0	0	0	0
Dc	0	0	0	1	0	-1	0	0	0	0	0	0	0	0
Ec	0	0	0	0	0	1	1	-1	0	0	0	0	0	0
Ee	0	0	0	0	0	0	0	1	0	0	0	0	-1	0
Fc	0	0	0	0	0	1	0	0	-1	0	0	0	0	0
Fe	0	0	0	0	0	0	0	0	1	0	0	0	0	-1
Gc	0	0	0	0	0	0	0	0	-1	1	0	0	0	0
Ge	0	0	0	0	0	0	0	0	1	-1	0	-1	0	0

$* \; [V1, V2, V3, V3, V5, V6, V7, V8, V9, V10, Vd1, Vd2, Vd3, Vd4]^T = [0, 0, 0, 0, 0, 0, 0, 0, 0, 0, 0, 0, 0, 0]^T$

$-\infty < V1 < \infty$
$0 \leq V2 < \infty$
$-\infty < V3 < \infty$
$0 \leq V4 < \infty$
$-\infty < V5 < \infty$
$-\infty < V6 < \infty$
$-\infty < V7 < \infty$
$0 \leq V8 < \infty$
$0 \leq V9 < \infty$
$0 \leq V10 < \infty$
$-\infty < Vd1 < \infty$
$-\infty < Vd2 < \infty$
$-\infty < Vd3 < \infty$
$-\infty < Vd4 < \infty$

Reaction	GPR
R2:	G1 **or** G2
R3:	G3
R4:	G4
R5:	G3
R6:	G6
R7:	G5 **and** G7
R10:	G8

Fig. 2 In the figure a small network is represented (**a**) composed by various components commonly used in constraint-based models such as metabolites, genes, reactions, and drains. The corresponding steady-state system (**b**), the bounds of the reactions (**c**), and gene protein rules (**d**) are shown

the biomass equation, where M represents the biomass components and the ∂m represents the contribution of these components for the biomass in mmol/(g of biomass).

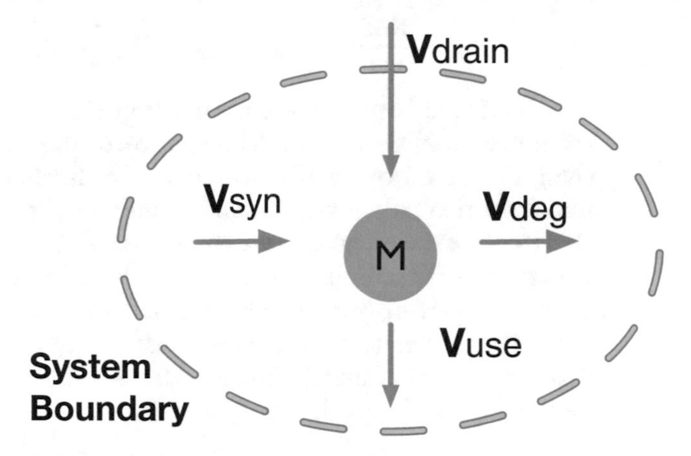

Fig. 3 Constraint-based models use the balance around each metabolite to represent a metabolic network. In the figure, all types of transformations that could be present in each CBM metabolite are represented. The V_{drain} represents the fluxes carrying in and out a metabolite to and from the system. These fluxes deal with the nutrients, as well as the outputs including the by-products of the system. V_{use} represents the fluxes required for the metabolic requirements, such as growth and maintenance. V_{syn} are the fluxes that synthesize each metabolite and V_{deg} are the fluxes that represent the consumption of the metabolite

$$\partial m \cdot M \rightarrow \text{biomass}$$

Such an equation is used to calculate the growth rate of the organism in h^{-1}, and thus the other fluxes on the model are in mmol/gDW/h units.

In order to define the availability of nutrients and excretion of products in/to the medium, some special reactions are built and included in the system, denominated by drains. Drains define which are the metabolites that could be inputs and/or outputs of the modeled organism. The next equation illustrates how those special reactions look like; the reaction is only composed by one metabolite M which is usually defined as a reactant. Thus, when its flux is positive, metabolite M is an output of the system, while, if the flux is negative it is an input of the system.

$$M \leftrightarrow$$

The mass balance of a specific metabolite M is illustrated in Fig. 3. In the figure, four different rates that can perturb the concentration of M are represented. V_{drain} represents the rate responsible for the translocation of metabolite M. Those rates will be dealing with requirements of the system, i.e., nutrients, as well as the by-products. V_{syn} are the fluxes that synthesize the metabolite and V_{deg} are the fluxes that consume metabolite M. V_{use} represents the rate of utilization of metabolite M for metabolic requirements, such as growth and maintenance. Thus, the conservation law [35] can be applied allowing the representation of the accumulation of metabolite M as

$$\frac{dM}{dt} = V_{syn} - V_{deg} - V_{use} - V_{drain}$$

CBMs are based on the assumption that metabolic transients are more rapid than both cellular growth rates and the dynamic changes in the organism's environment. Metabolism typically has transients that are shorter than a few minutes, and thus it is reasonable to place the metabolic system in a steady state when investigating aspects of metabolism related to growth. At the steady state, the concentrations of metabolites in a network are constant and the fluxes that generate a metabolite must be equivalent to the fluxes that consume that metabolite. So, the accumulation of an internal metabolite M is zero, as stated in

$$0 = V_{syn} - V_{deg} - V_{use} - V_{drain}$$

By applying the same principle to all metabolites, a homogeneous system of linear equations can be defined as

$$S^* v = 0$$

In the system, S is an $m \times n$ stoichiometric matrix, for a set of m metabolites and a set of n reactions, and v is the vector of n reaction fluxes. In this representation, it is assumed that all the reactions are mass balanced (Fig. 2b). For each reaction of a metabolic network, bounds for their minimum and maximum values can be configured to reflect thermodynamic feasibility, i.e., reaction directionality, and flux capacity of the reactions (Fig. 2c):

$$lb \leq vi \leq ub$$

Further details, such as translational/transcriptional representation in the form of Gene-Protein-Reaction (GPR) associations, are also typically included in the models [36]. The representation of GPR associations usually resorts to Boolean logic, where the relationships between reactions and their encoding genes are modeled as logical AND/OR operations. Typically, the AND operator represents the formation of protein complexes, and the OR operator usually represents genes encoding isoenzymes. This allows the identification of which genes are responsible for coding a reaction in the model (Fig. 2d).

3.1.1 Importing Constraint-Based Model in Optflux

Optflux allows several ways of loading a constraint-based model. Due to the absence of clear standards to collect this type of information prior to 2007, each author or CBM platform used its own specific format. With the release of the Systems Biology Markup Language (SBML) in 2003 [37] and subsequent availability of easy-to-use libraries for exchanging models in this format (libSBML) in 2008 [38], this format started to be adopted and became the most used standard for collecting CBM models.

Nonetheless, there are other legacy formats that the community still uses. Because of that, OptFlux is shipped with readers that provide support for loading models from several formats, namely:

1. SBML levels 2 and 3, with the option of using the flux balance constraints plug-in (FBC). OptFlux is able to import models using the SBML specification from level 2 version 1 onward;
2. Metatool reader specification[5] [39];
3. CellDesigner (SBML specification extended with layout information from CellDesigner);
4. Table format (useful to ease the burden of importing models from Excel datasheets—a format commonly used by authors to publish their new reconstructions).

Furthermore, OptFlux includes access to a model repository—an internal repository of different models and organisms, which makes the access to commonly used models much less cumbersome to OptFlux users. Currently, the repository provides access to nearly 50 models.

All of these readers are available via the same operation in OptFlux *menu: File/new Project*. This operation has four main steps, also depicted in Fig. 4:

1. Specify the project name and choose one of the available readers;
2. Define reader parameters; this step is different depending on the chosen reader and can have more than one sub-step;
3. Identify/define drains in the loaded model;
4. Identify biomass reaction.

One of the most important steps in the process of correctly importing a model is the definition of drains. Until this day, there are no standards to unambiguously define the drains on CBM models. Some authors opted for adding boundary metabolites in the model that should be removed during the import process. Other models have been created with no drains, being necessary to create them for all the external metabolites. Finally, some authors include all the necessary drains in the model, effectively eliminating the need for the user to do anything other than correctly importing the model matrix.

The user should understand how the drain reactions are dealt with in the model source file and select the applicable method in OptFlux:

- Do not create drains—if drains are already present in the model;
- Remove external metabolites—if it is necessary to delete boundary metabolites to create the drains;

[5] http://pinguin.biologie.uni-jena.de/bioinformatik/networks/metatool/metatool5.0/ecoli_networks.html

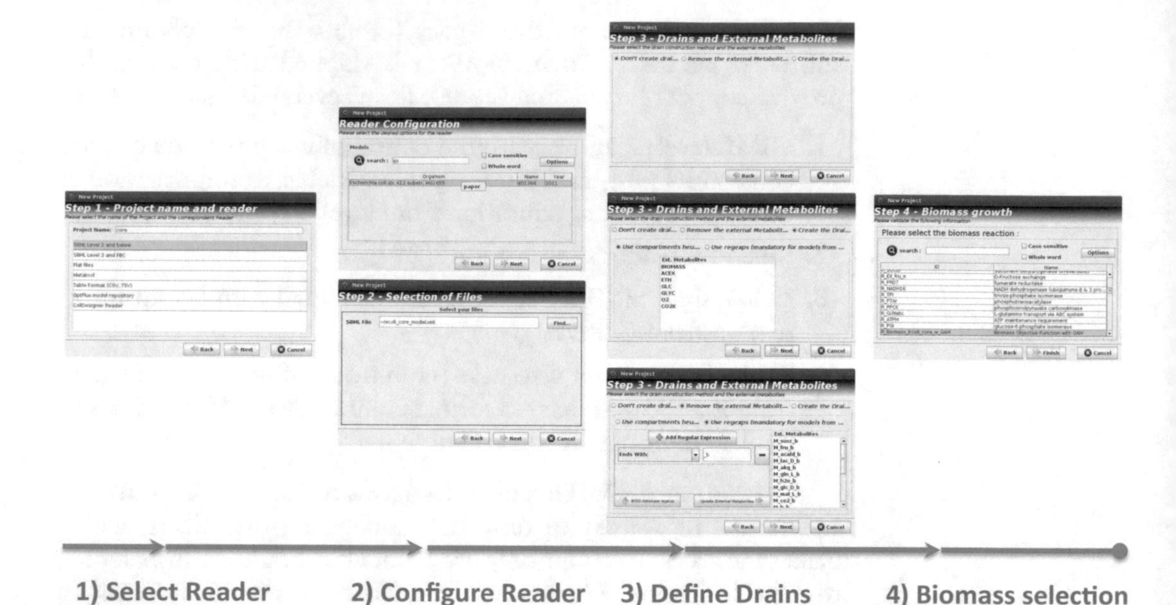

Fig. 4 The process of a project creation in OptFlux includes four main steps: (1) The Reader Selection where the user selects a specific reader to construct a project. OptFlux comes with readers for several formats such as SBML, CellDesigner, Table Format, and Model repository. (2) The reader configurations. Different readers can have different inputs and in this step the user is able to configure those inputs. In the figure, two different reader configurations are shown: the first one is the configuration for the model repository where the user selects one of the models present on the database (she/he can inspect the respective paper by clicking in the second button of the mouse), and the second is the SBML reader where the user only has to select a valid SBML file. (3) Define drains.—Different formats and different authors of models have different forms to define the drains on the models so it is necessary to select the method in this step. (4) Biomass selection, where the user is able to select a reaction to be the biomass artificial reaction; this will help on the usage of several operations on OptFlux

– Create drains—if it is necessary to create new drain reactions for the external metabolites.

When selecting either the "*Remove external Metabolites*" or the "*Create Drains*" *options*, the user must also select the method by which the external metabolites are detected. OptFlux provides two different ways of doing this:

– Use compartment heuristic—OptFlux begins by identifying the external compartment and subsequently mapping all the metabolites associated with it.

– Use regular expressions—the user is prompted to define a regular expression that OptFlux uses to compute the matching metabolite identifiers.

In Fig. 5, three examples are shown. On the first (line 1) the network already contains drain reactions so the option that should be chosen is "Don't create drains." On the second example (line 2), the network does not contain drain reactions, but there are

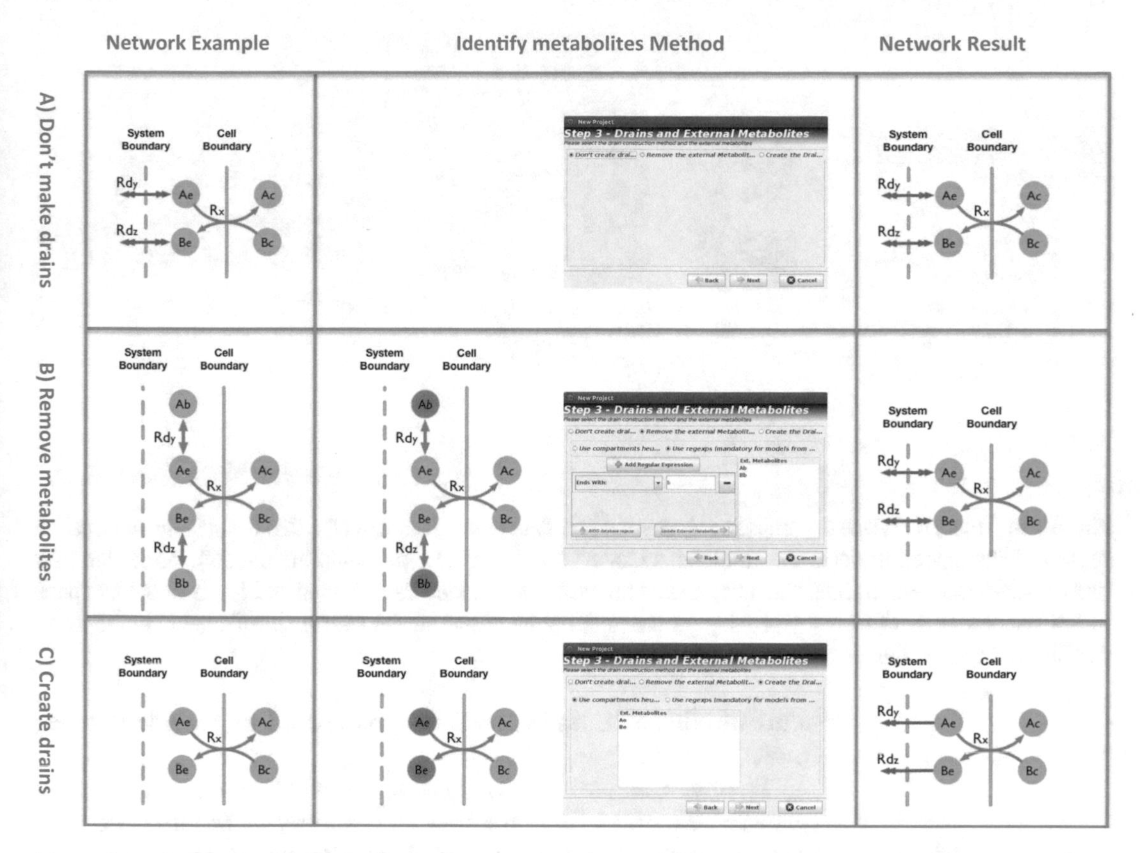

Fig. 5 In the figure, three standard examples of how drains could be stored in the models are schematized (first column), the best option to use in OptFlux to define drains (column 2) and the resulting network after OptFlux defines the drains (column 3)

reactions in the model that contain artificial metabolites that when removed will transform these reactions into drains. So, the option that should be selected in this case is "Remove external metabolites." Those artificial metabolites are called "boundary metabolites" and for OptFlux to remove them, the user should identify them first. On those examples, the boundary metabolites are in the same compartment as external metabolites so the user can use a regular expression in OptFlux to detect them (in this example "Ends with" "b").

In the third example (line 3), the network does not contain drains and no artificial metabolites are available, thus the drains need to be created. Since the metabolites where the drains should be added are the external metabolites, the user can use one of the two strategies to identify the metabolites: "External compartment," where all the metabolites that should have a drain are in the same external compartment, or use a regular expression ("Ends with" "e" on this example).

OptFlux can also use a heuristic method to identify the correct drain configuration, and select the best option by default. Being a

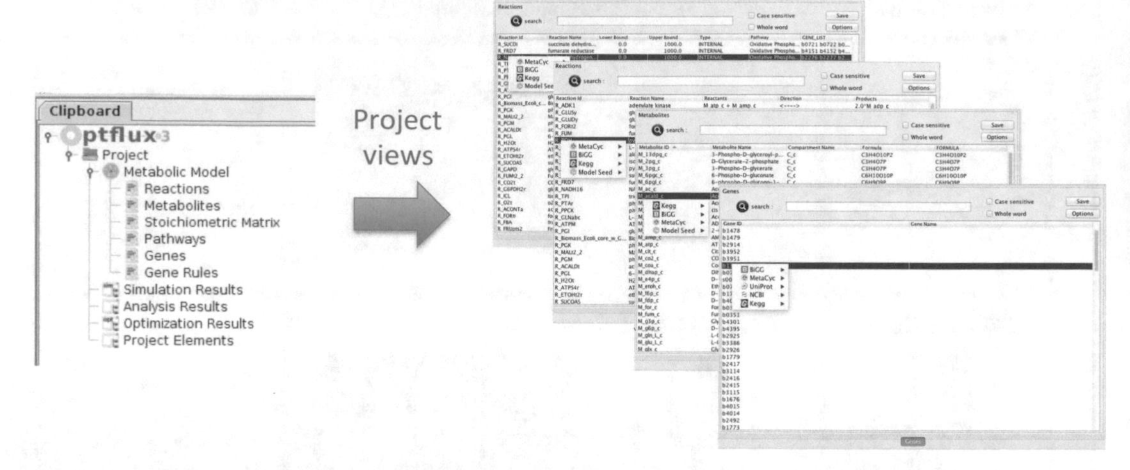

Fig. 6 The Project data type is composed by several sub-data types. The data type that collects the information regarding the loaded model is the "Metabolic Model" (the first data type inside the project). Inside, the user has detailed information about the components present in the model. Each component has a view that opens in the visual area of OptFlux (by clicking on the right button of the mouse over a specific entity in the table, OptFlux will search the entity in several databases)

heuristic, this method is not 100% guaranteed to provide the best result.

Finally, the user is also asked to select a default biomass (growth) equation from the model. This step is required because many operations in OptFlux, such as phenotype predictions and strain optimization, among others, require the biomass equation to be specified in the objective function. By setting the default biomass equation, the repeated use of many of these operations becomes greatly simplified. OptFlux also uses a heuristic to select a reaction that it believes to be the correct one, but the user should always confirm this and change it if necessary.

In the end of the "*New Project*" operation, a new project will appear in the clipboard section with the name chosen by the user in **step 1**. The core data type of the project is named "Metabolic Model" (Fig. 6). Inside, the user can access detailed information about reactions, metabolites, and also the stoichiometry of the system in a human-readable way. Furthermore, genes, gene rules, and pathway information are shown if available.

3.2 Visualization of Data Using CBMs

Within the field of Systems Biology, the analysis of different types of biological networks is an important task in understanding the underlying biological processes. Data can be associated with the biological components facilitating its visualization and interpretation. Visualizing data in this way contextualizes and enriches the dataset. Data-rich visualizations have been extremely valuable for viewing, interpreting, and communicating data. Two-dimensional pathway maps have long been a popular visual representation of metabolic pathways and other biological pathways.

CBM models are not easy to understand in their mathematical form, which turns their analysis and interpretation into a very hard and time-demanding task. To aid in this task, visualization tools have been developed to complement 2-dimensional pathways maps with CBM outputs. CellDesigner [40, 41] is one of the most popular tools for creating, editing, and visualizing biochemical networks, but it lacks specific methods for CBMs. Alternatively, Cytoscape [42] became a standard tool for the integrated analysis and visualization of biological networks, but faces the same problem of the CellDesigner platform, missing a native CBM specification. Regarding these problems, some platforms that integrate visualization technologies with the CBM methodology have been developed. The BioVisualizer platform [43], available as a plug-in for OptFlux, is one of them. BioVisualizer provides a visualization framework based on a well-defined abstract representation of metabolic pathways and CBMs. This plug-in provides a linkage between the constraint-based models and pathway layouts, natively understanding how to interact with each other and allowing the highlight of properties or results based on the use of the CBM.

3.2.1 Optflux Visualization Capabilities

OptFlux allows the creation of pathway layouts using the BioVisualizer plug-in [43]. In the current version of Optflux, this plugin comes already pre-installed.

The visualization plug-in provides all the functionalities related to the visualization and editing of the metabolic layout. One of these features is the specification of the default colors and shapes of the nodes. The graphical user interface (GUI) is composed of two major elements (Fig. 7): the network view, where it is possible to edit the network and click/drag the nodes (Fig. 7a), and the side panel where filters, overlaps, node information, zoom, and export functionalities are available (Fig. 7b). In this way, it is possible for the user to easily explore the network, using all the features the interface has to offer. The pathway layout used by the visualizer is the Force Directed Layout (FDL) [44] with adaptations to support fixed nodes. This was coupled with the possibility to fix/unfix nodes, allowing the user to fix a node to the specific position it is in, or drag it to a desired new position; unfixing a node will remove the position information of the node, making it susceptible to the FDL algorithm to adjust its position according to its surroundings. It is also possible to unfix and fix nodes by type, allowing a user to fix/unfix all reaction or metabolite nodes at the same time.

BioVisualizer is also able to customize loaded layouts, by clicking in the right mouse button over nodes on network view, where several procedures are available, namely:

1. Changing the node type. It allows the user to mark a metabolite node as a "currency metabolite," a special type of node. Often, highly connected metabolite nodes, the so-called

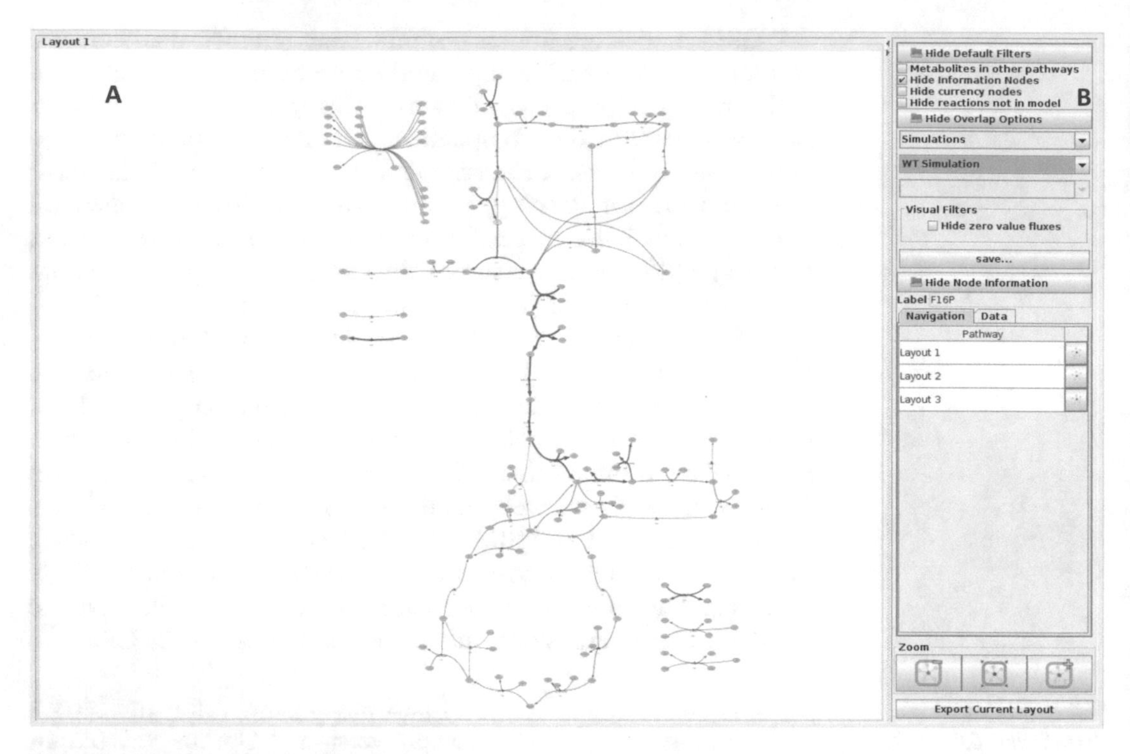

Fig. 7 BioVisualizer view. The view is divided into two different areas. The network viewer (**a**) where the users can access the nodes of the network and apply some functionality by clicking on the right button of the mouse. The side panel (**b**) with functionalities like node/edge filters, overlap selection, zooming options, navigation, and exporting the layout in several formats

"currency metabolites," like energy-carrying molecules (ATP, NAD(H), NADP(H), etc.) can be marked to clearly identify this type of node and possibly hide them to unclutter the pathway visualization.

2. Replicating a node. Highly connected nodes can be replicated to reduce the number of overlapping connections in the layout.
3. Highlight replicated nodes. Replicated nodes are highlighted when selected.
4. Merging equivalent nodes. Replicated nodes can be merged (one to one).

All these features, when combined with the import and export capabilities, allow the user to create and edit their own layouts, being able to export them for later use in this or other compatible software tools.

Filtering and overlaying capabilities are also provided. It is possible to filter the network, by hiding parts of it, based in the node type (e.g., hide all currency metabolites) or by reaction identifier.

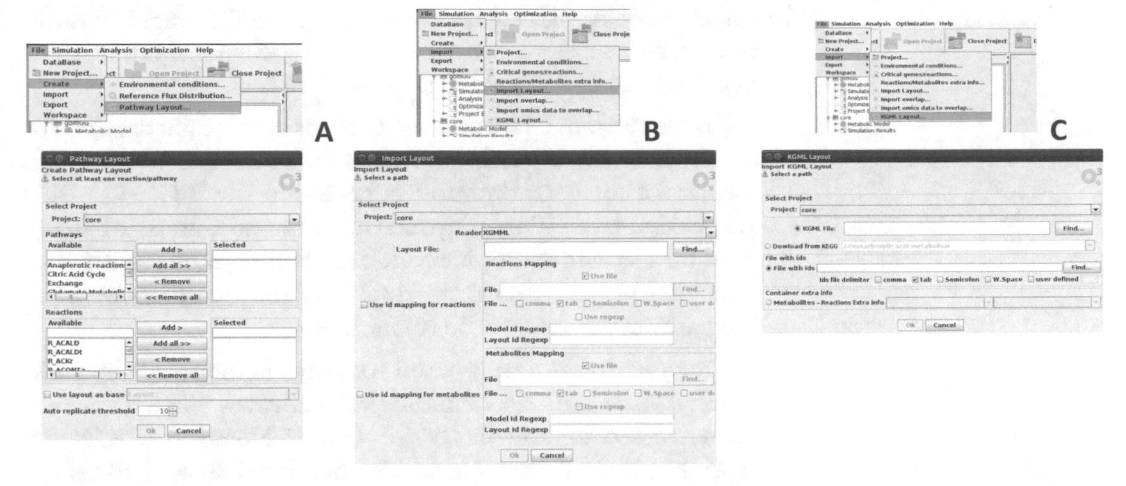

Fig. 8 Three different ways to create a layout for a model in OptFlux: creating a layout using model information (**a**), loading a layout file (**b**) or importing a KEGG layout (**c**)

It is also possible to overlay information over the network. The visualizer allows altering the visual aspect of the network components, supporting the change of the direction, thickness, and colors of the edges, while for nodes it is possible to change the color and shape. This feature allows, for instance, overlaying flux distributions in the metabolic pathway layouts.

3.2.2 Loading/Creating Layouts

The Biovisualizer plugin supports the creation of layouts using information existent in a model or the import of layouts in different formats. There are three different operations to import layouts in OptFlux (Fig. 8): the "Pathway Layout" operation in *menu:File/Create/Pathway Layout*, the "Import Layout" operation in *menu: File/Import/Import Layout*, and the "KGML Layouts" in menu: File/Import/KGML Layout.

Using the "Pathway Layout" operation (Fig. 8a) the user can build a layout using reaction stoichiometry information present in the model. This can be done by following two possible strategies: (1) choosing a list of reactions to be represented in the layout, or (2) in the case the model provides pathway information, builds the layout directly with the reactions from a selected set of pathways.

In the "Import Layout" operation (Fig. 8b) it is possible to select one of the following formats:

1. eXtensible Graph Markup and Modeling Language (XGMML): a format based on the Graph Modeling Language (GML), used for graph description using XML tags to describe nodes and edges. It is used in different tools such as Cytoscape [42].
2. CellDesigner SBML (CD-SBML): graphical notation system proposed by Funahashi and coworkers [40, 41], where layouts are stored using a specific extension of SBML.

3. Systems Biology Graphical Notation (SBGN): graphical notation system to represent networks of biochemical interactions in a standard and unambiguous way [45].
4. COBRA version 1 layouts: set of maps specifically created for the COBRA Toolbox. Several maps were constructed using this format for many of the models hosted in the BiGG-v1[6] [46] knowledge-base. These layouts have the advantage of being valid for use with multiple models containing similar pathways, since they provide a correct mapping of the identifiers between the pathway layout and BiGG models.
5. Escher[7]: Escher is a web-based tool for building, viewing, and sharing visualizations of biological pathways [47]. It uses a format based on the JavaScript Object Notation (JSON) to represent the layouts. Those files had been created for some of the models hosted in the BiGG-v2[8] [48].

The "KGML Layout" operation (Fig. 8c) allows the user to import a KEGG pathway[9] to OptFlux. To do so, the user can choose a KGML file that can be downloaded from the KEGG database or choose directly the pathway name to be retrieved from KEGG. This interface also provides the tools to map a metabolic model with a pathway layout. It is possible to load a 2-column file providing the mapping between the metabolic identifiers in the model and KEGG identifiers. If the model loaded in OptFlux has already that information, then it is possible to map it automatically.

3.3 Simulation Using CBM

3.3.1 Simulation Methods

On the previous chapter, it was presented how a system of homogeneous linear equations can be derived, assuming that there is no accumulation of metabolites, and thus all fluxes leading to their formation and degradation are mass-balanced [49, 50]. This system is representative of an entire metabolic phenotypic space of a specific organism strain. Such systems are typically underdetermined since the number of reactions (variables of the system) normally exceeds the number of metabolites (constraints of the system), so, a plurality of solutions exists, expressed in an infinite number of possibilities of distribution of metabolic fluxes through available reactions. These possibilities are constrained by the stoichiometric matrix and the flux limits, forming a domain of stoichiometrically feasible behaviors.

The feasible domain of a constraint-based model can be conceptualized as the "metabolic genotype," i.e., all reactions that can be catalyzed with enzymes codified by the genes in the genome of

[6] http://bigg1.ucsd.edu
[7] https://escher.github.io/
[8] http://bigg.ucsd.edu/
[9] http://www.genome.jp/kegg/pathway.html

the modeled strain. A particular point in this feasible domain is a specific flux distribution that can be conceptualized as a "metabolic phenotype" since this flux distribution is a characteristic of how the strain responds to an input, as a medium or stress. Therefore, constraint-based models are able to identify how a strain adjusts its metabolic distribution to an environmental perturbation, such as an alternative carbon source, or even to genetic modifications, such as gene deletions.

A particular solution for the distribution of metabolic fluxes may be found using constraint-based optimization by stating an objective and seeking its maximum/minimum value, within the stoichiometrically defined domain. Such objective encodes a biological rationale, mimicking how a cell choses the distribution of metabolic fluxes on the domain of feasible behavior. However, the mechanisms by which metabolic flux distributions are chosen by the cell are a complex interplay of enzymes, genetic regulatory, and signaling events and not all of those events are known in detail. Nonetheless, in terms of the evolutionary selection process it is expected that, for a given environment, a given cell chooses to express and regulate a specific set of metabolic enzymes, which act in concert to produce an "optimal metabolic flux distribution," which may be thought of as the "metabolic phenotype" of that strain under those conditions. Such "optimal flux distribution" depends on the objective used by the cell, such that in past years several phenotype prediction methods have been proposed on the top of the same steady-state assumptions, but differing on the objective function and/or additional constraints.

Flux Balance Analysis (FBA) was the first optimization-based phenotype prediction method developed for predicting the phenotypes using CBM models of metabolism. Briefly, FBA finds a distribution of fluxes that meets a certain objective formulated with Linear Programming (LP). The most common objective function used with FBA is the maximization of the specific growth rate, encoded in a biomass pseudo-equation, assuming that the metabolic phenotype of a wild-type strain is defined by a tendency to optimize its growth rate. In a practical way, FBA joins the underdetermined system of linear equations with the objective of maximizing the growth reaction flux. The growth reaction (explained in detail above) describes the consumption of biosynthetic precursors and energy requirements for synthesizing a specific amount of cellular material [49, 51–53].

One handicap of the flux distributions obtained with FBA is that they are not unique. The values of the reaction fluxes in the metabolic network can vary for the same optimal value of the objective function (i.e., multiple flux distributions can lead to the same maximum growth rate).

Parsimonious enzyme usage FBA (pFBA) [54, 55] is a two-step LP optimization problem, which tries to reach the simplest flux distribution under the maximum growth rate. In the first step, a normal FBA is issued to compute the optimal growth rate, while on the second step a minimization of the sum of the absolute values of all the reaction fluxes is performed, while the optimal growth rate calculated in the first level is maintained as a constraint. pFBA removes some artifacts in the flux distribution that do not contribute for the maximum growth and/or are physiologically unsound, such as fluxes related to futile cycles in the network. The assumption of this method is based on the efficiency of metabolic networks, which expects the cell to perform a certain task with the minimal amount of resources (amount of proteins and genes), thereby predicting the most resource-wise efficient flux distribution for the maximum growth rate.

While the maximum growth rate assumption has been well accepted by the scientific community as a good approximation to mimic the phenotypic behavior of a wild-type strain, the same cannot be said when the system is subjected to a perturbation, such as a gene deletion or a stress condition in the environment. To more accurately predict how a metabolic network reacts to a genomic perturbation, distinct phenotype prediction methods have been proposed.

The Minimization Of Metabolic Adjustment (MOMA) [56] was developed in 2002 and it was the first proposed formulation with the purpose of simulating the effect of genetic perturbations in a metabolic network. This methodology assumes that a mutant cell (i.e., a cell with a perturbation in its genome like a gene deletion) will try to minimize the adjustments of the flux values in comparison with its behavior on the wild-type strain. MOMA is formulated as a Quadratic Programming problem, assuming its objective function as the minimization of the Euclidean distance between the mutant set of fluxes and the reference wild-type fluxes. The growth predictions of gene knockouts simulated with MOMA are more conservative than the results obtained with FBA and it has been shown that MOMA can predict more accurately gene essentiality in some cases [56, 57].

Shlomi et al. pursued the same concept of minimal metabolic adjustment, but in a different perspective [58]. Instead of minimizing the flux differences between the mutant and the wild-type strains, the methodology entitled Regulatory On/Off Minimization of metabolic flux changes after genetic perturbations (ROOM) minimizes the number of reactions that are activated or deactivated in a mutant in comparison with a reference flux distribution. This approach requires the introduction of binary variables in the objective function, thus converting the problem into a Mixed-Integer Linear Programming problem, increasing its complexity. The assumption behind ROOM is that, when faced with a set of knockouts, a cell will adjust its internal fluxes by making the minimum amount of regulatory changes, i.e., the magnitude of the fluxes can change, but the set of active enzymes should be similar to the wild-type organism. The predictions obtained for ROOM were closer to FBA than MOMA and revealed that MOMA is better at estimating transient metabolic adaptations, while FBA and ROOM can better predict the phenotype of an evolved knockout mutant [58].

In addition to the regular MOMA formulation, two additional variations are available in the literature: linear MOMA (LMOMA) [59] and PSEUDO [60]. One common issue usually encountered in the flux distributions computed with MOMA is that it favors a large number of small flux changes in detriment of a few large changes in the metabolic network. This is caused by the quadratic formulation used to calculate the flux distance in MOMA and can be solved by using LMOMA, which uses the Manhattan distance between the reference and perturbed networks to find the flux phenotype of the knockout mutant. Another issue that might arise from using MOMA/LMOMA is the importance given to a reference flux distribution. Usually, the reference set of fluxes is calculated using FBA or pFBA [61–63] and any error in this flux distribution will be propagated to all the predictions. The methodology developed in PSEUDO can tackle this issue by not using a single flux distribution as a reference but a region of the flux space delimited by a minimum threshold imposed on the biomass yield [60]. The authors of this methodology reported some improvements over MOMA and FBA flux predictions [60].

Another issue encountered in the formulation of MOMA and LMOMA was the dependence of the mutant phenotype on the scale of the stoichiometry of the metabolic reactions [64]. By using different stoichiometric representations of a metabolic network that are biochemically equivalent, Brochado et al. showed that the simulation outcome of MOMA/LMOMA was sensitive to the stoichiometric representation chosen for the network [64]. Since biochemically equivalent networks should produce the same results, the authors proposed a new methodology entitled Minimization of Metabolites Balance (MiMBl). The formulation underlying MiMBl solves the stoichiometry dependence of other algorithms by using the metabolite turnovers as the variables in the objective function. Instead of minimizing the changes in the fluxes in comparison with a reference network, MiMBl minimizes the changes in the turnovers of all metabolites in the network. As a consequence, MiMBl provides more robust results, which are not dependent on the numerical stoichiometric representation chosen to describe a metabolic network.

3.3.2 Flux Variability Analysis

A key issue that may arise with the use of constraint-based phenotype prediction methods is the existence of alternative optimal solutions within the same maximal objective (e.g., growth rate), which can be achieved through different flux distributions. Flux Variability Analysis (FVA) is an efficient LP-based strategy used to calculate the full range of values of each flux that can be present, to achieve optimal or suboptimal objective states. This technique is one of the most used to investigate the limits of the feasible domain of a constraint-based model.

FVA can be a useful tool, not only for determining the ranges of the fluxes that could be achieved using a specific CBM model, but also to determine how a flux can vary in an optimal or suboptimal solution determined by a phenotype prediction method. In this latter case, the value of the objective function used by that specific phenotype prediction method is computed in a first step, and afterward its value is set as restriction, followed by two new optimizations to calculate the maximum and the minimum flux values for a specific reaction given those constraints.

Through the use of FVA it is possible to determine the domain of all the possible solutions (usually represented by an *n*-dimensional polyhedron called the flux cone) where each dimension refers to a flux) and to analyze how two fluxes influence each other.

To do so, it is necessary to select a target flux and a pivot flux. The operation begins by splitting the range of all possible values of the pivot flux in multiple steps. Afterward, for each step, the value of the pivot flux is fixed and the maximum and minimum values for the target flux are computed. These values can be plotted into a two-dimensional surface that can be thought of as the projection of the flux cone into the dimensions of the pivot and the target fluxes.

This technique can be very useful to understand how an input/output of the system influences the biomass flux (phase plane analysis), or to analyze the robustness of a production strain (production envelope analysis).

To help illustrate the first point (phase plane analysis), Fig. 9 is put in place, where the relationship between a pivot and a target

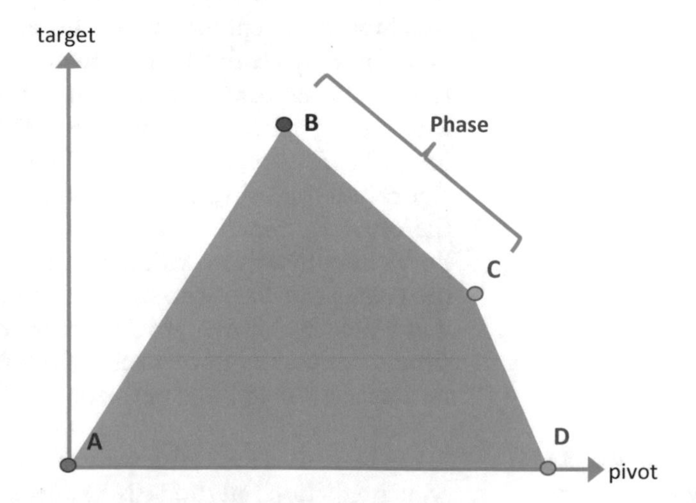

Fig. 9 Surface plot. The range of values of a pivot flux is split and the minimum and maximum values of the target flux are calculated generating in this way a poligon. In such poligon it is easy to identify the minimum and the maximum that the pivot flux could achieve (a and d) the maximum value of the target flux (b) and where there are changes in the phase (c)

flux are represented. In a closer inspection of the plot, it is possible to identify four different points—A, B, C, and D. Points A and D represent the minimum and maximum values of the pivot flux, point B represents the maximum achievable value of the target flux for the corresponding and optimal pivot flux value (b′). Finally, point C represents a phase change and c′ the corresponding pivot flux threshold value (prompting the phase change). Phase changes can be detected by analyzing the first derivatives of the points of the plot (i.e., the slope of the line) and correspond to the reduced cost of the pivot flux. The reduced costs are associated with the LP variables and represent the change in the objective function of the LP problem, promoted by a small change in the pivot flux. A positive reduced cost translates into an increase in the value of the objective function, while a negative one represents a decrease. These values are important, since they allow us to identify phenotypic phase changes. While no change is observed in the reduced cost, it can be assumed that the system does not need to change its behavior to comply with its restrictions. Alternatively, when a change in the reduced cost is observed, it can be assumed that a phase limit has been reached where the system needs to change its behavior to keep complying with its set of restrictions. These system behavior changes are translated to phenotypic changes. As an example, a phase change can be detected when simulating *E. coli* with oxygen limitations. By gradually reducing the oxygen uptake, it is possible to detect a phase where *E. coli* starts producing acetate and another where *E. coli* produces ethanol, formate, and acetate (via mixed-acid fermentation).

Another informative analysis that can be achieved via FVA is the robustness analysis of a production strain (production envelope analysis). In this case, FVA is also used to generate a plot similar to the one previously explained, usually called the production envelope. The production envelope is used to analyze the phenotype of a strain that is already producing a target product. The procedure to calculate this plot is the same as before, but in this case the pivot is usually the biomass flux and the target is the product flux. In a typical production envelope, there are three important points to keep in mind; A—the theoretical maximum production rate of the target metabolite, B—the maximum production rate of the target when the biomass is in its maximum value, and C—the threshold point of the pivot flux, where the minimum value of the target flux reaches 0. As exemplified in Fig. 10, analyzing the position of these three points in the plot allows the characterization of the solution robustness as:

1. Non-robust—when points B and C are aligned in the maximum of biomass. This means that there is a range of possible values for the target product, including 0 (no production).

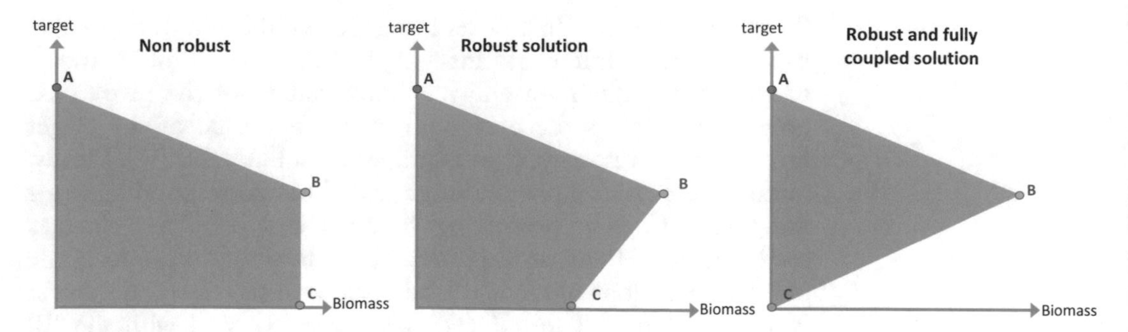

Fig. 10 Surface plots used to characterize productive phenotypes

2. Robust solution—when point C is not aligned with B, meaning that there is a range of biomass values between C and B, where there is a guaranteed minimum production of the target compound.
3. Robust and fully (biomass) coupled solution—when point C is overlapped with 0 in the biomass axis and not aligned with point B, meaning that the production of the target compound if guaranteed for the whole range of possible pivot (biomass) values.

The analysis of the results can reveal the best tradeoff between biomass formation and target product secretion.

In conclusion, FVA can be used to study the entire range of achievable metabolic functions, as well as the redundancy in optimal phenotypes [65].

3.3.3 OptFlux Operations

Environmental Conditions

When a project is loaded into OptFlux, the metabolic model is created with default lower and upper flux bounds that are defined in source information (e.g., SBML file). Those values can be analyzed in the reactions data type within the metabolic model. To override these default values, an environmental condition (EC) can be created. The EC is typically used to define the medium where the organism is growing, by defining lower/upper bounds in the drains available in the system. An EC could be also used to block reactions, change reversibilities or impose maximum capacities for some reactions. There are two different ways to create an EC in OptFlux (Fig. 11): importing a pre-existent environmental condition from file or creating it from scratch.

Importing Environmental conditions using text files can be performed using the operation on *menu:File/Import/Environmental conditions* (Fig. 11a). In this operation the user needs to select the project where the EC will be imported using a tabular data file. A tabular data file is a plain text file, where each line is a data record consisting of one or more fields separated by a delimiter character such as commas or tabs. In EC tabular files, each record is expected to have three fields, the first one is the reaction identifier (id), the

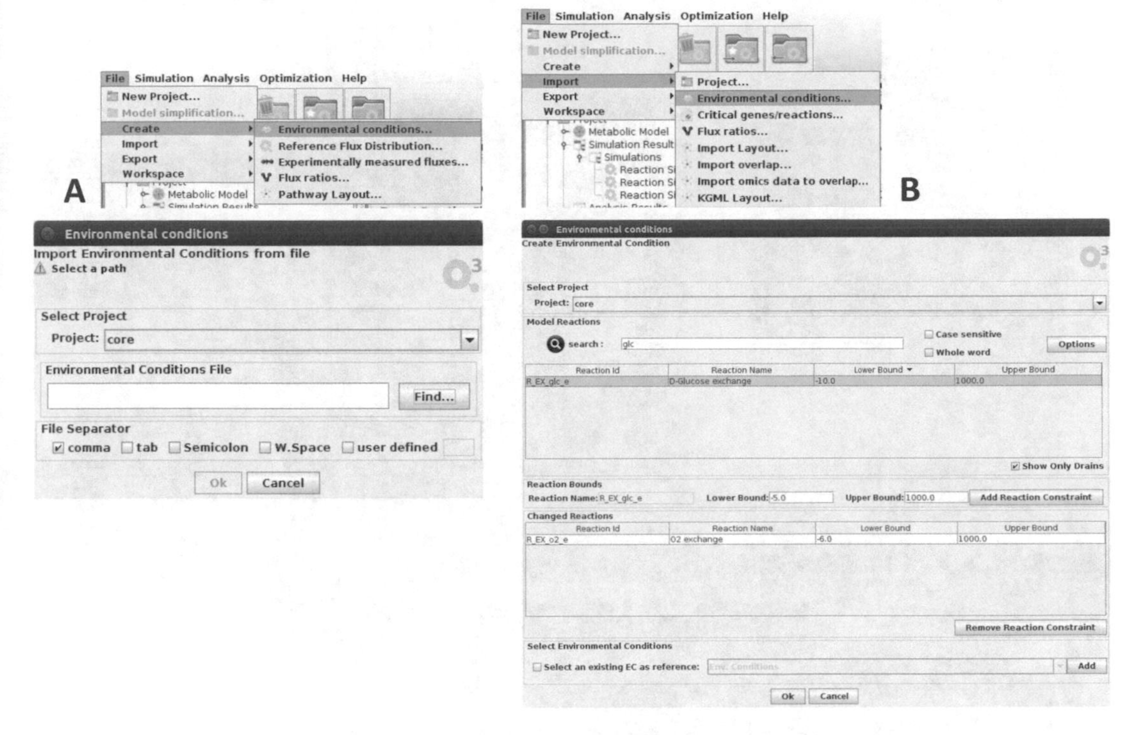

Fig. 11 Two different ways to create an environmental condition in OptFlux. Loading environmental conditions from a file (**a**), or creating an environmental condition by setting reaction bounds individually (**b**)

second the lower bound and the third the upper bound. The file separator can also be defined in the operation.

To create an EC from scratch the user must access the operation via *menu:File/create/Environmental conditions* (Fig. 11b). The operation allows the user to select reactions, one by one and define their limits. By default the checkbox for "Show Only Drains" is selected making only the drains visible, assuming that the user only has to define a growth medium. However, the user is able to define the bounds of any reaction in the model by unchecking that box. The upper part of the operation allows the selection of the reactions (easing its search), while the bottom part allows adding these constraints to the EC, specifying the lower and upper bounds. The user can also add all reaction bounds contained in a previously defined EC.

Phenotype Simulation

Optflux has implemented five operations to perform phenotype simulations each one with different assumptions to perform genetic modifications:

1. Wild-Type Simulations accessible in the *menu:simulation/wild type* (Fig. 12a).
2. Reaction Knockout simulations accessible in the *menu:simulation/knockouts/reaction* (Fig. 12b).

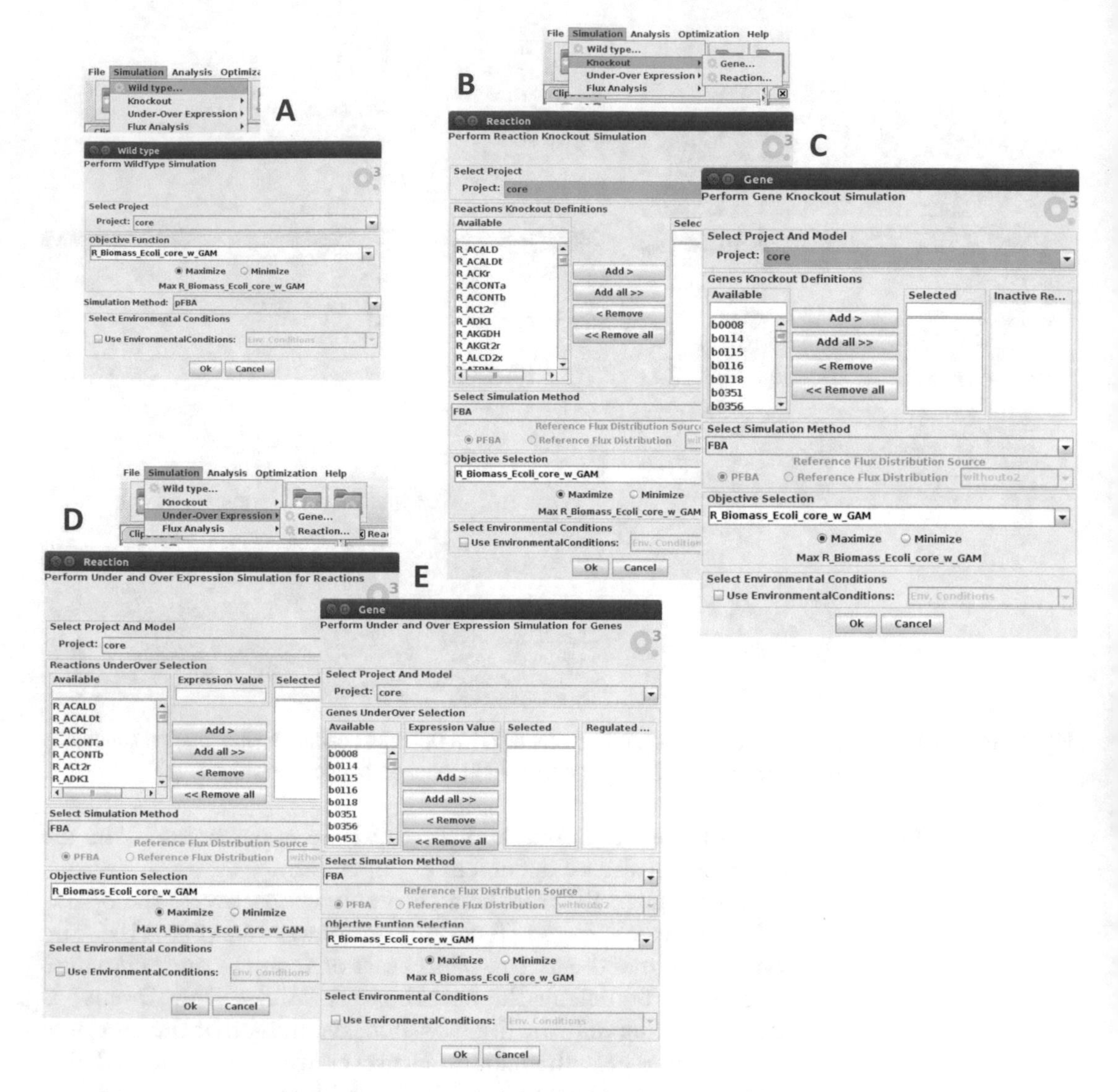

Fig. 12 Five different types of phenotype simulations to simulate in Optflux. Wild type (**a**), reaction knockouts (**b**), gene knockouts (**c**), reaction under/over-expression (**d**), and gene under/over-expression (**e**). In all the operations, the user is able to also select a simulation method, the biomass reaction, and the environmental condition

3. Reaction Under/Over-expression simulations accessible in the *menu:simulation/Under-Over Expression/reaction* (Fig. 12d).
4. Gene Knockout simulations accessible in the *menu:simulation/knockouts/gene* (Fig. 12c).
5. Genes Over/Under-expression simulations accessible in the *menu:simulation/Under-Over Expression/gene* (Fig. 12e).

For all of these operations it is possible to select the OptFlux project that will be used, an environmental condition (if it is necessary to override the default bounds existing in the model), the phenotype simulation method and some parameters for the objective function (method specific).

In terms of wild-type simulations, two simulation methods are available, FBA and pFBA; for the objective function it is required to specify a target reaction for optimization (by default the biomass equation is selected) and an objective function sense, i.e., maximizing or minimizing the selected flux (maximization is selected by default).

For other types of phenotype simulations (such as reaction/gene knockout/over-under expression—*see* **Note 1**), in addition to the FBA and pFBA methods, the user can also select MOMA, LMOMA, ROOM, and MiMBl. These additional methods share the common characteristic of requiring a reference flux distribution to run, and the way this reference is selected depends on the user preference. There are two possibilities; (1) the default option, where OptFlux performs a pFBA simulation and uses the resulting flux distribution as the reference; and, (2) select a previously computed phenotype prediction as the reference flux distribution.

The major difference between these operations is the type of genetic changes that will be simulated. When performing the wild-type simulation operation, no genetic changes are imposed upon the model, while for all the other operations perturbations at the reaction or at the gene level are allowed. For knockouts, two operations are available, allowing the selection of either a set of reactions or a set of genes to knockout. Finally, for over/under expression operations, a pFBA flux distribution is assumed as the reference and a set of reactions or genes is selected. For each entity (reaction or gene) the user is asked to define an expression value (over/under) relative to that of the reference flux distribution. For additional information consult the sections above and **Note 1**.

The result of any phenotype simulation operation is an object created in the clipboard, placed within the Simulation Results list, under the Simulations data type. The phenotype simulation result data type collects multiple information regarding the performed operation, which is displayed in several tabs (Fig. 13):

1. Simulation solution: In this tab, OptFlux presents some of the selected inputs for the simulation, such as the method name and the selected ECs, as well as the value for the objective function obtained with the selected phenotype prediction method, the biomass value, and the net conversion of the system. The net conversion represents the net rates of the metabolites that are consumed and produced, providing an overall perspective of the predicted consumptions and secretions for the current simulation.

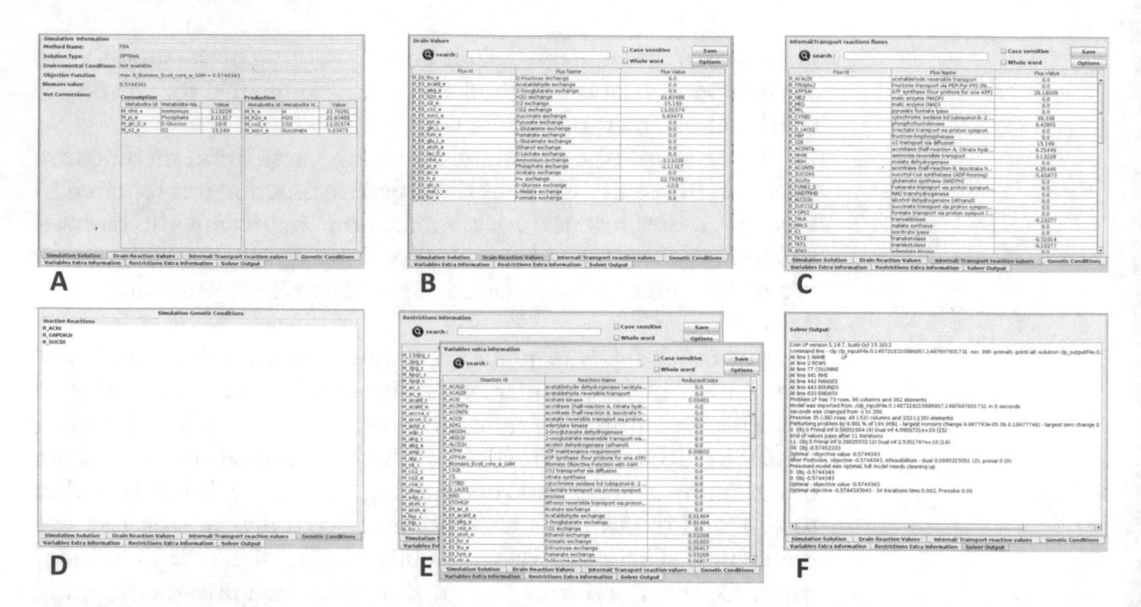

Fig. 13 Results for phenotype simulation in an example where three reactions were knocked out (R_ACKr, R_SUCDi, and R_G6PDH2r) using an E. coli core model [51]. (**a**) Phenotype simulation method and the net conversion showing the inputs and outputs of the system; (**b**) Drain reactions' fluxes; (**c**) Internal reactions' fluxes; (**d**) Genetic alterations; (**e**) Extra information regarding the variables and restrictions of the problem like shadow prices and reduced costs; (**f**) Raw solver output

2. The values of the fluxes are split between two views: one displays the drain reaction values and the other displays the Internal/Transport reaction values. In these views, the information is presented in a tabular format, allowing the user to search for specific information (via an intelligent search bar), hide the rows containing zero values, as well as exporting the information to a table format file.
3. Genetic conditions: allows inspecting the genetic conditions (including knockouts, over-under expression) applied. In the wild-type operations, no genetic conditions are shown.
4. Variables and restrictions extra information: extra information associated with the variables and the restrictions, such as shadow prices and reduced costs.
5. Solver output: textual output of the solver used in the constraint-based optimization problem.

Flux Variability Analysis

As previously mentioned to investigate the limits of the feasible domain of a CBM model or the alternative optimal solutions of the phenotype prediction methods, an FVA operation can be performed.

OptFlux provides two different operations concerning FVA analysis (Fig. 14):

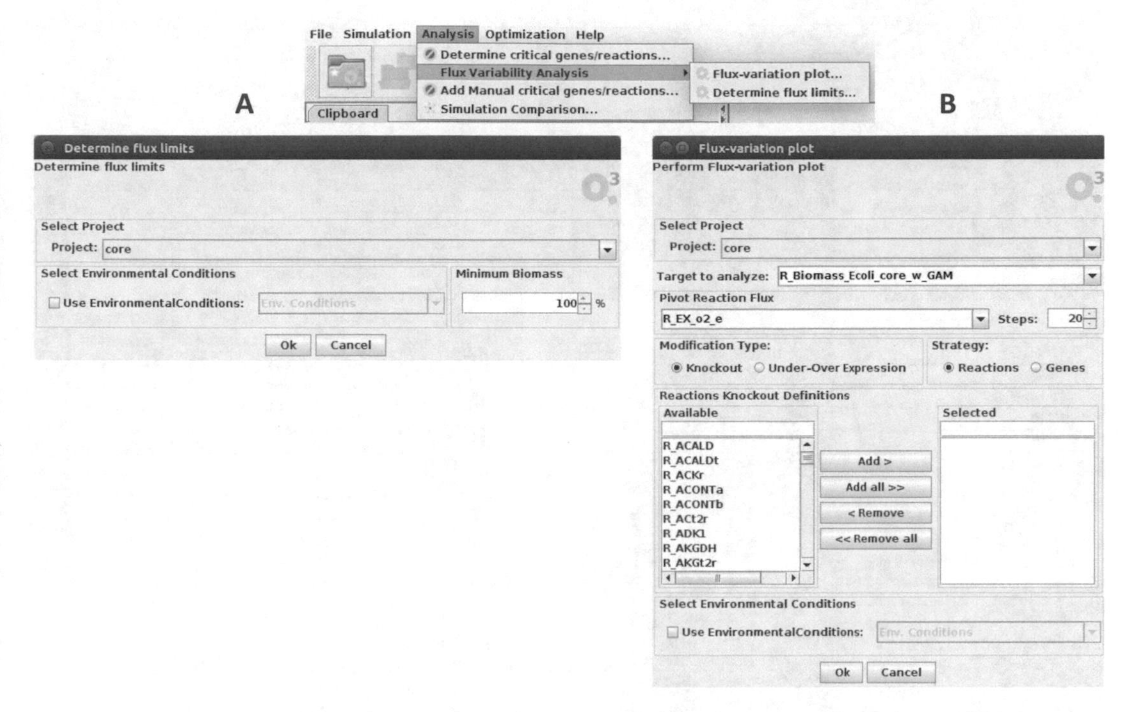

Fig. 14 OptFlux GUIs for accessing the two different ways of analyzing flux variability: determine all fluxes' ranges (**a**); determine a surface selecting a pivot and a target flux (**b**)

1. Determine flux limits in *menu:Analysis/Flux Variability Analysis/Determine flux limits* (Fig. 14).
2. Flux-variation plot in the *menu:Analysis/Flux Variability Analysis/Flux-variation plot* (Fig. 14).

The "Determine flux limits" operation is used to determine the domain of all fluxes present in the CBM model, by restricting the system to a defined level of the biomass flux. Such a level is a percentage of the wild-type value (maximum biomass value). In the configuration of the operation, the user selects the project, the biomass level used as a restriction, and can also select an EC. The result of the operation is a new object in the clipboard, placed within the "Analysis Results" list, under the "Flux Limits" data type. The result contains the minimum and maximum values possible for all the fluxes (Fig. 15).

The "Flux variation plot" operation is used to build a 2-dimentional projection of the solution space. To perform this operation the user must select a pivot and a target flux, and (optionally) can also impose a genetic perturbation (gene/reaction knockout or gene/reaction under/over expression). Using this operation the user is able to compute a phase plane analysis or an envelope production analysis depending on the chosen fluxes to pivot and the target (see above for more details). The result of the operation is

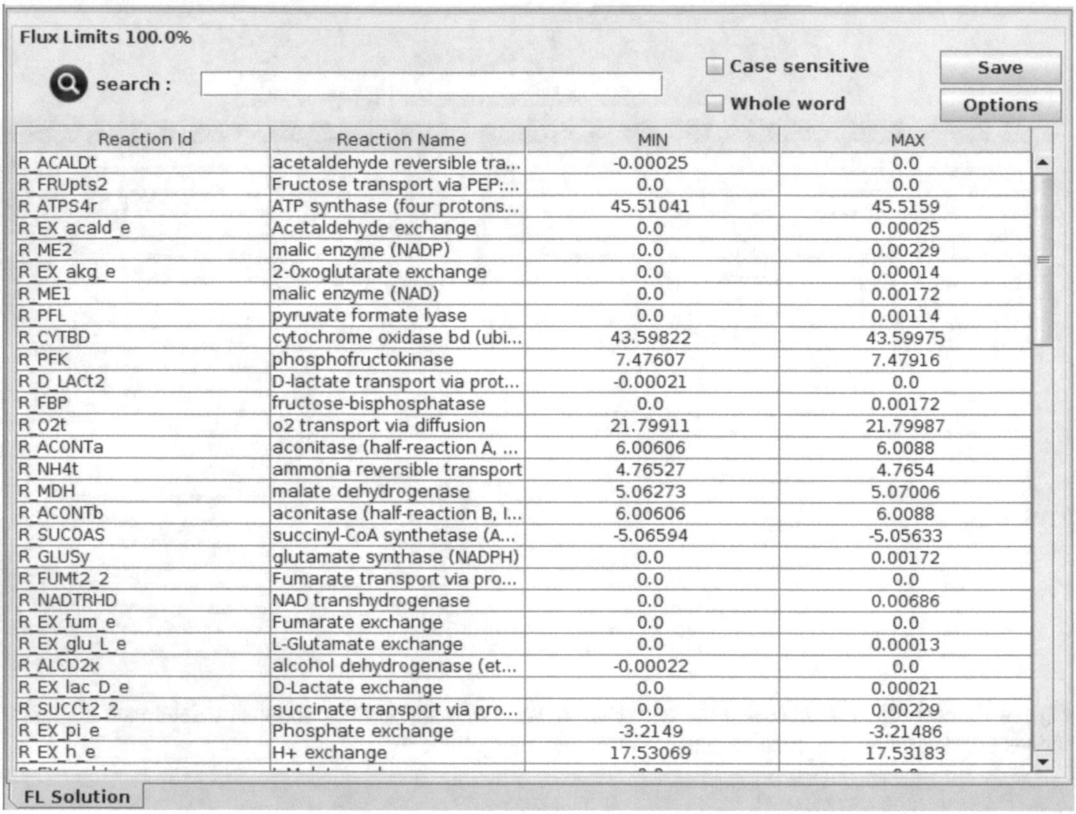

Flux Limits 100.0%

search : | Case sensitive | Whole word | Save | Options

Reaction Id	Reaction Name	MIN	MAX
R_ACALDt	acetaldehyde reversible tra...	-0.00025	0.0
R_FRUpts2	Fructose transport via PEP:...	0.0	0.0
R_ATPS4r	ATP synthase (four protons...	45.51041	45.5159
R_EX_acald_e	Acetaldehyde exchange	0.0	0.00025
R_ME2	malic enzyme (NADP)	0.0	0.00229
R_EX_akg_e	2-Oxoglutarate exchange	0.0	0.00014
R_ME1	malic enzyme (NAD)	0.0	0.00172
R_PFL	pyruvate formate lyase	0.0	0.00114
R_CYTBD	cytochrome oxidase bd (ubi...	43.59822	43.59975
R_PFK	phosphofructokinase	7.47607	7.47916
R_D_LACt2	D-lactate transport via prot...	-0.00021	0.0
R_FBP	fructose-bisphosphatase	0.0	0.00172
R_O2t	o2 transport via diffusion	21.79911	21.79987
R_ACONTa	aconitase (half-reaction A, ...	6.00606	6.0088
R_NH4t	ammonia reversible transport	4.76527	4.7654
R_MDH	malate dehydrogenase	5.06273	5.07006
R_ACONTb	aconitase (half-reaction B, I...	6.00606	6.0088
R_SUCOAS	succinyl-CoA synthetase (A...	-5.06594	-5.05633
R_GLUSy	glutamate synthase (NADPH)	0.0	0.00172
R_FUMt2_2	Fumarate transport via pro...	0.0	0.0
R_NADTRHD	NAD transhydrogenase	0.0	0.00686
R_EX_fum_e	Fumarate exchange	0.0	0.0
R_EX_glu_L_e	L-Glutamate exchange	0.0	0.00013
R_ALCD2x	alcohol dehydrogenase (et...	-0.00022	0.0
R_EX_lac_D_e	D-Lactate exchange	0.0	0.00021
R_SUCCt2_2	succinate transport via pro...	0.0	0.00229
R_EX_pi_e	Phosphate exchange	-3.2149	-3.21486
R_EX_h_e	H+ exchange	17.53069	17.53183

FL Solution

Fig. 15 Results of performing a "Determine flux limits" operation using E. coli core model using 100% of wild-type simulation as biomass restriction. On the result the user can inspect the minimum and maximum limits of all fluxes

a new object in the clipboard, placed within the "Analysis Results" list, under the "FVA Simulations" data type. The information of the new data type is displayed in several tabs:

1. "FVA Chart," where the user can visualize a plot that represents the projection of the two selected fluxes (pivot and target).
2. "FVA values," where the user can see the maximum and minimum values of the target flux restricted to a value of pivot flux, as well as the reduced costs of the pivot variable for the minimum and maximum simulations.
3. "FVA information," here the user can inspect the pre-configured parameters used to perform the operation.

In Figs. 16 and 17, an example of a phase plane analysis and an example of an envelope production analysis are represented, respectively.

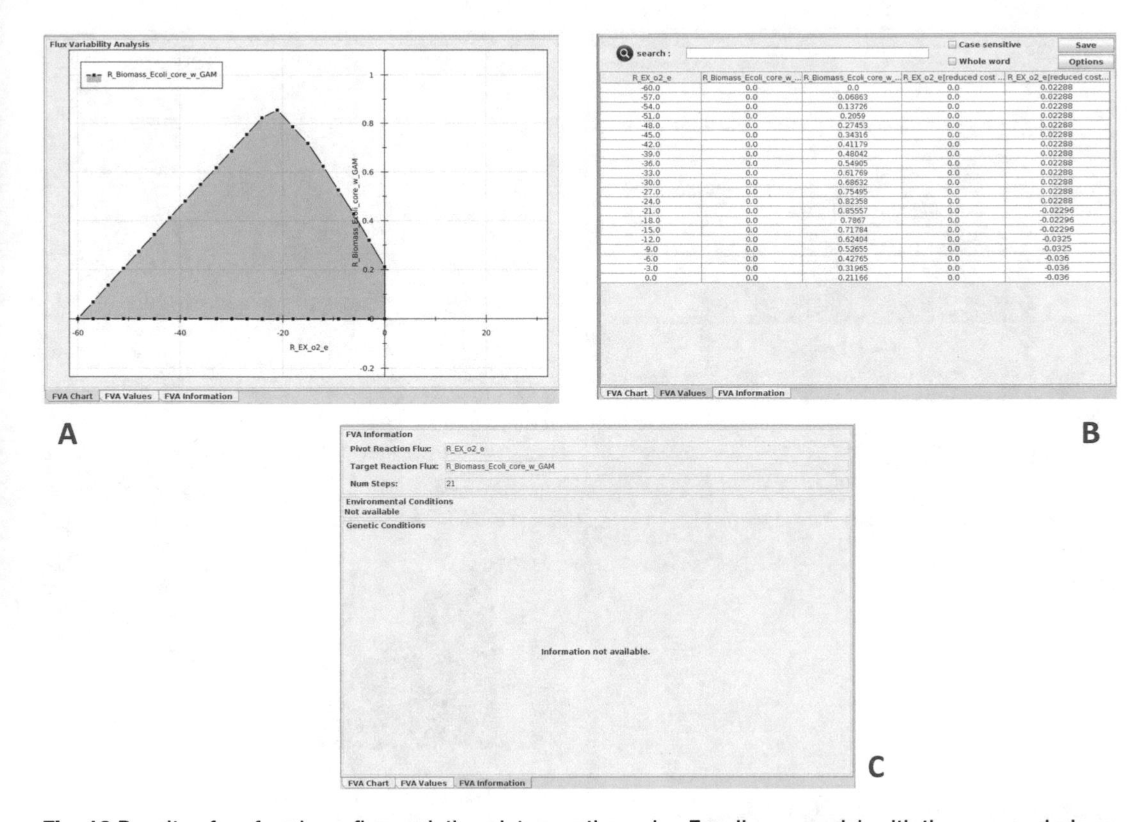

Fig. 16 Results of performing a flux-variation plot operation using E. coli core model, with the oxygen drain as pivot flux and the biomass as target flux. In the FVA chart view (**a**), the user is able to see the shape of the phase plan generated by the operation. In the FVA values view (**b**), the user is able to see all the steps calculated for the pivot flux, the minimum and maximum values of the target and also the reduced costs for the pivot target. In this example, by analyzing the reduced costs (last column) it is possible to detect four different phases: Phase 1 [−60, −21] mmol/gDWh of the oxygen input flux, where oxygen is decreasingly harmful for growth; Phase 2 [−21, −12] mmol/gDWh of the oxygen input flux, where E. coli produces Acetate as a co-product; Phase 3 [−12, −9] mmol/gDWh of the oxygen input flux, where E. coli produces Acetate and Formate as by-products and Phase 4 [−6, 0] mmol/gDWh of the oxygen input flux, where E. coli produces Acetate, Formate and Ethanol as by-products of the mixed-acid fermentation process. In the FVA information view (**c**), the user is able to see a summary of the configured parameters to reach the result.

Critical Genes/Reactions

A reaction or gene is considered critical (or essential) when its knockout leads to an unviable phenotype strain. Critical information is usually used not only to validate the models, but also to restrict the search when the objective is to construct a mutant strain to overproduce a compound of interest. In OptFlux, there are two operations that are able to build a critical reactions/genes set (Fig. 18):

1. *menu:Analysis/Determine Critical Genes/Reactions*—where OptFlux will perform a single knockout to genes or reactions assuming as critical all those where their knockout leads to a value of biomass less that 5% of the referenced value calculated

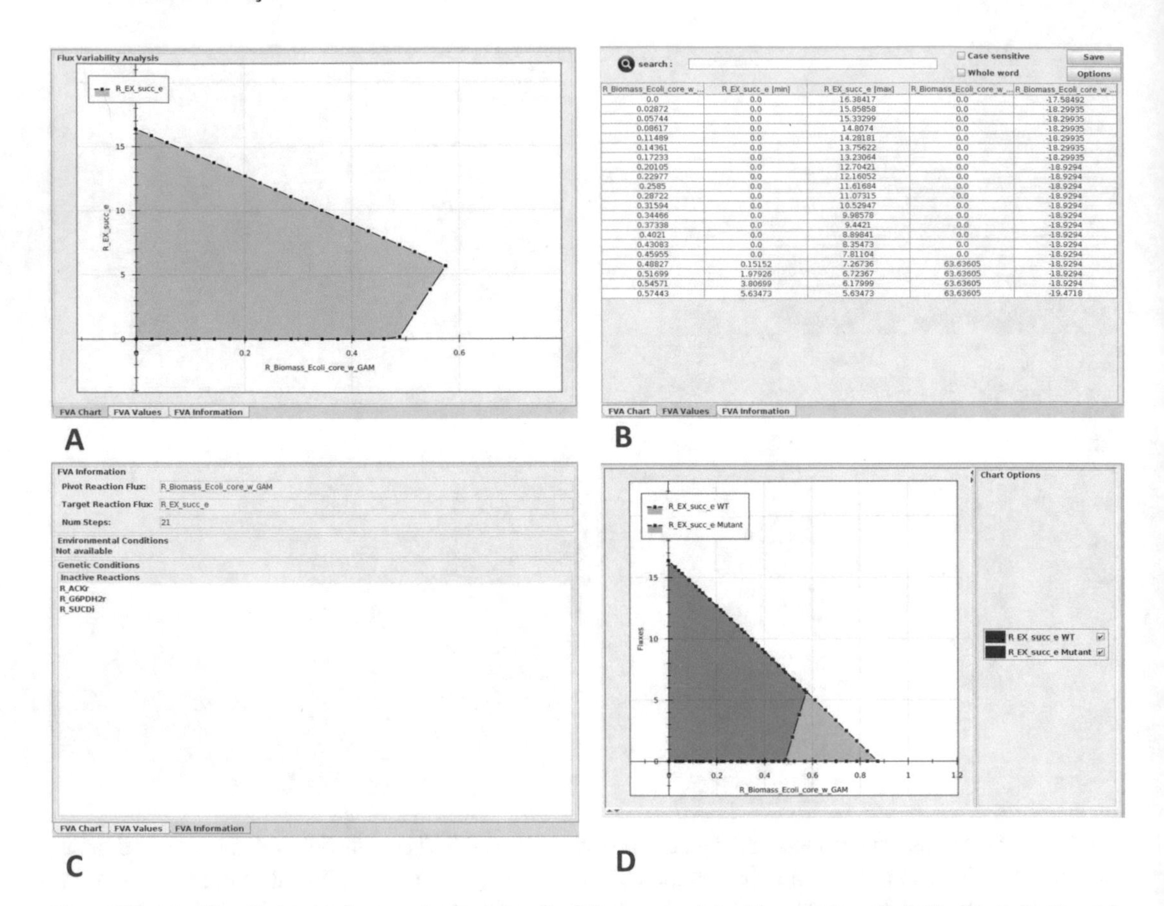

R_Biomass_Ecoli_core_w_...	R_EX_succ_e [min]	R_EX_succ_e [max]	R_Biomass_Ecoli_core_w_...	R_Biomass_Ecoli_core_w_...
0.0	0.0	16.38417	0.0	-17.58492
0.02872	0.0	15.85858	0.0	-18.29935
0.05744	0.0	15.33299	0.0	-18.29935
0.08617	0.0	14.8074	0.0	-18.29935
0.11489	0.0	14.28181	0.0	-18.29935
0.14361	0.0	13.75622	0.0	-18.29935
0.17233	0.0	13.23064	0.0	-18.29935
0.20105	0.0	12.70421	0.0	-18.9294
0.22977	0.0	12.16052	0.0	-18.9294
0.2585	0.0	11.61684	0.0	-18.9294
0.28722	0.0	11.07315	0.0	-18.9294
0.31594	0.0	10.52947	0.0	-18.9294
0.34466	0.0	9.98578	0.0	-18.9294
0.37338	0.0	9.4421	0.0	-18.9294
0.4021	0.0	8.89841	0.0	-18.9294
0.43083	0.0	8.35473	0.0	-18.9294
0.45955	0.0	7.81104	0.0	-18.9294
0.48827	0.15152	7.26736	63.63605	-18.9294
0.51699	1.97926	6.72367	63.63605	-18.9294
0.54571	3.80699	6.17999	63.63605	-18.9294
0.57443	5.63473	5.63473	63.63605	-19.4718

Fig. 17 Production Envelope for succinate using E. coli core model with reactions R_ACKr (Acetate kinase), R_SUCDi (succinate dehydrogenase), and R_G6PDH2r (G6P-dehydrogenase) suppressed. Production envelope plot (**a**); values for the biomass steps, minimum/maximum values of succinate and reduced costs to biomass (**b**); summary of the parameters used in the FVA operation (**c**); view overlapping the envelope production of wild type and mutant (**d**)

using wild-type simulation. On this operation, the user can also select an environmental condition.

2. *menu:Analysis/add manual critical genes/reactions*—where the user has the freedom to create a specific set of critical information. This operation is useful to the users that know that a given reaction or gene is essential or when the user wants to exclude some entities from the potential genetic alteration targets in strain optimization processes (i.e., ATPm is not a critical reaction but should not be considered as target of strain optimization process).

The result of both the operations is a new data type in the clipboard containing all critical genes or reactions.

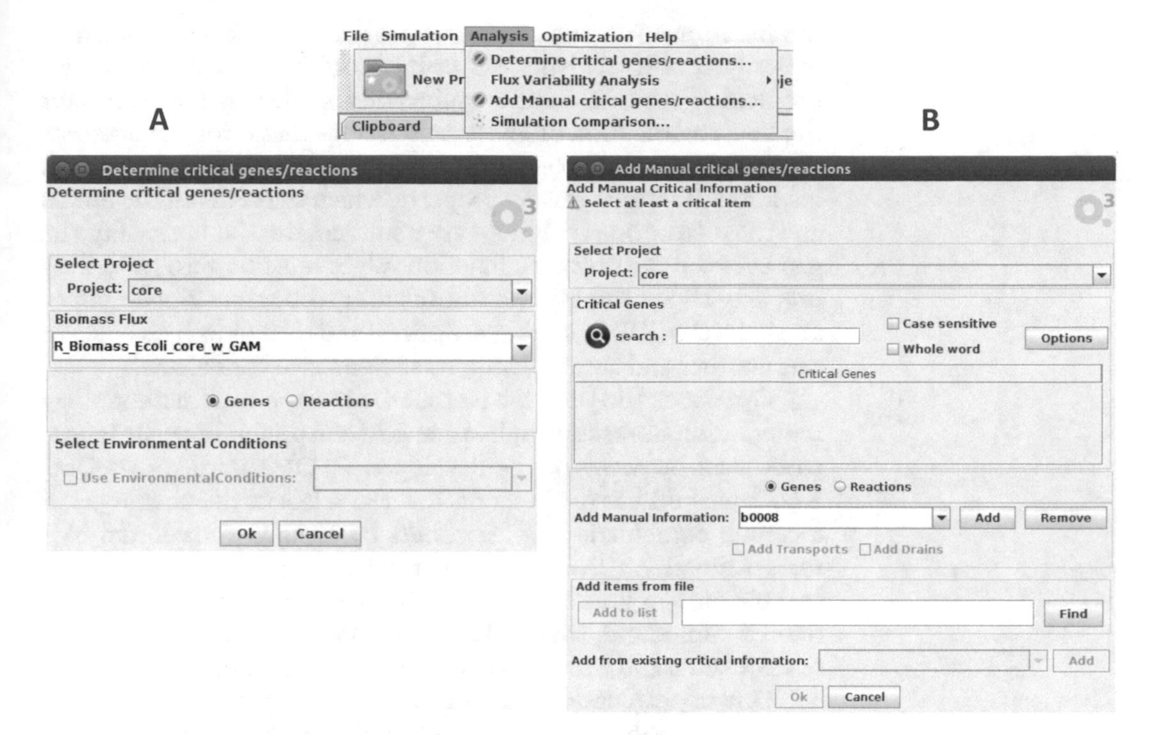

Fig. 18 Two different ways to create critical information on Optflux. (**a**) Determining by phenotype simulation approach, (**b**) Manual creation

3.4 Constraint-Based Strain Optimization Methods

While the in silico simulation of phenotypes can be of great help for analyzing the phenotype/genotype of a few rationally designed strains, it would take too long to manually discover combinations of genetic modifications that meet a certain metabolic engineering objective. In order to search for interesting genetic targets, several Computational Strain Optimization Methods (CSOM) have been developed to search among the large number of different strain engineering strategies for the ones that result in the desired phenotype [66].

CSOM links the abstraction of the constraint-based metabolic models and the assumptions of the constraint-based simulation methods to reveal insights into the best genetic design strategies to rationally address a ME objective.

In 2016, Maia and coworkers proposed a new taxonomy for the existent CSOMs, organizing them in three big clusters:

1. Bilevel mixed-integer programing methods;
2. Metaheuristics methods;
3. Elementary Mode Analysis based methods.

The bi-level mixed-integer programming methods are problems where two objective functions are simultaneously accounted for, usually the maximization of metabolite production and the maximum cellular growth. Such methods are formulated taking into account two distinct constraint-based optimization problems, the inner problem and the outer problem. The inner problem is typically concerned with the cellular objective, while the outer problem is focused on the engineering goal, i.e., the search for the best combination of genetic alterations that favor the overproduction of a desired compound. This mathematical formulation is based on the strong duality property, which states that if the primal and the dual optimal solutions are bounded, then, at optimality, the gap between the objective function values must be zero [67]. This property allows the bi-level formulation to be transformed into a single-level MILP by setting the primal and dual objectives equal to one another and accumulating their respective constraints.

OptKnock [68] was the first method to use such methodology representing a breakthrough in the field and establishing the framework used by many of the developed CSOMs until today. OptKnock uses FBA as the inner problem to calculate the phenotype of a certain combination of knockouts by assuming maximum biomass formation. The result returned by OptKnock is the best combination of knockouts that maximize the overproduction of a desired compound, while taking into account a maximum number of knockouts and a minimum biomass formation rate.

One important drawback of the target discovery methods that search for a global optimum solution like OptKnock is that, as the number of allowed genetic modifications increases, the total searchable space of strain designs grows exponentially, which makes the computation time required to solve the problem impractical.

This severely limits the maximum number of genetic modifications that can be included in the strain designs computed with OptKnock and similar methods. One of the possibilities to solve this limitation is to use metaheuristic methods, such as evolutionary algorithms or other nature-inspired heuristics to find strain designs with desired phenotypes.

Heuristic methods are usually computationally less expensive approaches for a myriad of optimization problems. Although, due to their nature, they do not guarantee that the overall optimal solutions are found, they allow the definition of optimization frameworks with an enriched set of objective functions, fostering a clear separation of the strain optimization from the phenotype simulation layers, while allowing optimization over larger search spaces (e.g., a higher number of genetic modifications). In Fig. 19, a generic workflow for a typical metaheuristic CSOM is presented.

The first effort to move in this direction was OptGene [69], presented by Patil and coworkers, which appeared shortly after the publication of OptKnock. Inspired by the Darwinian natural evolution theory, OptGene formulates a bi-level decoupled approach, supported by the use of a genetic algorithm [70]. The idea is to encode solutions as individuals in an evolving population. Here, each solution is represented as binary variables encoding the metabolic genome or a set of integer values encoding reaction deletions.

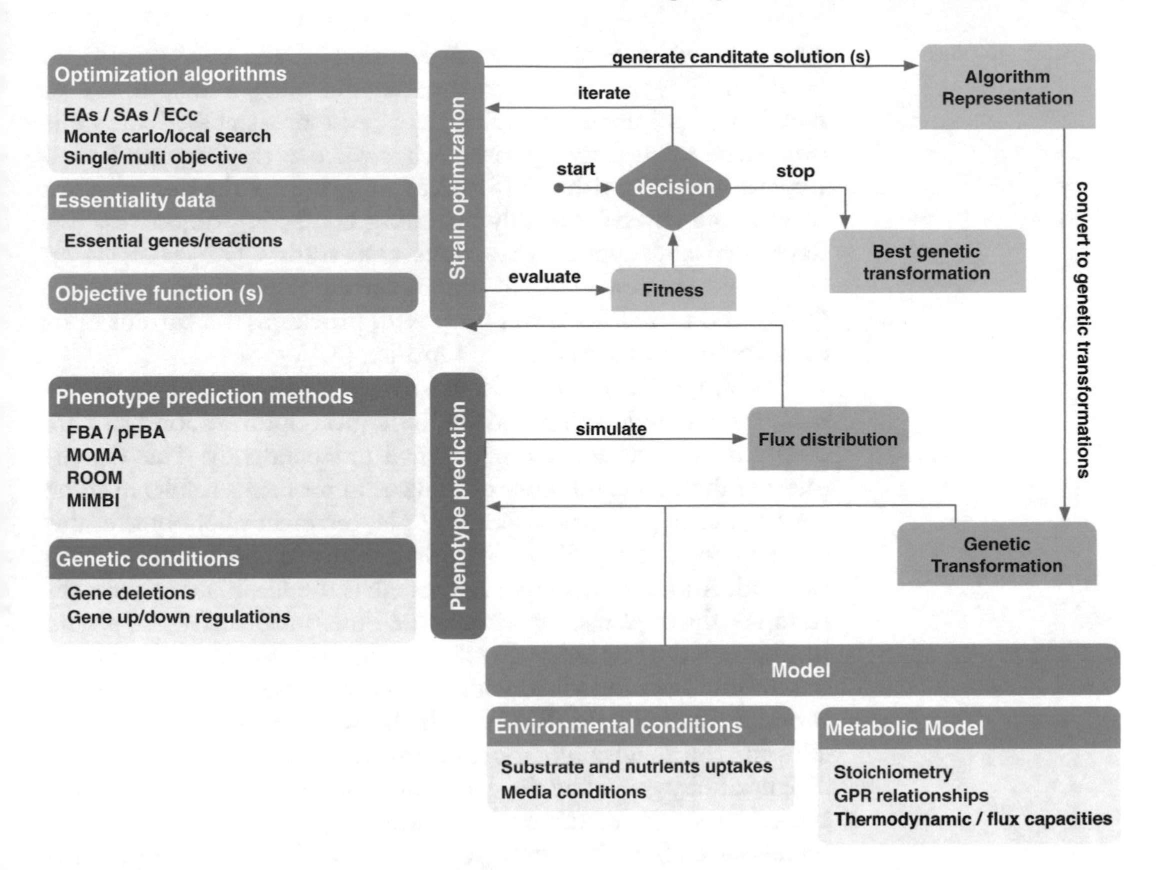

Fig. 19 Representation of the interactions between the meta-heuristic algorithms (blue), phenotype prediction (purple), and CBM (gray) on CSOM approaches

The algorithm starts by the random generation of an initial set of candidate solutions (the initial population), and each is decoded into a set of reaction deletions, which are translated into constraints. Each of these candidate solutions is simulated by means of one of the phenotype simulation methods described above, followed by the assignment of a fitness value by a user-defined objective function. Subsequently, the algorithm enters an iterative stage, starting with a selection step that chooses solutions as primary candidates for the reproduction in a stochastic way that depends on their assigned fitness (fitter individuals have a higher probability of generating offspring solutions). Finally, by combining these individuals via crossover or mutation operators, a new population is attained and re-evaluated. This cycle is repeated until a desired phenotype is achieved or another user-defined termination criterion is met (typically, a defined maximum number of generations or solutions evaluated).

The original implementation of OptGene was extended by the authors to support an optimized representation, which represents solutions as variable sized sets of deletions. Also, another nature inspired meta-heuristic—Simulated Annealing (SA)—was proposed, which evolves a single solution instead of a population, by mimicking the process of thermal annealing usually found in metallurgy and affine areas [71]. Later, an approach based on multi-objective evolutionary algorithms, in particular the Strength Pareto Evolutionary Algorithm 2 (SPEA2), allowing for the optimization of multiple criteria, possibly including conflicting objectives, was developed and applied to various case studies [72]. Finally, an integrated framework that homogenized these three heuristics (EAs, SAs, and SPEA2) was proposed, providing the current optimization back-end available in OptFlux [73].

Although these methods still follow a bi-level design, in this case, the bioengineering and the biological optimization tasks are clearly decoupled and are performed independently. This decoupling of the outer and inner optimization problems results in some very powerful properties. For example, the inner phenotype evaluation method can easily be swapped to any phenotype simulation method. Another important advantage is the flexibility in the definition of the objective function in the outer problem, which is here not bounded by linearity. Nonlinear objective functions (even discontinuous) can easily be included, as is the case with the biomass-product coupled yield (BPCY), which resembles productivity [69], allowing the definition of more meaningful and powerful functions. The flexibility gained by the decoupling of the two layers also allows the easier switch of the optimization heuristic used to search for metabolic engineering strategies and allows the different optimization tasks to be addressed with a similar framework.

3.4.1 Strain Optimization Using Optflux

OptFlux allows performing strain optimization operations using metaheuristic methods. This operation can be accessed on *menu: optimization/evolutionary* (Fig. 20). To perform this operation a user needs to define the following inputs:

1. Select the project that will be used to perform the optimization.
2. Select the outer-layer optimization method that will perform the ME optimization process. In this selection, the user is able to choose among one of the three available optimization algorithm: SPEA2, EA or SA, as well as the genetic modifications targets (genes or reactions) and strategy (knockouts or over/under expressions).
3. Select the objective functions. The objective functions will encode the objectives of the desired ME process. There are several objective functions available in OptFlux, including biomass-product coupled yield (BPCY), product yield with minimum biomass (YIELD), maximization/minimization of a flux value or maximization/minimization of the number of genetic modifications. Each objective function requires its own configuration, selecting which fluxes are used in the

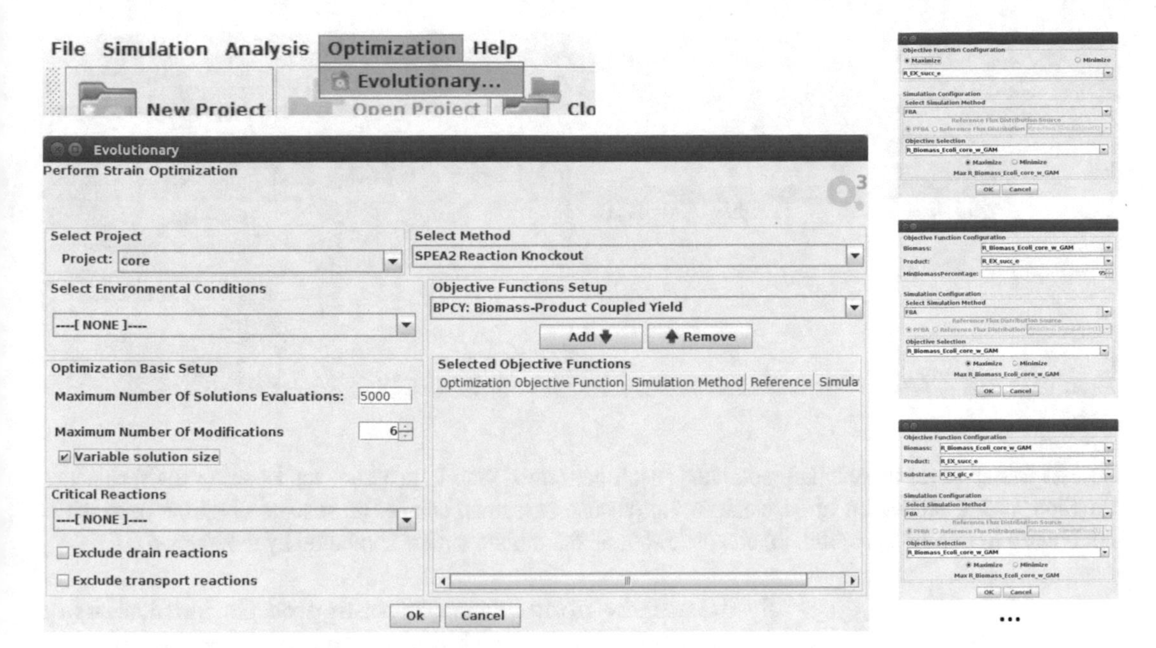

Fig. 20 Optflux Evolutionary strain optimization operation and some different objective functions setups

computation of the function value. A constraint-based strain simulation method must also be selected and associated with each objective function. This simulation configuration is similar to the simulation operation and allows selecting one of the implemented methods (FBA, pFBA, ROOM, MOMA, LMOMA, and MiMBl) and an objective function reaction, usually the biomass equation. This means that an optimization process can have multiple objective functions, each of them evaluated using different phenotype simulation configurations.

4. Select the environmental condition to be used.
5. It is also possible to select the essential information (genes or reactions), which excludes them from the optimization process, effectively reducing the search space.
6. The optimization basic setup allows the user to define the maximum number of solution evaluations (which is the number of phenotype simulations that will be performed—a termination condition), as well as the maximum number of genetic modifications. The user can also choose whether the solution sizes are fixed or if the algorithm is allowed to find solutions with different sizes. When the user decides for variable size solutions, this number will be used as the maximum possible size of the solutions.

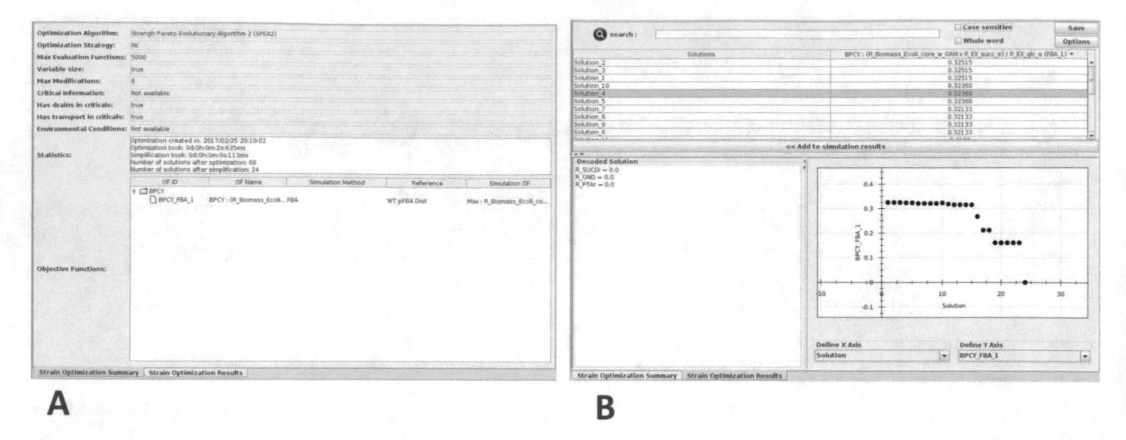

Fig. 21 Views of an evolutionary operation result performed with E. coli core model using BPCY objective function for the production of succinate using glucose as carbon source. (**a**) resume of all the parameters configured to run the operation. (**b**) view showing all the mutant strains computed by the method

By default the optimization is configured to find designs of up to six modifications, within a maximum of 5000 solution evaluations. These configurations are conservative and must be adapted for different models/conditions (e.g., by increasing the number of function evaluations, number of modifications, among others).

The result of the Evolutionary Algorithms is a new object in the clipboard, placed within the Optimization Results list, under the Evolutionary data type. The Optimization Result data type has two views associated (Fig. 21):

1. Strain optimization summary: where all the inputs used to produce the result are provided.
2. Strain optimization results, where the computed solutions are described. On this view, the user is also able to extract a specific simulation result to the clipboard for further analysis.

4 Notes

1. **Representation of Genetic modifications in CBMs**
   The suppression of a given metabolic function can be accomplished in vivo by disrupting the functioning of specific genes by targeted modifications through homologous recombination [74] or intron introduction [75]. With in silico CBMs, this task is commonly accomplished by imposing constraints that force the flux of the disabled reactions to zero, deterring the occurrence of flux over those reactions, followed by the evaluation of the effect of that perturbation.

   Constraint-based phenotype prediction methods can take advantage of gene-reaction rules information contained within the models to perform gene deletions instead of reactions deactivations. To attain that, it is necessary to use the Boolean rules to derive the off genes and understand which reactions must be deactivated. Such deactivation is performed by setting the reaction bounds of those reactions to 0. This type of gene deletions provides a closer resemblance to the in vivo scenario, since they inherently account for the occurrence of multifunctional and multimeric proteins, as well as isoenzymes.

Gonçalves and coworkers [76] suggested an alternative to use constraint-based phenotype prediction methods to represent over or under regulation of certain fluxes in the network. This methodology uses a wild-type flux distribution as a reference and a set of "mutated" reactions, where each reaction is associated with an expression level (p). If the expression value is smaller than 1, the reaction is under-expressed constraining its flux to be lower than the reference value, which will be multiplied by p. If the expression value is >1 the reaction will be overexpressed, and its flux is constrained to be higher than the reference value. The knockouts are formulated with an expression level of zero.

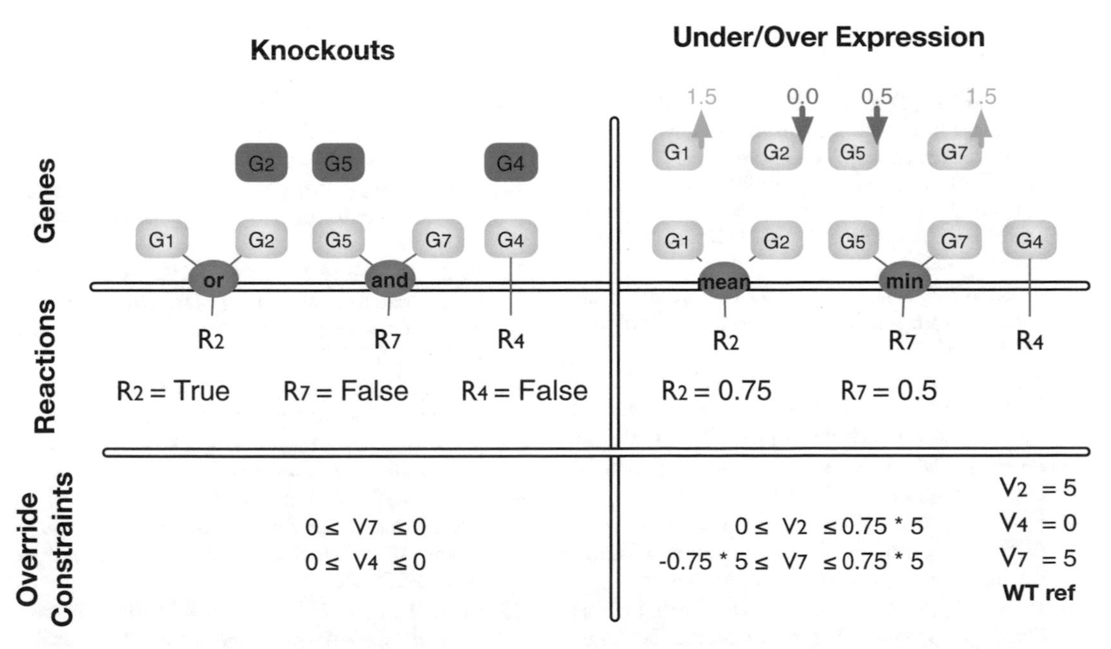

Fig. 22 Schema explaining how to represent genetic modifications within CBMs. In the left side, knockouts are shown (G2, G4, and G5), then the GPRs were applied to verify which reactions should be knocked out, and finally the bounds associated to deactivated reactions were set to 0. In the right side, under/over expressions are shown, where G1 and G7 were overexpressed, G5 was under-expressed and G2 was knocked out. GPRs were changed to support real values (Or operator changed to MEAN; AND operator changed to MIN), and these functions were applied to calculate the expression values of the reactions. Then, the bounds of the reactions were changed taking into account the wild-type flux distribution

To apply this methodology on the gene level, it was necessary to convert Boolean operators to functions able to handle numerical expression values by using the available GPR information. Typical examples of AND situations are the formation of protein complexes, where the expression of all the genes is necessary to produce the enzyme. With this in mind, the authors decided to translate the AND operator to the minimum value between the two expression values, implying that the expression of all the genes is necessary to produce the enzyme, and the gene for which the minimum number of copies are expressed will act as a bottleneck for the activity of the corresponding enzyme. On the other hand, typical examples of OR situations are genes encoding isoenzymes, where the genes involved work independently, so OR operators were transformed to the mean of all expression values. Thus, by using the converted GPR, it is possible to convert a set of genes associated with an expression value to a set of reactions associated with an expression value, allowing the flux bounds transformation explained before.

Figure 22 shows an example of how a genetic transformation is translated to the reactions bounds.

References

1. Jackson DA, Symons RH, Berg P (1972) Biochemical method for inserting new genetic information into {DNA} of Simian Virus 40: circular {SV40} {DNA} molecules containing lambda phage genes and the galactose operon of Escherichia coli. Proc Natl Acad Sci U S A 69:2904–2909
2. Lobban PE, Kaiser AD (1973) Enzymatic end-to-end joining of {DNA} molecules. J Mol Biol 78:453–471
3. Stephanopoulos G, Aristidou AA, Nielsen J (1998) Metabolic engineering: principles and methodologies. Academic press, New York, NY
4. Bailey J (1991) Toward a science of metabolic engineering. Science 252:1668–1675
5. Woolston BM, Edgar S, Stephanopoulos G (2013) Metabolic engineering: past and future. Annu Rev Chem Biomol Eng 4:259–288
6. Bailey JE, Birnbaum S, Galazzo JL et al (1990) Strategies and challenges in metabolic engineering. Ann N Y Acad Sci 589:1–15
7. Gavrilescu M, Maria G, Yusuf C (2005) Biotechnology—a sustainable alternative for chemical industry. Biotechnol Adv 23:471–499
8. Hatti-Kaul R, Rajni H-K, Ulrika T et al (2007) Industrial biotechnology for the production of bio-based chemicals – a cradle-to-grave perspective. Trends Biotechnol 25:119–124
9. Kitano H (2002) Systems biology: a brief overview. Science 295:1662–1664
10. Bork P (2005) Is there biological research beyond Systems Biology? A comparative analysis of terms. Mol Syst Biol 1:E1–E2
11. Schuster SC (2007) Next-generation sequencing transforms today's biology. Nat Methods 5:16–18
12. Papin JA, Tony H, Palsson BO et al (2005) Reconstruction of cellular signalling networks and analysis of their properties. Nat Rev Mol Cell Biol 6:99–111
13. Samaga R, Regina S, Steffen K (2013) Modeling approaches for qualitative and semi-quantitative analysis of cellular signaling networks. Cell Commun Signal 11:43
14. Covert MW, Knight EM, Reed JL et al (2004) Integrating high-throughput and computational data elucidates bacterial networks. Nature 429:92–96
15. Reed JL, Iman F, Ines T et al (2006) Towards multidimensional genome annotation. Nat Rev Genet 7:130–141
16. Fong SS, Palsson BØ (2004) Metabolic gene–deletion strains of Escherichia coli evolve to computationally predicted growth phenotypes. Nat Genet 36:1056–1058
17. Papin JA, Joerg S, Price ND et al (2004) Comparison of network-based pathway analysis methods. Trends Biotechnol 22:400–405

18. Rocha I, Förster J, Nielsen J (2008) Design and application of genome-scale reconstructed metabolic models. Methods Mol Biol 416:409–431
19. Thiele I, Palsson BØ (2010) A protocol for generating a high-quality genome-scale metabolic reconstruction. Nat Protoc 5:93–121
20. Hamilton JJ, Reed JL (2014) Software platforms to facilitate reconstructing genome-scale metabolic networks. Environ Microbiol 16:49–59
21. Notebaart RA, van Enckevort FHJ, Francke C et al (2006) Accelerating the reconstruction of genome-scale metabolic networks. BMC Bioinformatics 7:296
22. Agren R, Liu L, Shoaie S et al (2013) The {RAVEN} toolbox and its use for generating a genome-scale metabolic model for Penicillium chrysogenum. PLoS Comput Biol 9:e1002980
23. Henry CS, DeJongh M, Best AA et al (2010) High-throughput generation, optimization and analysis of genome-scale metabolic models. Nat Biotechnol 28:977–982
24. Dias O, Rocha M, Ferreira EC et al (2015) Reconstructing genome-scale metabolic models with merlin. Nucleic Acids Res 43:3899–3910
25. Smallbone K, Kieran S, Messiha HL et al (2013) A model of yeast glycolysis based on a consistent kinetic characterisation of all its enzymes. FEBS Lett 587:2832–2841
26. Smallbone K, Kieran S, Evangelos S et al (2010) Towards a genome-scale kinetic model of cellular metabolism. BMC Syst Biol 4:6
27. Bordbar A, Monk JM, King ZA et al (2014) Constraint-based models predict metabolic and associated cellular functions. Nat Rev Genet 15:107–120
28. Gombert AK, Jens N (2000) Mathematical modelling of metabolism. Curr Opin Biotechnol 11:180–186
29. Edwards JS, Palsson BO (1999) Systems properties of the haemophilus influenzaeRd metabolic genotype. J Biol Chem 274:17410–17416
30. Monk J, Nogales J, Palsson BO (2014) Optimizing genome-scale network reconstructions. Nat Biotechnol 32:447–452
31. Bailey JE (2001) Reflections on the scope and the future of metabolic engineering and its connections to functional genomics and drug discovery. Metab Eng 3:111–114
32. Rocha I, Maia P, Evangelista P et al (2010) {OptFlux}: an open-source software platform for in silico metabolic engineering. BMC Syst Biol 4:45
33. Glez-Peña D, Reboiro-Jato M, Maia P et al (2010) {AIBench}: a rapid application development framework for translational research in biomedicine. Comput Methods Programs Biomed 98:191–203
34. Schulze KL, Lipe RS (1964) Relationship between substrate concentration, growth rate, and respiration rate of Escherichia coli in continuous culture. Arch Mikrobiol 48:1–20
35. Bowen JH (1968) Basic principles and calculations in chemical engineering. Chem Eng Sci 23:191
36. Reed JL, Vo TD, Schilling CH, Palsson BO (2013) An expanded genome-scale model of Escherichia coli K-12 (i JR904 GSM/GPR). Genome Biol 4(9):R54
37. Hucka M, Finney A, Sauro HM et al (2003) The systems biology markup language ({SBML)}: a medium for representation and exchange of biochemical network models. Bioinformatics 19:524–531
38. Bornstein BJ, Keating SM, Jouraku A et al (2008) {LibSBML}: an {API} library for {SBML}. Bioinformatics 24:880–881
39. von Kamp A, Schuster S (2006) Metatool 5.0: fast and flexible elementary modes analysis. Bioinformatics 22:1930–1931
40. Funahashi A, Morohashi M, Kitano H et al (2003) {CellDesigner}: a process diagram editor for gene-regulatory and biochemical networks. BioSilico 1:159–162
41. Funahashi A, Matsuoka Y, Jouraku A et al (2008) {CellDesigner} 3.5: a versatile modeling tool for biochemical networks. Proc IEEE 96:1254–1265
42. Cline MS, Smoot M, Cerami E et al (2007) Integration of biological networks and gene expression data using Cytoscape. Nat Protoc 2:2366–2382
43. Noronha A, Vilaça P, Rocha M (2014) An integrated network visualization framework towards metabolic engineering applications. BMC Bioinformatics 15:420
44. Fruchterman TMJ, Reingold EM (1991) Graph drawing by force-directed placement. Softw Pract Exp 21:1129–1164
45. Le Novere N, Hucka M, Mi H et al (2009) The systems biology graphical notation. Nat Biotechnol 27(8):735–741
46. Schellenberger J, Park JO, Conrad TM et al (2010) {BiGG}: a Biochemical Genetic and Genomic knowledgebase of large scale metabolic reconstructions. BMC Bioinformatics 11:213
47. King ZA, Dräger A, Ebrahim A et al (2015) Escher: a web application for building, sharing, and embedding data-rich visualizations of biological pathways. PLoS Comput Biol 11: e1004321
48. King ZA, Lu J, Dräger A et al (2016) BiGG models: a platform for integrating, standardizing and sharing genome-scale models. Nucleic Acids Res 44:D515–D522

49. Varma A, Palsson BO (1994) Metabolic flux balancing: basic concepts, scientific and practical use. Nat Biotechnol 12:994
50. Papoutsakis ET (1984) Equations and calculations for fermentations of butyric acid bacteria. Biotechnol Bioeng 26:174–187
51. Orth JD, Thiele I, Palsson BØ (2010) What is flux balance analysis? Nat Biotechnol 28:245–248
52. Feist AM, Palsson BO (2010) The biomass objective function. Curr Opin Microbiol 13:344–349
53. Schuetz R, Kuepfer L, Sauer U (2007) Systematic evaluation of objective functions for predicting intracellular fluxes in Escherichia coli. Mol Syst Biol 3:119
54. Ponce de León M, Cancela H, Acerenza L (2008) A strategy to calculate the patterns of nutrient consumption by microorganisms applying a two-level optimisation principle to reconstructed metabolic networks. J Biol Phys 34:73–90
55. Lewis NE, Hixson KK, Conrad TM et al (2010) Omic data from evolved E. coli are consistent with computed optimal growth from genome-scale models. Mol Syst Biol 6:390
56. Segrè D, Vitkup D, Church GM (2002) Analysis of optimality in natural and perturbed metabolic networks. Proc Natl Acad Sci U S A 99:15112–15117
57. Snitkin ES, Dudley AM, Janse DM et al (2008) Model-driven analysis of experimentally determined growth phenotypes for 465 yeast gene deletion mutants under 16 different conditions. Genome Biol 9:R140
58. Shlomi T, Berkman O, Ruppin E (2005) Regulatory on/off minimization of metabolic flux changes after genetic perturbations. Proc Natl Acad Sci U S A 102:7695–7700
59. Becker SA, Feist AM, Mo ML et al (2007) Quantitative prediction of cellular metabolism with constraint-based models: the {COBRA} Toolbox. Nat Protoc 2:727–738
60. Wintermute EH, Lieberman TD, Silver PA (2013) An objective function exploiting suboptimal solutions in metabolic networks. BMC Syst Biol 7:98
61. Park JH, Lee KH, Kim TY et al (2007) Metabolic engineering of Escherichia coli for the production of L-valine based on transcriptome analysis and in silico gene knockout simulation. Proc Natl Acad Sci U S A 104:7797–7802
62. Alper H, Jin Y-S, Moxley JF et al (2005) Identifying gene targets for the metabolic engineering of lycopene biosynthesis in Escherichia coli. Metab Eng 7:155–164
63. Jung YK, Kim TY, Park SJ et al (2010) Metabolic engineering of Escherichia coli for the production of polylactic acid and its copolymers. Biotechnol Bioeng 105:161–171
64. Brochado AR, Andrejev S, Maranas CD et al (2012) Impact of stoichiometry representation on simulation of genotype-phenotype relationships in metabolic networks. PLoS Comput Biol 8:e1002758
65. Mahadevan R, Schilling CH (2003) The effects of alternate optimal solutions in constraint-based genome-scale metabolic models. Metab Eng 5:264–276
66. Maia P, Rocha M, Rocha I (2016) In silico {constraint-based} strain optimization methods: the quest for optimal cell factories. Microbiol Mol Biol Rev 80:45–67
67. Ignizio JP (1994) Linear programming. Pearson College Division, London
68. Burgard AP, Pharkya P, Maranas CD (2003) Optknock: a bilevel programming framework for identifying gene knockout strategies for microbial strain optimization. Biotechnol Bioeng 84:647–657
69. Patil KR, Rocha I, Förster J et al (2005) Evolutionary programming as a platform for in silico metabolic engineering. BMC Bioinformatics 6:308
70. Holland JH (1992) Adaptation in natural and artificial systems: an introductory analysis with applications to biology, control and artificial intelligence. MIT Press, Cambridge, MA
71. Rocha M, Maia P, Mendes R et al (2008) Natural computation meta-heuristics for the in silico optimization of microbial strains. BMC Bioinformatics 9:499
72. Maia P, Rocha I, Ferreira EC et al (2008) Evaluating evolutionary multiobjective algorithms for the in silico optimization of mutant strains. In: 8th IEEE International Conference on BioInformatics and BioEngineering, BIBE 2008
73. Maia P, Rocha I, Rocha M (2013) An integrated framework for strain optimization. In: 2013 IEEE Congress on Evolutionary Computation, CEC 2013, pp 198–205
74. Datsenko KA, Wanner BL (2000) One-step inactivation of chromosomal genes in Escherichia coli {K-12} using {PCR} products. Proc Natl Acad Sci U S A 97:6640–6645
75. Frazier CL, San Filippo J, Lambowitz AM et al (2003) Genetic manipulation of Lactococcus lactis by using targeted group {II} introns: generation of stable insertions without selection. Appl Environ Microbiol 69:1121–1128
76. Gonçalves E, Pereira R, Rocha I et al (2012) Optimization approaches for the in silico discovery of optimal targets for gene over/under-expression. J Comput Biol 19:102–114

Chapter 3

The MONGOOSE Rational Arithmetic Toolbox

Christopher Le and Leonid Chindelevitch

Abstract

The modeling of metabolic networks has seen a rapid expansion following the complete sequencing of thousands of genomes. The constraint-based modeling framework has emerged as one of the most popular approaches to reconstructing and analyzing genome-scale metabolic models. Its main assumption is that of a quasi-steady-state, requiring that the production of each internal metabolite be balanced by its consumption. However, due to the multiscale nature of the models, the large number of reactions and metabolites, and the use of floating-point arithmetic for the stoichiometric coefficients, ensuring that this assumption holds can be challenging.

The MONGOOSE toolbox addresses this problem by using rational arithmetic, thus ensuring that models are analyzed in a reproducible manner and consistently with modeling assumptions. In this chapter we present a protocol for the complete analysis of a metabolic network model using the MONGOOSE toolbox, via its newly developed GUI, and describe how it can be used as a model-checking platform both during and after the model construction process.

Key words Metabolic networks, Constraint-based analysis, Rational arithmetic, User interface, Reproducibility

1 Introduction

The rapid explosion of genomic technologies has led to a proliferation of complete genome sequences for a variety of organisms, from Archaea to Zooplankton [1]. The metabolism of many of these organisms is important, for reasons ranging from antimicrobial target discovery to metabolic engineering [2]. Given the complexity of metabolic processes, strategies such as systematic gene knockouts may not always be feasible, requiring the development of computational methods for rapid hypothesis generation and prioritization [3]. For this reason, a unified modeling framework enabling a genome-scale analysis of metabolism in these organisms becomes critical.

Such a framework, typically called constraint-based metabolic modeling, emerged in the early 1990s [4, 5]. Its main premise is that accurate predictions can be made simply from the knowledge

Marco Fondi (ed.), *Metabolic Network Reconstruction and Modeling: Methods and Protocols*, Methods in Molecular Biology, vol. 1716, https://doi.org/10.1007/978-1-4939-7528-0_3, © Springer Science+Business Media, LLC 2018

of the full complement of reactions available to an organism, without any detailed information on the kinetic parameters governing the rates of those reactions. By making the quasi-steady-state assumption, meaning that production and consumption balance one another for each internal metabolite, constraint-based models can be used to predict the distribution of reaction fluxes, as well as features such as growth rates, essential and synthetic lethal reactions, and minimal media [6]. This framework has been successfully used to analyze and make predictions about a number of organisms, in all the major domains of life [7].

However, many of the metabolic network models analyzed in the constraint-based framework have turned out to be poorly conditioned, meaning that a small perturbation to their coefficients could result in a large change in the model predictions [8]. Such perturbations are commonplace during analysis which is typically performed in floating-point arithmetic, an approximate representation of numbers on a computer that allows their fast manipulation. For this reason, the author has proposed an alternative way of analyzing these networks, using rational arithmetic, in which the coefficients of each reaction are represented as ratios of integers [8]. This representation, together with key insights into the structure of metabolic network models, forms the foundation of the MONGOOSE toolbox for metabolic network analysis described in this chapter.

2 Materials

In this section we describe the computational tools that need to be in place in order to make use of the MONGOOSE toolbox, and give detailed instructions on how to set them up. Since MONGOOSE was developed for a Unix-based environment, including Mac OS, we focus on those environments, and in particular use Mac OS as a running example. However, it is possible to run it in a Windows environment using a virtual machine—an approach based on a platform such as Cygwin is, however, unlikely to work.

2.1 Installing MONGOOSE and the GUI: The Easy Way

The easiest way to install MONGOOSE and its graphical user interface (GUI) is via the Docker [9] distribution available at https://hub.docker.com/r/ctlevn/mongoose/. If you follow the instructions on that page, you will be able to get MONGOOSE up and running in no time. The only disadvantage is that the files you create will need to be transferred from the Docker container back to your local hard drive at the end of each session. Alternatively, if you know Docker well you can update your container to include the results of the work you have performed, and restart it based on the same container identifier the next time around.

However, if Docker does not work for you for whatever reason, you will need to follow the more complex path outlined in the rest of this section.

2.2 Installing Python

The latest version of the Python software [10] can always be downloaded from the Python website. The first version of the toolbox, MONGOOSE 1.1, has been developed under Python 2.6, and is still available for download from GitHub, https://github.com/WGS-TB/MongooseGUI, for those users who choose to use Python 2. However, there is now a new version, MONGOOSE 2.0, ported to Python 3.5, available at https://github.com/WGS-TB/MongooseGUI3. MONGOOSE 2.0 has new features such as a much faster analysis thanks to the Python binding for *QSOpt_ex* [11] (*see* Subheading 2.5 below), which means that optimization problems can now be solved directly within Python instead of creating an input file, making a system call to run *QSopt_ex*, and then parsing the output file. For this reason, we use the Python 3.5 version in our running example in this section.

To install Python 3.5 on Mac OS, click on the appropriate Download button on the Python website, https://www.python.org/downloads/. This will download a.pkg file. Double-click on that file and follow the instructions to select the installation location (or use the default for simplicity) and any other options.

2.3 Installing Additional Python Modules

In order to make the most of the MONGOOSE toolbox you will need to install some additional Python modules, namely

- *setuptools* (needed to install some of the other modules listed below)
- *cython* [12] (needed for the Python interface to *QSopt_ex*)
- *python-libsbml* [13] (needed for reading metabolic models from SBML files)
- *xlrd* (needed for reading metabolic models from Excel files)
- *SIP* (needed for the GUI to MONGOOSE)
- *Qt* and *PyQt* [14] (needed for the GUI to MONGOOSE)

To install setuptools, simply open a Terminal window and run the command

```
pip3 install setuptools
```

To install *cython*, *python-libsbml*, *xlrd*, and *SIP* you can proceed in exactly the same way:

```
pip3 install cython
pip3 install python-libsbml
pip3 install xlrd
pip3 install sip
```

Finally, to install *Qt*, first download an installer package from the Qt website, https://www.qt.io/download-open-source/, then run it by double-clicking it and following the instructions (*see* also **Notes 1** and **2**). The installation folder, for instance /Applications/Qt, will be used in the installation of PyQt; we refer to it as PATH below.

When the installation is complete, you can install *PyQt*. To do so, download its source code from the Riverbank website, https://www.riverbankcomputing.com/software/pyqt/download (if you are following our running example, select the OS X source). Unpack the zipped file and run the following commands from the Terminal window (where PATH is the directory in which you installed *Qt*):

```
cd PyQt-mac-gpl-4.11.4
python configure-ng.py -qmake PATH/5.7/clang_64/bin/qmake
make
make install
```

See **Note 3** in case of any difficulties with installing *PyQt*.

2.4 Installing QSopt_ex

To install *QSopt_ex* in the most painless way possible, we highly recommend the fork of the project created by Jon Lund Steffensen, in support of the PSAMM toolbox [15].

Before you get it to work, however, you need to install several dependencies:

- *Libtool* [16] (needed to install *QSopt_ex* as a usable library)
- *GNU MP* [17] (needed to perform arbitrary precision computations)
- *libz* and *libbz2* (optional libraries to read and write compressed files)

This is actually a lot fewer dependencies relative to the original version, created by Daniel Espinoza [11]. Here is how to get them set up.

To get *Libtool*, the tool needed to install *QSopt_ex* as a library, download it from http://mirror.jre655.com/GNU/libtool/libtool-2.4.6.tar.gz. Unpack the zipped file and run the following commands from the Terminal window:

```
cd libtool-2.4.6
./configure -prefix /usr/local
make
make install
```

Also *see* **Note 4** for a possible additional dependency.

To get *GNU MP*, the arbitrary precision arithmetic library, download it from https://gmplib.org/download/gmp/gmp-6.0.

0a.tar.bz2. Unpack the zipped file and run the following commands from the Terminal window:

```
cd gmp-6.0.0
./configure
make
make check
make install
```

You are now ready to install *QSopt_ex*. You can obtain it from GitHub by opening a Terminal window and running the command

```
git clone https://github.com/jonls/qsopt-ex.git
```

Once that is done, run the following commands from the Terminal window:

```
cd qsopt-ex
./bootstrap
mkdir build && cd build
../configure -prefix /usr/local
make
make check
make install
```

See **Note 5** for interpreting the results of the tests performed during installation.

2.5 Installing python-qsoptex

The Python module *python-qsoptex*, created by Jon Lund Steffensen in support of the PSAMM toolbox [15], provides an additional tool to facilitate creating, editing, and solving linear programs in rational arithmetic. This saves a substantial amount of time compared to the original version of MONGOOSE, which required reading and writing files and using a system call for *QSopt_ex*.

To install *python-qsoptex*, start a Terminal window and run the commands

```
git clone https://github.com/jonls/python-qsoptex.git
cd python-qsoptex
python setup.py install
python test_qsoptex.py
```

If you get a line ending with "OK" at the end of this process, the installation was successful. If it fails during the installation process (penultimate step), you might be missing one of the dependencies and need to go back to one of the previous steps.

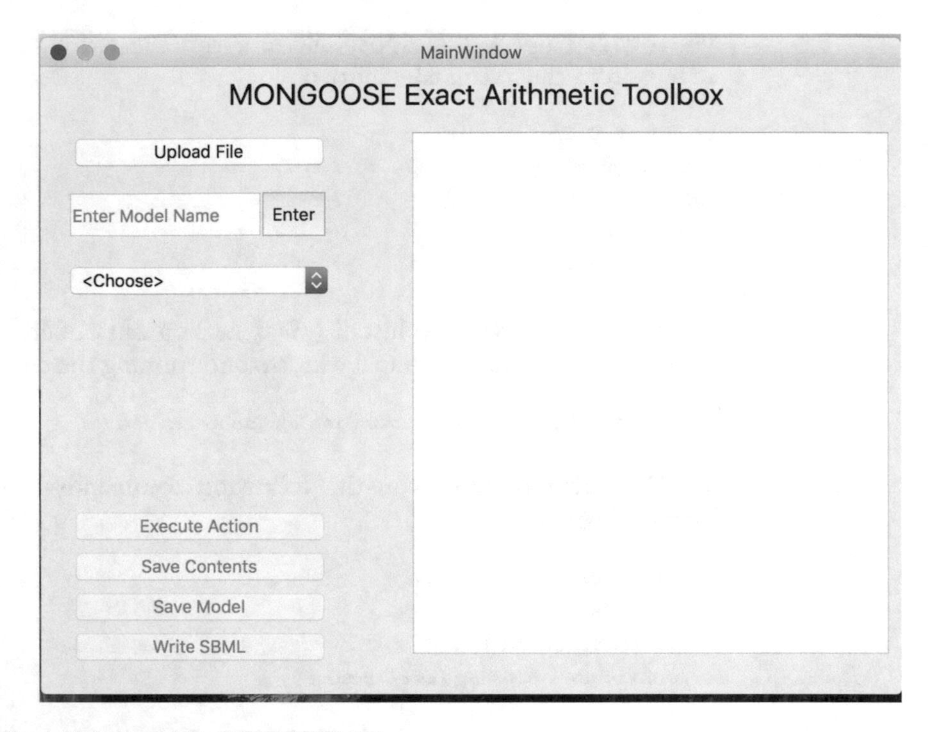

Fig. 1 The initial view of the MONGOOSE GUI

2.6 Installing MONGOOSE and Its GUI

After all this preparation, the installation of MONGOOSE itself, together with its GUI, is quite simple. You just need to run the commands

```
git clone https://github.com/WGS-TB/MongooseGUI3.git
cd MongooseGUI3
python MONGOOSEgui.py
```

This will start the MONGOOSE GUI, which looks as shown in Fig. 1.

3 Methods

In this section we describe the protocol to be followed to identify the most common structural features—or bugs—that metabolic network models have, using the constraint-based framework. When the detected model behavior is in accordance with the modeler's expectations, the reconstruction process should proceed to the next stage (such as model refinement). However, if model behavior violates the modeler's expectations, especially if those expectations are based on prior biological knowledge, it might be necessary to modify the model until the behavior changes.

3.1 Parsing a Metabolic Network Model

There are two formats supported by MONGOOSE, the mostly outdated Excel format and the newer SBML format [18]. As the vast majority of metabolic network models are published in the SBML format, we demonstrate the parsing on an SBML file, the one for *Acinetobacter baumannii* [19], abbreviated as AB1, chosen because it appears first in alphabetical order among the available reconstructions [20] and shows a number of interesting features detected by MONGOOSE. Unfortunately, since the parsing of a metabolic network model may require some flexibility and non-trivial user input, the parsing part of the MONGOOSE pipeline is not currently integrated with the GUI. However, we plan to integrate it in a future version.

MONGOOSE provides two parsing modes—the default mode is based on the original version of MONGOOSE, while the other one emulates the COBRA toolbox [21, 22]; the default is more suitable for model verification purposes, but both should give the same result on a properly designed model. The differences between the two modes are listed in Table 1. For more information, including details on how many of the existing models were affected by those differences, see the table at http://groups.csail.mit.edu/cb/mongoose/.

To parse a model in the default mode, start a Python session with the command

```
python -i ModelParsing.py
```

and then, once the input line starting with a " > > > " appears, run the command

```
model = parseSBML('Acinetobacter Baumannii.xml')
```

If you want to parse the model in the COBRA mode instead, run the command

```
model = parseSBML('Acinetobacter Baumannii.xml', True)
```

Table 1
Differences between the default mode and the COBRA mode for parsing

Decision	Default mode	COBRA mode
Is the metabolite constrained?	Not [external]	Not [boundaryCondition or name ends in _b]
Is the reaction forward-only?	Not [reversible]	[LOWER_BOUND $\geq$ 0]
Is the reaction reverse-only?	No[a]	[UPPER_BOUND $\leq$ 0]
Is the reaction the objective?	User-chosen[b]	[OBJECTIVE_COEFFICIENT == 1]

[a]Except after structural analysis
[b]Among reactions with "biomass" in their lowercased name

The parser may generate warnings, which should be regarded as signs that something needs to be corrected in the model. Here are the main ones, both of which are encountered in parsing AB1:

```
Warning: no candidate for extra, outside found in the list
Warning: no candidate for biomass found in the list
```

The first warning means that no external compartment specification could be identified. This can be fixed easily if there is a systematic naming convention that distinguishes external metabolites from internal ones. For instance, if metabolites whose compartment is specified as 'e' are to be considered external, run the command

```
model.adjustExternal('e')
```

On the other hand, there is also a way to specify external metabolites based on their names; in the AB1 model we are using as the example, any metabolite whose name ends in 'xt' is external, so we can run the command

```
model.adjustCompartments('xt', startPos = -2)
```

The second argument means that the position where the string we are looking for is expected to start is position −2, or 2 from the end, of the metabolite name.

The second warning means that no biomass reaction could be identified. This can also be fixed by specifying the particular reaction that should be used as the biomass reaction by searching for the reaction with the correct name or other defining properties. For instance, if the correct biomass reaction is reaction number 760, like in the AB1 model, run the command

```
model.biomassCoefficients[760] = 1
```

Another type of warning,

```
Warning: multiple candidates for biomass found in the list
```

occurs when multiple candidate reactions are identified; in this case, you will be given an explicit choice of reaction names and need to enter the number corresponding to the right one.

The internal representation that MONGOOSE uses for all parts of a metabolic network model are objects. For a reaction, this object consists of a name, a list of metabolite-coefficient pairs, a Boolean "reversible" value (determining whether it can proceed in both directions), and optional attributes. For a metabolite, this object contains its species, its compartment, a Boolean "external" value (determining whether it is subject to flux balance), and optional attributes. Multiple metabolites can share the same species, as long as they belong to different compartments. For the network as a whole, this object consists of a list of metabolites, a list of reactions, a stoichiometric matrix, and optional attributes. A lot of the optional attributes are created by the process of structurally analyzing the network, which we refer to as "reducing" it.

At the end of the parsing process, to avoid losing any of the changes you made, we recommend saving the model into a **shelve** file, by using the commands (recall that AB1 is an abbreviated name of the model organism)

```
s = shelve.open('MyModel')
s['AB1'] = model
s.close()
```

This will create a file storing a compressed representation of the model, which will be useful for future processing; in particular, this compressed representation can be directly loaded into and modified within the GUI, as we describe below.

3.2 Inspecting, Modifying, and Saving a Model

Once you have a shelve file containing a model, you can load it into the MONGOOSE GUI that you started by clicking on the Upload File button, then typing in the key corresponding to the model you want to load (in our example it will be AB1) and pressing the Enter button. Note that a shelve file may contain multiple models, and the full list of options will appear in the right-hand display (Fig. 2).

Once your choice has been confirmed, the drop-down menu will become active. It allows you to inspect the following elements of the model:

- a metabolite (its species name, compartment, and reduction status if known)
- a reaction (its name, metabolite-coefficient pairs, and reduction status if known)
- a reaction formula (in a readable format instead of metabolite-coefficient pairs)
- the biomass reaction (its number only)
- a reaction subset (if the network has been structurally analyzed; *see* Subheading 3.7)

In the snapshot below, we successively:

1. determine the biomass reaction
2. display its name and metabolite-coefficient pairs
3. print out its formula

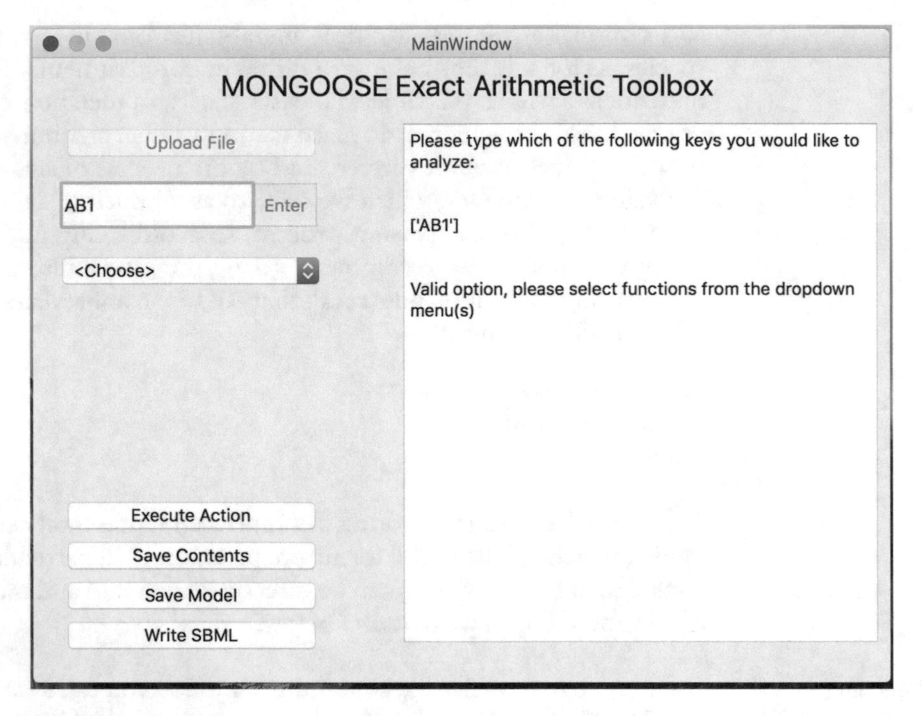

Fig. 2 Selecting the model in a file with the GUI

4. identify a metabolite involved in it, determine its species name and compartment

After making a choice from the drop-down menu, press the Execute Action button to see its result on the right-hand side display.

For **step 1**, we simply select findBiomassReaction from the drop-down menu and press the Execute Action button, which displays 760 as the index of the biomass reaction.

For **step 2**, we select reactions from the drop-down menu and enter the number 760 next to it, then select name from the second drop-down menu. After we press the Execute Action button, the name, "R761", appears. We then repeat the same process, but select pairs instead of name, which displays a list of pairs of the form [metabolite index, coefficient] (Fig. 3).

For **step 3**, we select printReactionFormula from the drop-down menu without changing the reaction number, which produces a human-readable version of the biomass reaction.

For **step 4**, we select metabolites from the drop-down menu, enter the number 707, and query its species by name, which produces the output "PHOSPHOLIPID". We also check whether it is an external metabolite, with the result—False—letting us know that it isn't (Fig. 4).

The GUI also allows you to modify a model, in the following ways:

MainWindow

MONGOOSE Exact Arithmetic Toolbox

Upload File

AB1 Enter

reactions 760

pairs

Execute Action

Save Contents

Save Model

Write SBML

```
>>> model.findBiomassReaction()

760

>>> model.reactions[760].name

R761

>>> model.reactions[760].pairs

[[372, Fraction(-3, 5)], [164, Fraction(-3, 100)], [874,
Fraction(-19, 100)], [707, Fraction(-1, 25)], [918,
Fraction(-1, 50)], [466, Fraction(-1, 200)], [398,
Fraction(-27, 1000)], [240, Fraction(-19, 500)], [63,
Fraction(-3, 100)], [363, Fraction(-46, 1)], [447,
Fraction(1, 1)], [821, Fraction(46, 1)], [87, Fraction(46, 1)]]
```

Fig. 3 The result of performing the first two queries described in the text

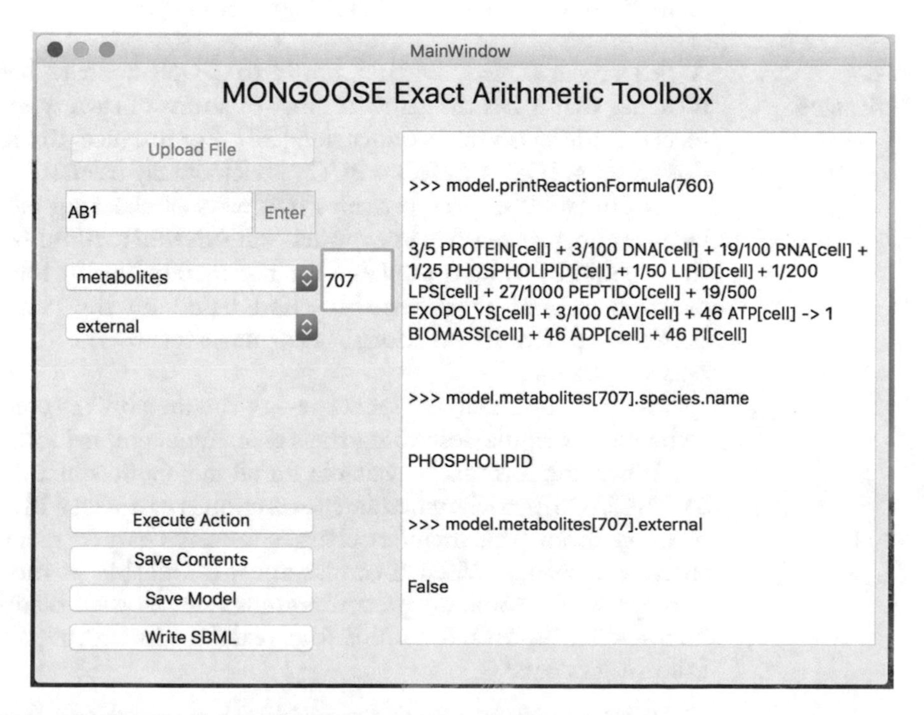

Fig. 4 The result of performing the second two queries described in the text

- add a reaction
- delete one or more reactions
- add a metabolite
- delete one or more metabolites
- reduce the network by performing structural analysis on it [8]

Here, we demonstrate the addition of a reaction (the deletion of reactions and the addition or deletion of metabolites work along similar lines), and discuss the network reduction in detail in the following sections.

For the model in our example, the biomass metabolite is only present in the biomass reaction, and thus will be a topologically blocked (dead-end) metabolite unless we add an export reaction for it. Just as we did for PHOSPHOLIPID in the previous example, we begin by listing its name to ensure that we have the correct one, and check that it is not external, so that the balance condition applies to it. Then we will add the reaction that exports it, giving it an appropriate name, and print out its formula to confirm that it is added correctly. Finally, we will perform a network reduction by using the `reduceNetwork` command (note that this can take several minutes for large models, so please be patient and don't close the GUI while it is processing). The elements of structural analysis performed during network reduction will be explained in detail in Subheadings 3.4–3.7, so we do not describe them here (Figs. 5 and 6).

3.3 Checking Elemental Balance

A properly specified reaction needs to be elementally balanced, meaning that it has the same number of atoms of each type on the reactant side as on the product side [23]. For instance, the reaction $C_2H_5OH + 3O_2 \rightarrow 3H_2O + 2CO_2$ is elementally balanced.

MONGOOSE now provides two ways of checking elemental balance for the reactions in a model—an *internal* method based on the chemical formulas available for the metabolites in the model description, and an *external* method based on the consistency between different reactions. Let us explain the difference between them.

Suppose first that all—or some—of the metabolites come with a chemical formula describing their elemental composition.

When the formula is available for all metabolites in a reaction, MONGOOSE checks whether the reaction is balanced with respect to every atom (this includes charges for any charged metabolites such as protons). When formulas are not available in the model description for some of the compounds (we call such compounds "unspecified"), MONGOOSE flags reactions in which one of the following occurs:

1. Only one metabolite is unspecified, but its inferred formula is not valid.

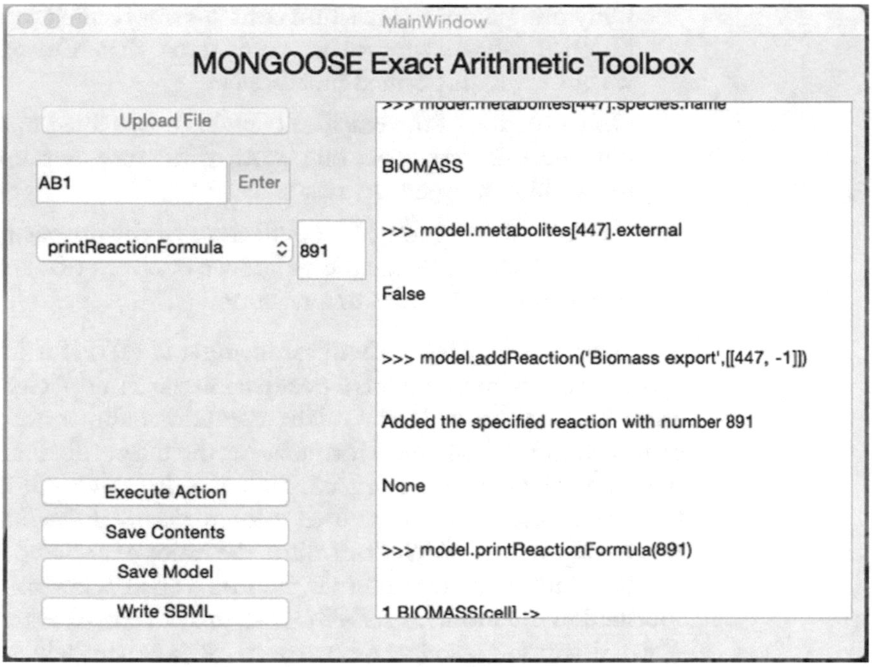

Fig. 5 The result of adding a biomass export reaction

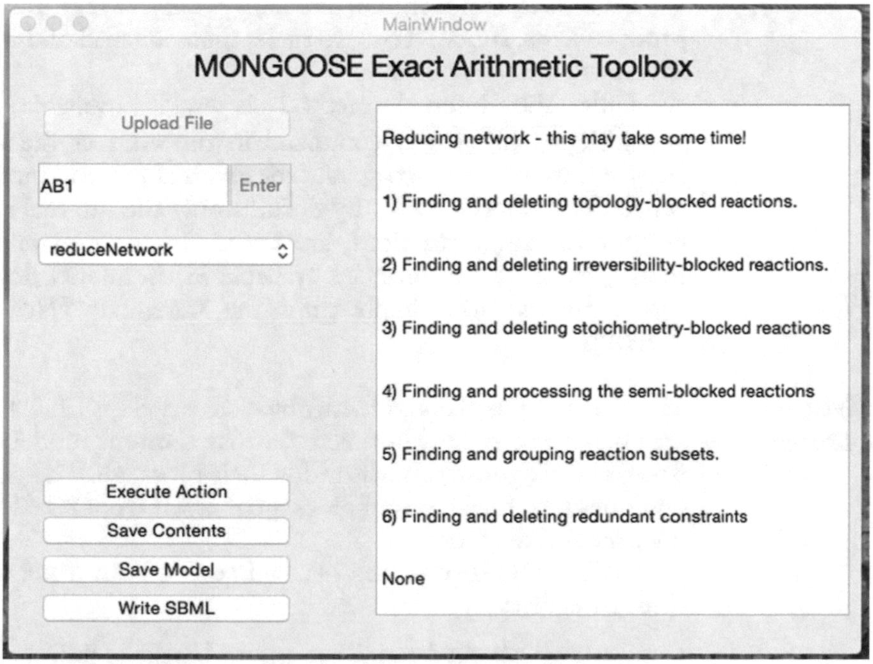

Fig. 6 The result of reducing the model

2. Only one side of the reaction contains unspecified metabolites, but has more atoms of a certain type than the other side, without the unspecified metabolites.
3. Only one side of the reaction contains unspecified metabolites, but there is less than one atom difference per unspecified metabolite between the two sides.
4. The gcd of stoichiometric coefficients of the unspecified metabolites does not divide the difference of atoms of a certain type between the two sides of a reaction.

Here, we say that a chemical formula is valid if it includes at least one atom and contains a positive integer multiplicity for every atom included in it. If any of the conditions above occurs, there cannot exist valid chemical formulas for the unspecified metabolites that make the reaction balanced. Note that MONGOOSE assumes that the molecules are described using a *chemical formula*, not an *empirical formula* which determines the ratios of each type of atom.

In addition to the internal method which relies on formulas specified in the model, MONGOOSE provides an external method which identifies groups of reactions, if any, for which a linear combination produces a "free lunch," a reaction with an empty reactant side and a non-empty product side [24]. In other words, if the net reaction of a group of reactions, taken with positive integer multiplicities, is of this type, this shows that at least one of the reactions in the group is elementally unbalanced.

MONGOOSE identifies and flags groups of reactions that can produce a free lunch, trying to make them as small and as disjoint from each other as possible.

Only the internal method is made available via the `checkElementalBalance` command in the GUI at the moment; however, we plan to incorporate the external method into a future version of MONGOOSE. By default, only the internal reactions, not the exchange reactions, are tested. In the case of the AB1 model, since no formulas are available in the model description, this command will simply print the statement "No formulas found!".

3.4 Identifying Blocked Reactions

MONGOOSE is able to identify blocked reactions [25], those that are unable to carry any non-zero flux based on the model specifications, and diagnose the reason for their blockage. The box below describes the three types of blockage that MONGOOSE is able to uncover and diagnose.

MONGOOSE classifies blocked reactions in a network into three categories:

- **topology-blocked** reactions, whose blockage follows from the topology of the metabolic network (e.g., reactions containing a dead-end metabolite)

- **stoichiometry-blocked** reactions, whose blockage follows from the flux balance constraints on the internal metabolites
- **irreversibility-blocked** reactions, whose blockage follows from the addition of irreversibility constraints to the flux balance constraints

The first step of the network reduction process consists in identifying all the blocked reactions in the model, and deleting them. It also deletes all the dead-end metabolites identified in conjunction with topology-blocked reactions. This does not alter the set of possible flux distributions of the network, since the blocked reactions can only carry zero flux by definition, and the dead-end metabolites do not add any effective constraints.

In order to identify the status of a particular reaction, if it is suspected that it might be blocked, you may use the reductionStatus option when querying the model about this reaction. However, there is also a way to access the list of all those reactions that suffer from a particular type of blockage—simply use one of the commands findTopoBlockedReactions , findStoichBlockedReactions, or findIrrevBlockedReactions in the GUI. This will produce the list of all the reactions blocked in this particular way. The command findSemiBlockedReactions produces the list of all semi-blocked reactions, discussed in Subheading 3.6. In addition, you can also see the list of all metabolites that are eliminated due to topology via the findTopoBlockedMetabolites command. *See* Fig. 7 for an illustration on the AB1 model.

The presence of blocked reactions is in general undesirable, and this is why in Subheading 3.5 we describe ways in which MONGOOSE can help unblock particular blocked reactions. However, the results of these automated methods cannot substitute for the expertise of someone with a detailed knowledge of the metabolism of the particular organism in question, and should be treated as suggestions.

3.5 Unblocking Blocked Reactions

The topology-blocked reactions are typically the simplest kind of reactions to unblock. First, one needs to compute the causal chain that has led to the blockage of this reaction (this is not always a single metabolite that is not used elsewhere in the model, but could be the result of an iterative dead-end elimination process). Then, starting with the first element of the causal chain, one needs to identify the reason for the blockage. Often, though not always, this will be the a typo in the model specification, where the name of a metabolite that should be identical to another one is spelled differently somewhere in the model. Another reason for topology-blocked reactions can simply be the lack of another reaction that contains a metabolite present in the first reaction of the causal chain for this blockage.

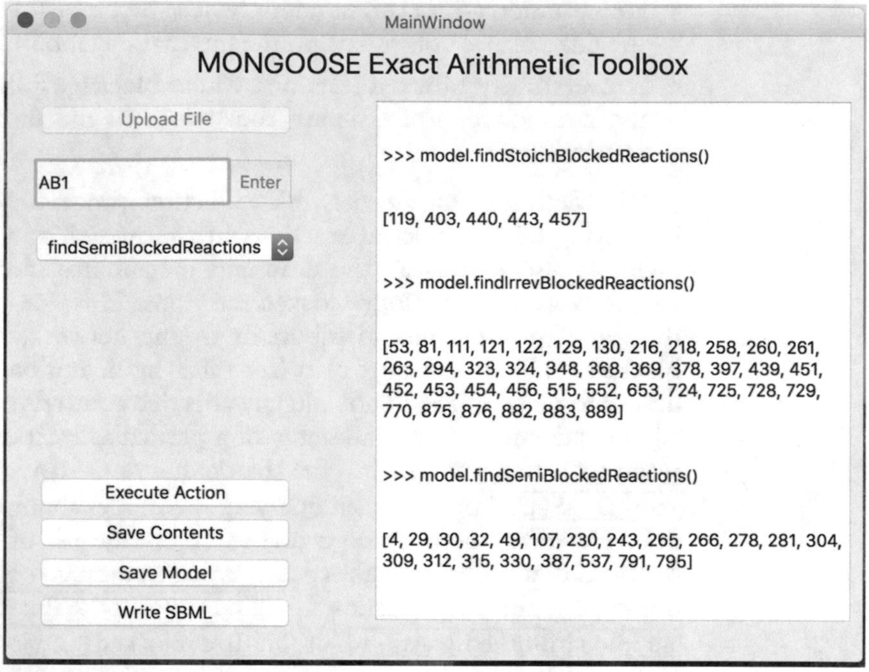

MainWindow

MONGOOSE Exact Arithmetic Toolbox

Upload File

AB1 Enter

findSemiBlockedReactions

Execute Action

Save Contents

Save Model

Write SBML

```
>>> model.findStoichBlockedReactions()

[119, 403, 440, 443, 457]

>>> model.findIrrevBlockedReactions()

[53, 81, 111, 121, 122, 129, 130, 216, 218, 258, 260, 261,
263, 294, 323, 324, 348, 368, 369, 378, 397, 439, 451,
452, 453, 454, 456, 515, 552, 653, 724, 725, 728, 729,
770, 875, 876, 882, 883, 889]

>>> model.findSemiBlockedReactions()

[4, 29, 30, 32, 49, 107, 230, 243, 265, 266, 278, 281, 304,
309, 312, 315, 330, 387, 537, 791, 795]
```

Fig. 7 The result of identifying stoichiometry-, irreversibility-, and semi-blocked reactions

In the case of a typo, the metabolite needs to be renamed, which can be done by identifying the reaction containing the incorrectly spelled metabolite name (the second link in the causal chain), adding a version of this reaction with the correct metabolite in place, deleting the original reaction, and deleting the incorrectly spelled metabolite. Once the changes have been made to the model, one should reduce the network once more and see if the problem has been resolved. In the case of a missing reaction, the blocked reaction can remain in the model, but a search should be initiated for a reaction that can produce (respectively, consume) the metabolite that the blocked reaction consumes (respectively, produces).

Unlike topology-blocked reactions, MONGOOSE can unblock stoichiometry- or irreversibility-blocked reactions by suggesting that certain model constraints be relaxed—flux balance constraints on a small number of metabolites in the former case, and irreversibility constraints on a small number of reactions in the latter case. Although these sets of constraints are typically quite small, there could be multiple solutions and they are not guaranteed to be the smallest possible (although they are always minimal in the sense that relaxing only some, but not all, of the constraints in them will not remove the blockage). For this reason, they should be treated as suggestions to the modeler rather than definitive changes to be made to the model [26].

The changes needed to unblock the biomass reaction can be identified using the unblockBiomassReaction command in the

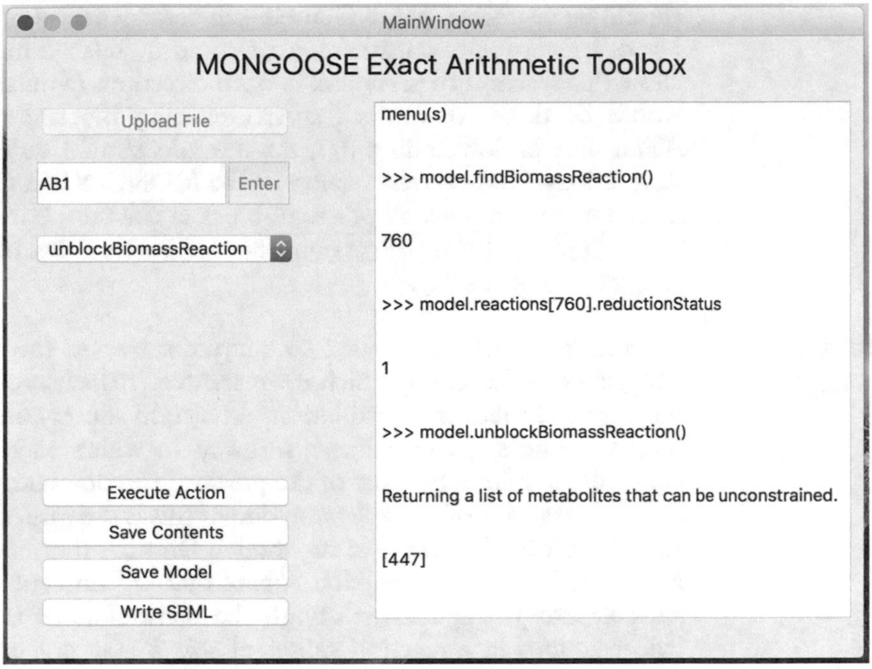

Fig. 8 The result of testing the reduction status of the biomass reaction and unblocking it

GUI. In the case of the AB1 model, this command will produce the message "The biomass reaction is not blocked; nothing to do here." However, if we run it on the parsed model before the addition of a biomass export reaction, it produces the list [447], containing the single BIOMASS metabolite, as an optional set of constraints to be relaxed (and indeed, the addition of a biomass export reaction effectively addresses the issue) (Fig. 8).

3.6 Identifying Semi-blocked Reactions

In addition to blocked reactions, MONGOOSE also identifies semi-blocked reactions [8], namely, reactions that are specified to be reversible, but for which only one direction (forward or reverse) is possible due to the model constraints. These are also potentially undesirable discrepancies between the model specification and its functionality, but may not be as severe as blocked reactions.

MONGOOSE identifies two types of semi-blocked reactions:

- **effectively forward** reactions, specified to be reversible but only able to carry positive (or zero) flux
- **effectively reverse** reactions, specified to be reversible but only able to carry negative (or zero) flux

Normally, any such reactions are labeled irreversible (and reversed in the case of effectively reverse reactions) by MONGOOSE before the next stage of the structural analysis. However, if the modeler strongly believes that some semi-blocked reaction should be reversible, MONGOOSE can help identify a small number of irreversibility constraints that could be relaxed in order to allow this reaction to carry flux in both directions (similar to what would be done to unblock an irreversibility-blocked reaction). Again, like in Subheading 3.6, these results should be treated as suggestions. The current version of the MONGOOSE GUI does not allow the unblocking of a semi-blocked reaction, but the command `findSemiBlockedReactions` does allow for a list of all such reactions to be displayed.

3.7 Identifying Reaction Subsets

A reaction subset (also called an enzyme subset in the literature [27]) is a set of reactions which are guaranteed to have proportional fluxes in any valid flux distribution. A reaction subset could correspond to the steps of a linear pathway, in which each reaction consumes a unique product of the previous reaction and produces a unique reactant of the following one. If this is the case, a reaction subset is a desired feature of the model. On the other hand, there can be reactions in a reaction subset that are unrelated to one another except implicitly, through the constraints of the model (the reactions in a reaction subset of size 2 may not even share any metabolites). In such a case, since inhibiting one model in a reaction subset inhibits all of the other ones, the behavior is undesirable and should be closely examined by the modeler.

MONGOOSE identifies two types of reaction subsets:

- **reversible** reaction subsets, in which all reactions can carry flux of either sign
- **irreversible** reaction subsets, in which all reactions can only carry positive flux

The advantage of identifying reaction subsets, from the point of view of network reduction, is that they can then be grouped into a single overall reaction without loss of information [27]. This is a helpful step in the subsequent computation of essential reactions, synthetic lethal pairs, and minimal media, discussed in Subheadings 3.8–3.9.

The last stage of the network reduction process also deletes any metabolites that define redundant constraints. At the end of the process, the network contains the information about the reduction status for each of the reactions and metabolites in the model, as well as all the information required to convert a flux vector in the reduced network to one in the original network (via the `reactionSubsets` attribute, which in this case can also contain a single reaction).

3.8 Identifying Essential Reactions and Synthetic Lethal Pairs

The concept of a **minimal cut set** is a very fruitful way of analyzing metabolic network models. A cut set is a set of reactions whose inhibition (by setting their flux to 0) disables the production of a target reaction, typically the biomass reaction. A minimal cut set [28, 29] is one such that no reaction can be removed from it while preserving the property defining a cut set. Minimal cut sets correspond to inhibition strategies and can point the way to discovering new drug targets or selecting metabolic engineering interventions.

MONGOOSE identifies two types of minimal cut sets:

- **essential reactions**, single reactions whose inhibition disables biomass production
- **synthetic lethal pairs**, pairs of reactions that are not essential on their own, but whose simultaneous inhibition disables biomass production

MONGOOSE performs the search for essential and synthetic lethal reactions [6] by using the reduced network, since its size is substantially smaller. It also samples a valid flux vector for each reaction in the reduced network, in a pre-processing step in order to further speed up the computation (since only those reactions or pairs of reactions that have a possibility of being minimal cut sets then need to be tested). In the final stage, MONGOOSE expands the minimal cut sets it found back to the original (non-reduced) network. The commands findEssentialReactions and findSyntheticLethalPairs implement this functionality in the MONGOOSE GUI (Fig. 9).

3.9 Identifying Minimal Media

A **medium** in a metabolic network model is a subset of the import reactions that enable biomass production. A minimal medium [30] is a medium from which no import reaction can be removed without disabling biomass production. The search for good minimal media is particularly informative for bacteria, where it can, for instance, lead to designing new growth media for culturing them in a laboratory setting.

MONGOOSE can identify a number of small minimal media based on the reduced network, then expand them to the original network. While these minimal media are not guaranteed to be as small as possible, they are typically quite small. The command findMinimalMedia implements this functionality in the MONGOOSE GUI (Fig. 10).

3.10 Ending Your Session

At the end of a session using the MONGOOSE GUI, press the SaveContents to save the history of the commands you ran on the model and their output, which will create a dialog box prompting you to enter a name and a location to save your file in. If you made changes to the model that you would like to save (in particular, if you used the reduceNetwork command and would like to save time

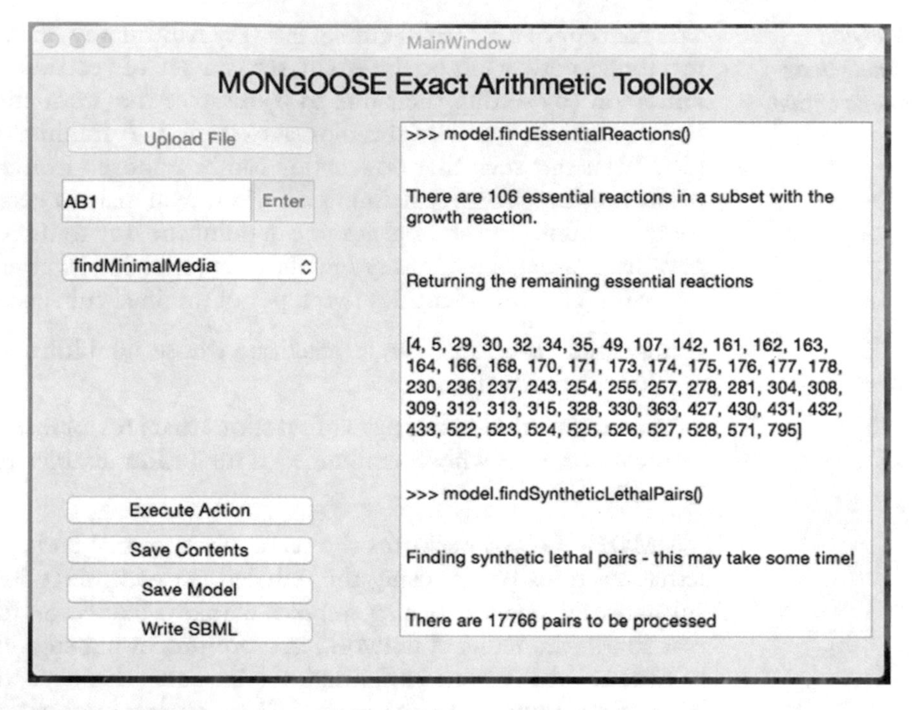

Fig. 9 The result of identifying the essential reactions and synthetic lethal pairs

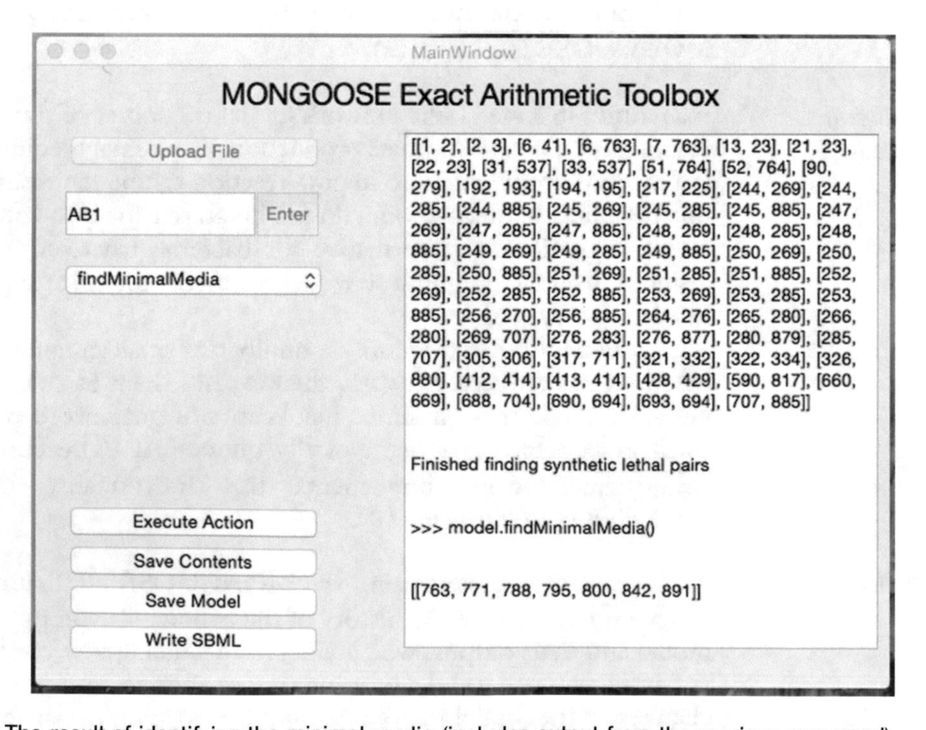

Fig. 10 The result of identifying the minimal media (includes output from the previous command)

by not having to run it again in a future analysis), you can also use the SaveModel button. Please note that this will overwrite the model in your original shelve file, so make sure that you are happy with the changes you made and want to overwrite the original model. It is always a good idea to create a backup version of the original shelve file that you can revert to if you change your mind later. Finally, the WriteSBML button enables you to write out the model into a text file in SBML format [18], making it suitable for analysis with most other toolboxes for metabolic network models.

We hope that you find MONGOOSE useful, and always welcome feedback, which will help us decide on the directions to pursue in developing it further.

4 Notes

1. If you have a fast Internet connection, it will be faster to use the offline installer for *Qt* than the online one, which can be slow due to the use of a mirror server.
2. When installing *Qt*, you may get this error: "You need to install Xcode version 5.0.0. Download Xcode from https://developer.apple.com/xcode". If you get this error you might need to install Xcode—not version 5.0.0, but the latest version that works on your operating system—go to the suggested website, download and install it. You will also need to accept the license via the command

   ```
   sudo /usr/bin/xcodebuild
   ```

   from a Terminal window. Otherwise you may get an error when installing *PyQt*.
3. Another problem that may arise during the installation of *PyQt* is a failure to locate SIP. In that case, instead of using *pip3* to install SIP, follow http://pyqt.sourceforge.net/Docs/sip4/installation.html's instructions and specify SIP's location via the `-sip` option when installing *PyQt*(by default, it is /Library/Frameworks/Python.framework/Versions/3.5/bin/sip).
4. When installing *QSopt_ex*, you may encounter a problem if you do not have the *Autoconf* [31] tool. It can be installed in essentially the same way as *Libtool*, from http://ftp.gnu.org/gnu/autoconf/autoconf-2.69.tar.gz. In this case, the first two commands in the Terminal window should be replaced by

   ```
   tar -xvf autoconf-2.69.tar.gz
   cd autoconf-2.69
   ```

5. When installing *GNU MP* and *QSopt_ex*, as you run *make check* before the final step, pay close attention to the result of the tests. If any of them fail, the software may not work correctly; however, warnings are (usually) perfectly fine.

Acknowledgements

The authors would like to thank Dan Park for creating the MONGOOSE website. In addition, the authors would like to acknowledge the invaluable input of Bonnie Berger, Aviv Regev, and Jason Trigg, as well as the help of Daniel Espinoza and Dan Steffy. This work is supported by an NSERC Discovery Grant as well as an Alfred P. Sloan Fellowship.

References

1. Benson DA, Clark K, Karsch-Mizrachi I et al (2015) GenBank. Nucleic Acids Res 43: D30–D35
2. Zhang C, Hua Q (2015) Applications of genome-scale metabolic models in biotechnology and systems medicine. Front Physiol 6:413
3. Long MR, Ong WK, Reed JL (2015) Computational methods in metabolic engineering for strain design. Curr Opin Biotechnol 34:135–141
4. Varma A, Palsson B (1994) Metabolic flux balancing: basic concepts, scientific and practical use. Nat Biotechnol 12:994–998
5. Varma A, Palsson B (1994) Stoichiometric flux balance models quantitatively predict growth and metabolic by-product secretion in wild-type Escherichia coli W3110. Appl Environ Microbiol 60:3724–3731
6. Suthers PF, Zomorrodi A, Maranas CD (2009) Genome-scale gene/reaction essentiality and synthetic lethality analysis. Mol Syst Biol 5:1
7. Gottstein W, Olivier BG, Bruggeman FJ, Teusink B (2016) Constraint-based stoichiometric modelling from single organisms to microbial communities. J R Soc Interface 13 (124):20160627
8. Chindelevitch L, Trigg J, Regev A, Berger B (2014) An exact arithmetic toolbox for a consistent and reproducible structural analysis of metabolic network models. Nat Commun 5:4893
9. Boettiger C (2015) An introduction to Docker for reproducible research, with examples from the R environment. In: ACM SIGOPS operating systems review, special issue on repeatability and sharing of experimental artifacts, vol 49, no 1. ACM, New York, pp 71–79
10. van Rossum G (1995) Python tutorial. Technical Report CS-R9526, Centrum voor Wiskunde en Informatica (CWI), Amsterdam
11. Applegate D, Cook W, Dash S, Espinoza D (2007) Exact solutions to linear programming problems. Oper Res Lett 35:693–699
12. Behnel S, Bradshaw R, Citro C, Dalcin L et al (2011) Cython: the best of both worlds. Comput Sci Eng 13:31–39
13. Bornstein BJ, Keating SM, Jouraku A, Hucka M (2008) LibSBML: an API library for SBML. Bioinformatics 24(6):880–881
14. PyQt Whitepaper. Riverbank Computing. http://www.riverbankcomputing.com/static/Docs/PyQt4/pyqt-whitepaper-a4.pdf Accessed October 9, 2017
15. Steffensen JL, Dufault-Thompson K, Zhang Y (2016) PSAMM: a portable system for the analysis of metabolic models. PLoS Comput Biol 12(2):e1004732
16. Matzigkeit G, Oliva A, Tanner T, Vaughan GV (2015) GNU Libtool reference manual. Samurai Media Limited, Surrey
17. Granlund T, The GMP Development Team (2016) GNU MP: the GNU Multiple Precision Arithmetic Library. http://gmplib.org/. Accessed October 9, 2017
18. Hucka M, Finney A, Sauro HM, Bolouri H et al (2003) The systems biology markup language (SBML): a medium for representation and exchange of biochemical network models. Bioinformatics 19(4):524–531
19. Kim HU, Kim TY, Lee SY (2010) Genome-scale metabolic network analysis and drug

targeting of multi-drug resistant pathogen Acinetobacter baumannii AYE. Mol BioSyst 6 (2):339–348

20. In Silico Organisms (2016) UCSD systems biology group. http://systemsbiology.ucsd.edu/InSilicoOrganisms/OtherOrganisms. Accessed October 9, 2017
21. Hyduke D, Schellenberger J, Que R, Fleming R et al (2011) COBRA Toolbox 2.0. Protoc Exch http://dx.doi.org/10.1038/protex.2011.234
22. Ebrahim A, Lerman JA, Palsson BO, Hyduke DR (2013) COBRApy: constraints-based reconstruction and analysis for python. BMC Syst Biol 7(74). http://dx.doi.org/10.1186/1752-0509-7-7
23. Ravikrishnan A, Raman K (2015) Critical assessment of genome-scale metabolic networks: the need for a unified standard. Brief Bioinform 16(6):1057–1068
24. Chindelevitch LA (2010) Extracting information from biological networks. Dissertation, Massachusetts Institute of Technology. http://hdl.handle.net/1721.1/64607
25. Schuster R, Schuster S (1991) Detecting strictly detailed balanced subnetworks in open chemical reaction networks. J Math Chem 6:17–40
26. Ponce-de-León M, Montero F, Peretó J (2013) Solving gap metabolites and blocked reactions in genome-scale models: application to the metabolic network of *Blattabacterium cuenoti*. BMC Syst Biol 7(114)
27. Gagneur J, Klamt S (2004) Computation of elementary modes: a unifying framework and the new binary approach. BMC Bioinf 5(175)
28. Klamt S, Gilles E (2004) Minimal cut sets in biochemical reaction networks. Bioinformatics 20(2):226–234
29. Acuña V, Chierichetti F, Lacroix V, Marchetti-Spaccamela A et al (2009) Modes and cuts in metabolic networks: complexity and algorithms. BioSystems 95(1):51–60
30. Suthers PF, Dasika MS, Kumar VS, Denisov G et al (2009) A genome-scale metabolic reconstruction of Mycoplasma genitalium, iPS189. PLoS Comput Biol 5(2):e1000285
31. Calcote J (2010) Autotools: a practioner's guide to GNU autoconf, automake, and libtool. No Starch Press, San Francisco, CA

Chapter 4

The FASTCORE Family: For the Fast Reconstruction of Compact Context-Specific Metabolic Networks Models

Maria Pires Pacheco and Thomas Sauter

Abstract

The FASTCORE family is a family of algorithms that are mainly used to build context-specific models but can also be applied to other tasks such as gapfilling and consistency testing. The FASTCORE family has very low computational demands with running times that are several orders of magnitude lower than its main competitors. Furthermore, the models built by the FASTCORE family have a better resolution power (defined as the ability to capture metabolic variations between different tissues, cell types, or contexts) than models from other algorithms.

Key words Context-specific, omics data integration, Algorithm, Metabolic modeling, Constraint-based modeling, FBA, Microarray

1 Introduction

The FASTCORE family comprises FASTCORE [1] and to the present date three published workflows: FASTCORMICS [2], FASTCC [1], and fastgapfill [3] that use the FASTCORE algorithm to perform model reconstructions, consistency testing, and gap-filling tasks.

FASTCORE completes a set of reactions with high confidence level (core reaction set) or a draft network for fastgapfill with a minimal number of reactions with a low confidence to obtain a consistent model. By definition, a consistent model does not include dead ends nor gaps and therefore every reaction can carry a nonzero flux. To this aim, FASTCORE forces the core set to carry a flux (practically a flux is pushed above a flux/no flux segregation threshold ε using an approximation of the cardinality function which allows employing a linear programming-based approach. To obtain a compact model, a L1-regularization is performed to minimize the flux through reactions with a low confidence level (non-core reactions) [1]. The search for compactness is motivated by the idea that the synthesis of enzymes or transporters consumes

Marco Fondi (ed.), *Metabolic Network Reconstruction and Modeling: Methods and Protocols*, Methods in Molecular Biology, vol. 1716, https://doi.org/10.1007/978-1-4939-7528-0_4,

energy and metabolites. Thus, an organism that is able to fulfil the required tasks to survive with minimal costs should have an evolutionary advantage.

Although the formalism of FASTCORE allows for the application of different tasks such as gapfilling, we will focus in this book chapter on how to use FASTCORE and FASTCORMICS for the building of compact context-specific models and FASTCC for the testing of consistency.

By a context-specific model, we understand a metabolic network based on the constraint-based modeling approach [4] that represents the metabolism of a given cell type or in a given condition such as disease state or hypoxia. Context-specific metabolic networks were first introduced by Shlomi et al. [4] and were motivated by the rationale that as different cell types express different genes due to epigenetic modifications or different external stimuli, the metabolism should also be cell-specific. To obtain context-specific models, omics data is used to extract a sub-network that only contains active reactions in the context of interest.

2 Material

To run the original implementation of FASTCORE, FASTCORMICS, or FASTCC (http://wwwen.uni.lu/recherche/fstc/life_sciences_research_unit/research_areas/systems_biology/software/fastcore), Matlab (www.mathworks.com) and the IBM CPLEX (https://www-01.ibm.com/software/commerce/optimization/cplex-optimizer) solver are required. Another implementation of FASTCORE and FASTCC, which can be run with other solvers, is available in the COBRA toolbox [5]. Further instructions on how to install the cobra toolbox can be found at https://opencobra.github.io. Moreover, the original code of FASTCORE can easily be modified to use other solvers. A version of FASTCORE, fastgapfill, and FASTCC is also available in python in the PSAMM metabolic modeling toolbox [6] (https://github.com/zhanglab/psamm).

FASTCORMICS requires the installation of R (https://cran.r-project.org/) for data preprocessing. Besides the packages required for standard microanalysis such as *affy* [7] or *oligo* [8] (depending on the type of arrays), and *ggplots* [9] for the plotting, FASTCORMICS needs the installation of the *frma* package [10, 11] and a reference vector used by the barcode function [11] such as *hgu133plus2barcodevecs* [12] for the HGU133plus2 platform. The packages used for the frma normalization and the barcode discretization can be downloaded from the Bioconductor website (https://www.bioconductor.org). Further packages needed for the quality control can also be downloaded from Bioconductor or from R-cran (https://cran.r-project.org/) mirrors using the *install.packages* function. And finally, a genome-scale reconstruction such as Recon X [13, 14], Recon2.2 [15], HMR, and HMR 2.0 for humans is required. Barcode [11] is currently available for human and mouse arrays. The extensive list of supported arrays and information on barcode can be found at http://barcode.luhs.org/.

3 Methods

3.1 FASTCORE

FASTCORE uses as input a consistent genome-scale reconstruction (such as Recon X or HMR) and a vector containing the indices of the reactions that are regarded as active with a high confidence level. The reactions present in the input network that are not included in the core set represent the non-core reaction set. The use of FASTCORE for data which can easily be segregated into active and inactive such as bibliographic data or proteomic is straightforward: The data IDs are translated into model gene IDs and matched to the reactions using the gene-protein-reaction (GPR) association rules. Reactions that are not under gene control are always assigned to the non-core set. In contrast, reactions associated with one unique gene are included in the core set if the data supports the expression of the associated genes with a high confidence level. For two or more genes, the impact of the genes on one reaction is given by Boolean rules. A Boolean AND requires all the genes in the rule to be expressed whereas OR means that at least one gene needs to be expressed. The association rules are sometimes a complex combination of ANDs and ORs. Therefore, an easy way to deal with these rules is to assign 0 and 1 to unexpressed and expressed genes, respectively. Then AND is replaced by the minimum of the 2 values and Or by the maximum function. After this mapping step, reactions associated with a 1 are included in the core reactions. The output of FASTCORE is a vector A that contains the indices of the reactions of the input model required to build a consistent model that contains every core reaction. Finally, to obtain a consistent output model, the reactions of the input model that are not present in the vector A are removed using, e.g., the *removeRxns* function of the COBRA toolbox.

Besides the hard core (the equivalent of core reactions) and low core reactions (the equivalent of non-core reactions of FASTCORE), MBA [16] introduced medium core reactions and CORDA [17] takes four different reactions sets: high (HC), medium (MC), and negative (NC) and others (OT) as input. FASTCORE can easily be modified to take different sets of reactions as input but the addition of sets for all the algorithms should be avoided as it requires the setting of additional thresholds that must be set based on solid biological or statistical evidences as the miss-classification of reactions has a large impact on the input networks. When setting a new threshold or modifying an existing one, the user must verify that the algorithm is giving accurate results on an independent dataset using for instance the benchmark procedures described in [18] or by splitting the dataset in a training set, used for the reconstruction and a validation set, used for the model testing as described in [2]. The integration of additional sets is often done to integrate data from a specific source such as the human protein atlas [19] where the data is classified in high, medium, low, and undetected. It is important to keep in mind that the classification used by databases is often heuristic if not arbitrary. Before setting a threshold, one should consider how the data is produced in order to determine if an undetected gene is really not expressed (like it is the case for RNA-seq data) or if the methods might not be sensitive enough. Put in other words, the user should carefully assess the specificity and sensitivity of the technique or the workflow used to produce the data.

For the following example, we assume that proteins tagged as undetected are controlled by unexpressed genes. In this case the bounds of the associated reactions can be set to 0 and FASTCC (see the following section) is run to eliminate the reactions that are no longer able carry a flux. The new consistent input model is fed to FASTCORE using the reactions associated with a high confidence level as core reactions. Here, medium and low core reactions are only included if they are required to build a consistent network. An alternative solution is to use the modified version of FASTCORE used by the FASTCORMICS workflow [2] and to set beside the transport reactions the medium core reactions to the set of not penalized reactions in order to favor the inclusion of medium over the low core reactions.

3.2 FASTCC

FASTCC is a variant of FASTCORE for the identification of blocked reactions that are unable to carry a flux due to the presence of dead ends or gaps. FASTCC takes a consistent network and the flux/no flux threshold ε.

By default, every reaction of the model is set as a core reaction. FASTCC tries to include all the irreversible reactions in the input model by maximizing an approximation of the cardinality function. Irreversible reactions that were not included in this step are identified as blocked reactions. Reactions that were shown to carry an absolute flux above epsilon are removed from the core set. Then reversible reactions are tested. To include reversible reactions that were not yet included, the sign in the stoichiometry matrix of the remaining reversible reactions is flipped. Finally, a one by one step, in which the sign of the remaining reversible reactions is changed again, is run. Reactions that were still included in the model are considered blocked reactions.

3.3 FASTCORMICS

For other types of data, such as microarray data where the expression is given in continuous values, thresholds that segregate between expressed and not expressed have to be set. Furthermore, for a microarray experiment, each probe is associated with a different, non-negligible noise level; therefore, a different threshold should ideally be set for each gene or probe. In the FASTCORMICS workflow, Barcode is used as a discretization method. Barcode [11] compares each probe to the intensity distribution of a given gene across thousands of arrays of the same platform to identify the lowest expression mode. The authors of barcode recommend a threshold of 5 Z-scores (standard deviations) from the lowest expression model to consider a gene to be expressed. The recommended 5 standard deviations correspond, for each probe, to a different intensity value, which depends on the median and the standard deviation of the not expressed population (lowest model). This allows taking different levels of noise into account. We introduced besides the 5 Z-score threshold, an inexpression threshold set at 0 standard deviation. The setting of an inexpression threshold prevents the inclusion of reactions that are associated with unexpressed genes. The use of two thresholds allows buffering the impact of the core reactions set. Reactions that are wrongly tagged as core reactions and part of an expressed branch will not be included in the model.

The Barcode workflow is implemented in R and requires the *frma* package and the vector containing the reference values specific to each array and species such as *hgu133plus2barcodevecs* [12] for the HGU133plus2 platform. At the moment, vectors only exist for humans and mouse and for a restrained number of platforms. It is advisable to use arrays from the same platform and besides the usual quality control that has to be performed in each microarray analysis, a principal components analysis (PCA) should be done in order to verify that the data are not subjected to large batch effects. A correlation plot allows also verifying if the samples are sufficiently dissimilar to capture variations at the metabolic level. Once the quality control is completed, the fRMA normalization is performed using "core" as the target option because currently metabolic models only include gene identifiers. Then, the barcode function is run with the output "z-score." The obtained z-scores are then further discretized in 1, 0, −1. Meanwhile, the data IDs are converted to model IDs. In the case that more than one probe ID corresponds to a given gene, the maximal expression value is considered. Note that, if information exists on the specificity of a probe set, this information can be taken into account to discard unspecific probes. The discretized values are matched to the model identifiers using the gene protein reaction (GPR) rules. As explained for FASTCORE, the activity state of a reaction is given by the GPR rules, the Boolean AND can be replaced by the minimum function, and the OR by the maximum function.

Further, if more than 1 array is used as input, only genes that are tagged as expressed or unexpressed in 90% of the arrays are taken into consideration. The bounds of the reactions which are associated with −1 (in 90% of the arrays) are set to zero so that they are removed from the model along with other reactions that are no longer able to carry a flux. Consequently, reactions associated with a 1 value (in 90% of the arrays) are included in the core set. Additional constraints, such as the medium constraints for cells in culture, allow increasing the accuracy of the models by reducing the space of possible solutions.

The models can also be forced to produce known metabolites or to produce biomass by adding these reactions to the core set. Note that in this case FASTCORE does not optimize for a function, which would assume that the metabolism is optimized for the fulfillment of this task, but only forces these reactions to carry a flux. Beside the transcriptomics data, other types of data can be added to the core set, such as literature data in order to further constrain the model. A modified version of FASTCORE is run that allows leaving a set of reactions not penalized. Usually, transport reactions supported by barcode are removed from the core set as being under control of promiscuous genes that control up to thousand reactions and are transferred to the not penalized set. Thus, their inclusion is favored above non-core reactions but the inclusion is not forced.

3.4 How Many Input Samples?

The number of arrays required for a reconstruction depends on the purpose of the model. If the aim is to build a generic liver model, a maximum of arrays from the same platforms and from different contexts or individuals, which passed the quality control, should be included in order to build a representative liver model. Other types of data might also be included such as proteomic data or literature evidence to further constrain the model. If the aim is to build a liver model for a specific condition only arrays matching these criteria should be included. However, if the aim is to build an image of a liver cell in a specific condition and time point, only one array should be selected and the inclusion of bibliomic data should be avoided. As the expression of a gene in one or more conditions does not mean that the gene is expressed in every condition, the number of included arrays has an impact on the resolution of a model, defined as the ability to capture metabolic variability between two conditions. As an example, some reactions that are induced by the environment might be excluded from the output model as these reactions are only expressed in some of the samples (below the 90% ubiquity threshold) that were submitted to the right stimulation. In contrast, the drawback of using one or only a few arrays is the difficulty to determine if the observed variability is related to the object of the study or to inter-sample variability.

3.5 Input Models or Reconstructions

Input models are obtained after the mapping of sequenced transcripts to the reactome retrieved from databases and after manual curation and testing. Generic models are far from being complete and contain gaps and dead ends that prevent around 30% of the reactions of the human Recon X [13, 14] models to carry a flux. But the main issue is related to the ambiguity between reconstructions that aim to be as complete as possible introducing sometimes the same reaction multiple times with different cofactors, causing the formation of loops. Loops affect the prediction power of models as the reactions in those loops can carry a flux even when not connected with the environment. The identification and removal of loops is a complex topic, especially as some loops are biologically relevant. Further, some reactions that are thermodynamically reversible need to be irreversible in physiological conditions to allow the ATP production rates observed in vitro [15]. The difference between reconstructions and model as well as the limitation of the traditional reconstruction processes were illustrated by a study that converted Recon2, a human reconstruction, to a model [15]. Although the integration of omics data improves the predictability of the output models, it does not correct all its flaws and manual curation of the input and output models might still be required. Manual curation is especially important because context-specific models are sensitive to the accuracy of the gene-reactions-rules, which required intensive curation in the Recon 2.2 study [15]. Therefore, it is advisable, before running FASTCORE, to verify that the main pathways, relevant to the subject of the study, are complete and correct in the input model and, if otherwise, to perform manual curation using literature data.

3.6 Manual Curation

Another reason why manual curation is required is because context-specific models fail to predict metabolic functions when the models are built uniquely based on omics data. The reason for this failure is the large number of reactions that are not under gene control that represent 30% of the input models networks [18]. These reactions are for instance diffusion reactions. The issue is that these reactions have the same weight so that if there are 2 possibilities to connect the core reactions with the same number of reactions, the two paths will be regarded as equivalent by the FASTCORE family and the other competing algorithms. Therefore, one or the other path can be included in the model. In the toy model (Fig. 1), A can be connected to B or A can be connected to D. The choice of one of the paths is random but a semi-manual curation strategy can be used in which models are built using the FASTCORE family, the function is tested and then non-core reactions that send a wrong direction are eliminated until the model is able to fulfil the required function. Note that medium composition information and the inclusion of more data in general allows improving the functionality of the model but does not correct all the flaws.

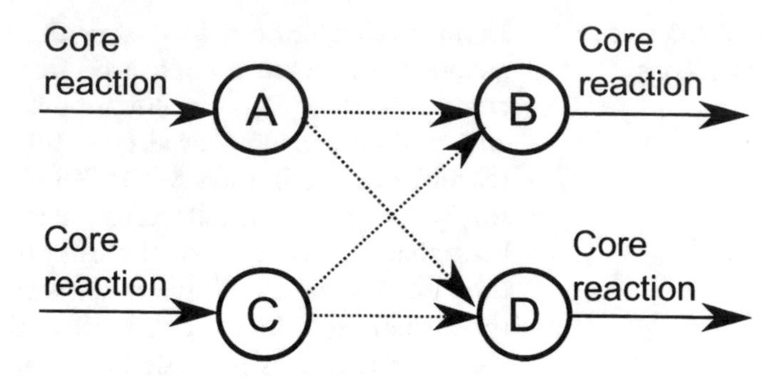

Fig. 1 In this toy network, the circles represent the metabolites and the arrows symbolize the reactions. Solid lines and dashed arrows mark the core and non-core reactions, respectively

If further manual curation is necessary, left-out cross-validation assays might help to narrow down the number of reactions that need to be checked. In left-out cross-validation assays, each core reaction is iteratively removed from the core set and the model is reconstructed using the usual FASTCORMICS workflow. The reactions that are present in all the output model of these reconstructions are core reactions that are supported by at least another core reaction that requires the activation of the left-out core reaction (Hard core reaction) or are non-core reactions that are supported by at least two core reactions. In the same way, left-out unexpressed reactions are iteratively removed from the non-expressed set. Reactions that are inactive in all models (hard core inactive reactions) are inactive because a least another not expressed reaction that is in the same branch prevents the reactions of the branch to carry a flux or are non-core reactions that are not required by the core set to carry a flux. The focus should then be set on reactions that are not hard core, hard core inactive reactions, and core reactions.

4 Notes

1. *Solvers*: The FASTCORE family was implemented using the CPLEX optimization software package from IBM as after testing several solvers. CPLEX was the solver that showed the best performances. Nevertheless, the FASTCORE family can easily be modified to be used with other solvers. Note also that most of the solvers used in metabolic modeling are float point solvers which have a tolerance to the violation of constraints of a linear program optimization problem. For CPLEX the tolerance is of $1e{-}6$, therefore the epsilon threshold should be set above this value. Some competing solvers have a higher tolerance and therefore epsilon should be adapted accordingly. Although CPLEX has a low violation tolerance, some models might only

allow fluxes that are below this threshold (due to too narrow bounds or too high stoichiometry coefficient). Fluxes below this threshold cannot be handled by float point solver and therefore the use of fixed solver as MOONGOOSE [20] like recommended. Another alternative is to change the unit of the biomass and to adopt the bounds accordingly in order to exit the tolerance zone of the solvers.

2. *Parameters*: The code of FASTCORMICS and FASTCORE as well as several parameters like the expression thresholds, the ubiquity criteria, or epsilon can be modified to correspond to the needs of the user, but it is advised to take into consideration that the workflows were validated using the given parameter setting. Therefore, the user should verify that the introduced modifications do not alter the accuracy of the FASTCORE family by using for, i.e., the benchmark procedures [18] on an independent dataset. Further, the introduction of arbitrary thresholds should be avoided.

   Furthermore, in the next months, we will release algorithms or workflows to integrate other types of data such as RNA-seq or to perform other tasks.

Acknowledgments

We would like to thank Dr. Nikos Vlassis for his feedback.

References

1. Vlassis N, Pacheco MP, Sauter T (2014) Fast reconstruction of compact context-specific metabolic network models. PLoS Comput Biol 10:e1003424
2. Pacheco MP et al (2015) Integrated metabolic modelling reveals cell-type specific epigenetic control points of the macrophage metabolic network. BMC Genomics 16:809. https://doi.org/10.1186/s12864-015-1984-4
3. Thiele I, Vlassis N, Fleming RMT (2014) fastGapFill: efficient gap filling in metabolic networks. Bioinformatics 30:2529–2531
4. Shlomi T, Cabili MN, Herrgard MJ, Palsson BO, Ruppin E (2008) Network-based prediction of human tissue-specific metabolism. Nat Biotechnol 26:1003–1010
5. Schellenberger J et al (2011) Quantitative prediction of cellular metabolism with constraint-based models: the {COBRA} {T}oolbox v2.0. Nat Protoc 6:1290–1307
6. Steffensen JL, Dufault-Thompson K, Zhang Y (2016) PSAMM: a portable system for the analysis of metabolic models. PLoS Comput Biol 12:1–29
7. Gautier L, Cope L, Bolstad BM, Irizarry RA (2004) affy — analysis of Affymetrix GeneChip data at the probe level. Bioinformatics 20:307–315
8. Carvalho BS, Irizarry RA (2010) A framework for oligonucleotide microarray preprocessing. Bioinformatics 26:2363–2367
9. Wicklham H (2009) ggplot2: elegant graphics for data analysis. Springer, New York, NY
10. McCall MN, Bolstad BM, Irizarry RA (2010) Frozen robust multiarray analysis (fRMA). Biostatistics 11:242–253
11. McCall MN, Uppal K, Jaffee HA, Zilliox MJ, Irizarry RA (2011) The Gene Expression Barcode: leveraging public data repositories to begin cataloging the human and murine transcriptomes. Nucleic Acids Res 39: D1011–D1015
12. McCall MN, Irizarry RA (2016) hgu133plus2-barcodevecs: hgu133plus2 data for barcode. R package version 1.10.0

13. Duarte NC et al (2007) Global reconstruction of the human metabolic network based on genomic and bibliomic data. Proc Natl Acad Sci 104:1777–1782
14. Thiele I et al (2013) A community-driven global reconstruction of human metabolism. Nat Biotechnol 31:419
15. Swainston N et al (2016) Recon 2.2: from reconstruction to model of human metabolism. Metabolomics 12:109. https://doi.org/10.1007/s11306-016-1051-4
16. Jerby L, Shlomi T, Ruppin E (2010) Computational reconstruction of tissue-specific metabolic models: application to human liver metabolism. Mol Syst Biol 6:401
17. Schultz A, Qutub AA (2016) Reconstruction of Tissue-Specific Metabolic Networks Using CORDA. PLoS Comput Biol 12:e1004808
18. Pacheco MP, Pfau T, Sauter T (2016) Benchmarking procedures for high-throughput context specific reconstruction algorithms. Front Physiol 6:410
19. Uhlen M (2005) A human protein atlas for normal and cancer tissues based on antibody proteomics. Mol Cell Proteomics 4:1920–1932
20. Chindelevitch L, Trigg J, Regev A, Berger B (2014) An exact arithmetic toolbox for a consistent and reproducible structural analysis of metabolic network models. Nat Commun 5:4893

Chapter 5

Reconstruction and Analysis of Central Metabolism in Microbes

Janaka N. Edirisinghe, José P. Faria, Nomi L. Harris, Benjamin H. Allen, and Christopher S. Henry

Abstract

Genome-scale metabolic models (GEMs) generated from automated reconstruction pipelines often lack accuracy due to the need for extensive gapfilling and the inference of periphery metabolic pathways based on lower-confidence annotations. The central carbon pathways and electron transport chains are among the most well-understood regions of microbial metabolism, and these pathways contribute significantly toward defining cellular behavior and growth conditions. Thus, it is often useful to construct a simplified core metabolic model (CMM) that is comprised of only the high-confidence central pathways. In this chapter, we discuss methods for producing core metabolic models (CMM) based on genome annotations. With its reduced scope compared to GEMs, CMM reconstruction focuses on accurate representation of the central metabolic pathways related to energy biosynthesis and accurate energy yield predictions. We demonstrate the reconstruction and analysis of CMMs using the DOE Systems Biology Knowledgebase (KBase). The complete workflow is available at http://kbase.us/core-models/.

Key words Central metabolism, Core metabolic models, Metabolic model reconstruction, Flux balance analysis, Biochemical pathways, Model comparison

1 Introduction

Central carbon metabolism is a key component in the metabolic network of living organisms as these pathways harbor many of the most important mechanisms for energy biosynthesis, as well as producing the precursor compounds for most essential biomass building blocks. The energy production strategies defined in the central metabolic pathways have a significant impact on the behavior and growth conditions of microorganisms, thus playing a crucial role in the quantitative prediction of biomass and energy yields [1, 2]. Energy production strategies in microbes are highly diversified, unlike those in higher eukaryotes. These strategies primarily depend on environmental factors such as: (1) carbon source

Marco Fondi (ed.), *Metabolic Network Reconstruction and Modeling: Methods and Protocols*, Methods in Molecular Biology, vol. 1716, https://doi.org/10.1007/978-1-4939-7528-0_5,

utilization; (2) ability to respire by reducing numerous electron acceptors; and (3) fermentation capabilities.

It continues to be challenging to make accurate computational predictions based on metabolic models and in silico simulations interpreting complex microbial behavior. Tools for automated metabolic model reconstruction such as ModelSEED [3–5] can rapidly generate draft genome-scale metabolic models from annotated genome sequences [6]. However, these draft models, and in some cases even curated published models, can lack accuracy in predicting growth yields, ATP production yields, and central carbon flux profiles. This poor accuracy stems primarily from three common problems: (1) poor representation of energy biosynthesis pathways; (2) a lack of diverse electron transport chain (ETC) variations; and (3) addition of extensive gapfilling reactions that can sometimes misrepresent an organism's behavior [7].

Many of these problems can be avoided by using a simplified model comprised of only the most confidently annotated and biologically critical pathways for energy biosynthesis [8] (Fig. 1). We define these models as Core Metabolic Models (CMM), and they consist primarily of the sugar oxidation pathways, the fermentation pathways (Fig. 2), and the ETC variations. We previously developed an approach for the reconstruction and analysis of CMMs based on annotated genome sequences [9], which we implemented as a pipeline in the DOE Systems Biology Knowledgebase (KBase). In this chapter, we demonstrate how this analysis workflow can be run in KBase. The complete workflow, including example data and commentary, can be accessed from http://kbase.us/core-models. The pipeline is comprised of four main steps: (1) genome annotation by RAST [10]; (2) CMM reconstruction [9]; (3) gapfilling [7]; and (4) flux balance analysis (FBA) [11]. We also discuss methods for exploring metabolic diversity by studying the variations in central metabolic pathways in a phylogenetic context.

2 Materials

In this section, we describe the data and tools required to build CMMs using the KBase Narrative Interface (https://narrative.kbase.us). Methods that use the data and tools listed in this section are described in detail in Subheading 3.

2.1 KBase Narrative Interface

In KBase, reproducible workflows called *Narratives* can be created and shared. Narratives can include data, analysis steps, results, visualizations, and commentary. Narratives can be shared with collaborators as "active papers" that let others repeat the analysis workflows and even alter parameters or input data to achieve different or improved results. We encourage readers to view and copy the Core Model Construction Narrative (see http://kbase.us/core-

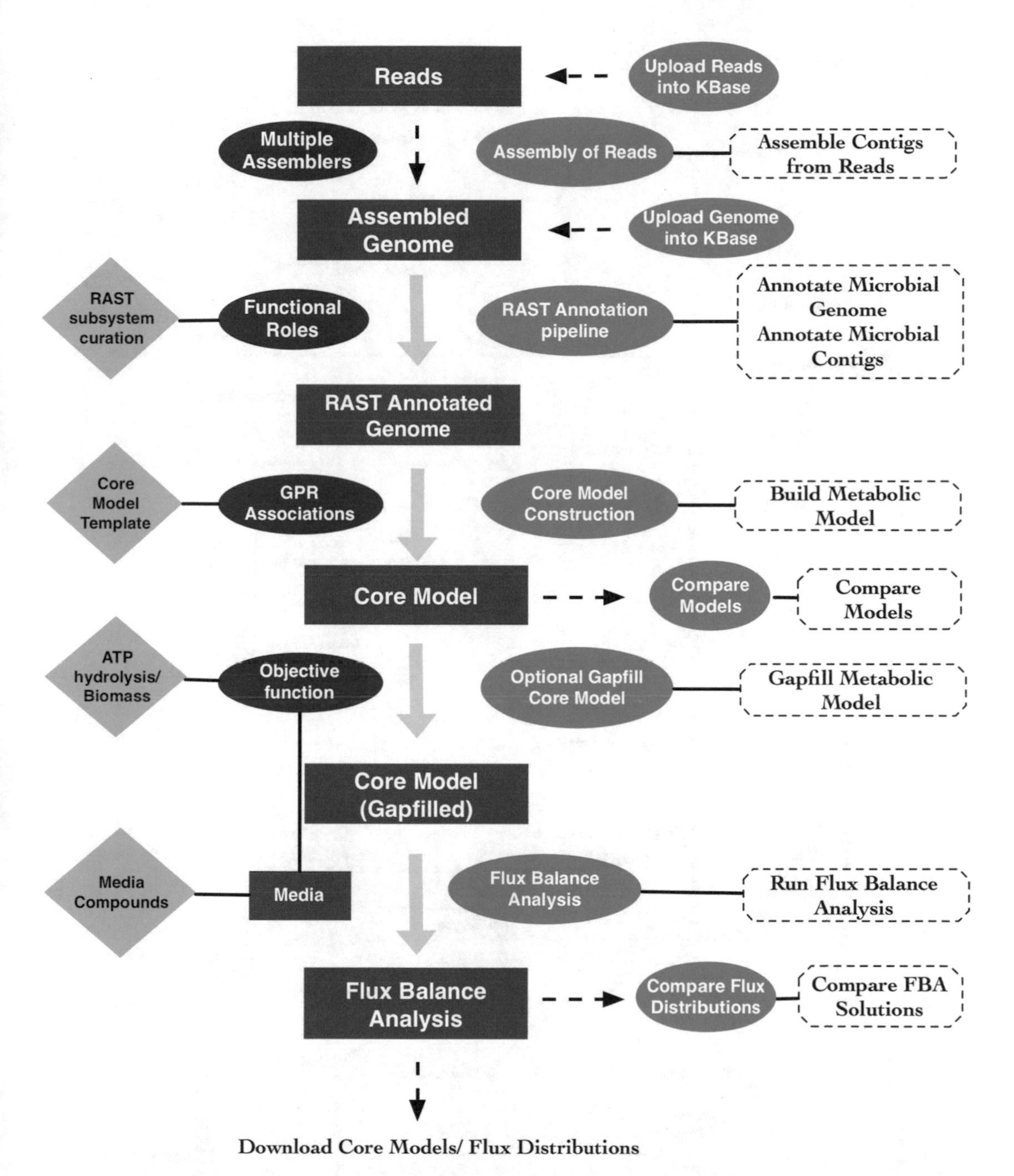

Fig. 1 Seven-step pipeline used to construct and analyze core metabolic models in KBase. The core model reconstruction pipeline is comprised of seven apps (rounded rectangles with dashed borders), which operate on specific data types (magenta rectangles). These apps are driven by several curated reference data sources (green diamonds), including RAST subsystems, the template model pathways, the template model objective functions, and the compounds that make media formulations used for gapfilling and FBA. The purple ovals identify the essential components/data (explained in the text) required for apps; the blue ovals show the steps performed by the apps. Dashed arrows show optional steps while turquoise arrows show major steps of the pipeline. The resulting data can be downloaded as explained in **Note 1**

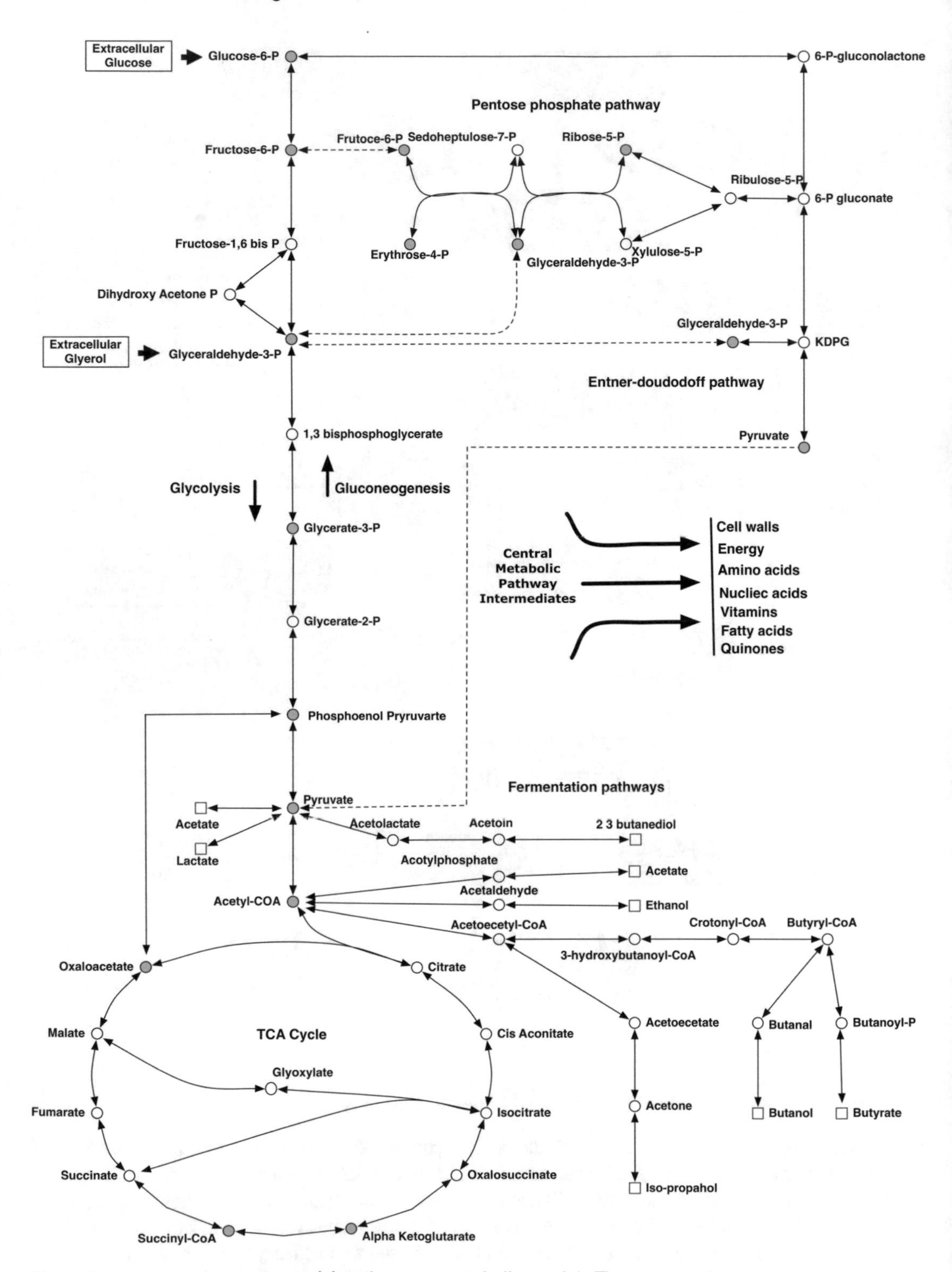

Fig. 2 Metabolic pathways comprising the core metabolic model. The core model template encompasses 12 central metabolic pathways including sugar oxidation (glycolysis, gluconeogenesis, Entner-Doudoroff, pentose phosphate), TCA cycle and fermentation pathways (fermentation end products displayed as squares with blue borders). These pathways produce 16 biomass precursor molecules (green circles) (Table 1)

models) and try running the steps on the example data or using their own data.

2.2 Core Metabolic Model Reconstruction Pipeline (Apps)

The CMM reconstruction pipeline in KBase permits a user to progress from raw genome sequencing reads through assembly and annotation to core model reconstruction and then perform model analysis and comparison. The pipeline is comprised of seven apps (centered on the four main steps previously mentioned), which are described in this section (Fig. 1).

The first step of the pipeline is the *Assemble Contigs from Reads* app, which accepts short reads from Next-Generation Sequencing (NGS) as input and produces assembled contigs as output. KBase includes numerous apps for genome assembly, but the *Assemble Contigs from Reads* app is the most sophisticated, as it enables users to run multiple assemblers at once, then aids in selecting the best set of contigs produced by all the assemblers.

The second step of the pipeline is the *Annotate Microbial Contigs* app. This app is based on the RAST (Rapid Annotations using Subsystems Technology) pipeline for microbial genome annotation [12]. The app accepts assembled contigs as input, performs gene calling using a combination of Glimmer [13] and Prodigal [14], then functionally annotates genes from the SEED subsystems ontology [15] using a kmer-based approach [16]. Alternatively, if one already has a genome with existing gene calls in GenBank format, the *Annotate Microbial Genome* app can be used to simply re-annotate the existing genome while keeping the gene calls intact. This app also uses the RAST approach for functional annotation. Note that when an existing genome is imported into KBase, its original annotations are kept intact. Unless the imported genome was generated by RAST or PATRIC [17], it is likely that the annotations do not conform to the SEED subsystems ontology. As a result, it is currently necessary to re-annotate these genomes using the *Annotate Microbial Genome* app prior to building a metabolic model. Both annotation apps produce an annotated genome as output, which includes data on all genome contigs, genes, proteins, and functional annotations.

The third step of the pipeline is the *Build Metabolic Model* app. In this app, the functional annotations generated by the RAST-based genome annotation apps are used to generate a draft metabolic model. A draft model consists of three parts: (1) a network of metabolic reactions (including both gene-associated reactions and spontaneous reactions); (2) a set of gene-protein-reaction (GPR) associations that dictate how each reaction activity depends on associated gene activity; and (3) a biomass composition reaction that defines the small molecule building blocks that comprise 1 g of biomass (e.g., amino acids, nucleotides, lipids, cofactors, cell-wall components, and energy). This app produces genome-scale metabolic models by default, but it is possible to select the core template to build a core model instead. Core models contain far fewer reactions, and their biomass composition reaction uses only central-carbon precursor molecules. The reactions included in the models produced by the *Build Metabolic Model* app are selected from the ModelSEED [18] biochemistry database. This curated database contains mass and charge balanced reactions, standardized to aqueous conditions at neutral pH. The Model SEED reaction database integrates biochemistry from KEGG [19, 20], MetaCyc [21], EcoCyc [22], Plant BioCyc, Plant Metabolic Networks, and Gramene [23]. The database is available for download from GitHub (https://github.com/ModelSEED/ModelSEEDDatabase/blob/master/Biochemistry/).

The fourth step of the pipeline is the *Gapfill Metabolic Model* app. Draft metabolic models (built using the *Build Metabolic Model* app) usually have missing reactions (gaps) due to incomplete or incorrect functional genome annotations. As a result, these models are unable to produce biomass using media on which the organism typically is capable of growing. Gapfilling algorithms can overcome this problem by identifying the minimum number of new reactions that must be added to the model, or existing reactions that must be made reversible to enable the production of biomass. The gapfilling app in KBase uses Model SEED reaction database for gapfilling. It works equally well on core or genome-scale metabolic models. When gapfilling a core model, only the reactions present in the core model template (*see* Subheading 2.3) are considered for gapfilling. When gapfilling a genome-scale model, all 13,000 reactions from the ModelSEED [18] biochemistry database are considered for gapfilling.

The fifth step of the pipeline is the *Run Flux Balance Analysis* app. This app predicts the flow of metabolites through the metabolic network of an organism by optimizing for the selected cellular objective function, which is typically the production of biomass. Flux Balance Analysis (FBA) is a constraint-based approach that estimates growth-optimal fluxes through all the reactions in the metabolic network, thereby making it possible to estimate the growth rate of an organism (the rate of biomass production) or the rate of production of a given metabolic output on a specified media. This app makes it possible to analyze an organism's growth on different substrates and to evaluate the reactions and metabolites that carry fluxes in each growth condition. In addition to optimizing the biomass, one can choose to optimize a certain reaction (e.g., transporter reaction) so that the model optimizes to produce flux through that reaction. The *Run Flux Balance Analysis* app requires the user to specify a media formulation in which the growth will be simulated. In KBase, the media contains a list of the chemical compounds that are available for consumption in the flux simulation. KBase currently maintain more than 500 commonly used media conditions. In addition, users are able to build and upload their own custom media formulations. The *Run Flux Balance Analysis* app includes a range of FBA algorithms, including flux variability analysis, gene essentiality prediction, and expression data analysis.

The sixth step of the pipeline is the *Compare Models* app. This app provides comparative analysis of two or more models based on reactions, compounds, biomass, and proteins families. The app provides overall statistics of conserved reactions, conserved compounds, and conserved biomass precursors across metabolic models.

The seventh and final step of the pipeline is the *Compare FBA Solutions* app. KBase permits the use of flux balance analysis to predict how an organism will behave metabolically in a wide range of growth conditions. With this capability, it quickly becomes important to be able to compare the flux profiles predicted by FBA side-by-side in order to understand how an organism's behavior changes from one condition to the next, or how the behavior of two different organisms differs within a single condition. The *Compare FBA Solutions* app enables this comparison. FBA solutions are compared on three levels: (1) the objective value for each FBA solution; (2) the flux through each reaction in each FBA solution; and (3) the uptake and excretion of metabolites in each FBA solution. For the flux comparison, reaction fluxes are categorized into four possible states: not in model; no flux; forward flux; and reverse flux. Metabolite fluxes are categorized into similar states: not in model; no flux; uptake; and excretion. FBA solutions are compared based on these states, and solutions with similar states are compared based on magnitude of flux.

2.3 Metabolic Pathways in the Core Model Template

All metabolic model reconstruction in KBase is built upon a set of model templates, each of which integrates the three types of data needed to build a model from an annotated genome: (1) the full set of reactions that comprise the metabolic pathways across a wide range of organisms; (2) the SEED functional roles associated with the enzymes that perform all metabolic reactions; and (3) the default objective functions to be used in the reconstructed models (*see* Subheading 2.5). Different model templates are used to construct different types of models (e.g., plants, gram negative genomes, gram positive genomes), and for core models, a specific core model template (CMT) is applied.

The CMT integrates 200 highly curated reactions (https://github.com/ModelSEED/ModelSEEDDatabase/blob/master/Templates/Core/Reactions.tsv) encompassing 12 key energy biosynthesis pathways (Fig. 2) linked to central metabolism including: sugar degradation pathways (Glycolysis, Entner-Doudoroff, Citric acid cycle and Pentose phosphate), fermentation pathways (producing end products: lactate, acetate, formate, ethanol, 2,3-butanediol, butyrate, butanol, and acetone) that are derived

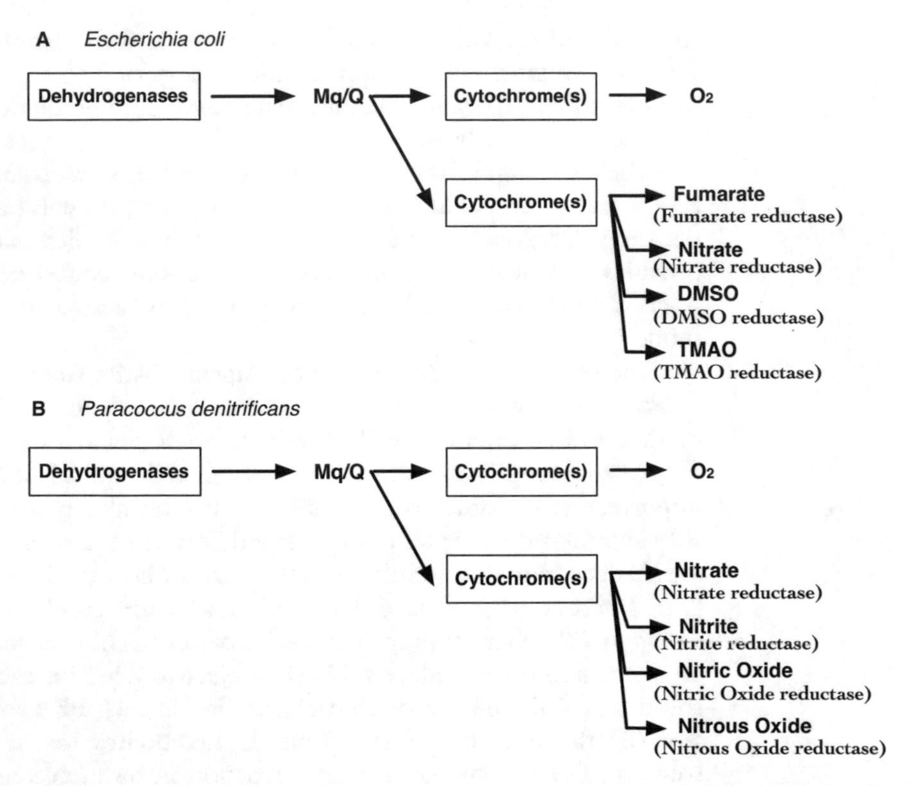

Fig. 3 Diverse electron transport chains in bacteria. Escherichia coli (**a**) and *Paracoccus denitrificans* (**b**) are able to respire aerobically and anaerobically by reducing nitrate. *E. coli* (**a**) is able to reduce organic electron acceptors fumarate, DMSO, and TMAO. *P. denitrificans* (**b**) is able to reduce more inorganic electron acceptors including nitrite, nitric oxide, and nitrous oxide

from central metabolism as well as various aerobic and anaerobic Electron Transport Chains (*see* Subheading 2.4). These pathways are considered the building blocks of the central metabolism that is represented by our CMT, and they were derived from an analysis of a phylogenetically diverse set of well-studied model organisms, including *Escherichia coli*, *Bacillus subtilis*, *Pseudomonas aeroginosa*, *Clostridium acetobutylicum*, and *Paracococcus denitrificans* [9]. We also added a number of manually curated ETC reactions to the CMT. These reactions reflect the diverse ETC variations in aerobic respiration as well as facilitating the reduction of number of anaerobic electron acceptors (*see* Subheading 3.4). The CMT maps its 200 reactions to over 400 SEED functional roles through complexes (*see* Subheading 3.3, Fig. 3). These mappings are used to associated genes to the CMT reactions when building a CMM from an annotated genome.

2.4 Encoding of ETC Diversity in the Core Model Template

Unlike the electron transport chains of higher eukaryotes, bacterial ETCs are highly diversified. As a result, they are able to grow in a variety of aerobic and anaerobic environments reducing anaerobic electron acceptors such as nitrate, nitrite, fumarate, dimethyl sulfoxide (DMSO), and trimethylamine N-oxide (TMAO). Given the importance of the ETCs in governing cell behavior and growth conditions, significant curation was invested to encode various ETCs as a part of the CMT. For instance, *Escherichia coli* (Fig. 3a) is known to respire aerobically and anaerobically reducing nitrate, fumarate, TMAO, and DMSO. *Paracoccus denitrificans* (Fig. 3b) is also able grow aerobically and anaerobically reducing a variety of nitrogen-based compounds including nitrate, nitrite, nitrous oxide, and nitric oxide. Better annotation of ETCs aids identification of complex respiration types and makes energy yield predictions more accurate.

2.5 Default Objective Functions in the Core Model Template

The CMT integrates two default objective functions for the CMMs: (1) a biomass production objective function, modeled by maximizing the simultaneous production of 16 central carbon precursors needed to produce 1 g of biomass (green circles in Fig. 2); and (2) an energy production objective function modeled by maximizing flux through an ATP hydrolysis reaction. The biomass biosynthesis objective function in our CMT was constructed based on the biomass precursor stoichiometry that was derived by Varma and Palsson [24] and used in one of the earliest models of *E. coli* (Table 1) [25]. In our analysis of the biomass objective function, we found that gapfilling was occasionally required to enable synthesis of all essential biomass precursors in our biomass object function [9]. For this reason, we also include the energy object function in the CMT, which permits a focused study of energy biosynthesis in our core models without any gapfilling. Using this objective function, we computed ATP production yields in all models without any gapfilling; hence, these computations were based solely on reactions derived from existing RAST annotations.

3 Methods

3.1 Construction of a Draft Core Metabolic Model from an Annotated Genome

Here, we apply our core model reconstruction pipeline (*see* Subheading 2.2 and Fig. 1) in KBase to build and analyze a core model for the genome *Escherichia coli* K12 (see http://kbase.us/core-models/). Because we are starting with an imported genome, we skip the genome assembly step of our pipeline and apply the *Annotate Microbial Genome* app to re-annotate our genome with functions from the SEED subsystems ontology [15]. In this re-annotation step, RAST assigns 3889 genes with 3797 distinct functions. 1804 of these functions appear in the SEED subsystem ontology.

Now that the genome is annotated with SEED functions, the *Build Metabolic Model* app can be used with the CMT (*see* Subheading 2.3) selected to build a draft CMM. In addition to constructing

Table 1
Central carbon precursors of small-molecule building blocks of biomass

Biomass compound	Coefficient
NADPH	−1.8225
D-Erythrose4-phosphate	−0.8977
NADH	3.547
Phosphoenolpyruvate	−0.5191
NADP	1.8225
NAD	−3.547
H_2O	−41.257
Acetyl-CoA	−3.7478
ADP	41.257
CoA	3.7478
ATP	−41.257
Pyruvate	−2.8328
3-Phosphoglycerate	−1.496
Oxaloacetate	−1.7867
Phosphate	41.257
D-fructose-6-phosphate	−0.0709
ribose-5-phosphate	−0.8977
H+	46.6265
Glyceraldehyde3-phosphate	−0.129
2-Oxoglutarate	−1.0789
D-glucose-6-phosphate	−0.205

Compound names and associated coefficients of the biomass biosynthesis objective function used in CMMs. This biomass stoichiometry originally derived by Varma and Palsson [8] and used in CMMs with modifications [9]

a core model, this app has an optional "Gapfill metabolic model" checkbox. If this option is selected, then when the *Build Metabolic Model* step finishes, the *Gapfill Metabolic Model* step will start automatically. For the sake of our example, this checkbox is left "unchecked" so these steps can be run separately. When the *Build Metabolic Model* step completes, a draft model is created, which is comprised of 158 reactions, 168 compounds, and 478 genes. Because gapfilling was not run automatically, this draft model only includes reactions that are associated with genes.

The gene associations were generated based on a two-step process. In the *Annotate Microbial Genome* app, the genes in our genome were assigned biological functions (e.g., Pyruvate kinase

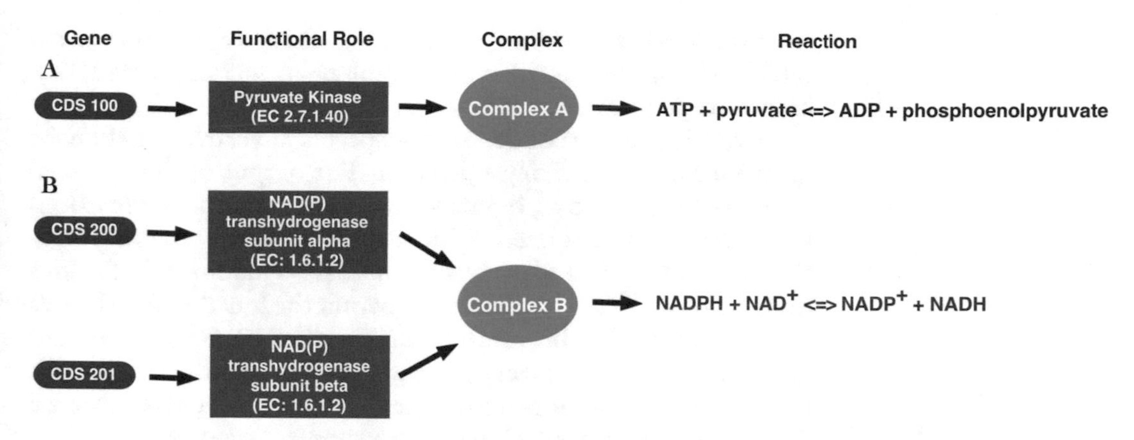

Fig. 4 Organization of genes, gene annotations, complexes, and the biochemical reactions in gene protein reaction (GPR) mappings. Panel (**a**) shows a gene assigned Pyruvate Kinase (EC 2.7.1.40) as a function. This gene is mapped first to a complex (Complex A), then to a biochemical reaction. Panel (**b**) shows two genes that were assigned the NAD(P) transhydrogenase (EC 1.6.1.2) alpha and beta subunits as functions. These genes are mapped first to a single complex (Complex B), then to the appropriate reaction

(EC 2.7.1.40)), and in our CMT, these functions are mapped to the appropriate biochemical reactions. Thus, the *Build Metabolic Model* app maps the reactions associated with function *A* in the CMT to the gene(s) associated with function *A* in the genome (Fig. 4a). As some metabolic enzymes have multiple functional subunits encoded by separate genes, this mapping process also integrates information about such complexes, so the genes encoding separate subunits are mapped to the appropriate reaction as a group (Fig. 4b). If only one subunit is annotated in the genome, the reaction is still added to the model, although a note is made that the other subunits appear to be missing.

The draft model also has two different objective functions, as defined by our CMT (*see* Subheading 2.5): an energy production function (called bio2) and a biomass production function (called bio1). These objective functions play a role in the gapfilling of the model performed in the next step of our model reconstruction pipeline, as well as in how the model is analyzed during flux balance analysis.

3.2 Gapfilling Core Metabolic Model for Energy and Biomass Production

The next step of our pipeline is to gapfill the CMM to enable the production of energy and biomass in a specified growth condition. We must specify a growth condition when gapfilling because the nutrients present in the growth condition have a major impact on the reactions required to permit growth and energy production. In KBase, growth conditions are specified as media formulations, which specify the concentration and uptake ranges of all metabolites known to be available in the growth condition. By default, gapfilling will use a special growth condition called *Complete* media, which includes all metabolites for which there is a transport reaction in KBase (*see* Subheading 2.2). In the case of our *E. coli* K12 CMM, we will gapfill in glucose minimal media as *E. coli* K12 is known to grow in this condition.

Gapfilling also requires that a specific objective function be specified for the gapfilling operation. The output of this specified objective function (e.g., biomass reaction or a transporter reaction) is constrained to a nonzero value, while linear programming algorithms are applied to identify a minimal set of additional reactions that must be added to the model to permit the function to achieve a nonzero value. In the case of our *E. coli* K12 CMM, we have specified the biomass reaction, which produces all the compounds involved in the biomass biosynthesis (Table 1), as the objective function resulting in an objective value greater than zero.

Because we have two separate objective functions in our CMM, we run the gapfilling with each function. As it turns out, our *E. coli* K12 CMM requires no additional reactions to reach a nonzero value with either of our objective functions. This result was expected for our energy production objective function, which was specifically designed to require minimal or no gapfilling. However, some genomes do require at least some gapfilling to permit standard biomass production. This lack of significant required gapfilling highlights one of the major strengths of using CMMs: CMMs generally require far less (or no) gapfilling compared to genome-scale models, meaning the predictions they make will be based primarily if not entirely on the genome annotations. Now that our *E. coli* K12 CMM has been demonstrated to be capable of producing both energy and biomass, we can use flux balance analysis to predict the flux profile in *E. coli* that optimizes each of these objective functions.

3.3 Analysis of Core Metabolic Model with Flux Balance Analysis

In the fifth step of our pipeline, we use the *Run Flux Balance Analysis* app in KBase to optimize the production of biomass and energy (*see* Subheading 2.5) in our CMM, while also predicting metabolite uptake, intracellular flux profile, and growth/ATP production yields. As with gapfilling, FBA requires that a growth condition (media) be specified for the analysis, and as before, we select glucose minimal media under aerobic conditions as our desired growth condition for analysis. Our FBA reveals a biomass yield of 0.12 g biomass/mmol glucose uptake and an energy yield of 26.5 mmol ATP per mmol glucose uptake. In KBase, the *Run Flux Balance Analysis* app automatically also runs flux variability analysis (FVA), which enables the classification of model reactions during predicted growth or energy production. FVA reveals that 37 (27%) of reactions in *E. coli* are essential for biomass production when growing in glucose minimal media, while 31 (19%) are essential for energy production. As expected, simple energy production requires fewer pathways than biomass production.

3.4 Comparative Analysis of CMMs and Flux Distributions

Now that our *E. coli* CMM has been built and analyzed, it is interesting to apply this same pipeline to examine other genomes and/or growth conditions. KBase includes tools that support the comparison of models and/or FBA solutions when studying multiple genomes in multiple growth conditions. Comparative analysis of the models helps to reveal metabolic pathways that are common to all models compared and also to identify unique parts of metabolism for each individual model. Comparing flux distributions allows the identification of high flux pathways and pathways that are not being utilized under certain environmental conditions.

We demonstrate this capability by applying our CMM reconstruction pipeline to build a model of *Paracoccous denitrificans* PD1222. We then compare the models (*E. coli* and *P. denitrificans*) and their predicted flux profiles using the *Compare Models* and *Compare FBA Solutions* apps respectively. This analysis reveals that *P. denitrificans* and *E. coli* have 50 reactions in common, with only 4 reactions unique to *P. denitrificans* and 112 reactions unique to *E. coli.* Comparing the FBA predictions, we find that the reactions common to both models were largely essential for biomass production, while the reactions unique to each model were active but not essential. We also compared the flux profiles of *E. coli* that optimize the energy yield in glucose minimal media under aerobic and anaerobic conditions. This comparison reveals that ATP yield is much higher (26.5 ATPmmol/mmol of glucose) in the aerobic condition where 31 reactions have nonzero fluxes compared to in the anaerobic condition (2.75 ATPmmol/mmol of glucose) where only 21 reactions have nonzero fluxes. The difference in energy yield is due to the fact that under aerobic conditions, *E. coli* is able to fully oxidize glucose-utilizing aerobic ETCs, yielding more energy, whereas in anaerobic conditions with no anaerobic electron acceptors present, energy is produced solely through fermentation by substrate level phosphorylation.

3.5 Determining Metabolic Pathways in CMMs and Phylogenetic Distribution

To study and evaluate the metabolic potential of an organism, it is useful to ascertain the existence of classical metabolic pathways. We have used CMMs to determine the presence or absence of key energy biosynthesis-related pathways (Fig. 2). We have developed a set of Boolean rules to determine the presence and absence of each pathway based on reactions present in each of the CMMs [9]. This methodology allows for alternative reactions within an individual step of each pathway, but every step of each defined pathway must be annotated in order for the pathway to be classified as present. Boolean rules that were used to determine the existence for Glycolysis and Gluconeogenesis are listed in Table 2.

Once the presence and absence of pathways have been determined, there are multiple ways to analyze the pathway data. In Fig. 5, we have painted pathway presence and absence data for Glycolysis, Gluconeogenesis, and Entner-Doudoroff on a

Table 2
Booleans rules used to govern pathway presence/absence in CMMs (rules that are displayed in this table were used to determine presence/absence of glycolysis and gluconeogenesis)

Section 1	
Enzyme names	*Reaction ID*
phospho_glucose_isomerase	rxn00558
ATP-dependent-pfk	rxn00545
ADP-dependent-pfk	rxn04043
ppi-dependent-pfk	rxn00551
NAD-dependent_phosphoglycerate_dehydrogenase	rxn00781
NADP-dependent_phosphoglycerate_dehydrogenase	rxn00782
ATP-dependent_phosphoglycerate_kinase	rxn01100
GTP-dependent_phosphoglycerate_kinase	rxn01105
phosphoglycerate_mutase	rxn01106
Enolase	rxn00459
pyruvate_kinase	rxn00148
fructose_bis_phosphotase	rxn00549
f1,6_bisphosphate_aldolase	rxn00786
ATP_pyruvate_water_phosphotransferase	rxn00147
Section 2	
Enzyme names/pathway segments	*Rule*
gdh	means 1 of {NAD-dependent_phosphoglycerate_dehydrogenase,NAD-dependent_phosphoglycerate_dehydrogenase}
pgk	means 1 of {ATP-dependent_phosphoglycerate_kinase, ADP-dependent_phosphoglycerate_kinase}
pgm	means phosphoglycerate_mutase
pyrk	means pyruvate_kinase
pfk	means 1 of {ATP-dependent-pfk,ADP-dependent-pfk,ppi-dependent-pfk}
G3P-PYR	means gdh and pgk and pgm and enolase and pyrk
G3P-PEP	means gdh and pgk and pgm and enolase
F6P-PYR	means pfk and f1,6_bisphosphate_aldolase and G3P-PYR
G6P-PYR	means phospho_glucose_isomerase and F6P-PYR

(continued)

Table 2
(continued)

glycolysis_t1	means G6P-PYR
glycolysis_t2	means F6P-PYR and not G6P-PYR
glycolysis	means glycolysis_t1 or glycolysis_t2
gluconeogeneis	means fructose_bis_phosphotase and G3P-PEP or (fructose_bis_phosphotase and G3P-PEP and ATP_pyruvate_water_phosphotransferase)
glycolysis_is_supported	means glycolysis
glycolysis_is_not_supported	means not glycolysis
glycolysis_is_ADP-dependent	means glycolysis and (ADP-dependent-pfk or ADP-dependent_phosphoglycerate_kinase)
glycolysis_is_ppi-dependent	means glycolysis and ppi-dependent-pfk

Section 1 of the table displays the reaction names and the corresponding reaction ids for the pathways that are considered for establishing Boolean rules. Section 2 displays assigned rules for reactions that are mentioned in Section 1. Rules able to facilitate: (1) pathway steps that may have more than one enzymatic reaction (e.g., *gdh* means 1 of {NAD dependent_phosphoglycerate_dehydrogenase,NAD-dependent_phosphoglycerate_dehydrogenase}), (2) partial pathway segments (e.g., G3P-PEP means *gdh* and *pgk* and *pgm* and enolase), and (3) complete pathways (e.g., glycolysis means glycolysis_t1 or glycolysis_t2)
Boolean rules were established for 12 central metabolic pathways including pathways that are mentioned in this table and were originally published in Edirisinghe et al. [9])

phylogenetic tree where it depicts phylogenetic distribution of a given biochemical pathway. In Fig. 6, we have organized all CMMs by their taxonomic groups against pathway presence and absence data. Taxonomic groups that are displayed along the horizontal axis of Fig. 6 were sorted sequentially as they appear in a 16S rRNA-based phylogenetic tree [9].

3.6 Overview and Discussion

In this chapter, we present a detailed protocol for the reconstruction and analysis of core metabolic models (CMMs) in KBase. In comparison to genome-scale models, CMMs are simpler and can accurately determine: (1) ATP yields based on different growth/environmental conditions, (2) ETC variations and respiration types, (3) ability to produce fermentation products, (4) presence and absence of classical biochemical pathways in central metabolism, and (5) ability to produce key metabolic pathway intermediates in central metabolism which are precursors of essential biomass components of the cell.

We have implemented the CMM construction and analysis pipeline using KBase apps (*see* Subheading 2.2) with commentary (*see* Subheading 2.1), where the following major steps are demonstrated: (1) annotation of microbial genomes, (2) reconstruction of CMM, (3) gapfilling of CMM, and (4) perform flux balance analysis (Fig. 1). Comparative analysis of CMMs and flux distributions based on different media conditions is also demonstrated.

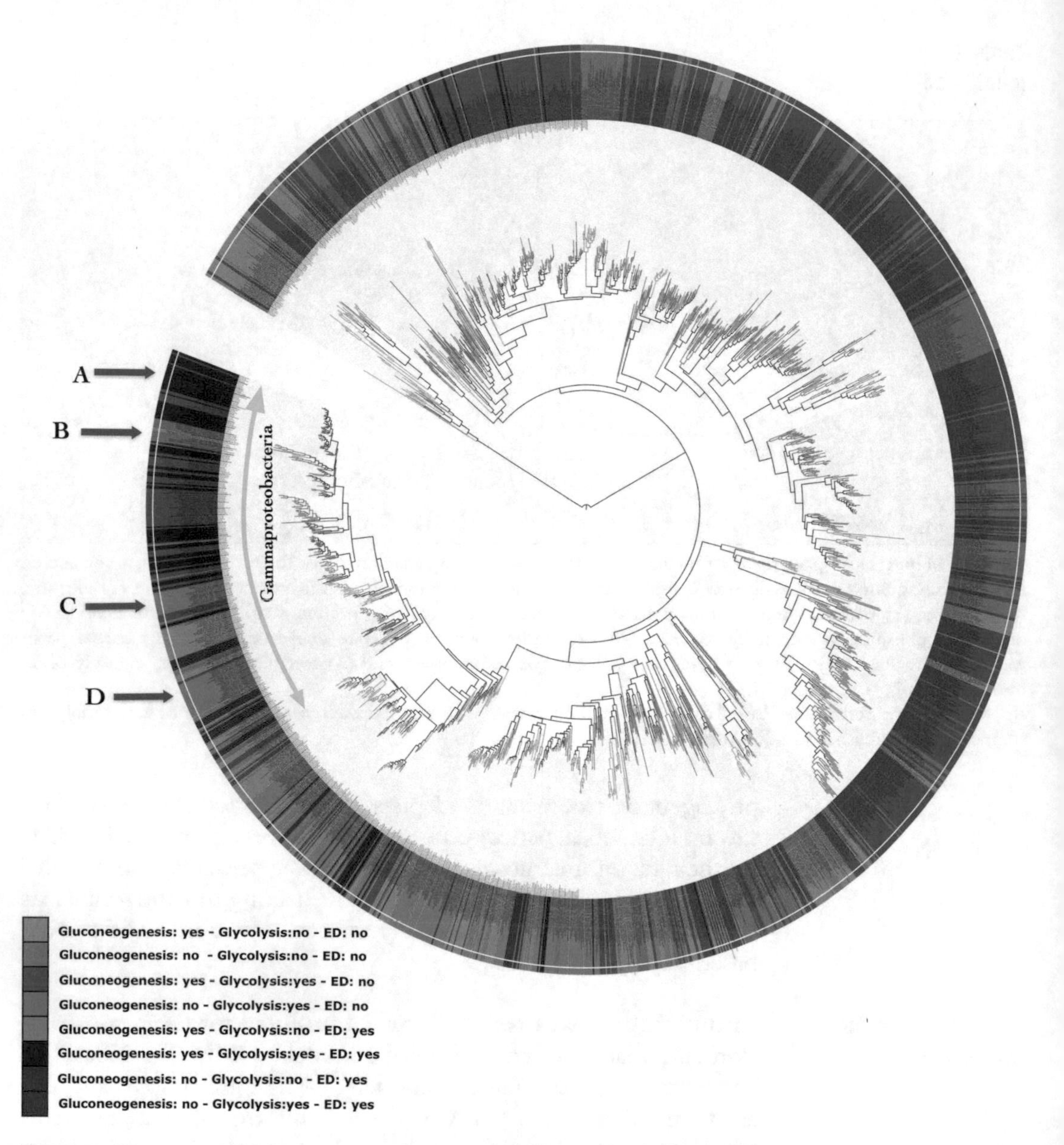

Fig. 5 Phylogenetic distribution of central metabolic pathways (originally published in Edirisinghe et al. [9]). Microbial life tree (16S $OTU_{98.5}$) depicting the presence and absence of sugar degradation pathways glycolysis, gluconeogenesis, and Entner-Doudoroff. The name of the organism and the phylum can be found at the leaf of the tree in the high-resolution image. The colored branches depict which clades gained or lost certain metabolic pathways. The curved arrow shows the range of the group Gammaproteobacteria, and the straight arrows indicate the regions where species belongs to several different genera have different phenotypes with the same taxonomic group: Escherichia and Salmonella (purple) (A), (B) *Buchnera* (green), (C) *Shewanella* (light blue) and (D) *Pseudomonas* (light blue). A high-resolution image of this figure can be accessed at http://bioseed.mcs.anl.gov/~janakae/coremodel/springer/fig5.pdf

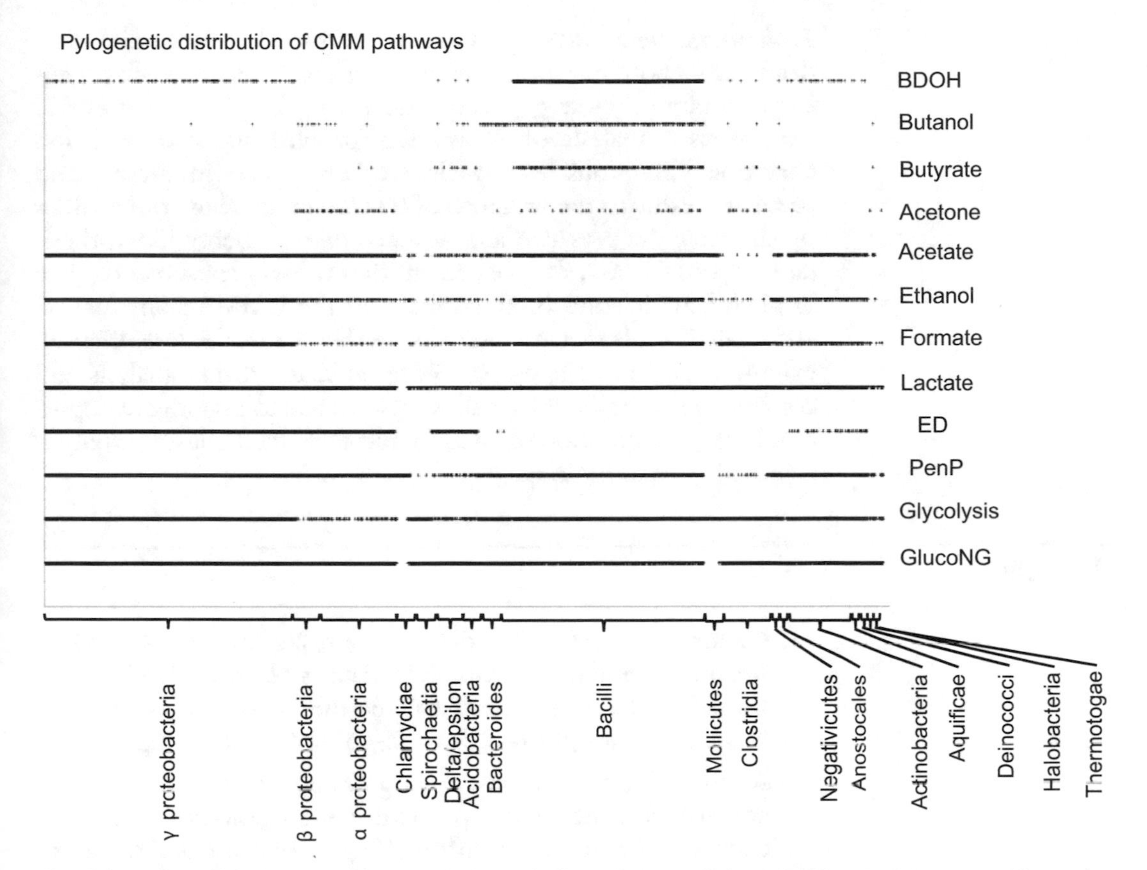

Fig. 6 Presence and absence of key central metabolic pathways of about 8100 organisms sorted by major phylogenetic groups originally published in Edirisinghe et al. [9]). Taxonomic groups that are displayed in the horizontal axis of the graph were sorted sequentially as they appear in a 16SrRNA-based phylogenetic tree (*GlucoNG* gluconeogenesis, *ED* Entner-Doudoroff, *PenP* Pentose Phosphate)

Along with the described reconstruction and analysis tools, KBase also offers a large amount of public data including microbial genomes and media formulations that aid in the CMM reconstruction process across the microbial tree of life (*see* Subheadings 2.2 and 3.5).

In our specific example where we used the *Escherichia coli* K12 annotated genome as the starting point, we demonstrated the CMM's ability to predict energy yields and biomass without requiring any gapfilling reactions, thus the CMM predictions are solely based on genome annotations. We performed flux balance analysis (FBA) coupled with flux variability analysis (FVA) (*see* Subheading 2.2) using the *E. coli* K12 CMM in glucose minimal media to predict metabolite uptake and excretion, intracellular flux profiles, and growth/ATP production yields. These analyses reveal the essential reactions required for *E. coli* K12 to predict energy yields or biomass/growth under a specific media/environmental condition. A comparative analysis of the CMM of *E. coli* and

P. denitrificans showed conservation of essential metabolic reactions across both organisms, while reactions reflecting each organism's unique biology were deemed mainly nonessential. Comparative analysis of *E. coli* flux profiles under aerobic and anaerobic conditions has revealed the differences in energy yield predictions due to the presence of ETCs. We conclude our analysis by showing the presence and absence of key energy biosynthesis pathways in CMMs, and we present the pathway conservation data in phylogenetic context. In addition to the CMM reconstruction and analysis tools that are discussed in this chapter, KBase offers an extensive catalog of apps (*see* **Note 2**) that provide analysis and comparison capabilities that allow researchers to investigate important biological questions related to microbial metabolism and other topics in systems biology.

4 Notes

1. Genomes, CMMs, Flux distributions, comparative analysis of the models and flux distributions data can be downloaded from the KBase Narrative interface (see the instructions at http://kbase.us/data-upload-download-guide/downloading-data/).
2. KBase offers an extensive catalog of apps for metabolic model construction and for comparative analysis genomes. The list of apps can be found at https://narrative.kbase.us/#catalog/apps/

References

1. Gottschalk G (1988) Bacterial metabolism. Springer, New York, NY
2. Gottschalk G (1989) How *Escherichia coli* synthesizes ATP during aerobic growth of glucose. In: Bacterial metabolism. Springer, New York, NY, pp 13–35
3. Henry CS, DeJongh M, Best AA, Frybarger PM, Linsay B, Stevens RL (2010) High-throughput generation, optimization and analysis of genome-scale metabolic models. Nat Biotechnol 28(9):977–982. https://doi.org/10.1038/nbt.1672
4. Karp PD, Paley S, Romero P (2002) The Pathway Tools software. Bioinformatics 18(Suppl 1):S225–S232
5. Becker SA, Feist AM, Mo ML, Hannum G, Palsson BO, Herrgard MJ (2007) Quantitative prediction of cellular metabolism with constraint-based models: the COBRA Toolbox. Nat Protoc 2(3):727–738
6. Monk J, Palsson BO (2014) Genetics. Predicting microbial growth. Science (New York, NY) 344(6191):1448–1449. https://doi.org/10.1126/science.1253388
7. Kumar VS, Dasika MS, Maranas CD (2007) Optimization based automated curation of metabolic reconstructions. BMC Bioinformatics 8:212
8. Varma A, Palsson BO (1994) Stoichiometric flux balance models quantitatively predict growth and metabolic by-product secretion in wild-type Escherichia-coli W3110. Appl Environ Microbiol 60(10):3724–3731
9. Edirisinghe JN, Weisenhorn P, Conrad N, Xia F, Overbeek R, Stevens RL, Henry CS (2016) Modeling central metabolism and energy biosynthesis across microbial life. BMC Genomics 17:568. https://doi.org/10.1186/s12864-016-2887-8
10. Aziz RK, Bartels D, Best AA, DeJongh M, Disz T, Edwards RA, Formsma K, Gerdes S, Glass EM, Kubal M, Meyer F, Olsen GJ, Olson R, Osterman AL, Overbeek RA, McNeil LK, Paarmann D, Paczian T, Parrello B, Pusch

GD, Reich C, Stevens R, Vassieva O, Vonstein V, Wilke A, Zagnitko O (2008) The RAST Server: rapid annotations using subsystems technology. BMC Genomics 9:75. https://doi.org/10.1186/1471-2164-9-75

11. Orth JD, Thiele I, Palsson BO (2010) What is flux balance analysis? Nat Biotechnol 28 (3):245–248. https://doi.org/10.1038/nbt.1614. nbt.1614 [pii]

12. Aziz RK, Bartels D, Best AA, DeJongh M, Disz T, Edwards RA, Formsma K, Gerdes S, Glass EM, Kubal M, Meyer F, Olsen GJ, Olson R, Osterman AL, Overbeek RA, McNeil LK, Paarmann D, Paczian T, Parrello B, Pusch GD, Reich C, Stevens R, Vassieva O, Vonstein V, Wilke A, Zagnitko O (2008) The RAST server: rapid annotations using subsystems technology. BMC Genomics 9:15. https://doi.org/10.1186/1471-2164-9-75. 75 [pii]

13. Delcher AL, Bratke KA, Powers EC, Salzberg SL (2007) Identifying bacterial genes and endosymbiont DNA with Glimmer. Bioinformatics 23(6):673–679. https://doi.org/10.1093/bioinformatics/btm009, btm009 [pii]

14. Hyatt D, Chen GL, Locascio PF, Land ML, Larimer FW, Hauser LJ (2010) Prodigal: prokaryotic gene recognition and translation initiation site identification. BMC Bioinformatics 11:119. https://doi.org/10.1186/1471-2105-11-119

15. Overbeek R, Olson R, Pusch GD, Olsen GJ, Davis JJ, Disz T, Edwards RA, Gerdes S, Parrello B, Shukla M, Vonstein V, Wattam AR, Xia F, Stevens R (2014) The SEED and the Rapid Annotation of microbial genomes using Subsystems Technology (RAST). Nucleic Acids Res 42(Database issue):D206–D214. https://doi.org/10.1093/nar/gkt1226

16. Brettin T, Davis JJ, Disz T, Edwards RA, Gerdes S, Olsen GJ, Olson R, Overbeek R, Parrello B, Pusch GD, Shukla M, Thomason JA III, Stevens R, Vonstein V, Wattam AR, Xia F (2015) RASTtk: a modular and extensible implementation of the RAST algorithm for building custom annotation pipelines and annotating batches of genomes. Sci Rep 5:8365. https://doi.org/10.1038/srep08365

17. Snyder EE, Kampanya N, Lu J, Nordberg EK, Karur HR, Shukla M, Soneja J, Tian Y, Xue T, Yoo H, Zhang F, Dharmanolla C, Dongre NV, Gillespie JJ, Hamelius J, Hance M, Huntington KI, Jukneliene D, Koziski J, Mackasmiel L, Mane SP, Nguyen V, Purkayastha A, Shallom J, Yu G, Guo Y, Gabbard J, Hix D, Azad AF, Baker SC, Boyle SM, Khudyakov Y, Meng XJ, Rupprecht C, Vinje J, Crasta OR, Czar MJ, Dickerman A, Eckart JD, Kenyon R, Will R, Setubal JC, Sobral BW (2007) PATRIC: the VBI PathoSystems Resource Integration Center. Nucleic Acids Res 35(Database issue): D401–D406. https://doi.org/10.1093/nar/gkl858. gkl858 [pii]

18. Tran TT, Dam P, Su Z, Poole FL II, Adams MW, Zhou GT, Xu Y (2007) Operon prediction in Pyrococcus furiosus. Nucleic Acids Res 35(1):11–20. https://doi.org/10.1093/nar/gkl974, gkl974 [pii]

19. Kanehisa M, Araki M, Goto S, Hattori M, Hirakawa M, Itoh M, Katayama T, Kawashima S, Okuda S, Tokimatsu T, Yamanishi Y (2008) KEGG for linking genomes to life and the environment. Nucleic Acids Res 36 (Database issue):D480–D484

20. Kanehisa M, Goto S (2000) KEGG: Kyoto encyclopedia of genes and genomes. Nucleic Acids Res 28(1):27–30

21. Caspi R, Altman T, Dale JM, Dreher K, Fulcher CA, Gilham F, Kaipa P, Karthikeyan AS, Kothari A, Krummenacker M, Latendresse M, Mueller LA, Paley S, Popescu L, Pujar A, Shearer AG, Zhang P, Karp PD The MetaCyc database of metabolic pathways and enzymes and the BioCyc collection of pathway/genome databases. Nucleic Acids Res 38(Database issue):D473–D479. https://doi.org/10.1093/nar/gkp875. gkp875 [pii]

22. Karp PD, Riley M, Saier M, Paulsen IT, Collado-Vides J, Paley SM, Pellegrini-Toole A, Bonavides C, Gama-Castro S (2002) The EcoCyc database. Nucleic Acids Res 30(1):56–58

23. Ware D, Jaiswal P, Ni J, Pan X, Chang K, Clark K, Teytelman L, Schmidt S, Zhao W, Cartinhour S, McCouch S, Stein L (2002) Gramene: a resource for comparative grass genomics. Nucleic Acids Res 30(1):103–105

24. Varma A, Palsson BO (1993) Metabolic capabilities of Escherichia-coli. 2. Optimal-growth patterns. J Theor Biol 165(4):503–522

25. Varma A, Palsson BO (1993) Metabolic capabilities of Escherichia-coli.1. Synthesis of biosynthetic precursors and cofactors. J Theor Biol 165(4):477–502

Chapter 6

Using PSAMM for the Curation and Analysis of Genome-Scale Metabolic Models

Keith Dufault-Thompson, Jon Lund Steffensen, and Ying Zhang

Abstract

PSAMM is an open source software package that supports the iterative curation and analysis of genome-scale models (GEMs). It aims to integrate the annotation and consistency checking of metabolic models with the simulation of metabolic fluxes. The model representation in PSAMM is compatible with version tracking systems like Git, which allows for full documentation of model file changes and enables collaborative curations of large, complex models. This chapter provides a protocol for using PSAMM functions and a detailed description of the various aspects in setting up and using PSAMM for the simulation and analysis of metabolic models. The overall PSAMM workflow outlined in this chapter includes the import and export of model files, the documentation of model modifications using the Git version control system, the application of consistency checking functions for model curations, and the numerical simulation of metabolic models.

Key words PSAMM, YAML model format, Metabolic modeling, GEMs, Model consistency checking, Constraint-based flux simulation, Gap filling, Minimal network, Model version tracking

1 Introduction

Advances in sequencing technology have led to the increased availability of omics data from a wide variety of organisms. Through annotations of the omics data, in silico models could be constructed to simulate the functions and activities of individual organisms [1, 2]. These genome-scale models (GEMs) of metabolism specifically represent the metabolic potentials encoded by complete genomes. GEMs have broad applications in bioengineering [3], in vitro experimental design [4], and studying organism to organism interactions [5]. Despite this technological development, it is still challenging to construct and apply GEMs as accurate representations of genotype-phenotype associations. For example, a critical assessment of almost one hundred GEMs has revealed that inconsistencies could be introduced to GEMs during the curation and simulation of metabolic models [6]. These inconsistencies are difficult to identify due to the complexity of models, and moreover, due

Marco Fondi (ed.), *Metabolic Network Reconstruction and Modeling: Methods and Protocols*, Methods in Molecular Biology, vol. 1716, https://doi.org/10.1007/978-1-4939-7528-0_6,

to the lack of a standardized model format and standardized model curation and model consistency checking software. To help resolve these challenges, this chapter provides an introduction to an integrated software system, named PSAMM, for the integrative curation and construction of GEMs [7].

PSAMM stands for the Portable System for the Analysis of Metabolic Models. It is an integrated platform that supports iterative curations and mathematical simulations of GEMs. At the core of PSAMM is a number of model consistency checking and model simulation functions. These functions can be applied to identify potential model inconsistencies [8–10] and simulate reaction fluxes under variable environmental settings [11, 12]. With these functions various aspects of the organism's physiology can be explored, including fluxes through metabolic reactions and pathways, gene essentiality, and biomass production. PSAMM represents GEMs in a YAML-based model format, which provides flexible and human readable model files, and is compatible with version-tracking software, such as Git [13], Subversion [(14),] or Mercurial [15]. This YAML-based format seeks to alleviate some of the common inconsistencies that could be introduced during the editing of metabolic models. In addition, it provides a modular representation of the complex components in GEMs. Hence, PSAMM supports the curation of model components with version tracking, and it allows for the independent manipulation of mathematical simulation settings (i.e., media definition, flux limits, objective functions, etc.) and static model components (i.e., definition of genes, metabolites, and reactions).

2 Materials

The PSAMM software package includes diverse functions that can be applied for model import/export, model curation, and model simulation. Figure 1 illustrates an integrated workflow of PSAMM functions, including model representations, model curations, and mathematical simulations. The YAML model files can be imported from existing models or exported following conventional model representations, and model edits can be documented through version tracking systems. An extensive list of PSAMM-related resources including software programs, model collections, documentations, and tutorial materials is provided in Table 1 and detailed below.

2.1 PSAMM Modeling Operations

PSAMM is available as an open source software. The source code can be accessed on GitHub and the Python Package Index (PyPI) (*see* **Note 1**). The software is portable across all major platforms (Windows, Linux, and macOS). Once installed all PSAMM functions can be accessed through a central command-line interface using the *psamm-model* command. This central command calls

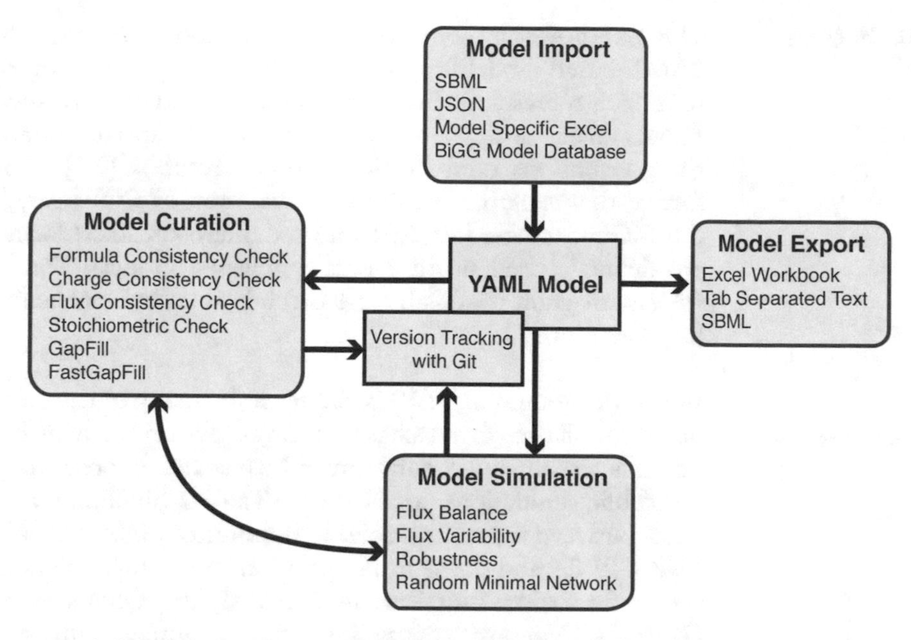

Fig. 1 A representation of the functions supported in the PSAMM software package. Aspects of the workflow related to model imports and exports are shown in red boxes, the model files and version documentation are shown in black boxes, while the model checking and model simulation functions are shown in blue boxes

Table 1
A list of PSAMM resources

Materials	Description	URL	Type
PSAMM	PSAMM overview	https://zhanglab.github.io/psamm/	Website
PSAMM-model	Base modeling platform	https://github.com/zhanglab/psamm	Software
PSAMM-import	Conversions of models to YAML format	https://github.com/zhanglab/psamm-import	Software
PSAMM-model-collection	Collection of published models in YAML format	https://github.com/zhanglab/psamm-model-collection	Git-tracked model repository
PSAMM Tutorial	Examples of PSAMM functions performed on a toy model	https://psamm.readthedocs.io/en/latest/tutorial.html	Tutorial
PSAMM Documentation	Documentation on all PSAMM functions	https://psamm.readthedocs.io/en/latest/	Documentation

the diverse programs in PSAMM to perform model consistency checking and simulation functions. In addition to the command line interface, an application programming interface is provided for PSAMM functions to allow for their use in other programs (*see* **Note 2**).

2.2 PSAMM Model Import

The *psamm-import* command is used for converting models to the YAML-based model format. It enables the application of PSAMM to models represented in other commonly used formats, such as the SBML format as defined either by the SBML specification [16] or the conventions used by the COBRA Toolbox [17], the JSON format that is derived from the conventions of COBRApy [18] or the BiGG database [19, 20], and the Microsoft Excel formats that are defined based on individual published models. The *psamm-import* program is available on GitHub as an independent repository (*see* **Note 1**).

2.3 Linear Programming Solver

Linear programming (LP) is the basis for many of the constraint-based simulation approaches that can be performed with PSAMM. For this reason, an LP solver must be installed in order to support metabolic simulations (*see* **Note 1**). Table 2 highlights the differences between the various solvers supported by PSAMM. While the IBM CPLEX and the Gurobi optimizers are compatible with all of the optimization functions in PSAMM, the QSopt_ex and the GLPK solvers are compatible with a limited subset of the PSAMM functions because PSAMM does not support mixed integer linear programming (MILP) interface with these solvers. For example, the *tfba* loop removal and the *gapfill* functions are not supported by the QSopt_ex or the GLPK solver.

2.4 PSAMM Python Dependencies

In addition to an LP solver, PSAMM relies on a number of required and optional dependencies for full functionality. The PyYAML package is required for parsing model files in the YAML format and is installed automatically when PSAMM is installed using pip (*see* **Note 1**). The packages xlrd and XlsxWriter are optional and are used only when the Excel import and export functions are used, respectively.

Table 2
Supported linear programming solvers along with details on their availability, supported linear programming functions, and a description of what PSAMM functions can be used with each

LP solver	License	PSAMM interface	PSAMM functions supported
CPLEX	Commercial (paid) or academic (free)	LP and MILP	All
Gurobi	Commercial (paid) or academic (free)	LP and MILP	All
QSopt_ex	Open source	LP only	*fba*, *fva*, *robustness* without the "*--loop-removal tfba*" option
GLPK	Open source	LP only	*fba*, *fva*, *robustness* without the "*--loop-removal tfba*" option

3 Methods

3.1 Building Metabolic Models in the YAML Format

The first step in creating a metabolic model involves generating a collection of metabolic reactions from genome annotations. This process can be done through various automated platforms such as KBase [21], KASS [22], ModelSEED [23], and MetaCyc [24] or through manual annotation. The end result of this process is a set of metabolic reactions that are associated with genes coding for enzymes that catalyze those reactions.

In the next step, a GEM can be represented in YAML format based on the identified associations of genes and proteins with metabolic reactions. The YAML models are represented as text files based on the YAML specification [25], which uses line breaks and whitespaces as meaningful structural information and because of this, is compatible with version-tracking systems. A YAML model is typically stored as separated modules in a model directory, where each module is used to represent a discrete component of the full model. An example of a toy metabolic model that contains a single transport reaction and two exchange reactions is provided in Fig. 2. Individual modules were formulated to describe the different components, including the static features of metabolites and reactions, as well as the dynamic features of model flux bounds, media conditions, and other settings related to mathematical simulations. Details of the individual model components are described below.

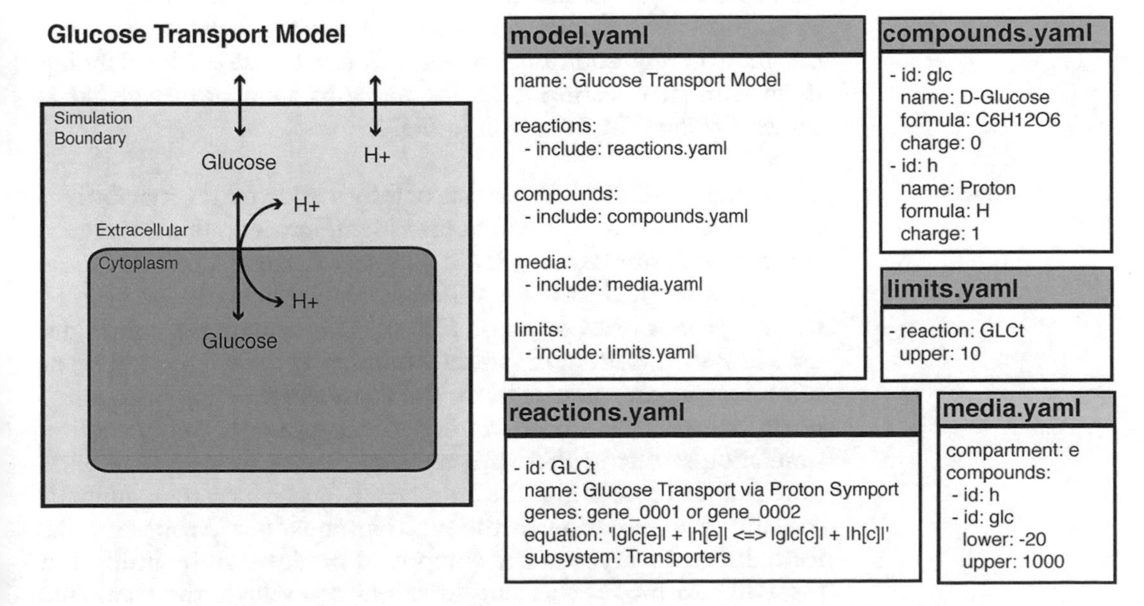

Fig. 2 A schematic representation of a toy model that contains only one transport reaction and two exchange reactions. The model is represented in a YAML format, which includes a central model file (*model.yaml*) and separated file modules (*reactions.yaml*, *compounds.yaml*, *limits.yaml*, and *metia.yaml*) for representing the different model components. References to the file modules are defined in the central model file using the *include* labels

3.1.1 Central Model File

The YAML format metabolic models are organized around a central *model.yaml* file that acts as the access point to the other files in the model (Fig. 2). This file consists of references to the reactions, compounds, limits, and media modules, which can be stored in separate files and referenced using their relative paths in the model directory (*see* **Note 3**). This central file is also where model properties are defined, including the model name, default flux limits, and the reaction identifier of a biomass equation. The biomass equation designated in this file is used as a default objective function in metabolic simulations.

3.1.2 Reactions Module

The reactions module contains information pertaining to the metabolic reactions in a model. Each reaction entry in this module must have an identifier and associated metabolic equation. The reaction entries can also contain additional data defining enzyme names, gene associations, enzyme commission numbers, or any model-specific features that are defined by the model curator. Reaction equations are representations of biochemical conversions, usually catalyzed by an enzyme, written as an equation of metabolites, and their corresponding stoichiometry as can be seen in the reaction GLCt in Fig. 2. Gene associations for a reaction represent the association of gene products as enzyme complexes or alternative enzymes. The gene associations can be written as logical statements, which use "and" to indicate enzyme complexes composed of multiple gene products, and use "or" to indicate alternative enzymes. The gene logic can be used in multiple PSAMM functions to perform gene deletions and look at minimal gene networks in the model. Any additional properties can be added by defining them with new feature keys like the subsystem property that is shown for the GLCt reaction in Fig. 2.

3.1.3 Compounds Module

The compounds module consists of information on the metabolites in a model such as the compound identifiers, compound names, compound formulas, and ionic charges. Examples of how these properties are included for different metabolites can be seen in the *compounds.yaml* panel of Fig. 2. The compound names are used in PSAMM for automatically translating the compound identifiers in reaction equations into the compound names. Such automatic translations occur during the output of mathematical simulation results in PSAMM, making it easier to analyze simulation results by relating the numerical flux values to a human-readable representation of the reaction equations. Properties like formulas and charges in the compound module can be utilized in the PSAMM model checking functions to evaluate the elemental and charge balances of reaction equations.

3.1.4 Media Module

The media module is used to define the bounds of nutrient uptake and metabolic byproduct diffusion. It is composed of a list of extracellular compounds, which represent the exchange of these compounds between the model and the simulated environment. The exchange fluxes of extracellular compounds are bounded by the upper and lower flux values set in the media module. Lower bounds indicate the maximum flux of compound uptake, while upper bounds determine the maximum flux of byproduct diffusion. If no specific bounds are set in this file, like with the "h" compound in Fig. 2, the bounds will be set to a maximum range of $[-x, x]$, where x is the default bounds of reaction fluxes, meaning that the compounds can be freely consumed or produced by the model.

3.1.5 Limits Module

The limits module is used to define flux bounds of metabolic reactions in a model. This module is optional and is required only when the flux bounds deviate from the standard settings in PSAMM, which assigns $[0, x]$ to irreversible reactions and $[-x, x]$ to reversible reactions, where x by default is equal to 1000, or when a default flux limit is set in the central model file, x will be equal to the default limits value. Reactions can be limited in this module by constraining their fluxes with the upper, lower, or fixed properties to alter the maximum bounds, minimum bounds, or fix the flux at a specific value, respectively. An example of this can be seen in the limits property that has been set for the GLCt reaction in the Fig. 2, panel *limits.yaml*, where the reaction GLCt was limited to a maximum flux of 10.

3.1.6 Model Definition

By default, PSAMM includes all the reactions in the reaction module into the construction of metabolic simulations. Alternatively, a model definition file, consisting of a list of reaction identities, can be used to define a specific subset of reactions from the reactions module that will be included in model simulations. This feature is useful when utilizing the *gapfill* and *fastgapfill* functions (Subheading 3.3.5).

3.1.7 Customization of the YAML Model Format

The YAML-based model format can be customized by including user-defined properties in any modules. User-defined properties can be used to include notes from annotations, define subgroups of entries, and provide additional information that could be useful for the interpretation of modeling results. Besides the YAML format, individual modules can also be represented in a tab-separated text format to allow for further flexibility in the presentation of the model data. As an example, a text format representation of the *compounds.yaml* panel in Fig. 2 is shown below:

compounds.tsv			
id	name	formula	charge
glc	D-Glucose	$C_6H_{12}O_6$	0
h	Proton	H	1

3.2 Model Import/Export

Besides YAML models, PSAMM supports the import and export of models represented in other commonly used model formats (Subheading 2.2). This feature increases the compatibility of PSAMM with a wide range of existing models and modeling databases. From the imported models, a PSAMM workflow can be applied to support the simulation, consistency checking, manual curations, and a resulting curated version of the model can be exported to different formats, including the SBML format, the tab-separated table format, or the Microsoft Excel format (Fig. 1).

3.2.1 Model Import

The model import functions in PSAMM are supported by the *psamm-import* command, which can be run as follows:

```
psamm-import <TYPE> --source <INPUT_PATH> --dest <OUTPUT_PATH>
```

The "--source" and "--dest" parameters are used to define the paths to the original model files and the imported YAML model directory. The supported import types include SBML, JSON, or Excel formats. A detailed list of supported import types can be viewed by running the following command:

```
psamm-import list
```

Additionally, the *psamm-import-bigg* command can be used to download and convert metabolic models contained in the online BiGG metabolic model database [19, 20].

3.2.2 Model Export

PSAMM provides three model exporting options, through the *psamm-model* functions, for the export of SBML, Excel, and tab-separated table formats. These can be accessed using the following commands:

```
# SBML export
psamm-model sbmlexport > output.sbml
# Excel export
psamm-model excelexport output.xlsx
# Table export
psamm-model tableexport {TYPE} > output.tsv
```

The SBML export function converts YAML model files to a SBML file following conventions of SBML specifications [16] and the flux balance constraints (FBC) extension [26]. All data in the model, including reactions, compounds, gene definitions, flux bound settings, objective function settings, and exchange flux settings, are all exported to the SBML model. If the model contains user-defined properties that are not part of the SBML specification, those will be stored in the SBML notes section. Similarly, the Excel export function converts all data files in a model into individual worksheets of the Excel file, which include both standard properties and user-defined properties of the model components. Finally, the table export function converts designated components of the model, such as the reactions, compounds, medium, limits, or the model metadata, into a tab-separated text format.

3.3 Model Consistency Checking

Model consistency checking is an important step in metabolic modeling. It can be applied to evaluate mass and charge balances of biochemical equations, identify knowledge gaps in a metabolic reconstruction, and resolve potential inconsistencies introduced from manual editing of the models. PSAMM provides diverse approaches for the consistency checking and the subsequent curation of metabolic models (Fig. 3). The *chargecheck*, *formulacheck*, and *masscheck* functions can be used to examine the consistency within metabolic equations, the *fluxcheck* function can be used to identify blocked reactions, and the gap filling functions, *gapfill* and *fastgapfill*, can be used to propose new reactions for unblocking reactions in a model.

3.3.1 Formula Balance Checking

The *formulacheck* function in PSAMM is used to check the stoichiometric consistency of individual reaction equations (Fig. 3). This function checks the balance of individual reaction equations and can be applied only when compound formulas are annotated with the "*formula*" property in the compounds module. Figure 3 illustrates the application of *formulacheck* on an imbalanced fumarase reaction, where a water molecule was missing from the right hand side of the equation. The *formulacheck* command identified

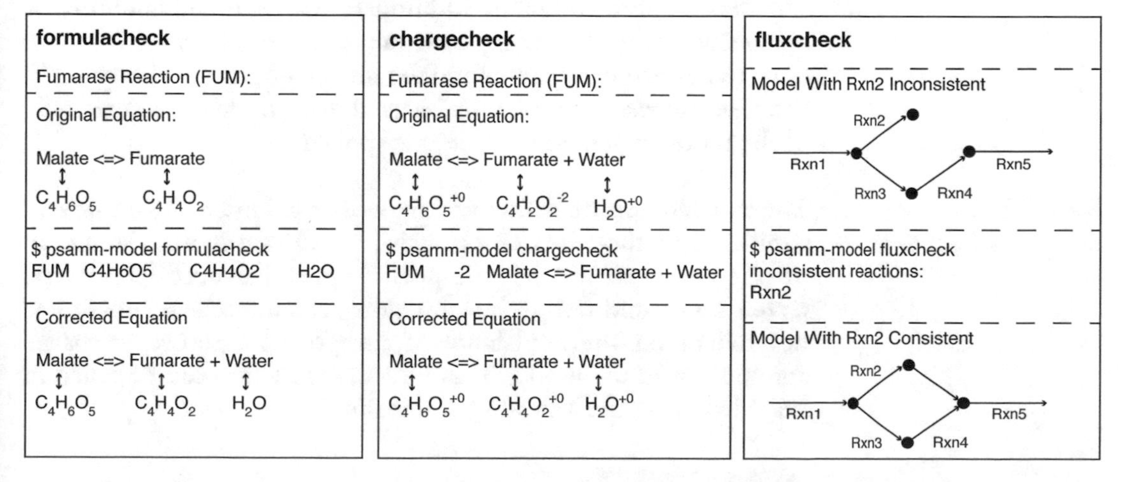

Fig. 3 Examples of consistency checking functions in PSAMM. The applications of three functions, *chargecheck*, *formulacheck*, and *fluxcheck*, were illustrated. The analysis of original models by each function could point to potential inconsistencies in the model and provide references for subsequent model curations and corrections

the inconsistency and provided information about the missing elements. The *formulacheck* is limited to the given compound formula in the compounds module, and hence its accuracy relies on the formula annotation of the metabolites. Additionally, if compound formulas are missing for some metabolites, reactions involving those metabolites will not be analyzed by the *formulacheck* function.

3.3.2 Charge Balance Checking

The *chargecheck* function in PSAMM is used to check for balance of charges in a reaction equation. The *chargecheck* function relies on the charge information that is provided in the compounds module with the "*charge*" label and can be influenced by the accuracy of the charge annotation. In Fig. 3 the *chargecheck* command is demonstrated on a fumarase reaction, where differences in protonation state caused a charge imbalance on the two sides of the reaction equation. The *chargecheck* command identified the charge inconsistency, and provided support to manual curations by listing the charge differences.

3.3.3 Flux Consistency Checking

The *fluxcheck* function in PSAMM is used to identify blocked reactions that cannot carry any flux in a model. This function minimizes and maximizes the fluxes through each reaction in the model and reactions with a flux range of [0,0] are reported as blocked. An example of using *fluxcheck* can be seen in Fig. 3, where *RXN2* was blocked because it leads to a "dead end" in the metabolic pathway. This blockage was identified by the *fluxcheck* function. Subsequently, the pathway of *RXN2* can be manually investigated to unblock the reaction, either by including new reactions or by changing the directionality of existing reactions. When *fluxcheck* is run without any additional options, it will minimize or maximize flux values using constraints set in the current model (e.g., as defined in the limits and the media modules). Alternatively, the "*--unrestricted*" option can be used to examine the connectivity of the model when no constraint is applied.

3.3.4 Stoichiometric Consistency Checking

The stoichiometric consistency can be checked in PSAMM using an additional method, *masscheck*, which provides a global estimation of whether all reactions in a model are mass balanced. Unlike the *formulacheck* and *chargecheck* functions, the masscheck function is not reliant on the availability of compound formula or charge annotations in the model. This *masscheck* function can be called in PSAMM using the following command:

```
psamm-model masscheck
```

It checks if compounds in the model can be assigned a nonzero mass value so that no residue mass is present between the reactants and the products of each reaction. For example, given a model with two reactions:

```
RXN1: A + B => C
RXN2: C => B
```

The *masscheck* function will report an inconsistency in this case because the two reactions, *RXN1* and *RXN2*, cannot be both balanced when compound *A* is assigned a nonzero mass. This indicates that a metabolite is either missing in one reaction, or in excess in the other reaction. Due to the interconnected nature of many reactions in metabolic networks, the assignment of mass values may not be unique, and hence the residue mass values may be assigned to any of the reactions that are connected to the imbalanced reaction [7, 8]. In such cases, *masscheck* can be run in an iterative fashion, with the balanced reactions verified after each iteration marked with the "*--checked*" parameter, until all the potentially imbalanced reactions are identified.

3.3.5 Gap Filling

PSAMM contains two gap filling functions implemented based on two distinct algorithms, GapFind/GapFill [10] and fastGapFill [9]. The GapFind/GapFill algorithm is implemented in the *gapfill* function. It first identifies blocked compounds that cannot be produced, and then it identifies the smallest set of changes to the model that could enable the production of the compounds. Possible changes proposed by *gapfill* include: adding new metabolic reactions, adding transport or exchange reactions, and changing existing irreversible reactions to be reversible. Similarly, the fastGapFill algorithm is implemented in the *fastgapfill* function, which searches for a consistent subset of reactions from a broader reaction database. The fastGapFill algorithm is faster than the GapFind/GapFill algorithm, so it can be applied to larger metabolic models. Unlike *gapfill*, the *fastgapfill* function will not propose changes to the reversibility of model reactions.

Both of the gap filling functions support the exploration of new gap filling reactions from an extended reaction module that consists of not only reactions defined in the model definition file (Subheading 3.1.6), but also new reactions that could contribute to unblocking reactions in the model. Additionally, the gap filling functions can also explore the addition of new transporting and exchange reactions to help facilitate the production or removal of blocked compounds. The *fastgapfill* function provides additional controls to set the penalty (or "cost") of adding new reactions into the model. This can be achieved by setting up a two-column table, with each reaction identifier mapped to a designated penalty score, and by using the "*--penalty*" option:

```
psamm-model fastgapfill --penalty penalty.tsv
```

Additionally, default penalties can be set for different types of reactions using the "*--db-penalty*", "*--tp-penalty*", and "*--ex-penalty*" options, which define the penalty of adding reactions from the extended reaction database, the transporter reactions, or the exchange reactions, respectively. By default, a penalty value of 1 is assigned for each reaction added to the model.

3.4 Version Tracking with YAML Models

The model curation process can be tracked with version control systems, such as Git [13], Subversion [14], or Mercurial [15]. The version-tracked models can be set up as local projects, or can be put on remote servers to support collaborative annotations. Using Git as an example, the following sections provide an overview of how the version control systems can be used for documenting the collaborative model curation process.

3.4.1 Git Setup

Before Git is used for the first time, the author name and email has to be configured. This is so that Git can keep track of the author of each change to the model.

```
git config --global user.name "John Doe"
git config --global user.email johndoe@example.com
```

3.4.2 Model Repository Initiation

If a model repository is available, the git clone command can be used to make a local copy of the model files. For example, the PSAMM model repository contains over 50 curated, YAML formatted models from the current literature. These models can be cloned to a user's local repository using the following command:

```
git clone https://github.com/zhanglab/psamm-model-collection.
git
```

The *git clone* command will create a local copy of the remote repository called *psamm-model-collection*, which contains multiple models curated in the PSAMM project [7]. This local copy mirrors the remote repository that is stored in the GitHub server. It can be updated to incorporate new changes in the remote repository, and it can be used to upload curated new changes from local repository to the server.

A new model repository can also be initialized using local files of a model. This can be achieved using the *git init* command. For example, command below would initiate a new git repository that incorporates model files in a local directory named "new-model":

```
cd new-model
git init
```

3.4.3 Tracking Changes in the Model Repository

After initializing the Git repository, model files can be added to and tracked by the repository through the *git add* and *git commit* commands. In a collaborative project, model files can be edited simultaneously by different curators in their individual copies of the model repository. This would require the tracking of all model files in the local model repository, which can be set up as follows:

```
git add *.yaml
git commit -m "Initiate tracking of YAML model files"
```

When changes are made to the model, the *git diff* command can be used to review the changes since a previous commit. Inspecting the diff output is an important step to make sure that no unintentional changes were made to the model files.

```
git diff
```

The new changes to model files can be committed to the repository using the *git commit* command after inspection of model differences. A commit in Git version control system is a record of file changes labeled with the author who made the changes, a timestamp of when the changes were committed, and a short description of what were changed in the model files. The commit can be made by first staging the changes using the *git add* command on the modified model files, and then performing the *git commit* command to commit the staged changes. A message should be provided with each commit to briefly describe the nature of changes made to the model using the "*-m*" option.

```
#-stage the local files that have been changed
git add <filename>
#-commit changes to the Git version control system
git commit -m "<description>"
#-push committed changes to the remote server
git push
```

The version history of a model can be inspected with the *git log* command. The users can also obtain detailed information on the commits using the "*-p*" option.

```
git log
git log -p
```

3.4.4 Advanced Features for Collaboration

A number of commands in Git are useful for supporting the collaborative annotation of metabolic models. For example, the "*--branch*" option can be used to support independent testing cases where specific aspects of a model are modified and examined:

```
git checkout --branch complete-b6-synthesis
```

In the above example, the model curator created a new branch, named "*complete-b6-synthesis*", to support the targeted curation of the vitamin B6 synthesis pathway. The use of branches is also the key for using these repositories in collaborative projects. In those cases, multiple researchers could create branches from a base version of a model and perform their edits on those branches without generating conflicts in the files. After these branches are fully tested, they can be merged back into the master branch of the model using the *git merge* command:

```
git checkout master
git merge complete-b6-synthesis
```

Other tools in Git can be used to support a more efficient analysis of the tracked model versions. The *git blame* command can be used to point out the last time a specific line of a model file was changed. This can be used to pinpoint changes to a given model file. The *git bisect* command is another powerful tool for inspecting the model history. This can be applied to quickly identify the context of specific changes to the model files.

3.5 Model Simulations

PSAMM supports diverse functions for performing mathematical simulations with metabolic models. Most of the simulation functions in PSAMM provide outputs in a tab-separated text format, which can be further analyzed using other command-line tools or imported into programs like Excel or R for downstream analyses and visualizations. The *psamm-model* command provides the interface for calling model simulation functions in PSAMM, and the options for each function are detailed below. Additionally, some global options are available for the definition of simulation conditions in multiple functions (*see* **Notes 5–9**).

3.5.1 Flux Balance Analysis

The flux balance analysis (FBA) is broadly applied for the constraint-based analysis of metabolic models [11, 27]. This function is performed using the following command:

```
psamm-model fba
```

By default, the *fba* function in PSAMM maximizes the biomass reaction, which can be defined in the central model file (Subheading 3.1.1). The output of the *fba* function includes flux values for each individual reaction in the metabolic model that contribute to optimization of the objective function. The flux values are provided in a tab-separated format along with original equations of the metabolic reactions, as well as a list of genes associated with the reactions. The *fba* command by default will only output reactions that carry nonzero fluxes in a given simulation. To obtain fluxes for all the reactions in the model, an "*--all-reactions*" option can be set to force the output of all reactions and fluxes.

3.5.2 Flux Variability Analysis

The flux variability analysis (FVA) is applied to determine the lower and upper bounds of metabolic reactions under specific simulation conditions [28]. It can be carried out using the following command:

```
psamm-model fva
```

By default, the *fva* function is performed on every reaction of the model when the objective function is constrained to its optimized value. This allows for an exploration of variabilities in the model and provides insights into alternative solutions that could contribute to an optimized objective flux. Results from the *fva* function are presented as tab-separated values, including reaction identifiers, lower flux bounds, upper flux bounds, and the reaction equations.

3.5.3 Robustness Analysis

The robustness analysis can be used to determine how a model responds to changes of specific reaction fluxes [29]. A robustness analysis can be performed in PSAMM using the following command:

```
psamm-model robustness EX_glc-D
```

The example above explores the fluxes of the objective function of a model when perturbing the uptake flux of D-glucose (EX_glc-D). By default, the *robustness* command in PSAMM optimizes the biomass objective function defined in the central model file when varying the flux of a perturbation function (e.g., EX_glc-D in the above example). The objective function can also be assigned using the "*--objective*" option to permit optimization of other reactions in the system. Additionally, the "*--minimum*" and "*--maximum*" options can be used to set the lower and upper bounds of the perturbation, respectively, and the "*--steps*" option can be used to set the number of simulation steps over the range of the varying reaction flux. The output of the *robustness* function is a tab-separated table containing the flux of the varying reaction and the corresponding objective flux. The fluxes of all the reactions can be printed for each step of the simulation using the "*--all-reaction-fluxes*" option.

3.5.4 Minimal Network Analysis

The *randomsparse* function in PSAMM identifies the viable minimal networks of a metabolic model. It performs successive random deletions of the genes or reactions in a model until any more deletion would make the model nonviable. The randomsparse function can be carried out through the follow command:

```
psamm-model randomsparse --type genes 90%
```

In the above example, the "*--type*" option indicates the unit of deletions in the minimal network identification. Possible values of the "*--type*" option include genes, reactions, or exchange, where the exchange option would limit the successive deletions to a list of exchange reactions and is useful for identifying minimal medium conditions. The number 90 in the above command defines a threshold, in percentage, of the required objective flux as compared to the wild-type objective flux. The model will be considered nonviable when the optimized objective flux falls below this given threshold. The result of the *randomsparse* function is a tab-separated table, where the first column lists the identifiers of genes or reactions, depending on the definition of the "*--type*" option, and the second column lists a binary value, with 1 indicating the presence of an element in the minimal network and 0 indicating the absence of an element in the minimal network.

4 Notes

1. Installing PSAMM and Linear Programming Solvers in the Python Virtual Environment
   It is recommended that PSAMM should be installed in a virtual Python environment. Virtual environments are isolated instances of the Python environment, which allow users to install packages without requesting system administrator access to a computer and provide users with more control over the version management of the software package. The Python virtual environment can be set up using the *virtualenv* command. In the example below, a virtual environment, named "*psamm-env*," was initiated and activated:

   ```
   virtualenv psamm-env
   source psamm-env/bin/activate
   ```

   After initiating the Python virtual environment, the *pip* command can be used for the installation of PSAMM functions:

   ```
   #Installation of the PSAMM package
   pip install psamm
   #Installation of psamm-import
   pip install git+https://github.com/zhanglab/psamm-import.
   git
   ```

When using the Python virtual environment, a Python interface should be installed for each linear programming solver so that the solver can be accessed by PSAMM. After activating the virtual environment, the *pip* command can be used to install the linear programming solver interface.

CPLEX (available at https://www-01.ibm.com/software/commerce/optimization/cplex-optimizer/)
pip install <path-to-cplex>/cplex/python/<python-version>/<platform>/ Notes: 1. <path-to-cplex> is the location where CPLEX was installed on the system; <python-version> is the version number of Python (e.g., 2.x or 3.x); <platform> indicates the system platform, e.g., x86_64-osx when running on a macOS system. 2. Users of Python 2.7 should use CPLEX for Python 2.6 and similarly, users of Python 3.5 or 3.6 should use CPLEX for Python 3.4.

Gurobi (available at http://www.gurobi.com/products/gurobi-optimizer)
pip install <path-to-gurobi-python> Note: <path-to-gurobi-python> depends on the system. For example, on macOS the path is /Library/<gurobi-version>/mac64.

GLPK (available at https://www.gnu.org/software/glpk/)
pip install swiglpk Note: The GLPK library must be installed on the system before the pip command is run. On Linux platforms the GLPK library may be available through the system package manager, for example on Ubuntu GLPK can be installed on the system using "*apt-get install libglpk-dev.*"

QSopt_ex (available at http://www.dii.uchile.cl/~daespino/ESolver_doc/main.html)
pip install python-qsoptex Note: The QSopt_ex library must already be installed on the system as explained in the python-qsoptex documentation. On Linux platforms the QSopt_ex library may be available through the system package manager, for example on recent versions of Ubuntu it can be installed with "*apt-get install libqsopt-ex-dev*".

2. Application Programming Interface of PSAMM
   The PSAMM application programming interface (API) provides access to all PSAMM functions in the development of new Python programs. This allows for the integration of PSAMM into customized pipelines that can be used for new modeling applications or for the analysis of specific models.

3. Model References

   The "*--model*" option in PSAMM can be used with any function to indicate the path to the central model file of a YAML formatted model. When this option is not used, PSAMM looks for a file named *model.yaml* in the user's current directory. The "*--model*" option provides an alternative way to call a model when the central model file is not following the standard naming convention, or when the user needs to call a model from outside of the current directory. An example of using the "*--model*" option is shown below:

```
psamm-model --model ./model.yaml fba
```

4. Global Parameters of PSAMM

   PSAMM provides a number of global parameters that can be applied in multiple functions to further define the approaches used for analyzing metabolic models. These parameters increase the flexibility of PSAMM applications and can be used to compare the results from different simulation algorithms.

5. Solver Selection

   The "*--solver*" option in PSAMM allows for the selection of different solvers for a given function. Values supported for this option includes *cplex*, *gurobi*, *glpk*, and *qsoptex*. While the *cplex* and *gurobi* support all functions in PSAMM, the *glpk* and *qsoptex* only support limited options of the *fba*, *fva*, and *robustness* functions (Subheading 2.2).

```
psamm-model masscheck --solver name=cplex
```

6. Objective Function Definition

   The "*--objective*" parameter can be used to designate a reaction as the objective function for a metabolic simulation when using the *fba*, *fva*, *robustness,* and *randomsparse* functions. If a reaction is not designated, then the default biomass reaction set in the central model file will be maximized during the simulation.

7. Loop Removal

   The "*--loop-removal*" option can be used to eliminate artificial flux loops in the model simulations [30]. PSAMM implements two algorithms for the loop removal function, including a global L1 minimization of the flux values and an application of the thermodynamic constraints [12, 31]. The "*--loop-removal*" option can be assigned using the following command:

```
psamm-model fba --loop-removal=l1min
```

8. Epsilon Option
   Due to small rounding errors in the linear programming solvers, it is possible that some fluxes will have small nonzero values that are very close to zero. In order to help interpretation of the simulation results the "*--epsilon*" option can be used to set a minimum flux value that the simulation will consider to be a nonzero number. For example, if the epsilon value is set to be 0.001 then any value below 0.001 will be considered to be equal to zero. By default, the epsilon value is set to 10^{-5}. In the *gapfill* function, epsilon is important for the stability when solving the gap-filling problem. If *gapfill* fails to run on a certain model, sometimes a slight increase of the epsilon value may resolve this problem. If the *gapfill* problem remains to be unsolvable, the *fastgapfill* function can be used as an alternative to the *gapfill* function.
9. Parallelization Options
   The "*--parallel*" option can be used in *fva* or *robustness* functions to define the number of parallel processes to launch when running these functions. The parallelization of PSAMM functions can increase the efficiency of simulation processes. By default, the *fva* and *robustness* functions will try to use as many processes as possible to speed up the function. Setting "*--parallel*" option would provide more control over the resource allocation of a system.

References

1. Feist AM, Palsson BØ (2008) The growing scope of application od genome-scale metabolic reconstructions: the case of E. coli. Nat Biotechnol 26:659–667
2. Durot M, Bourguignon PY, Schachter V (2009) Genome-scale models of bacterial metabolism: reconstruction and applications. FEMS Microbiol Rev 33:164–190
3. Burgard AP, Pharkya P, Maranas CD (2003) OptKnock: a bilevel programming framework for identifying gene knockout strategies for microbial strain optimization. Biotechnol Bioeng 84:647–657
4. M a O, Palsson BØ, Papin JA (2009) Applications of genome-scale metabolic reconstructions. Mol Syst Biol 5:320
5. Biggs MB, Medlock GL, Kolling GL, Papin JA (2015) Metabolic network modeling of microbial communities. Wiley Interdiscip Rev Syst Biol Med 7:317
6. Ravikrishnan A, Raman K (2015) Critical assessment of genome-scale metabolic networks: the need for a unified standard. Brief Bioinform 16:1057–1068
7. Steffensen JL, Dufault-Thompson K, Zhang Y (2016) PSAMM: a portable system for the analysis of metabolic models. PLoS Comput Biol 12:e1004732
8. Gevorgyan A, Poolman MG, Fell DA (2008) Detection of stoichiometric inconsistencies in biomolecular models. Bioinformatics 24:2245–2251
9. Thiele I, Vlassis N, Fleming RMT (2014) FASTGAPFILL: efficient gap filling in metabolic networks. Bioinformatics 30:2529–2531
10. Satish Kumar V, Dasika MS, Maranas CD (2007) Optimization based automated curation of metabolic reconstructions. BMC Bioinformatics 8:212
11. Orth JD, Thiele I, Palsson BØ (2010) What is flux balance analysis? Nat Biotechnol 28:245–248
12. Müller AC, Bockmayr A (2013) Fast thermodynamically constrained flux variability analysis. Bioinformatics 29:903
13. Git (2016) Git. https://git-scm.com/. Accessed 10 Nov 2016

14. Apache Subversion (2004) Apache Software Foundation. https://subversion.apache.org. Accessed 10 Nov 2016
15. Mercurial (2016) Mercurial Source Control Management. https://www.mercurial-scm.org. Accessed 10 Nov 2016
16. Hucka M, Bergmann F, Hoops S, Keating S, Sahle S, Schaff J, Smith L, Wilkinson D (2015) The Systems Biology Markup Language (SBML): language specification for level 3 version 1 core. J Integr Bioinform 12(2):266
17. Hyduke D, Hyduke D, Schellenberger J, Que R, Fleming R, Thiele I, Orth J, Feist A, Zielinski D, Bordbar A, Lewis N, Rahmanian S, Kang J, Palsson B (2011) COBRA Toolbox 2.0. Protoc Exch 33:1–35
18. Ebrahim A, Lerman JA, Palsson BO, Hyduke DR (2013) COBRApy: COnstraints-Based Reconstruction and Analysis for Python. BMC Syst Biol 7:74
19. Schellenberger J, Park JO, Conrad TM, Palsson BØ (2010) BiGG: a Biochemical Genetic and Genomic knowledgebase of large scale metabolic reconstructions. BMC Bioinformatics 11:213
20. King ZA, Lu J, Dräger A, Miller P, Federowicz S, Lerman JA, Ebrahim A, Palsson BO, Lewis NE (2016) BiGG Models: a platform for integrating, standardizing and sharing genome-scale models. Nucleic Acids Res 44: D515–D522
21. KBase: The Department of Energy Systems Biology Knowledgebase (2015) https://kbase.us. US Department of Energy. Accessed 10 Nov 2016
22. Moriya Y, Itoh M, Okuda S, Yoshizawa AC, Kanehisa M (2007) KAAS: an automatic genome annotation and pathway reconstruction server. Nucleic Acids Res 35:W182
23. Overbeek R, Olson R, Pusch GD, Olsen GJ, Davis JJ, Disz T, Edwards RA, Gerdes S, Parrello B, Shukla M, Vonstein V, Wattam AR, Xia F, Stevens R (2014) The SEED and the Rapid Annotation of microbial genomes using Subsystems Technology (RAST). Nucleic Acids Res 42:D206
24. Caspi R, Billington R, Ferrer L, Foerster H, Fulcher CA, Keseler IM, Kothari A, Krummenacker M, Latendresse M, Mueller LA, Ong Q, Paley S, Subhraveti P, Weaver DS, Karp PD (2016) The MetaCyc database of metabolic pathways and enzymes and the BioCyc collection of pathway/genome databases. Nucleic Acids Res 44:D471–D480
25. Ben-kiki O, Evans C (2009) YAML Ain't Markup Language.
26. Olivier BG, Bergmann FT (2015) The Systems Biology Markup Language (SBML) level 3 package: flux balance constraints. J Integr Bioinform 12:269
27. Fell DA, Small JR (1986) Fat synthesis in adipose tissue. An examination of stoichiometric constraints. Biochem J 238:781–786
28. Mahadevan R, Schilling CH (2003) The effects of alternate optimal solutions in constraint-based genome-scale metabolic models. Metab Eng 5:264–276
29. Edwards JS, Palsson BO (2000) Robustness analysis of the Escherichia coli metabolic network. Biotechnol Prog 16:927–939
30. Schilling CH, Letscher D, Palsson B (2000) Theory for the systemic definition of metabolic pathways and their use in interpreting metabolic function from a pathway-oriented perspective. J Theor Biol 203:229–248
31. Schellenberger J, Lewis NE, Palsson B (2011) Elimination of thermodynamically infeasible loops in steady-state metabolic models. Biophys J 100:544–553

Chapter 7

Integration of Comparative Genomics with Genome-Scale Metabolic Modeling to Investigate Strain-Specific Phenotypical Differences

Jonathan Monk and Emanuele Bosi

Abstract

Genome-scale metabolic reconstructions are powerful resources that allow translation biological knowledge and genomic information to phenotypical predictions using a number of constraint-based methods. This approach has been applied in recent years to gain deep insights into the cellular phenotype role of the genes at a systems-level, driving the design of targeted experiments and paving the way for knowledge-based synthetic biology.

The identification of genetic determinants underlying the variability at the phenotypical level is crucial to understand the evolutionary trajectories of a bacterial species. Recently, genome-scale metabolic models of different strains have been assembled to highlight the intra-species diversity at the metabolic level. The strain-specific metabolic capabilities and auxotrophies can be used to identify factors related to the lifestyle diversity of a bacterial species.

In this chapter, we present the computational steps to perform genome-scale metabolic modeling in the context of comparative genomics, and the different challenges related to this task.

Key words Niche adaptation, Microbial evolution, Constraint-based metabolic modeling, Flux balance analysis, Comparative genomics

1 Introduction

One of the prominent features of bacteria is their ability to adapt and thrive in changing conditions [1]. This versatility is underpinned by a remarkable genomic variability even between closely related strains [2–4], which is often driven by genetic elements facilitating the horizontal exchange of DNA, also known as *Horizontal Gene Transfer* (HGT) [5]. The evolution of bacteria, and in particular that of pathogens, often has important implications for humans (e.g., emergence of multi-drug-resistant strains), Thus, it has become important to develop computational strategies to gain insights into the evolutionary trajectories of different bacterial species.

Marco Fondi (ed.), *Metabolic Network Reconstruction and Modeling: Methods and Protocols*, Methods in Molecular Biology, vol. 1716, https://doi.org/10.1007/978-1-4939-7528-0_7, © Springer Science+Business Media, LLC 2018

The introduction of the Pangenome concept [6, 7] revolutionized the field of comparative genomics for bacteria, overcoming the problems related to their high genomic variability and the lack of a reference genome. According to this paradigm, the whole genetic repertoire of the taxa, also called *Pangenome*, can be used to describe the genomic diversity within a taxa. The Pangenome can be divided into *Core genome*, the set of universally shared genes which mostly have a housekeeping role, and *Dispensable genome*, the set of clade-specific genes which contributes to strain differentiation and adaptation to different conditions.

This approach has been fruitfully used to gain evolutionary insights, with practical applications such as the design of targeted vaccines in the framework known as *reverse-vaccinology* [8–10]. Despite its utility, the Pangenome cannot be used to quantify the phenotypical effects of genome variability within a species. In this context, *Genome-scale metabolic network reconstructions* (GENREs) have been used to define conserved and strain-specific metabolic capabilities of *E. coli* and *S. aureus* species [11, 12], relating differences in metabolic capabilities with the lifestyle diversity of strains within these species. This knowledge can effectively be used to define the metabolic potential of a bacterial species and to identify metabolic features characterizing the (more) virulent strains.

In this chapter, we will illustrate a protocol to study the genomic variability within a bacterial species, predicting phenotypical differences by means of constraint-based metabolic modeling.

2 Materials

2.1 Genome Sequence

The sequence of an organism's genome can be analyzed to identify the genetic elements encoding functional products. These, in turn, can be annotated to identify their role in cellular biology, and mapped on other genomes to identify homologous genes. The publicly available genomes can be downloaded from the NCBI's ftp site in different formats such as FASTA and Genbank, together with their annotated elements (when present). It should be stressed that, although the analyses reported here can be equally performed on complete and draft sequence, it is preferable to use the complete versions, if available. In the case that a genome is lacking an annotation (i.e., de novo sequenced genomes), this gap can be easily filled using a tool suited for this task (such as Prokka [13]).

2.2 Sequence Comparison Tools

The homology relationship between genes (i.e., orthology, paralogy, inparalogy) can be assessed by means of sequence similarity, which is evaluated with tools such as BLAST, BLAT, or MUMmer [14–16]. Since the identification of orthologous genes between two genomes is not trivial, a number of approaches have been implemented to address this task. Recently [17], a method for benchmarking the present tools has been proposed, helping the identification of the most effective methods for this kind of analyses.

2.3 Reference Metabolic Reconstructions

The reconstruction of good-quality metabolic reconstructions is a demanding task, often comprising multiple computational and experimental steps, as described by Thiele and Palsson [18]. However, such an approach is not feasible to be applied on a large number of strains. To overcome this scalability issue, curated reference GENREs from closely related organisms have been used to derive strain-specific reconstruction on large-scale datasets via inference of homology relationships [11, 12]. While it is possible to consider a single GENRE as a reference, this could lead to the presence of biases in the strain-specific GENREs. Therefore, it is preferable to integrate different reconstructions in a single-reference GENRE.

The reference GENREs should be SBML compliant (Systems Biology Markup Language) and must be consistent, i.e., the metabolites, reactions, and compartments should adhere to standardized identifiers. If not, they should be reconciled to achieve consistency. Moreover, the sequences of the corresponding genomes should be available, with the names of the genes consistent with those present in the reconstructions.

2.4 Metabolic Modeling Framework

There are many tools that can be used to predict strain-specific phenotypes with constraint-based modeling [19]. Of these, the most widely used is the COBRA toolbox [20, 21], which has been integrated with the computational environments of MatLab and Python. This framework, regardless of the programming language implementation, includes many important features to import the SBML files and perform in silico simulations and analyses within the paradigm of Object-Oriented Programming (OOP).

This chapter will report code examples of the Python version of COBRA. Therefore, to replicate the analyses here reported, the following resources should be installed:

1. Python v2.7 or greater.
2. COBRApy [21] (https://github.com/opencobra/cobratoolbox).
3. Gurobi (or another linear programming solver compatible with COBRApy) (http://www.gurobi.com/).
4. Pandas v0.7 or above.

3 Methods

3.1 Obtain Publicly Available Genome Sequences

The first step of the protocol is the construction of a dataset of representative organisms of a taxa (**query dataset**) for which the corresponding genomic sequence is available. The NCBI's ftp website can be accessed with different protocols to automatize the download of these genomes. In **Note 1** we report a bash script to be used on a machine with a UNIX operative system (such as Ubuntu or Debian) to accomplish this task for a user-defined taxa.

3.2 Annotate Bacterial Genomes

Although the majority of the publicly available genomes are annotated, it is possible that some sequences may require annotation. If this is the case, it is preferable to have the annotation method consistent for the used dataset, that is, all the genomes should be annotated with the same tool to prevent annotation pipeline-biased results.

Generally, genome annotation for bacteria is considered to be a "solved" problem, which can be performed using one of the many tools available. We suggest Prokka, a user-friendly, customizable software whose predictions have been shown to be accurate [13]. Moreover, Prokka can produce GFF3 files that can be easily passed to Roary to perform high-throughput Pangenome analyses. A sample command line for Prokka is:

```
prokka --outdir mydir --prefix mygenome genome_sequence.fa
```

where *genome_sequence.fa* is the fasta file reporting the nucleotide sequence of the genome of interest, *mydir* is a directory to be generated containing the output files, and *mygenome* is the prefix of the output files (whose extensions and descriptions are reported in Table 1).

3.3 Obtain and Reconcile Reference GENREs

As previously said, one or more curated reference GENREs are required to derive strain-specific reconstructions. Although GENREs in SBML format are usually provided as supplementary files in the corresponding research papers, they can also be found in repositories such as MetaNetX, MemoSys2, and BiGG, or under this github repository: https://github.com/opencobra/m_model_collection. The organisms with established biological knowledge (such as presence of physiological and expression data, well-defined biomass composition, manually curated metabolic reconstructions, etc.) should be prioritized during the choice of reference GENREs.

As previously said in Subheading 2, the use of multiple GENREs (e.g., those reconstructed by different groups) is preferable to avoid biases due to undersampling of the metabolic potential of the species under investigation. However, the GENREs should share the naming standard of metabolites, reactions, and compartments.

Table 1
The table reports the extension and description of the files generated with Prokka

Extension	Description
.gff	This is the master annotation in GFF3 format, containing both sequences and annotations. It can be viewed directly in Artemis or IGV.
.gbk	This is a standard Genbank file derived from the master .gff. If the input to prokka was a multi-FASTA, then this will be a multi-Genbank, with one record for each sequence.
.fna	Nucleotide FASTA file of the input contig sequences.
.faa	Protein FASTA file of the translated CDS sequences.
.ffn	Nucleotide FASTA file of all the annotated sequences, not just CDS.
.sqn	An ASN1 format "Sequin" file for submission to Genbank. It needs to be edited to set the correct taxonomy, authors, related publication etc.
.fsa	Nucleotide FASTA file of the input contig sequences, used by "tbl2asn" to create the .sqn file. It is mostly the same as the .fna file, but with extra Sequin tags in the sequence description lines.
.tbl	Feature Table file, used by "tbl2asn" to create the .sqn file.
.err	Unacceptable annotations—the NCBI discrepancy report.
.log	Contains all the output that Prokka produced during its run. This is a record of what settings you used, even if the --quiet option was enabled.
.txt	Statistics relating to the annotated features found.
Extension	Description
.gff	This is the master annotation in GFF3 format, containing both sequences and annotations. It can be viewed directly in Artemis or IGV.
.gbk	This is a standard Genbank file derived from the master .gff. If the input to prokka was a multi-FASTA, then this will be a multi-Genbank, with one record for each sequence.
.fna	Nucleotide FASTA file of the input contig sequences.
.faa	Protein FASTA file of the translated CDS sequences.
.ffn	Nucleotide FASTA file of all the annotated sequences, not just CDS.
.sqn	An ASN1 format "Sequin" file for submission to Genbank. It needs to be edited to set the correct taxonomy, authors, related publication, etc.
.fsa	Nucleotide FASTA file of the input contig sequences, used by "tbl2asn" to create the .sqn file. It is mostly the same as the .fna file, but with extra Sequin tags in the sequence description lines.
.tbl	Feature Table file, used by "tbl2asn" to create the .sqn file.
.err	Unacceptable annotations—the NCBI discrepancy report.
.log	Contains all the output that Prokka produced during its run. This is a record of what settings you used, even if the --quiet option was enabled.
.txt	Statistics relating to the annotated features found.

If not, the name should be converted to the same standard to achieve consistency (we will refer to this process as *reconciliation*). One way to reconcile two GENREs is to use the *MNXref namespace* from the MetaNetX database, which assign an identifier (*MNXid*) to compounds (basing on chemical structure and nomenclature), reactions, and compounds from KEGG, MetaCyc, ChEBI, BKM/BRENDA, BiGG, RAST, UniPathway, BioPath, HMDB, LipidMaps, and Reactome. The identifier starts with the MNX prefix, followed by a single letter indicating the type of object: R for reaction; M for metabolite; C for cellular compartment, and followed by an integer number, e.g., MNXR5678. In other words, different GEMREs with names following the standard of these databases can be reconciled by mapping their names to the MNXref namespace (directly on the SBML files, or within the Python/MATLAB environment). In order to perform this, the cross-ref files should be downloaded from MetaNetX (*see* Table 2) and used to replace the original names in the GENREs with the corresponding MNXids.

After the GENREs have been reconciled, the models should be manually curated to identify (and fix) eventual inconsistencies related to the reactions embedded in the reconstructions. For

Table 2
The cross-reference files for each model object type (Chemicals, Compartments, Reactions) are reported here along with a brief description of their contents

Object	File name	Link	Columns
Chemicals	chem_xref.tsv	http://www.metanetx.org/cgi-bin/mnxget/mnxref/chem_xref.tsv	(1) The identifier of a chemical compound in an external resource [XREF] (2) The corresponding identifier in the MNXref namespace [MNX_ID] (3) The evidence tag for the mapping [STRING] (4) The description given by the external resource [STRING]
Compartments	comp_xref.tsv	http://www.metanetx.org/cgi-bin/mnxget/mnxref/comp_xref.tsv	(1) The identifier of a subcellular compartment in an external resource [XREF] (2) The corresponding identifier in the MNXref namespace [MNX_ID] (3) The description given by the external resource [STRING]
Reactions	reac_xref.tsv	http://www.metanetx.org/cgi-bin/mnxget/mnxref/reac_xref.tsv	(1) The identifier of a biochemical reaction in an external resource [XREF] (2) The corresponding identifier in the MNXref namespace [MNX_ID]

instance, it is possible that the same reaction is present in two GENREs with (apparently) different stoichiometry (e.g., "$H_2 + 1/2O_2 \Leftrightarrow H_2O$" vs. "$2H_2 + O_2 \Leftrightarrow 2H_2O$"), compound name/protonation state (e.g., "malic acid" vs. "malate + H^+"), reaction equilibrium and its (reversible vs. irreversible), and gene-protein-reaction rule (e.g., "(Gene 1 AND Gene 2)" vs. "(Gene 1 OR Gene 2)").

At this point, the reference GENREs are ready to be merged in a single reconstruction (which will be referred to as *rGENRE* from now on) which will be used in the following steps.

3.4 Identification of Orthologous Genes and Pangenome Formulation

The identification of homologous genes originated from events of speciation (orthology), duplication (paralogy), or a combination of these (inparalogy) between two organisms has become a routine task in computational biology which can be performed with a number of tools. Recently, a systematic benchmark platform has been developed (available at http://orthology.benchmarkservice.org) and used to compare the performance of 15 well-established methods. The results show that the performance of MetaPhOrs, Inparanoid, Hieranoid, and OrthoInspector [22–26] is pretty strong in most of the benchmarks. Therefore, for the purpose of this protocol, we suggest the use of Inparanoid for the identification of orthologous genes, given its good accuracy and usability.

Regardless of the tool used, the coding sequences of the query dataset should be used to identify, for each strain, the orthologs of the genes present in the rGENRE.

To formulation of the Pangenome for a given taxa relies on the identification of groups of orthologous genes (OGs) between a set of representative genomes within the taxa. Although Inparanoid can, in principle, be used to accomplish this task, the number of pairwise comparisons that must be performed increases exponentially with the size of the considered dataset. Therefore, it is preferable to use software specifically optimized for the Pangenome construction, such as Get_homologues or Roary [27, 28]. The latter, in particular, is specifically suited for large-scale datasets since its running time is orders of magnitude lower than the other methods.

3.5 Construction of Strain-Specific GEMREs

As already mentioned, the rGENRE can be exploited to easily obtain metabolic models for a large number of closely related strains (*target strains*), which would otherwise be not feasible with the classical framework used for producing high-quality reconstructions.

For each strain, the following resources are required: (1) the rGENRE in a computable format (i.e., SBML); (2) a table reporting the ortholog pairs between the target strain and the genes present in rGENRE; (3) a platform to import, modify, and save reconstructions embedded in SBML files (e.g., CobraPy).

More details:

1. The rGENRE will allow creating a "*copy Model*" object for the target strain.
2. The table of ortholog pairs will be used to identify genes of rGENRE that are absent in the target strain. These genes (and the corresponding reactions) will be removed from the copy Model. The use of a function specific for this task (e.g., delete_model_genes) will ensure that the corresponding metabolic reactions are removed according to their Gene-Protein-Reaction (GPR) rule.
3. Optionally, the gene names of the copy Model can be changed according to the name of the corresponding orthologous genes/OGs.

3.6 Constraint-Based Modeling

Finally, the GENREs obtained for each target strain can be analyzed using Constraint-based methods (such as Flux Balance Analysis, FBA) to evaluate the phenotypical differences predicted in the in silico growth simulations.

3.6.1 Auxotrophies Prediction

Since different strains might colonize environments with diverse nutrient compositions, they will likely differ in the ability of synthesizing compounds normally present in their environments. From an evolutionary viewpoint, these strains will be subjected to selective pressure which will favor the loss of the genes involved in these anabolic pathways. The effect, at the phenotypical level, is the presence of strains requiring the addition of specific compounds to the minimal medium in order to grow (auxotrophies).

Translating this to the context of Constraint-based Modeling, auxotrophic strains are not able to grow on the minimal media defined for the reference strains (e.g., M9 medium for *E. coli* K12), therefore the nutrients necessary for the growth of these strains must be identified and supplied to their models. This is done by using *Gap-filling methods* (such as GrowMatch and Smiley [29, 30]), which use mixed-integer linear programming to identify the minimal subset of nutrients which should be added to each model to allow for growth. As soon as the minimal growth-supporting conditions are determined for each strain, a systematic investigation of their growth capabilities in different conditions can be carried out.

3.6.2 Strain-Specific Growth Capabilities

In this analysis the GEMREs will be tested to identify which compounds can be used as a carbon source. This will require, for each strain, to switch out of the "standard" carbon source present in the minimal medium with all the possible compounds to identify, with FBA, which can be used as a carbon source by the strain.

The same approach can be used to evaluate the differences in the compounds used as nitrogen, phosphorous and sulfur sources, and/or in different growth conditions (i.e., anaerobiosis).

4 Notes

This section reports practical examples of the methods previously described, divided into two workflow packages.

4.1 Workflow 1: Dataset Construction and Pangenome Analysis

In this workflow, we will illustrate how to: (1) systematically download and annotate all the genomes available in GenBank for a bacterial species and (2) use them to obtain a Pangenome with Roary. The examples here reported should be executed in a UNIX environment (e.g., Ubuntu).

Resources required

- The UNIX utilities *curl* and *wget*
- Prokka v1.11+ and its dependencies
- Inparanoid v4.1+ and its dependencies
- Roary v3.7+ and its dependencies

Procedure

The *wget* utility from the UNIX environment represents a powerful tool to automatically download data deposited in GenBank. The bash code reported below (*Script 1*), given a species (in this case *E. coli*), will create a new folder called *coliGenomes*, containing a sub-folder for each strain of that species with its sequenced genome present in GenBank. In this tutorial, we will limit our analysis to the first 10 *E. coli* strains. If you want to remove this limitation, change the comments in the indicated lines below.

Script 1

```
genbank=ftp://ftp.ncbi.nlm.nih.gov/genomes/genbank/bacteria/

targetTaxa=Escherichia_coli

outDir=coliGenomes

mkdir -p $outDir

genomes=`curl -ls

ftp://ftp.ncbi.nlm.nih.gov/genomes/genbank/bacteria/$targetTaxa/latest_assembly_

versions/ | head` # this line takes the first 10 E. coli genomes. Uncomment the line

below to download all the genomes

# genomes=`curl -ls

ftp://ftp.ncbi.nlm.nih.gov/genomes/genbank/bacteria/$targetTaxa/latest_assembly_ver

sions/`

for g in $genomes

do

    mkdir $outDir/$g

    wget

ftp://ftp.ncbi.nlm.nih.gov/genomes/genbank/bacteria/$targetTaxa/latest_assembly_ve

rsions/$g/* -P $outDir/$g

done
```

The same code can be used for any bacterial species, by changing the content of the *targetTaxa* variable with a string indicating the species of interest, formatted as it follows: *Genus_species.*

After all the sequences of the species have been downloaded (which might take up to several hours, depending on the number of genomes available for a species), the resulting directory should contain a sub-folder for each strain (named according to the GenBank assembly accession and name), containing files reported in Table 3 (named according to the pattern [assembly accession.version]_[assembly name]_content.[format]).

Table 3
The name of the files present in a GenBank directory are reported here, along with a brief description of their content

File name	Description
assembly_status.txt	A text file reporting the current status of this version of the assembly ("latest," "replaced," or "suppressed"). Any assembly anomalies are also reported
*_assembly_report.txt	Tab-delimited text file reporting the name, role, and sequence accession.version for objects in the assembly. The file header contains meta-data for the assembly including: assembly name, assembly accession.version, scientific name of the organism and its taxonomy ID, assembly submitter, and sequence release date
*_assembly_stats.txt	Tab-delimited text file reporting statistics for the assembly including: total length, ungapped length, contig & scaffold counts, contig-N50, scaffold-L50, scaffold-N50, scaffold-N75 & scaffold-N90
*_assembly_regions. txt	Provided for assemblies that include alternate or patch assembly units. Tab-delimited text file reporting the location of genomic regions and the alt/patch scaffolds placed within those regions
*_assembly_structure directory	Contains AGP files that define how component sequences are organized into scaffolds and/or chromosomes. Other files define how scaffolds and chromosomes are organized into non-nuclear and other assembly-units, and how any alternate or patch scaffolds are placed relative to the chromosomes. Only present if the assembly has an internal structure
*_cds_from_genomic. fna.gz	FASTA format of the nucleotide sequences corresponding to all CDS features annotated on the assembly, based on the genome sequence
*_feature_table.txt.gz	Tab-delimited text file reporting locations and attributes for a subset of annotated features. Included feature types are: gene, CDS, RNA (all types), operon, C/V/N/S_region, and V/D/J_segment. Replaces the .ptt and .rnt format files that were provided in the old genomes FTP directories
*_genomic.fna.gz	FASTA format of the genomic sequence(s) in the assembly. Repetitive sequences in eukaryotes are masked to lower-case. The genomic.fna.gz file includes all top-level sequences in the assembly (chromosomes, plasmids, organelles, unlocalized scaffolds, unplaced scaffolds, and any alternate loci or patch scaffolds). Scaffolds that are part of the chromosomes are not included because they are redundant with the chromosome sequences; sequences for these placed scaffolds are provided under the assembly_structure directory
*_genomic.gbff.gz	GenBank flat file format of the genomic sequence(s) in the assembly. This file includes both the genomic sequence and the CONTIG description (for CON records), hence, it replaces both the .gbk and .gbs format files that were provided in the old genomes FTP directories
*_genomic.gff.gz	Annotation of the genomic sequence(s) in Generic Feature Format Version 3 (GFF3). Additional information about NCBI's GFF files is available at ftp://ftp.ncbi.nlm.nih.gov/genomes/README_GFF3.txt
*_protein.faa.gz	FASTA format of the accessioned protein products annotated on the genome assembly

(continued)

Table 3
(continued)

File name	Description
*_protein.gpff.gz	GenPept format of the accessioned protein products annotated on the genome assembly
*_wgsmaster.gbff.gz	GenBank flat file format of the WGS master for the assembly (present only if a WGS master record exists for the sequences in the assembly)
annotation_hashes.txt	Tab-delimited text file reporting hash values for different aspects of the annotation data. The hashes are useful to monitor for when annotation has changed in a way that is significant for a particular use case and warrants downloading the updated records
md5checksums.txt	File checksums are provided for all data files in the directory

The asterisk symbol should be replaced with the following prefix: [assembly accession.version]_[assembly name], where "assembly accession.version" and "assembly_name" are unique identifiers for a given assembly

The amino acid sequence of the annotated protein products, reported in the **_protein.faa.gz* file, will be used to infer the homology relationships with the genes from other organisms.

Although unlikely, it is possible that some of the downloaded genomes should be annotated. Here, we report how to obtain the sequence of the coding elements using Prokka. The bash code reported below (*Script 2*) scans the directory tree previously obtained, identifies the strains lacking an annotation, decompresses their genome sequence files, and calls Prokka to annotate them.

Script 2

```
inpDir=coliGenomes # this name must match that of the "outDir" variable from the code

previously reported

accessions=`ls $inpDir`

for a in $accessions

do

  if [[ ! -e $inpDir/$a/$a"_protein.faa.gz" ]] # if accession $a lacks an annotation

    then genome=$inpDir/$a/$a"_genomic.fna.gz" # find the genome file

      gunzip -d $genome # unzip it

      unzippedGenome=`echo $genome | sed 's/\.gz//'`

      prefix=`echo $genome | grep -o GCA_[0-9]*\.[0-9] | head -1`

      prokka --force --prefix $prefix --locustag $prefix --outdir prokka_tmp

$unzippedGenome # use prokka with the accession id as locus tag

      cp prokka_tmp/* $inpDir/$a # move the output files in the right folder

      rm -r prokka_tmp # remove the temporary directory

  fi

done
```

Finally, gene sequences can be compared to identify the homology relationships between two genomes. We propose different approaches to carry out the homology analysis, depending on the complexity (number) of the comparisons that must be performed. There are basically two cases that can be considered.

In the first case, the task is to compare a number of genomes to a reference. The computational complexity of this analysis scales linearly with the number of genomes considered, making it feasible to use a tool such as Inparanoid.

Here, we report an example that should work on the directory structure previously. Assuming that Inparanoid has been successfully installed, and that a reference genome (*refGenome.faa*) is present, the following code should automatize the pairwise comparisons between each genome and the reference.

Script 3

```
inpDir=coliGenomes # this name must match that of the "outDir" variable from the code
previously reported
refGenome=refGenome.faa # this variable should contain the path to the reference
genome
inparanoidDir=./inparanoid-4.1 # this variable should contain the path to the inparanoid
directory
accessions=`ls $inpDir`
cp $inparanoidDir/* . # as said by the inparanoid authors, the directory content shuld
be copied in the project's directory
for a in $accessions
do
  echo "I'm comparing $a accession with the reference genome $refGenome"
  a_faa=$inpDir/$a/$a"_protein.faa.gz"
  if [[ ! -e $a_faa ]]; then a_faa=$inpDir/$a/`ls $inpDir/$a | grep .faa`; fi # if there is
no *_protein.faa.gz get whatever *.faa is there
  if [[ -n `file $a_faa | grep "gzip compressed data"` ]] # if the *.faa file is zipped, unzip
it
    then gunzip -d $a_faa
    a_faa=$inpDir/$a/`ls $inpDir/$a | grep .faa`
  fi
  cp $a_faa .
  a_faa=`ls -rt | tail -1`
  perl inparanoid.pl $a_faa $refGenome
  rm $a_faa
done
```

After the script has finished running, the working directory should contain four output files for each genome that has been compared to the reference. In more detail, the patterns of file naming and their descriptions are listed below (*NameQ* is the FASTA file name of the query genome, while *NameR* is the FASTA file name of the reference genome).

1. **Output.[NameQ]-[NameR]**: reports a summary for the genome pair comparison, as well as the composition of each orthologs group and the score.
2. **orthologs. [NameQ]-[NameR].html**: basically contains the same information as the previous file, encoded in a html format.
3. **table.[NameQ]-[NameR]**: is a tab-separated text file reporting the id of the orthologous group (OrtoID), the corresponding score (Score), the embedded genes from the query genomes (OrtoA), and those from the reference (OrtoB).
4. **sqltable.[NameQ]-[NameR]**: basically contains the same information as the previous file in a format compatible with relational database systems (e.g., SQL).

In the second case, we want to compute the Pangenome of a species. To achieve this, a large number of gene comparisons, roughly the square of the number of genomes, must be performed. Since this task requires a considerable computational effort, a method optimized for large-scale analyses, like Roary, should be preferred. This tool takes as input the assembly annotation of different strains and computes the Pangenome of a species using a limited quantity of time and memory. As reported by the authors, 128 strains can be analyzed in 1 h using 1 Gb of RAM, representing an incredible advancement over the other methods which would take weeks to analyze the same inputs. However, the authors also mention the fact that Roary is not intended for the analysis of diverse datasets, therefore this tool is not suited for comparisons at the genus-level (and above).

To work, Roary requires the annotation files in GFF3 format with unique locus tags, including the nucleotide sequences at the end of the files. Since the GFF3 files from NCBI do not fulfill this requirement, the user can, either, download the GenBank files plus nucleotide sequence and convert them in compliant GFF3 files, or re-annotate the genome with Prokka, which produces files compatible with Roary. After obtaining valid input files, the user should select the proper options to tweak the software, according to the analysis that must be performed (we strongly suggest the reader to check the software manual). For example, Roary can align and concatenate the core genes to produce a phylogeny with the *–e* option.

The example reported below (*Script 4*) is intended to work with the directory structure obtained with *Script 1*.

Script 4

```
inpDir=coliGenomes # this name must match that of the "outDir" variable from Script 1

accessions=`ls $inpDir`

# STEP 1: annotate all the genomes with Prokka

for a in $accessions

do

  rm $inpDir/$a/*.gff # remove any possible gff previously present

  genome=$inpDir/$a/$a"_genomic.fna.gz" # find the genome file

  gunzip -d $genome # unzip it

  unzippedGenome=`echo $genome | sed 's/\.gz//'`

  prefix=`echo $genome | grep -o GCA_[0-9]*\.[0-9] | head -1`

  prokka --force --prefix $prefix --locustag $prefix --outdir prokka_tmp

$unzippedGenome # use prokka with the accession id as locus tag

  cp prokka_tmp/* $inpDir/$a # move the output files in the right folder

  rm -r prokka_tmp # remove the temporary directory

done

# STEP 2: use Roary

  roaryInputs=`ls $inpDir/*/*.gff`

  extraOptions="" # insert any extra option for roary here

  roaryCl="roary $extraOptions $roaryInputs"

  echo "Roary is being used with this command line: "$roaryCl

  eval roaryCl
```

4.2 Workflow 2: Strain-Specific GENREs Analyses

In this workflow, we look at generating strain-specific metabolic models of the diverse *E. coli* species to examine strain-specific auxotrophies and highlight adaptation to diverse environments.

Monk et al. constructed genome-scale models of metabolism (GEMs) for all *E. coli* strains with fully sequenced genomes [11]. The GEMs were used to predict strain-specific auxotrophies. Several of the strains modeled were predicted to be auxotrophic for vitamins niacin (vitamin B3), thiamin (vitamin B1), or folate (vitamin B9). Other strains modeled have lost biosynthetic pathways for essential amino acids methionine, tryptophan, or leucine. The GEMs were also used to systematically analyze growth capabilities in over 650 different growth-supporting environments. The results show that unique strain-specific metabolic capabilities correspond to pathotypes and environmental niches.

In this workflow, we guide the readers step by step through a few of the analyses in the study by Monk et al. [11]. We aim to:

1. Build strain-specific metabolic models based on a well-curated reference model
2. Predict strain-specific auxotrophies for each modeled strain using gap filling methods
3. Systematically investigate growth capabilities in different nutritional environments

This workflow demonstrates using a highly curated reference metabolic model to create strain-specific models of 55 closely related strains using genome sequences alone. The models are highly predictive and shed light on pathogenesis and adaptation to specific micro-environments.

Resources required

- Python v2.7
- COBRApy (https://github.com/opencobra/cobratoolbox)
- Gurobi (or another linear programming solver compatible with COBRApy) (http://www.gurobi.com/)
- iJO1366 [31]
- List orthologous genes to *E. coli* K-12 MG1655 (included)
- *E. coli* K-12 MG1655 metabolic model iJO1366 (included)
- Pandas v0.7+

Procedure

Installation of the COBRApy has been described elsewhere [21]. Following installation and initialization of the toolbox, analyses relevant to this study can be conducted, as follows. Commands entered into the python prompt are signified with "In [xx]:", and "#" signifies a comment in python code. Details of the function calls in python can be obtained from python tutorials and documentation.

Determining Orthologous Genes to *E. coli K-12 MG1655*

There are many methods for determining orthologous genes between *E. coli* K-12 MG1655 and the other 55 selected strains of *E. coli* as discussed above. For this example we used the RAST corresponding genes tool that combines Bi-directional best BLAST hits to find best hits between two strains of *E. coli*.

Once a tool has been used to call orthologs one must select a threshold for similarity. We picked 80% similarity as our cutoff—any gene that was under this threshold was determined to be missing. For ease in following this tutorial we have included a matrix of orthologous genes for each *E. coli* strain compared to *E. coli* K-12 MG1655.

The matrix of orthologous genes (provided in the file *ortho_matrix_pegs.csv*) reports the sequence similarity of the orthologs from other strains (columns) with the *E. coli* K-12 MG1655 genes (rows). *See* Fig. 1 for an example of this matrix. The values are percentage identity values (PID) determined using the BLAST software package to determine the percentage of amino acid

	316385.7	469008.1	481805.6	566546.3	216592.3	444450.8	198215.6	574521.7	198214.7	344609.1
b0007	100	99.37	99.37	99.16	98.95	98.95	99.13	99.16	99.13	98.74
b3821	100	99.31	100	100	99.65	100	100	98.96	100	99.65
b2519	100	99.72	99.86	99.08	98.94	98.41	pseudo	98.02	97.42	99.08
b2518	100	100	100	100	100	100	100	98.6	100	100
b3824	100	100	100	100	100	100	100	100	100	100
b3825	100	100	100	100	99.71	99.71	99.71	99.71	99.71	100
b3350	100	99.83	99.67	99.83	99.5	pseudo	97.02	99	97.02	99.67
b3458	100	100	100	100	99.73	99.73	99.73	98.92	99.73	100
b3451	100	100	99.29	99.29	98.93	99.29	99.29	98.22	99.29	99.29
b3829	100	99.47	98.8	98.94	99.2	99.2	98.94	99.2	98.94	99.2
b3453	100	100	100	100	99.77	100	99.09	99.75	99.09	99.54
b3452	100	99.66	100	99.66	100	98.64	99.66	99.49	99.66	99.66
b3455	100	100	100	99.61	100	100	100	99.22	100	100
b2920	99.8	100	99.59	99.59	100	99.39	99.4	-	99.11	99.24
b2515	100	100	100	100	99.73	100	99.46	100	99.46	99.19
b3359	100	100	99.75	99.26	99.51	99.51	98.77	98.52	98.77	98.52
b3823	100	99.03	99.51	100	99.03	100	99.51	98.95	99.51	99.51
b1488	100	100	99.48	100	97.41	99.48	97.3	-	97.3	98.45
b1539	100	100	99.19	100	100	99.6	99.19	99.19	98.79	99.6
b4072	100	100	100	100	99.55	100	100	99.1	100	99.55
b0003	100	99.68	99.68	99.68	99.68	99.35	99.35	99.68	99.35	99.68
b4070	100	100	100	100	99.79	99.79	99.79	99.77	99.79	100
b4071	98.37	100	99.47	100	99.47	100	98.4	99.42	98.4	99.47
b4077	100	99.54	99.77	99.54	99.77	99.77	99.54	99.77	99.54	99.77
b2877	100	98.44	99.48	96.88	96.88	95.83	-	98.44		-
b0002	100	99.58	99.44	99.44	99.3	99.58	99.72	100	99.58	99.44

Fig. 1 Example of the matrix of orthologous genes (provided in the file *ortho_matrix_pegs.csv*), which reports the sequence similarity of the orthologs from other strains (columns) with the *E. coli* K-12 MG1655 genes (rows)

```
In [12]: data = pd.read_csv('ortho_matrix_pegs.csv',index_col=0)
         data[data.columns[:5]][:10]

Out[12]:        316385.7 469008.14 481805.6 566546.3 216592.3
         b0007     100.0     99.37    99.37    99.16    98.95
         b3821     100.0     99.31    100.0    100.0    99.65
         b2519     100.0     99.72    99.86    99.08    98.94
         b2518     100.0     100.0    100.0    100.0    100.0
         b3824     100.0     100.0    100.0    100.0    100.0
         b3825     100.0     100.0    100.0    100.0    99.71
         b3350     100.0     99.83    99.67    99.83     99.5
         b3458     100.0     100.0    100.0    100.0    99.73
         b3451     100.0     100.0    99.29    99.29    98.93
         b3829     100.0     99.47     98.8    98.94     99.2
```

Fig. 2 Command to import the matrix of orthologous genes as a python dataframe

similarity between each gene. Genes with lower than 80% PID are marked with "-." Genes that are annotated as pseudogenes are marked "pseudo."

The matrix of orthologous genes can be read into python (Fig. 2):

Next, we can use this matrix of orthologs to build strain-specific models by removing reactions from the reference *E. coli* K-12 MG1655 model.

Build strain-specific models

Using the orthologous genes we can remove those genes expected to be missing from each strain of *E. coli.* We do this by using the delete_model_genes module, part of pyCobra. This module uses the gene-reaction-relation rule to determine if a reaction should be removed from the model based on a missing gene. We loop through each strain and use delete_model_genes to remove genes and reactions from the reference model—thereby constructing strain-specific models of each strain (Fig. 3).

Predict strain-specific auxotrophies

After removing reactions with no corresponding encoded gene in each strain-specific model, many of them will no longer be able to grow on M9 minimal media (the default model inputs). This is because genes that are essential in this condition may not be present in all strains of *E. coli.* Thus the strains are auxotrophic for different nutrients.

To discover which nutrients must be added to each strain-specific model we can use gap-filling methods to determine the minimal set of nutrients that must be added to each model to allow them to grow again in M9+glucose minimal media.

```
# loop through each E. coli strain in the dataset:
for org in data.columns:
    # missing genes are marked with a '-':
    missing_genes = data[data[org] == '-'][org].index.tolist()
    # pseudogenes are marked as 'pseudo':
    pseudogenes = data[data[org] == 'pseudo'][org].index.tolist()

    # combine all missing genes and pseudogenes into a list:
    genes_to_remove = missing_genes + pseudogenes

    # load the base E. coli K-12 MG1655 model
    model = load_matlab_model('ijo1366.mat')
    # change the models name to correspond with its strain:
    model.name = strains[float(org)]
    # chage the model's id to be the strain's taxonomy id:
    model.id = org

    # if there are more than 0 genes to delete:
    if len(genes_to_remove) > 0:
        # use the delete_model_genes command to \
        # delete the genes from the model:
        delete_model_genes(model, genes_to_remove)

    # if there are no genes to delete make empty \
    # list of deleted genes and reactions:
    else:
        model._trimmed_genes = []
        model._trimmed_reactions = []
    # print the model name and how many genes & reactions were removed
    print "%s (%s): deleted: %i genes, %i reactions"% \
        (strains[float(org)], org,len(model._trimmed_genes), \
         len(model._trimmed_reactions))
    # save the new strain-specific model:
    dump(model, open('models/'+org+'.pickle','wb'))
```

```
Escherichia coli str. K-12 substr. DH10B (316385.7): deleted: 30 genes, 23 reactions
Escherichia coli BL21(DE3) AM946981 (469008.14): deleted: 58 genes, 52 reactions
Escherichia coli ATCC 8739 (481805.6): deleted: 45 genes, 41 reactions
Escherichia coli W (566546.3): deleted: 54 genes, 46 reactions
Escherichia coli 042 (216592.3): deleted: 63 genes, 60 reactions
Escherichia coli O157:H7 str. EC4115 (444450.8): deleted: 96 genes, 80 reactions
Shigella flexneri 2a str. 2457T (198215.6): deleted: 150 genes, 168 reactions
Escherichia coli O127:H6 str. E2348/69 (574521.7): deleted: 116 genes, 97 reactions
Shigella flexneri 2a str. 301 (198214.7): deleted: 143 genes, 162 reactions
```

Fig. 3 List of python commands to create strain-specific models

Gap-filling methods use mixed integer linear programming to optimize growth of the model subject to minimizing the addition of new components to the in-silico media. Therefore, these methods can predict the minimal set of metabolites that must be added to each model to allow them to grow again.

We can loop through each strain-specific model to test if they are capable of producing biomass. If they are not we can run the gap filling algorithm on them (*see* Fig. 4).

```
In [58]: models = glob('models/*.pickle') # get all of the strain specific models
         solutions = pd.DataFrame()
         for m in models:
             model = load(open(m))
             model.optimize() # optimize the model for growth
             if model.solution.f < 0.00001: # if the model doesn't grow, gap fill it!
                 print "gap filling %s"%strains[float(model.id)]
                 # run grow match algorithm with only exchange reactions
                 sol = growMatch(model, dm_rxns=0, ex_rxns=1,iterations=1)
                 # open the exchange reactions determined using grow match:
                 for rxn in sol[0]:
                     rxn = rxn.id.replace('_reverse','')
                     model.reactions[model.reactions.index(rxn)].lower_bound = -1
         #save the results:
         gaps = load(open('gap_fill.pickle','r'))
         gaps = gaps.fillna(0)

         gap filling Shigella boydii CDC 3083-94
         gap filling Shigella sonnei Ss046
         gap filling Escherichia coli str. K-12 substr. DH10B
         gap filling Shigella flexneri 2a str. 301
         gap filling Shigella boydii Sb227
         gap filling Escherichia coli CFT073
```

Fig. 4 List of python commands to perform gap-filling on strains not growing on the defined medium

Table 4
The additional growth requirements are reported for each strain (rows)

Strain	Glycit	Indole	leu	metsox	nac	thf	thm
Shigella flexneri 2a str. 301	0	0	0	1	0	1	0
Escherichia coli CFT073	0	1	0	0	0	0	0
Shigella boydii Sb227	1	0	0	0	0	0	1
Shigella sonnei Ss046	0	0	0	0	1	1	0
Escherichia coli st. K-12 substr. DH10B	0	0	1	0	0	0	0
Shigella boydii CDC 3083-94	1	0	0	0	0	0	1
Escherichia coli IAI39	0	0	0	0	0	1	0
Shigella flexneri 2002017	0	0	0	0	0	1	0

The compounds (columns) have a value of 0 if they are not essential for the growth of the corresponding strain, whereas a value of 1 means that the strain is auxotroph for that compound

The results of this gap filling are presented in Table 4. We can see that 8 of the 55 strains are predicted to be auxotrophic for different amino acids and vitamins. Note that gap-filling results are non-unique and growth may be restored with different combinations of metabolites.

```
In [*]: models = glob('models/gap_filled/*.pickle')
        all_growth = {}
        print 'testing all C sources for:'
        for m in models:
            model = load(open(m))
            print "\t", strains[float(model.id)]
            print strains[float(model.id)]
            rxn = model.reactions[model.reactions.index('EX_glc_LPAREN_e_RPAREN_')]
            rxn.lower_bound = 0
            growth = {}
            for r in model.reactions:
                if r.id.startswith('EX'):
                    lb = r.lower_bound
                    r.lower_bound = -10
                    model.optimize()
                    if model.solution.f < 0.0001 or model.solution.f == np.nan:
                        model.solution.f =0
                    growth[r.name.split('exchange')[0]] = model.solution.f
                    r.lower_bound = lb
            all_growth[strains[float(model.id)]] = growth
        all_growth = pd.DataFrame(all_growth).transpose()
        all_growth.to_csv('all_growth_predictions.csv')

        testing all C sources for:
                Escherichia coli S88
        Escherichia coli S88
                Shigella boydii CDC 3083-94
```

Fig. 5 List of python commands to verify the growth capabilities of each strain

Determine strain-specific growth capabilities

Now that we have determined growth-supporting minimal media conditions for all of our 55 models we are ready to systematically investigate their growth in different nutrient conditions. We will switch out the base carbon source in our minimal media to determine growth-supporting carbon sources for each strain. We do this by removing glucose from our minimal media and then looping through all of the possible inputs to determine if they support growth for each strain (Fig. 5).

These same steps can be performed again for different nutrient conditions, for example all potential carbon sources could be profiled in an anaerobic environment. Further, all potential nitrogen, phosphorous, and sulfur sources could also be examined. We leave this exercise to the interested reader.

We can cluster these results to look for patterns in the growth capabilities and to ask questions regarding whether growth capabilities correspond to any given properties of these strains, for example their phylogeny, lifestyle, or pathogenic properties (Fig. 6).

The most immediate observation is that the *Shigella* strains have lost the capability to grow in many different nutrient environments compared to the other strains of *E. coli*. Note that these

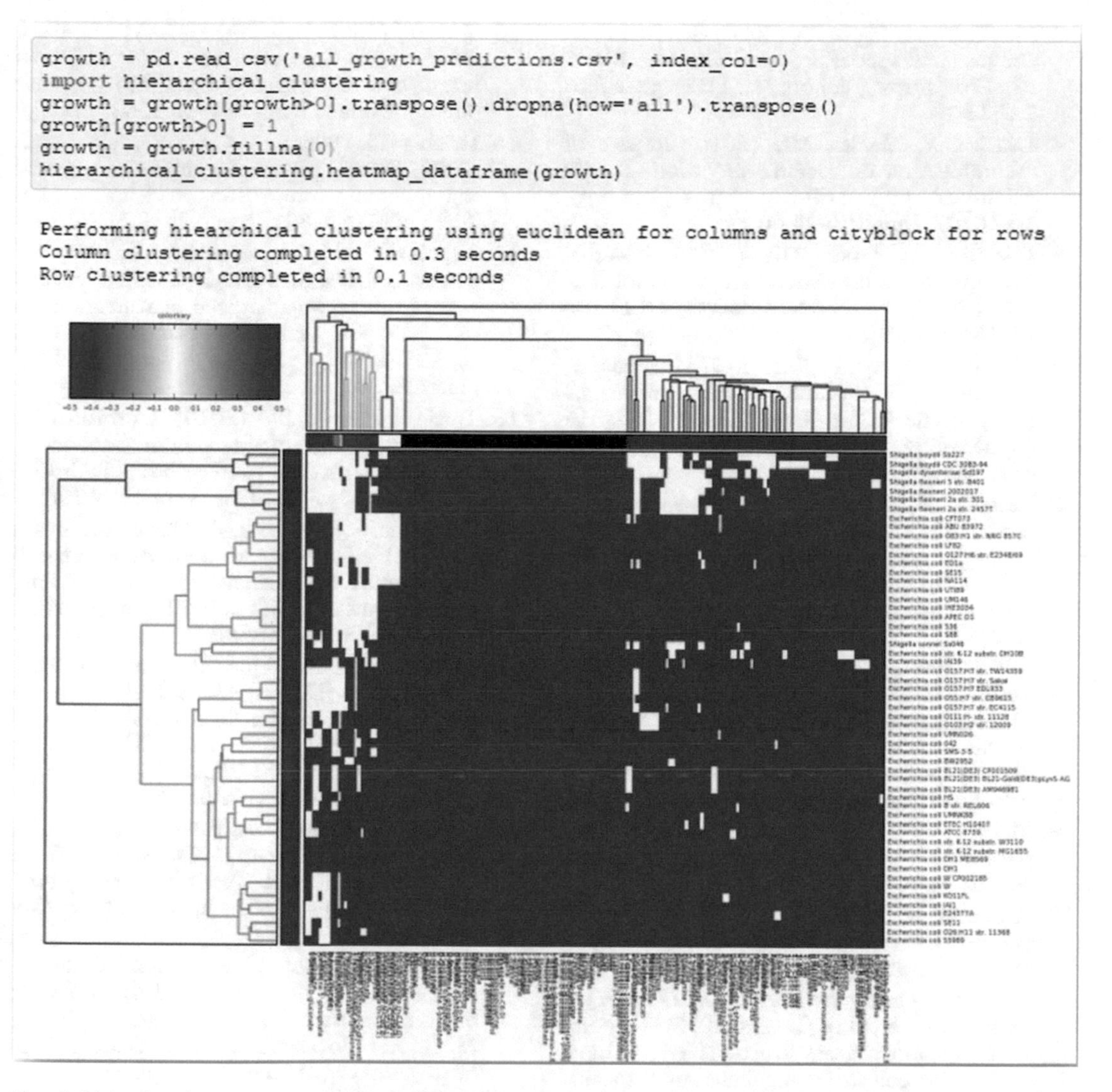

growth = pd.read_csv('all_growth_predictions.csv', index_col=0)
import hierarchical_clustering
growth = growth[growth>0].transpose().dropna(how='all').transpose()
growth[growth>0] = 1
growth = growth.fillna(0)
hierarchical_clustering.heatmap_dataframe(growth)
Performing hiearchical clustering using euclidean for columns and cityblock for rows
Column clustering completed in 0.3 seconds
Row clustering completed in 0.1 seconds

Fig. 6 List of python commands to visualize the growth capabilities of different strains. In the heatmap reported, the rows represent the strains while the columns represent the different compounds used for testing the growth capabilities. Each heatmap cell colored in red represent a compound in which a given strain is able to grow, whereas a white color stands for no growth

results are preliminary because the models have not undergone any comprehensive manual curation—but they are a good start and interesting results are already apparent!

References

1. Brooks AN, Turkarslan S, Beer KD, Lo FY, Baliga NS (2011) Adaptation of cells to new environments. Wiley Interdiscip Rev Syst Biol Med 3(5):544–561. https://doi.org/10.1002/wsbm.136

2. Telford JL (2008) Bacterial genome variability and its impact on vaccine design. Cell Host Microbe 3(6):408–416. https://doi.org/10.1016/j.chom.2008.05.004

3. Doolittle WF, Papke RT (2006) Genomics and the bacterial species problem. Genome Biol 7 (9):116. https://doi.org/10.1186/gb-2006-7-9-116
4. Didelot X, Maiden MC (2010) Impact of recombination on bacterial evolution. Trends Microbiol 18(7):315–322. https://doi.org/10.1016/j.tim.2010.04.002
5. Darmon E, Leach DR (2014) Bacterial genome instability. Microbiol Mol Biol Rev 78(1):1–39. https://doi.org/10.1128/MMBR.00035-13
6. Medini D, Donati C, Tettelin H, Masignani V, Rappuoli R (2005) The microbial pan-genome. Curr Opin Genet Dev 15 (6):589–594. https://doi.org/10.1016/j.gde.2005.09.006
7. Tettelin H, Masignani V, Cieslewicz MJ, Donati C, Medini D, Ward NL et al (2005) Genome analysis of multiple pathogenic isolates of Streptococcus agalactiae: implications for the microbial "pan-genome". Proc Natl Acad Sci U S A 102(39):13950–13955. https://doi.org/10.1073/pnas.0506758102.
8. Tettelin H, Medini D, Donati C, Masignani V (2006) Towards a universal group B Streptococcus vaccine using multistrain genome analysis. Expert Rev Vaccines 5(5):687–694. https://doi.org/10.1586/14760584.5.5.687
9. Mora M, Veggi D, Santini L, Pizza M, Rappuoli R (2003) Reverse vaccinology. Drug Discov Today 8(10):459–464
10. Rappuoli R (2001) Reverse vaccinology, a genome-based approach to vaccine development. Vaccine 19(17-19):2688–2691
11. Monk JM, Charusanti P, Aziz RK, Lerman JA, Premyodhin N, Orth JD et al (2013) Genome-scale metabolic reconstructions of multiple Escherichia coli strains highlight strain-specific adaptations to nutritional environments. Proc Natl Acad Sci U S A 110(50):20338–20343. https://doi.org/10.1073/pnas.1307797110.
12. Bosi E, Monk JM, Aziz RK, Fondi M, Nizet V, Palsson BO (2016) Comparative genome-scale modelling of Staphylococcus aureus strains identifies strain-specific metabolic capabilities linked to pathogenicity. Proc Natl Acad Sci U S A 113(26):E3801–E3809. https://doi.org/10.1073/pnas.1523199113.
13. Seemann T (2014) Prokka: rapid prokaryotic genome annotation. Bioinformatics 30 (14):2068–2069. https://doi.org/10.1093/bioinformatics/btu153
14. Altschul SF, Gish W, Miller W, Myers EW, Lipman DJ (1990) Basic local alignment search tool. J Mol Biol 215(3):403–410. https://doi.org/10.1016/S0022-2836(05)80360-2
15. Kent WJ (2002) BLAT--the BLAST-like alignment tool. Genome Res 12(4):656–664. https://doi.org/10.1101/gr.229202. Article published online before March 2002.
16. Delcher AL, Phillippy A, Carlton J, Salzberg SL (2002) Fast algorithms for large-scale genome alignment and comparison. Nucleic Acids Res 30(11):2478–2483
17. Altenhoff AM, Boeckmann B, Capella-Gutierrez S, Dalquen DA, DeLuca T, Forslund K et al (2016) Standardized benchmarking in the quest for orthologs. Nat Methods 13 (5):425–430. https://doi.org/10.1038/nmeth.3830
18. Thiele I, Palsson BO (2010) A protocol for generating a high-quality genome-scale metabolic reconstruction. Nat Protoc 5(1):93–121. https://doi.org/10.1038/nprot.2009.203
19. Dandekar T, Fieselmann A, Majeed S, Ahmed Z (2014) Software applications toward quantitative metabolic flux analysis and modeling. Brief Bioinform 15(1):91–107. https://doi.org/10.1093/bib/bbs065
20. Schellenberger J, Que R, Fleming RM, Thiele I, Orth JD, Feist AM et al (2011) Quantitative prediction of cellular metabolism with constraint-based models: the COBRA Toolbox v2.0. Nat Protoc 6(9):1290–1307. https://doi.org/10.1038/nprot.2011.308
21. Ebrahim A, Lerman JA, Palsson BO, Hyduke DR (2013) COBRApy: COnstraints-Based Reconstruction and Analysis for Python. BMC Syst Biol 7:74. https://doi.org/10.1186/1752-0509-7-74
22. Berglund AC, Sjolund E, Ostlund G, Sonnhammer EL (2008) InParanoid 6: eukaryotic ortholog clusters with inparalogs. Nucleic Acids Res 36(Database issue):D263–D266. https://doi.org/10.1093/nar/gkm1020
23. Kaduk M, Riegler C, Lemp O, Sonnhammer EL (2016) HieranoiDB: a database of orthologs inferred by Hieranoid. Nucleic Acids Res 45:D687. https://doi.org/10.1093/nar/gkw923.
24. Linard B, Allot A, Schneider R, Morel C, Ripp R, Bigler M et al (2015) OrthoInspector 2.0: software and database updates. Bioinformatics 31(3):447–448. https://doi.org/10.1093/bioinformatics/btu642
25. O'Brien KP, Remm M, Sonnhammer EL (2005) Inparanoid: a comprehensive database of eukaryotic orthologs. Nucleic Acids Res 33 (Database issue):D476–D480. https://doi.org/10.1093/nar/gki107
26. Pryszcz LP, Huerta-Cepas J, Gabaldon T (2011) MetaPhOrs: orthology and paralogy predictions from multiple phylogenetic

evidence using a consistency-based confidence score. Nucleic Acids Res 39(5):e32. https://doi.org/10.1093/nar/gkq953

27. Contreras-Moreira B, Vinuesa P (2013) GET_HOMOLOGUES, a versatile software package for scalable and robust microbial pan-genome analysis. Appl Environ Microbiol 79 (24):7696–7701. https://doi.org/10.1128/AEM.02411-13
28. Page AJ, Cummins CA, Hunt M, Wong VK, Reuter S, Holden MT et al (2015) Roary: rapid large-scale prokaryote pan genome analysis. Bioinformatics 31(22):3691–3693. https://doi.org/10.1093/bioinformatics/btv421
29. Kumar VS, Maranas CD (2009) GrowMatch: an automated method for reconciling in silico/in vivo growth predictions. PLoS Comput Biol 5(3):e1000308. https://doi.org/10.1371/journal.pcbi.1000308
30. Reed JL, Patel TR, Chen KH, Joyce AR, Applebee MK, Herring CD et al (2006) Systems approach to refining genome annotation. Proc Natl Acad Sci U S A 103 (46):17480–17484. https://doi.org/10.1073/pnas.0603364103.
31. Orth JD, Conrad TM, Na J, Lerman JA, Nam H, Feist AM et al (2011) A comprehensive genome-scale reconstruction of Escherichia coli metabolism--2011. Mol Syst Biol 7:535. https://doi.org/10.1038/msb.2011.65

Chapter 8

Template-Assisted Metabolic Reconstruction and Assembly of Hybrid Bacterial Models

Tiziano Vignolini, Alessio Mengoni, and Marco Fondi

Abstract

Intraspecific genomic exchanges happen frequently between bacteria living in the same natural environment and can also be performed artificially in the laboratory for basic research or genetic/metabolic engineering purposes. In silico metabolic reconstruction and simulation of the metabolism of the hybrid strains that result from these processes can be used to predict the phenotypic outcome of such genomic rearrangements; this can be especially helpful as a designing tool in the purview of synthetic biology. However, reconstructing the metabolism of a bacterium with a hybrid genome through in silico approaches is not a trivial task, as it requires taking into account the complex relationships existing between metabolic genes and how they change (or remain unchanged) when new genes are placed in a different genomic context. Furthermore, in order to "mix" the metabolic models of different bacterial strains one needs at least two different metabolic models to begin with, and reconstructing a genome-scale model from the ground up is a challenging task itself, requiring an intensive manual effort and a great deal of information. In this chapter, we propose two general protocols to address the aforementioned issues of: (1) quickly generating strain-specific metabolic models, given the relevant genomic sequence and an already existing, high-quality metabolic model of a different strain belonging to the same species, and (2) reconstructing the metabolic model of a hybrid strain containing genomic elements from two different parental strains.

Key words Metabolic modeling, Template-based reconstruction, Gene homology, Synthetic biology, Horizontal gene transfer, Genome evolution

1 Introduction

Genome manipulation and genome-wide rewiring of cellular networks are important issues in synthetic biology as well as in evolutionary genomics. The possibility of creating large synthetic gene constructs, which can allow completely modifying many cellular metabolic pathways is now mature [1]. Quantitative predictions of the corresponding changes in the overall metabolic network have to be performed in order to optimize genes constructs and well integrate their activities into the cellular metabolic context. Moreover, especially in prokaryotes, genome evolution is mainly represented by the pangenome model [2], where a core set of

Marco Fondi (ed.), *Metabolic Network Reconstruction and Modeling: Methods and Protocols*, Methods in Molecular Biology, vol. 1716, https://doi.org/10.1007/978-1-4939-7528-0_8,

common (shared) genes is associated with a plethora of many different dispensable (exclusive to some strains only) genes, which often constitute a large fraction of the genome and have horizontal origins [3]. The dispensable genome fraction could be functionally (transcriptionally, metabolically) related to the core fraction [4]. Consequently, it could be not trivial to predict the possible phenotypic (metabolic) changes and the adaptive evolutionary values of the dispensable gene fractions present in prokaryotic species.

In this context, metabolic modeling has become a helpful tool for the rational design of genomic constructs and for predicting the phenotypic outcome of engineered (or native) organisms [5–7] or even communities of organisms [8–10]. Reconstructing accurate models of existing organisms is still a complex and time-consuming task [11] although, over the years, metabolic models of many model organisms have been published [12–16]. This has important drawbacks since, nowadays, the process of building a metabolic model may not necessarily start from scratch. Indeed, when trying to reconstruct the metabolism of an organism evolutionarily related to another whose metabolism has already been successfully formalized into a model, a great deal of information can be taken from the existing one. By using the widely accepted assumption of functional conservation between orthologous genes harbored by closely related organisms [17], the process of building strain-specific metabolic models can be largely simplified and automated. The same principle can be used to rapidly reconstruct the metabolism of "hybrid" organisms, whose genome derives from different sources due to horizontal gene transfer (HGT) events or synthetic biology.

In this protocol we propose two general approaches (and related code) to address two main procedures: (1) the generation of strain-specific metabolic models using both genomic information and an already existing, high-quality metabolic model of a different bacterial strain belonging to the same species; (2) the reconstruction of a hybrid metabolic model embedding metabolic capabilities encoded by genomic elements (i.e., operons, gene clusters, replicons) from two different parental strains.

2 Materials

In this section, we describe what is needed to: (1) rapidly reconstruct a metabolic model, given a pre-existing reference model of a closely related strain, and (2) reconstruct the metabolic model of a hybrid strain containing genomic elements from two different parental strains.

2.1 Reference Model

The quality of the reconstructed model is strongly dependent on the quality of the available reference model. Thus, the reference model should be of high quality, capable of predicting metabolic phenotypes in strong agreement with experimental data, and possibly having already passed peer review. The reference model should be encoded as an .xml file in SBML format.

2.2 Genome Sequence

The genomic sequence is the primary source of information used in drafting a metabolic model. Therefore, a FASTA file containing either the complete genome, a set of contigs, or a list of the coding sequences of the strain whose metabolism is intended to be reconstructed has to be available.

2.3 Online Reconstruction Tools

Several tools exist for drafting the metabolic model of an organism, given the relevant list of metabolic genes. Here, we propose a combination of genome annotation and model reconstruction pipelines that are easy to use and readily accessible via the KBase online platform (https://kbase.us/).

2.3.1 KBase

KBase (the Department of Energy Systems Biology Knowledgebase, https://kbase.us/) is an online platform which integrates open-source software and data banks to be used in systems biology. KBase provides several tools to reconstruct and analyze biological models, all within a unified graphical interface (a so-called narrative interface), so that users do not need to access them from multiple sources, making it easier to develop and run complex workflows.

2.3.2 RAST

RAST (Rapid Annotation using Subsystems Technology, http://rast.nmpdr.org/) is a freely accessible, fully automated server for the annotation of bacterial and archaeal genomes. It requires the users to upload either a complete genome or a set of contigs in FASTA format, and then performs an extensive process of gene calling and functional annotation of protein coding genes, rRNAs, and tRNAs within a few hours [18].

2.3.3 ModelSEED

ModelSEED (http://www.theseed.org/) is an open-source, freely accessible web application for generating, optimizing, and analyzing genome-scale metabolic models [19]. It utilizes information derived from the functional annotation of a genome to generate a list of metabolic reactions, and to map them into a network in a fully automated manner. Developed by the same authors, it works harmonically with the RAST server and makes use of the SEED metabolic database [20].

2.4 Software for the Identification of Orthologous Genes: InParanoid

InParanoid is an automated software for the identification of orthologous genes [21]. Homologous genes that originate from a speciation event are called orthologs [22], and homologs that originate from a duplication event within the same genome are called paralogs [23]. If the duplication event was followed by a speciation, the paralogous genes are called outparalogs and can be found in different species. If instead the duplication event happened after the speciation, the paralogous genes are called inparalogs. Since a pair of outparalogs may have acquired a more diversified function than an inparalog pair, it is useful to distinguish between the two. Furthermore, clustering together the inparalogs of a genome allows for a more accurate identification of both one-to-one and many-to-many orthology cases between different genomes. The InParanoid software is capable of performing both inparalog clusterization and orthology groups identification by an extensive utilization of BLAST (Basic Local Alignment Search Tool) [24].

2.5 Modeling Framework

There are several available tools for performing constraint-based metabolic modeling. SBML-formatted models are generally recognized by these tools, which can be used to manipulate them and to perform simulations. Among them, the COBRA toolbox [25] is probably the most widely adopted, and the one we are going to deal with in this chapter. The version of this package we are going to refer to is to be used within the Matlab (The Mathworks Inc.) numerical computation and visualization environment (a Python version, COBRApy [26], is also available). Also needed are libSMBL (an API library for manipulation of systems biology models) [27] and a Linear Programming (LP) solver supported by the COBRA Toolbox as, for example, glpk (http://www.gnu.org/software/glpk) or Gurobi (Gurobi Optimization, http://www.gurobi.com). Please refer to specific literature/manuals/websites for information on the installation and configuration of these tools (*see* **Note 2**).

3 Methods

3.1 Reconstructing a Strain-Specific Metabolic Model

In this section, we are going to describe a protocol for rapidly generating the metabolic model of a bacterial strain which is closely related to another strain, the relevant metabolic model of which is already available (template model). Much of the information that would normally be acquired through several months of literature analysis and manual model curation, such as accurate gene annotations or metabolic pathway information needed to fill gaps in the network, will be derived directly from the template model. In particular, we are going to show how comparative genomics information regarding different strains can be exploited to correctly re-annotate and expand an automatically drafted model, by assuming that orthologous genes perform the same function in closely related organisms.

3.1.1 Obtain a List of Orthologous Genes

First, it is necessary to derive a list of the orthologous genes shared between the organism used as a template and the one that has to be reconstructed. This list shall be a text file organized in two columns, containing a list of gene IDs from the template genome in the first column, and a list of the relevant orthologs in the organism to be reconstructed in the second column, the orthologous couples being on the same row (*see* **Note 3**).

To generate this list, the InParanoid software can be used to perform an orthology search between the two genomes, to be loaded onto the program in the form of two FASTA files containing the translated gene sequences with their relevant IDs.

The output of an InParanoid search consists in a number of files, containing information regarding the identified orthology groups and the relevant sequence similarity scores. Assuming the orthology search has been performed between two FASTA files named "template.fasta" and "draft.fasta"—the first being the template organism's set of proteins and the second being the set of proteins of the organism whose metabolism is to be reconstructed—the needed output file is a text file named "table.template.fasta-draft.fasta," which will be located in the InParanoid folder. The third column of this file consists in a list of gene IDs from the template genome, and each row of the fourth column contains the closest ortholog for the corresponding gene in column 3. These two columns need to be extracted and written into a separate text file for further use. In a Unix environment, this can be done quickly from the command line:

```
cut -f '3' table.template.fasta-draft.fasta | cut -d \  -f 1 >
column1
cut -f '4' table.template.fasta-draft.fasta | cut -d \  -f 1 >
column2
pr -mt column1 column2 > orthologList
```

An example of the list obtained through this process is provided below:

GenomeA_*geneA*	GenomeB_*geneA*
GenomeA_*geneB*	GenomeB_*geneB*
GenomeA_*geneC*	GenomeB_*geneC*

3.1.2 Obtain a Draft Metabolic Model

Once the orthologs list has been prepared, the process of drafting and then refining the model of the strain under analysis can start. Obtaining a draft model is a simple and straightforward operation, and can be easily carried out via the previously described online services accessible from the KBase narrative interface. This can be done by uploading on the interface a FASTA file containing the complete genome (or a set of contigs) of the organism to be reconstructed, and then using the "Annotate Microbial Contigs" method from the RAST application, which uses the input FASTA file to generate a "Genome" file with functional annotations.

The next step is using this file as input for the "Build Metabolic Model" method from the ModelSEED application to automatically generate a draft metabolic reconstruction in SBML format. This preliminary model won't likely be functional, incapable of producing biomass, and will be far from an accurate reproduction of the real organism.

3.1.3 Expand and Correct the Draft Metabolic Model

Having obtained both the draft metabolic model and the list of orthologous genes, and having the template model at hand, everything is in order to begin expanding the newly generated draft metabolic model by filling any gaps it might contain, and correcting the inevitable annotation errors that result from the automatic gene calling process. The procedure to be described in this paragraph works by comparing the functional annotations of the genes in the draft model with the ones in the template model by means of the orthologs list, and correcting any discrepancy by re-annotating the genes in the draft model (*see* **Note 1**). It generally happens that many metabolic genes are not inserted in a model during the drafting process, so that draft reconstructions usually are some hundreds of genes smaller than the manually curated model of a closely related organism. In the following procedure, the orthologs list is used to expand the draft model, by adding metabolic genes found both in the orthologs list and in the template model.

It is important to note that, since this procedure relies heavily on the accuracy of the model used as a template to expand and correct the draft model, any error eventually present in the former is going to be propagated to the latter.

3.1.4 Load Both Models and the Orthologs list in the Modeling Framework

To carry out the procedure several tools from the COBRA toolbox will be needed, and the operations will be performed in a Matlab environment. Thus, we will assume that Matlab has been properly installed on the workstation, as well as the COBRA toolbox with all the relevant dependencies and a valid LP solver.

First of all, start Matlab and initialize the COBRA toolbox with

```
initCobraToolbox
```

Assuming that the SBML-formatted models have been saved in the working directory, and they are called "template_model.xml" and "draft_model.xml," import them into the Matlab workspace with

```
template = readCbModel('template_model.xml');
draft = readCbModel('draft_model.xml');
```

Then import the orthologs list. On the "Home" tab, in the "Variable" section, select "Import Data." Make sure the column separator is set as "tab" and both columns are flagged as "text," and then import the data as a 2-column cell array.

3.1.5 Begin the Enhancing Process

Once all of the necessary variables are loaded in the workspace, the process of correcting and expanding the draft model can begin. A Matlab function called "enhanceDraftModel" that does the job is freely downloadable from the GitHub repository https://github.com/TVignolini/replicon-swap. The function calls, among several functions already present in the standard Matlab packages and in the COBRA toolbox, two more Matlab functions called "findCommonRxns" and "simplifyRules" which are downloadable from the same GitHub repository, and thus they should be downloaded together (*see* **Note 4**). The "simplifyRules" function relies also on another script written in Python language called "booleanSimplifier.py" which is downloadable from the same directory and requires having Python installed on the workstation in order to work.

In the following paragraphs, we will explain the main operations performed in this pipeline.

First, all the metabolic reactions of the two input models need to be divided into several subsets, based on their gene association rules and on their presence in one or both of the models:

Subset 1: reactions that are present in the template model and that, according to their gene association rules, can be carried out by genes which have an ortholog in the draft model.

This subset is identified by knocking out, in the template model, every gene that does not have an ortholog in the draft model, and then selecting the reactions that are still functional (i.e., they still carry a flux value greater than zero). Assuming that the ortholog list has been called "dictionary," the Matlab code for this operation is the following:

```
[~,~,constrRxnNames,~] = deleteModelGenes (template, setdiff
(template.genes, dictionary (:,1)));
subset1 = setdiff (template.rxns, constrRxnNames);
```

where *subset1* is a vector of cells containing the IDs of the reactions in subset 1.

Subset 2: reactions belonging to subset 1 which are not present in the draft (target) model.

To isolate this subset it is sufficient to subtract all of the reactions in the draft model from the *subset1* list:

```
subset2 = setdiff (subset1, draft.rxns);
```

Subset 3: reactions belonging to subset 1 which are present also in the draft model.

To isolate this subset it is sufficient to intersect *subset1* with the reactions in the draft model.

```
subset3 = intersect (subset1, draft.rxns);
```

Subset 4: reactions that are present in the draft model alone and which, according to their gene association rules, can be carried out by genes that have an ortholog in the template model.

This subset is isolated in a fashion similar to subset 1, with the only difference that the gene knockouts are performed in the draft model and that every reaction also present in the template model is removed from subset 4.

```
[~,~,constrRxnNames,~] = deleteModelGenes (draft, setdiff
(draft.genes, dictionary (:,2)));
subset1 = setdiff (setdiff (draft.rxns, constrRxnNames), tem-
plate.rxns);
```

Subset 5: reactions that are present in the draft model and do not fall in any of the previous subsets. This subset does not include exchange and biomass reactions.

This subset is isolated by subtracting all of the other subsets from the reactions in the draft model.

```
subset5 = setdiff (draft.rxns, cat (1, subset1, subset4,
EXlist, biomass_draft));
```

where *EXlist* is a vector of cells containing the IDs of the exchange reactions in the draft model, and *biomass_draft* is a cell or a vector of cells containing the ID(s) of the biomass reaction(s) in the draft model. Subsets 2 and 3 are not included, as they are both part of subset 1.

3.1.6 Define the Gene Association Rules for Each Subset

Now that the reactions in both the models have been categorized, two additional tasks need to be performed: (1) determine which of them are to be kept/discarded/added in the "enhanced" model, and (2) define the relevant gene association rules. Each of the previously defined subsets will be approached in a different way. Since subset 1 is the union of subsets 2 and 3, only these last two will be independently considered in the next steps.

Subset 2: as in our previous definition, this subset contains all of the reactions present in the template model (but not in the draft target model) which can be carried out by genes that have an ortholog in the draft model. Thus, by assuming that orthologous genes perform the same function in different organisms, the same reactions should be added in the "enhanced" model. The gene association rules for these reactions should be the same they have in the template model; but, since it is possible that these rules contain some other genes (linked with an "or" logical statement to orthologous genes, and as such nonessential for the reaction to carry flux) which are not part of any orthology group, these genes should be removed from the relevant rules (which are Boolean expressions of varying complexity) by simplifying them.

This can be achieved by renaming every gene that needs to be removed from the rules as "False" and then running the "booleanSimplifier" Python script from within Matlab:

```
% Rename genes as "False"
nonOrthologs_template = setdiff (template.genes, dictionary
(:,1));
for n = 1:length (nonOrthologs_template)
     subset2_rules (:) = strrep (subset2_rules (:), nonOrtho-
logs_template (n), 'False');
end
% Simplify the rules
for n = 1:length (subset2_rules)
    if ~isempty (subset2_rules {n})
        tmpRule = sprintf ('''%s''', subset2_rules {n});
        [~,tmpRule] = system (sprintf ('python booleanSimpli-
fier.py %s', tmpRule));
        subset2_rules (n) = cellstr (tmpRule);
    end
end
```

where *subset2_rules* is a cell array containing the rules associated with every reaction in subset 2 (in the same order).

Subset 3: as said, this subset contains all of the reactions present in both the template and the draft models which can also be carried out by orthologous genes alone. To determine the gene association rules of this subset the information coming from the template model should be prioritized (since it is assumed to be more accurate), but also consider any unique gene present in the draft model that has been predicted to be associated with that reaction. Thus, the rules for this particular subset of reactions should be composed as follows:

1. Take the relevant rules from the template model and simplify them by removing every non-ortholog gene.
2. Take the relevant rules from the draft model and simplify them by removing every ortholog gene, thus leaving only the genes that are unique to the draft model.
3. Concatenate the previous two Boolean expressions with an "or" statement.

In Matlab code:

```
% Rename genes as "False"
for n = 1:length (nonOrthologs_template)
     subset3_rules_template (:) = strrep (subset3_rules_tem-
```

```
plate (:), nonOrthologs_template (n), 'False');
end
orthologs_draft = intersect (draft.genes, dictionary (:,2));
for n = 1:length (orthologs_draft)
    subset3_rules_draft (:) = strrep (subset3_rules_draft (:),
orthologs_draft (n), 'False');
end
% Simplify the rules
for n = 1:length (subset3_rules_template)
    if ~isempty (subset3_rules_template {n})
          tmpRule = sprintf ('''%s''', subset3_rules_template
{n});
        [~,tmpRule] = system (sprintf ('python booleanSimpli-
fier.py %s', tmpRule));
        subset3_rules_template (n) = cellstr (tmpRule);
    end
end
for n = 1:length (subset3_rules_draft)
    if ~isempty (subset3_rules_draft {n})
        tmpRule = sprintf ('''%s''', subset3_rules_draft {n});
        [~,tmpRule] = system (sprintf ('python booleanSimpli-
fier.py %s', tmpRule));
        subset3_rules_draft (n) = cellstr (tmpRule);
    end
end
% Concatenate the boolean expressions
subset3_rules = cell (length (subset3), 1);
for n = 1:length (subset3_rules_template)
    if ~isempty (subset3_rules_draft {n})
            subset3_rules_template (n) = strcat ({'('},
subset3_rules_template (n), {' or '}, subset3_rules_draft
(n), {')'});
    end
end
```

where *subset3_rules_template* is a cell array containing the rules associated with every reaction in subset 3 (in the same order), as they are in the template model. The same goes for *subset3_rules_-draft*, which contains the rules for the same reactions, but taken from the draft model.

Subset 4: this subset contains all of the reactions present in the draft model (but not in the template model) which can be carried out by genes that find an ortholog in the template model. As the template model is taken as reference and it does not contain this subset of reactions even though the relevant genes are present, these reactions are assumed to be the product of errors during the automated genome annotation process. As such, they should be excluded from the final model.

Subset 5: this subset contains all of the reactions in the draft model which do not belong in any of the previously defined categories (i.e., they are present in both models but with completely different gene association rules, or they are solely present in the draft model). These reactions should maintain the original rules they had in the draft model, as no additional information can be obtained from the template.

3.1.7 Generate the Final "Enhanced" Model

After all of the gene association rules have been defined, the function "createModel" can be used to generate a model containing the reactions in subsets 2, 3, and 5 and the relevant rules. However, this function needs some additional input: first and foremost, the chemical equations for every reaction, defining the identity and the stoichiometry of the metabolites involved, as well as the directionality of the reaction. Moreover, to obtain a more complete model, the upper and lower flux bounds for every reaction and also their complete names need to be specified. These further sets of information (reaction equations, upper and lower bounds, complete names of the reactions) should be taken from the template model for reactions in subsets 2 and 3.

Since some of the reactions in subset 5 are exclusive to the draft model, they should be checked on a case-by-case basis. If the reaction is present also in the template model, the equation, flux bounds, and complete name should be taken from that source; otherwise, they must be taken from the draft model.

To do this in Matlab:

```
for n = 1:length (subset5)
if ismember (subset5 (n), template.rxns)
        tmpID = strmatch (subset5 (n), template.rxns,'exact');
        subset5_equations (n) = template.rxnEquations (tmpID);
        subset5_lb {n} = template.lb (tmpID);
        subset5_ub {n} = template.ub (tmpID);
        subset5_names (n) = template.rxnNames (tmpID);
    else tmpID = strmatch (subset5 (n), draft.rxns,'exact');
        subset5_equations (n) = draft.rxnEquations (tmpID);
        subset5_lb {n} = draft.lb (tmpID);
        subset5_ub {n} = draft.ub (tmpID);
        subset5_names (n) = draft.rxnNames (tmpID);
    end
```

After having obtained all of the needed input, the final model can be generated by concatenating the different subsets in single lists (being careful to maintain the same order in all of the lists) and then feeding them to the "createModel" function:

```
% Create single lists
enhanced_reactions = cat (1, subset2, subset3, subset5);
```

```
enhanced_names = cat (1, subset2_names, subset3_names, sub-
set5_names);
enhanced_equations = cat (1, subset2_equations, subset3_equa-
tions, subset5_equations);
enhanced_lb = cat (1, subset2_lb, subset3_lb, subset5_lb);
enhanced_ub = cat (1, subset2_ub, subset3_ub, subset5_ub);
enhanced_rules = cat (1, subset2_rules, subset3_rules, subse-
t5_rules);
% Generate model
enhancedModel = createModel (enhanced_reactions, enhanced_-
names, enhanced_equations, '', enhanced_lb, enhanced_ub, '',
enhanced_rules, '', '');
```

3.2 Reconstructing the Metabolic Model of a Hybrid Strain

In this section, we are going to describe a protocol for reconstructing a metabolic model containing genes from two different existing models (which will be referred to as "parental models") of closely related bacterial strains. This process relies on the previously described assumption of functional superposition between orthologous genes (*see* Subheading 3.1), and starts from here to predict the metabolic interactions that might arise when metabolic genes from different organisms are placed in the same genomic context. A Matlab function called "repliconSwap" that performs this operation is freely downloadable from the GitHub repository https://github.com/TVignolini/replicon-swap, and—like the function described in the previous section—it relies on the two ulterior Matlab functions "findCommonRxns" and "simplifyModel" and on the Python script "booleanSimplifier.py," all of which are downloadable from the same repository. In the following paragraphs, we will explain the main operations performed in this pipeline.

The following steps are performed in a Matlab environment and rely on the COBRA toolbox, with the same requirements described for the procedure in Subheading 3.1.

3.2.1 Obtain a List of Orthologous Genes

First, a list analogous to the one described in Subheading 3.1.1 is needed, containing the orthologous genes between the two parental organisms. Please refer to Subheading 3.1.1 on how to obtain such list.

When the ortholog list is ready, import it in the Matlab workspace as a 2-column cell array of text. We will refer to the resulting Matlab variable as *dictionary* from now on.

3.2.2 Define the Content of the Hybrid Genome

The next step is defining the set of genes that are going to make up the genome of the hybrid model. In the procedure we are going to describe, the generation of a hybrid model is carried out as a "transplantation" process: a set of genes is removed from a "receiving" model (assumed to be the model providing the largest contribution, in terms of number of genes, to the hybrid model), which are replaced by another set of genes coming from a "donor" model. Thus, the genes to be removed from the receiving model need to be defined together with the genes coming from the donor model that are to be transplanted.

For this purpose, it is required to write down two separate lists containing the two sets of genes, and import them into the Matlab workspace as two cell vectors of text. From now on, we will refer to the corresponding Matlab variables as *removedGenes* (for the list of genes to be removed from the receiving model) and *transplantedGenes* (for those coming from the donor model).

3.2.3 Generate a Hybrid Model

Assuming that the two models have been imported into the workspace with the names of *receiving* and *donor*, the reconstruction of the hybrid model can begin. The process involves resolving three main instances: (1) figuring out which reactions in the receiving model are turned off after removing the intended portion of DNA, (2) figuring out which of them are turned back on after inserting the external set of genes (functional complementation effects), and (3) which reactions that were previously only present in the donor model may end up in the hybrid.

First, every gene in the portion of the genome that is going to be removed from the receiving model has to be replaced with the corresponding ortholog (if any) in the donor model.

In Matlab:

```
for n=1:length (dictionary)
    if ismember (dictionary (n,1), removedGenes)
    receiving.grRules = strrep (receiving.grRules, dictionary
{n,1}, dictionary {n,2});
    receiving.genes  =  strrep  (receiving.genes,dictionary
{n,1}, dictionary {n,2});
end
```

In the donor model, consider the portion of genome that is not going to be transplanted. Replace every gene in this subset with the corresponding ortholog harbored by the not-to-be-removed portion of the receiving genome (if any ortholog is present).

In Matlab language:

```
for n=1:length (dictionary)
    if ~ismember (dictionary (n,2), transplantedGenes)
    donor.grRules = strrep (donor.grRules, dictionary {n,2},
dictionary {n,1});
    donor.genes  =  strrep  (donor.genes,dictionary  {n,2},
dictionary {n,1});
end
```

In both models, if any gene is going to end up in the hybrid model along with an ortholog, replace it in the relevant gene association rule(s) as a Boolean expression consisting in the two genes separated by an “or” statement.

In Matlab:

```
for n=1:length (dictionary)
     if (ismember (dictionary (n,2), transplantedGenes) &&
~ismember (dictionary (n,1), removedGenes))
   receiving.grRules = strrep (receiving.grRules, dictionary
{n,1}, strcat ({'('}, dictionary (n,1), {' or '}, dictionary
(n,2), {')'}));
   donor.grRules = strrep (donor.grRules, dictionary {n,2},
strcat ({'('}, dictionary (n,2), {' or '}, dictionary (n,1),
{')'}));
end
```

Now that this preliminary remodeling of the parental models' gene association rules has been performed, a receiving and a donor submodels that will then be merged in the final hybrid model are required.

First, a knockout of all of the genes that are going to be removed from the receiving model is to be performed in order to check which of the reactions are still functional. Any gene that has been predicted to have an ortholog in the incoming set of transplanted genes should not be removed—but, since such genes should already have been replaced with the corresponding orthologs in the previous step, it will be sufficient to knock out all of the genes in the *removedGenes* list.

```
[~, ~, constrRxnNames, ~] = deleteModelGenes (receiving,
intersect (receiving.genes, removedGenes));
```

And then extract a receiving submodel with the “extractSubNetwork” function.

```
subReceiving = extractSubNetwork (receiving, setdiff (receiv-
ing.rxns, constrRxnNames));
```

The same approach should be applied to the donor model to extract a donor submodel (being careful to delete all the exchange and biomass reactions beforehand, in order to prevent duplicates).

```
[~, ~, constrRxnNames, ~] = deleteModelGenes (donor, setdiff
(donor.genes, transplantedGenes));
subDonor = extractSubNetwork (donor, setdiff (donor.rxns, cat
(1, constrRxnNames, EXlist, biomass));
```

where *EXlist* is a vector of cells containing the IDs of the exchange reactions in the donor model, and *biomass* is a cell or a vector of cells containing the ID(s) of the biomass reaction(s) in the donor model.

Now the gene association rules of the two submodels have to be double-checked in order to ensure that they do not contain any gene that won't be present in the final hybrid model. This can be done by renaming them as "False" and using the "booleanSimplifier.py" script to simplify the relevant rules. Please note that, automated gapfilling pipelines often insert genes named "Unknown" in the metabolic reconstructions. Care must be taken in order not to eliminate such genes during the aforementioned procedure.

```
unwantedGenesReceiving = setdiff (intersect (receiving.genes,
cat (1, removedGenes, setdiff (donor.genes, transplant-
edGenes))), 'Unknown');
unwantedGenesDonor = setdiff (intersect (donor.genes, cat
(1, removedGenes, setdiff (donor.genes, transplantedGenes))),
'Unknown');
for n=1:length (unwantedGenesReceiving)
      subReceiving.grRules = strrep (subReceiving.grRules,
unwantedGenesReceiving (n), 'False');
end
for n=1:length (unwantedGenesDonor)
     subDonor.grRules = strrep (subDonor.grRules, unwantedGe-
nesDonor (n), 'False');
end
subReceiving = simplifyModel (subReceiving);
subDonor = simplifyModel (subDonor);
```

Since the two parental submodels do not contain a "reaction equations" field, it must be added before merging them.

```
subReceiving_rxnEquations = printRxnFormula (subReceiving);
subReceiving = setfield (subReceiving, 'rxnEquations', subRe-
ceiving_rxnEquations);
subDonor_rxnEquations = printRxnFormula (subDonor);
subDonor = setfield (subDonor, 'rxnEquations', subDonor_rxnE-
quations);
```

Now the hybrid model from the two parental submodels can be assembled: the hybrid model will contain the sum of the reactions contained in the other 2. The two submodels will be likely composed of mostly unique reactions but will also embed several duplicates (the number varying with the extent of the genomic transfer that is being simulated): the reactions that are solely present in one of the two parental submodels will of course maintain the relevant gene association rules, but a different case is that of duplicate reactions. First, these duplicates need to be identified by comparing the stoichiometric matrix in the two submodels, by the use of the "findCommonRxns" function.

```
doubleRxns = findCommonRxns (subReceiving, subDonor);
```

This function will return a 2-column cell array containing the IDs of the duplicate reactions. If the two parental models have been obtained via the "enhanceDraftModel" function discussed in Subheading 3.1.4, the two columns of the array will be identical. Otherwise, there may be some discrepancies between the IDs of identical reactions in the two submodels. In any case, it is useful to isolate any name discrepancy as follows:

```
wrongRxns = setdiff (doubleRxns (:,2), doubleRxns (:,1));
```

Here, we arbitrarily choose the receiving submodel as a reference for the reaction IDs that will end up in the hybrid model for the duplicate reactions. The gene association rules for these reactions will be discussed later. Here, we simply assign them the rules as they are in the receiving model.

The next step involves creating single lists of reaction IDs, reaction equations, upper and lower flux bounds, and gene association rules to use as input to build the hybrid model:

```
hybridRxns = unique (setdiff (cat (1, subReceiving.rxns,
subDonor.rxns), wrongRxns));
for n = 1:length (hybridRxns)
    if ismember (hybridRxns (n), subReceiving.rxns)
         tmpID = strmatch (hybridRxns (n), subReceiving.rxns,
'exact');
         hybridRxnEquations (n,1) = subReceiving.rxnEquations
(tmpID);
        hybridRxnNames (n,1) = subReceiving.rxnNames (tmpID);
        hybridLB (n,1) = subReceiving.lb (tmpID);
        hybridUB (n,1) = subReceiving.ub (tmpID);
        hybridGrRules (n,1) = subReceiving.grRules (tmpID);
    elseif ismember (hybridRxns (n), subDonor.rxns)
          tmpID = strmatch (hybridRxns (n), subDonor.rxns,
'exact');
          hybridRxnEquations (n,1) = subDonor.rxnEquations
(tmpID);
        hybridRxnNames (n,1) = subDonor.rxnNames (tmpID);
        hybridLB (n,1) = subDonor.lb (tmpID);
        hybridUB (n,1) = subDonor.ub (tmpID);
        hybridGrRules (n,1) = subDonor.grRules (tmpID);
    end
end
```

Now gene association rules for the duplicate reactions can be taken into consideration. They should be composed as follows: take the rules from the donor submodel and simplify them by removing every ortholog gene—thus leaving only the genes that are unique to the donor model—and concatenate them with the previously defined rules (contained in the *hybridGrRules* vector) with an "or" statement.

In Matlab:

```
donorOrthologs = intersect (dictionary (:,2), subDonor.genes);
tmpRules = subDonor.grRules;
for n = 1:length (donorOrthologs)
    tmpRules = strrep (tmpRules, donorOrthologs (n), 'False');
end
for n = 1:length (doubleRxns (:,1))
    tmpID1 = strmatch (doubleRxns (n,1), hybridRxns, 'exact');
      tmpID2 = strmatch (doubleRxns (n,2), subDonor.rxns,
'exact');
    tmpRule = tmpRules {tmpID2};
    if ~isempty (tmpRule)
         % Add brackets before and after the string prior to
semplification
        tmpRule = regexprep (tmpRule, '(.*)', '\($1\)');
        tmpRule = sprintf ('''%s''', tmpRule);
        [~, tmpRule] = system (sprintf ('python booleanSimpli-
fier %s', tmpRule));
        % Remove extra brackets
        tmpRule = regexprep (tmpRule, '\((.*)\)', '$1');
        hybridGrRules (tmpID1) = strcat ({'('}, hybridGrRules
(tmpID1), {' or '}, tmpRule, {')'});
    end
end
```

Finally, the "createModel" function can be fed with all of the previously defined input to generate the hybrid model:

```
[hybrid] = createModel (hybridRxns, hybridRxnNames, hybridRx-
nEquations, '', hybridLB, hybridUB, '', hybridGrRules, '',
'');
```

4 Notes

1. If the latest version of the COBRA toolbox for Matlab is being used, make sure that the models have their metabolites compartmentalized with a bracketed tag (e.g., [c] and [e]) and not

with an underscore (e.g., _c0 and _e0)—as is sometimes the case—since the latter won't be recognized as a proper compartment tag by the relevant COBRA functions.

2. In both of the pipelines, there are steps in which some gene association rules are parsed by Matlab and some strings containing gene IDs are changed with different strings. Problems may arise when using models in which gene IDs have different lengths, since portions of longer gene ID strings can be mistaken for shorter IDs which fit within them (e.g., the first part of a gene called "abc1234" could be mistaken for a gene called "abc123" when parsed by Matlab), which can lead to errors. To prevent this, particular characters can be added to each gene ID, such as an alphabetic character to mark the end of genes IDs which end in numbers. Make sure that these characters are removed at the end of the pipeline.

   Here is a sample Matlab script that adds an "X" at the end of every gene ID contained in the .genes and .grRules fields of a model (in this exemplified case, all of the gene IDs begin with the characters "id"):

```
% Add an "X" at the end of every gene ID
for n=1:length (model.genes)
 model.genes {n} = regexprep (model.genes {n}, '(?<gene>.
*)', '$<gene>X');
end
for n=1:length (model.grRules)
 model.grRules {n} = regexprep (model.grRules {n}, '(id
[^ )]*)', '$1X');
end
for n=1:length (model.grRules)
 model.grRules {n} = strrep (model.grRules {n}, 'Unknown
[^X]', 'UnknownX');
end
```

   Change/add the same characters to the gene IDs present in the orthologs list under consideration, since the correspondence between the IDs in the list and those in the models must be exact.

3. When trying to use either of the two pipelines described in this chapter, please make sure that the two starting models considered (draft and template for the first pipeline, or the two parental models for the second one) are as homogeneous as possible regarding the names given to reactions and metabolites. Discrepancies between reaction or metabolite IDs or the

suffix used to define the cellular compartment of each metabolite (e.g., [c] and [e] versus _c0 and _e0) could result in unwanted reaction superpositions and/or possible gaps in the models that come out of the pipeline.

4. To check for any discrepancy in the reaction IDs of two different models, the "findCommonRxns" Matlab function mentioned in Subheading 3.1.4 can be used, as it compares the stoichiometric matrices in the two models to identify the equivalent reactions. The function returns a 2-column cell array containing a list of reaction IDs from the first model in the first column and the IDs of equivalent reactions from the second model in the second column (arranged in the same order). Assuming that the 2-column cell array has been named *reactions*, name discrepancies can be identified by using the setdiff function:

```
discrepancies = setdiff (reactions (:,1), reactions
(:,2));
```

References

1. Tang N, Ma S, Tian J (2013) New tools for cost-effective DNA synthesis. In: Synthetic biology. Academic Press, New York, NY, pp 3–21
2. Medini D, Donati C, Tettelin H, Masignani V, Rappuoli R (2005) The microbial pan-genome. Curr Opin Genet Dev 15:589–594. https://doi.org/10.1016/j.gde.2005.09.006
3. Wiedenbeck J, Cohan FM (2011) Origins of bacterial diversity through horizontal genetic transfer and adaptation to new ecological niches. FEMS Microbiol Rev 35:957–976. https://doi.org/10.1111/j.1574-6976.2011.00292.x
4. Galardini M, Brilli M, Spini G, Rossi M, Roncaglia B, Bani A, Chiancianesi M, Moretto M, Engelen K, Bacci G, Pini F, Biondi EG, Bazzicalupo M, Mengoni A (2015) Evolution of intra-specific regulatory networks in a multipartite bacterial genome. PLoS Comput Biol 11:e1004478. https://doi.org/10.1371/journal.pcbi.1004478
5. Wiechert W (2002) Modeling and simulation: tools for metabolic engineering. J Biotechnol 94:37–63. https://doi.org/10.1016/S0168-1656(01)00418-7
6. Zhang C, Hua Q (2016) Applications of genome-scale metabolic models in biotechnology and systems medicine. Front Physiol 6:1–8. https://doi.org/10.3389/fphys.2015.00413
7. Zou W, Liu L, Zhang J, Yang H, Zhou M, Hua Q, Chen J (2012) Reconstruction and analysis of a genome-scale metabolic model of the vitamin C producing industrial strain Ketogulonicigenium vulgare WSH-001. J Biotechnol 161:42–48. https://doi.org/10.1016/j.jbiotec.2012.05.015
8. Biggs MB, Medlock GL, Kolling GL, Papin JA (2015) Metabolic network modeling of microbial communities. Wiley Interdiscip Rev Syst Biol Med 7:317–334. https://doi.org/10.1002/wsbm.1308
9. Henry CS, Bernstein HC, Weisenhorn P, Taylor RC, Lee JY, Zucker J, Song HS (2016) Microbial community metabolic modeling: a community data-driven network reconstruction. J Cell Physiol 231:2339–2345. https://doi.org/10.1002/jcp.25428
10. Sung J, Hale V, Merkel AC, Kim PJ, Chia N (2016) Metabolic modeling with Big Data and the gut microbiome. Appl Transl Genomics 10:10. https://doi.org/10.1016/j.atg.2016.02.001
11. Thiele I, Palsson BØ (2010) A protocol for generating a high-quality genome-scale metabolic reconstruction. Nat Protoc 5:93–121. https://doi.org/10.1038/nprot.2009.203

12. Fleischmann RD, Adams MD, White O, Clayton RA, Kirkness EF, Kerlavage AR, Bult CJ, Tomb JF, Dougherty BA, Merrick JM, Mckenney K, Sutton G, Fitzhugh W, Fields C, Gocayne JD, Scott J, Shirley R, Liu LI, Glodek A, Kelley JM, Weidman JF, Phillips CA, Spriggs T, Hedblom E, Cotton MD, Utterback TR, Hanna MC, Nguyen DT, Saudek DM, Brandon RC, Fine LD, Fritchman JL, Fuhrmann JL, Geoghagen NSM, Gnehm CL, Mcdonald LA, Small KV, Fraser CM, Smith HO, Venter JC (1995) Whole-genome random sequencing and assembly of haemophilus-influenzae Rd. Science 269:496–512. https://doi.org/10.1126/science.7542800
13. The C. elegans Sequencing Consortium (1998) Genome sequence of the nematode *C. elegans*: a platform for investigating biology. Science 282:2012–2018. https://doi.org/10.1126/science.282.5396.2012
14. Edwards JS, Palsson BØ (2000) The Escherichia coli MG1655 in silico metabolic genotype: its definition, characteristics, and capabilities. Proc Natl Acad Sci U S A 97:5528–5533. https://doi.org/10.1073/pnas.97.10.5528
15. Förster J, Famili I, Fu P, Palsson B, Nielsen J (2003) Genome-scale reconstruction of the Saccharomyces cerevisiae metabolic network. Genome Res 13:244–253. https://doi.org/10.1101/gr.234503
16. Jamshidi N, Palsson BØ (2007) Investigating the metabolic capabilities of Mycobacterium tuberculosis H37Rv using the in silico strain iNJ661 and proposing alternative drug targets. BMC Syst Biol 1:26. https://doi.org/10.1186/1752-0509-1-26
17. Rentzsch R, C a O (2009) Protein function prediction--the power of multiplicity. Trends Biotechnol 27:210–219. https://doi.org/10.1016/j.tibtech.2009.01.002
18. Aziz RK, Bartels D, Best AA, DeJongh M, Disz T, Edwards RA, Formsma K, Gerdes S, Glass EM, Kubal M, Meyer F, Olsen GJ, Olson R, Osterman AL, Overbeek RA, McNeil LK, Paarmann D, Paczian T, Parrello B, Pusch GD, Reich C, Stevens R, Vassieva O, Vonstein V, Wilke A, Zagnitko O (2008) The RAST Server: rapid annotations using subsystems technology. BMC Genomics 9:75. https://doi.org/10.1186/1471-2164-9-75
19. Aziz RK, Devoid S, Disz T, Edwards RA, Henry CS, Olsen GJ, Olson R, Overbeek R, Parrello B, Pusch GD, Stevens RL, Vonstein V, Xia F (2012) SEED servers: high-performance access to the SEED genomes, annotations, and metabolic models. PLoS One 7:e48053. https://doi.org/10.1371/journal.pone.0048053
20. Henry CS, DeJongh M, Best AA, Frybarger PM, Linsay B, Stevens RL (2010) High-throughput generation, optimization and analysis of genome-scale metabolic models. Nat Biotechnol 28:977–982. https://doi.org/10.1038/nbt.1672
21. Sonnhammer ELL, Östlund G (2015) InParanoid 8: orthology analysis between 273 proteomes, mostly eukaryotic. Nucleic Acids Res 43:D234–D239. https://doi.org/10.1093/nar/gku1203
22. Fitch WM (1970) Distinguishing homologous from analogous proteins. Syst Zool 19:99–113. https://doi.org/10.2307/2412448
23. Koonin EV (2005) Orthologs, paralogs, and evolutionary genomics. Annu Rev Genet 39:309–338. https://doi.org/10.1146/annurev.genet.39.073003.114725
24. Madden T (2013) The BLAST sequence analysis tool. BLAST Seq Anal Tool 1–17.
25. Schellenberger J, Que R, Fleming RMT, Thiele I, Orth JD, Feist AM, Zielinski DC, Bordbar A, Lewis NE, Rahmanian S, Kang J, Hyduke DR, Palsson BØ (2011) Quantitative prediction of cellular metabolism with constraint-based models: the COBRA Toolbox v2.0. Nat Protoc 6:1290–1307. https://doi.org/10.1038/nprot.2011.308
26. Ebrahim A, Lerman JA, Palsson BO, Hyduke DR (2013) COBRApy: COnstraints-Based Reconstruction and Analysis for Python. BMC Syst Biol 7:74. https://doi.org/10.1186/1752-0509-7-74
27. Bornstein BJ, Keating SM, Jouraku A, Hucka M (2008) LibSBML: an API library for SBML. Bioinformatics 24:880–881. https://doi.org/10.1093/bioinformatics/btn051

Chapter 9

Integrated Host-Pathogen Metabolic Reconstructions

Anu Raghunathan and Neema Jamshidi

Abstract

The science and art of Genome scale metabolic network reconstructions have been explicitly documented in the literature for organisms across all the three kingdoms of life. Constraints-based models derived from such reconstructions have been used to assess metabolic phenotypes of their complex connections to genotype accurately. The problem of infectious disease is complex due to the multifactorial response of the host to the pathogen. Systems biology approaches and modeling allow one to study, understand, and predict emergent properties of such complex responses. The integration of the host and pathogen metabolic networks and the subsequent merger of their stoichiometric matrices is nontrivial and requires understanding of both pathogen and host metabolism and physiologies. The protocol here describes the detailed process of network and stoichiometric matrix merger using a salmonella-mouse macrophage model. The protocol also discusses the interfacial and objective functions required to actually embark on the analysis of host-pathogen interaction models.

Key words Host pathogen integrated model, Flux balance analysis, Constraints-based analysis, Stoichiometric merge

1 Introduction

Infection, a leading cause of human mortality and morbidity, has been a bane for mankind since the dawn of civilization. A huge variation in individual outcomes of host infection following exposure to potentially life-threatening pathogens shows the complex and functional diversity of pathogen biology and host response [1]. Systems biology, in this context popularized as systems medicine, allows one to gain insights into host-pathogen interaction and outcomes like infection [2]. These can later be applied for controlling pathogen behavior through treatment strategies in the host. There has been significant expectation from systems medicine to rapidly translate into novel therapies leading to personalized medicine. Much of this expectation has yet to be realized [3], due to limitations in not only conceptual understanding of the respective pathogens and hosts but also the need for concepts and methodologies to understand, abstract, and interpret/predict their

Marco Fondi (ed.), *Metabolic Network Reconstruction and Modeling: Methods and Protocols*, Methods in Molecular Biology, vol. 1716, https://doi.org/10.1007/978-1-4939-7528-0_9, © Springer Science+Business Media, LLC 2018

further interaction. Systems analysis in medicine is not new with one of the first examples [4] resulting in the emergent concept of regulation and control by Claude Bernard put forth as the concept of constancy of the internal environment (*le milieu intérieur*); who also foresaw the need for the application of mathematics to biology. Although there are several layers of hierarchy at the molecular level that dictate causality and functional phenotypic outcomes in the host and pathogen, metabolism can summarily be a common denominator integrating and bridging the gap between genotype and phenotypes. The concept of network reconstruction and translation to models is central to metabolism [5]. These models have been available for computing, understanding, and predicting complex phenotypes for individual organisms that could be potential pathogens and hosts.

Genome-scale manually curated metabolic reconstructions (GEMs) summarize existing knowledge of cellular pathways in organisms in a well-structured, mathematical manner [2, 6–11]. The processes to assemble these in a bottom-up fashion and algorithms that semi-automate them have been established and proven to be accurate for predicting function of many pathogens [12–15]. These network reconstructions are then converted into condition-specific models, formulated as optimization problems based on biochemical reaction stoichiometry and physicochemical constraints such as mass and energy conservation. The constraint-based approach allows integration of several data-types and allows computation of physiological properties of cell function despite incomplete knowledge about reaction rates, kinetic constants, enzyme activities, and metabolite concentrations [16–18]. There are many models that exist for both extracellular and intracellular pathogens in the literature [19]. Host models exist that can range from being the full organism like mouse or human to organelle like macrophage to multi-tissue representations [20–23]. However, there are very few models that describe host-pathogen interaction. A human alveolar macrophage-mTB model was the first host-pathogen interaction model assembled [21]. Further models that explore interactions of plasmodium falciparum in RBC to represent malarial infection have been reconstructed and probed to understand multiple metabolic states [24, 25]. These models have been used to identify essential metabolic connections between host and pathogen. The data explosion in biology facilitates the development of context-specific pathogen-host interaction models that represent the multifactorial response of infection.

This chapter is intended as a protocol for host pathogen interaction (HPI) model reconstruction and will describe the main steps of this process, focusing on infection with an intracellular microbe. The details of development of the pathogen or the host model can be found in legacy data on the COBRA approach. A comprehensive stepwise protocol for the systematic construction of a host-pathogen model has been previously described [26], comprised of four general steps, (1) pre-integration model check, (2) model integration, (3) testing of integration, and (4) simulation. Herein, we highlight specific aspects of the integration procedure in the context of real-world, genome-scale models, focusing on the model integration step and aspects of integration testing (steps 2 and 3 in [26]). We reiterate the importance of constraints and objective function in simulations.

2 Materials

In this section, we describe the essential components to build a HPI model. These include:

1. A high-quality pathogen GEM.
2. A high-quality host GEM.
3. Constraints and Objectives.
4. Modeling and simulation platform.

2.1 A High-Quality Model for Pathogen

A vast literature exists on building high-quality metabolic reconstructions and their translation to models for organisms that are potential pathogens [16]. Many algorithms exist online for drafting such models in an automated fashion as well [27]. The content, coverage, and predictive sensitivity of reconstructions differ based on the experimental data available for that organism. Hence, it is essential to check whether they conform to all the rules of metabolic network reconstruction and ultimately represent a high-quality, curated model capable of computing cell function accurately. Many models of potential pathogens exist in the literature [15–18]; however, running a functionality test suite is highly recommended [13] to avoid pitfalls in interpreting results from HPI modeling and simulation. In this protocol, we use a model, iRR1083 of the intracellular pathogen *Salmonella typhimurium* LT2.

2.2 A High-Quality Model for Host

Similar to the discussion in Subheading 2.1, a carefully curated host model is essential to developing a HPI model. The host models generally derived from global reconstructions need to be verified for their ability to accomplish the requisite tissue-specific metabolic functions. Often expansion of global reconstructions is required to represent desired infection states. Condition-specific tissue specific host models can be derived from their global counterparts by integrating OMICs data-types [21]. These methods are highlighted in Subheading 3 and details can be found elsewhere. In this protocol, we use a model, iMM1415, of the murine host *Mus musculus* [20].

2.2.1 Tailoring Host Models via Incorporating OMICs Data

There are many algorithms that have been developed to incorporate gene expression constraints in models to represent different tissues or disease states [28–30]. Also, more recently proteomic data have also been used to build metabolic reconstructions [31]. Algorithms to integrate metabolomic and fluxomic data [25, 32] into networks to assess flux states and metabolic pathway analysis are also on the rise. The interested readers are referred to available review articles outlining some of these methods. In this protocol, gene expression data is used to tailor the mouse model to represent the macrophage using the GIMME algorithm [33].

2.3 Constraints

The appropriate application of constraints is critical successful of host-pathogen interaction modeling [26]. For HPI modeling, we primarily assume a quasi-homeostatic state (rather than a strict steady state). Thus, mass balance or conservation is implicit over the time scale of interest. The constraints range from multiple OMICS data types to macro and micro nutrient demand to observed coupling within and across the host and pathogen. The OMICS data that have so far been integrated include transcriptomic and metabolomics data [23, 25]. Since the constraints are to be understood in the context of the model, some of these constraints will be discussed later.

2.4 Modeling and Simulation Platform

There have been a growing number of tools developed for performing constraint-based metabolic modeling (reviewed in [10, 34]). In recent years, there have been an increasing number of freeware programs (Fig. 1) as well as packages written for commercial tools that have been designed specifically for constraint-based modeling simulation and analysis [35]. The construction and use of these models has to a large degree been standardized, although improvements continue to be made [35–37]. The Systems Biology Markup Language (SBML) has been one of the more successful efforts to improve sharing and standardization of these models [38]. Models in the SBML format are recognized and can be converted to many other formats necessary for computation. In addition, a supported Linear Programming (LP) solver and a libSMBL API library are needed [13, 38]. For further information on the installation and configuration of these tools, users are referred to the respective manuals and literature [13].

3 Methods

The process of Host Pathogen Interaction Model development is delineated (Fig. 2). The details of the process are discussed in the following sections. As noted above, we will not focus on every single step, but rather a few of the critical steps required for integration and some basic testing of the model. As noted in

OptFlux
SBRT
MetaFluxNet
BioOpt
SurreyFBA
FASIMU
GEMSiRV
CellNetAnalyzer/FluxAnalyzer
COBRA Toolbox
SNA Toolbox
FBA-SimVis
MetaFlux
CycSim
WEbcoli
GSMN-TB
Acorn
Model SEED
FAME
MicrobesFlux

Fig. 1 List of modeling platforms, tools, and software available for constraints-based modeling and flux balance analysis. The ones in red are standalone, while those in purple represent web-based tools. The ones listed out in blue are toolbox based (compiled from [50])

Subheading 2, we consider infection of a mouse macrophage by salmonella using available expression data [39] and previously published manually curated and reconstructed GEMS for the mouse (iMM1415) and *Salmonella typhimurium* LT2 (iRR1083) [17, 20] [using Matlab and the COBRA toolbox [13]. A merger toy network is also discussed to illustrate the details (*see* **Note 1**).

3.1 Pre-Integration Model Check

The Pre-integration model check is one of the first steps involved in HPI model development. This is mandatory even for the integration of two high-quality GEMS of the pathogen and the host, due to potential discrepancies resulting after the merge [26].

The curated, validated models iRR1083 and iMM1415 have previously been tested and we leave it to the interested reader to perform the mass balance checks and any functionality testing of interest. It is essential to make sure that the principle of mass conservation is not violated. For that reason, mass balance and stoichiometric consistency check must be performed. Flux variability should also be performed to identify any “free energy” cycles and confirm that there is no net production of any metabolite when no substrates are available for uptake [26].

The pre-integration check technically could involve as many tests the modeler desires in order to ascertain model accuracy and functionality. In the COBRA toolbox some of these tests come prepackaged as a functionality test suite. However, the FTS can theoretically comprise of any number of desired simulations (that are in silico physiological tests) to ensure correct biological

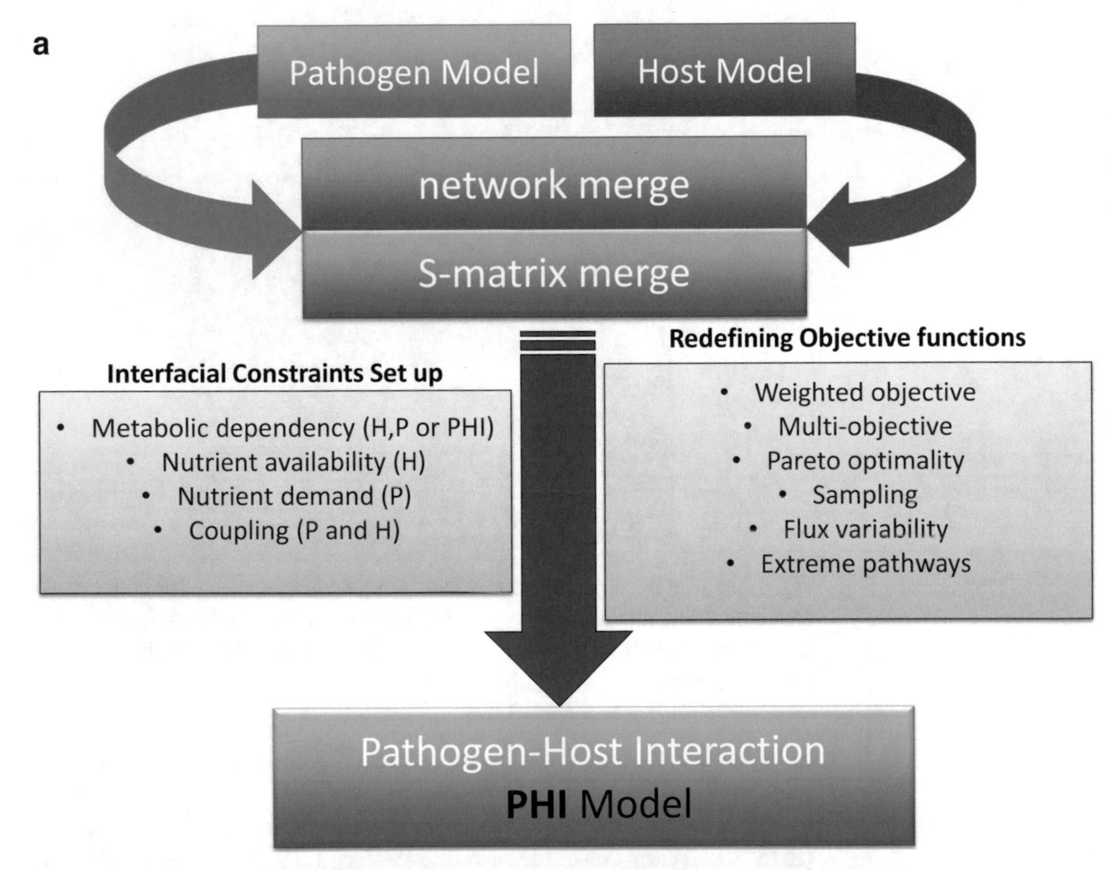

Fig. 2 (**a**) Overall schematic of development of Pathogen-Host Interaction Models. The first step is the merging networks followed by a compartment-specific row wise stoichiometric matrix merge. The process is followed by setting constraints and redefining objectives appropriate for the interaction. The final result is a host-pathogen interaction (HPI) model. (**b**) Flow chart of Network and Stoichiometric Matrix merge steps. The process is discussed using the salmonella-mouse interaction model. Merging networks and stoichiometric matrices to develop the HPI model is a nontrivial process. The detailed execution of the steps is delineated on the RHS of the flow chart. The overlapping metabolites in iRR1083 [17] directly exchange with iMM1415 [40] cytoplasm. Exchanges for metabolites in iRR1083 not shared by host directly exchange with the extracellular environment

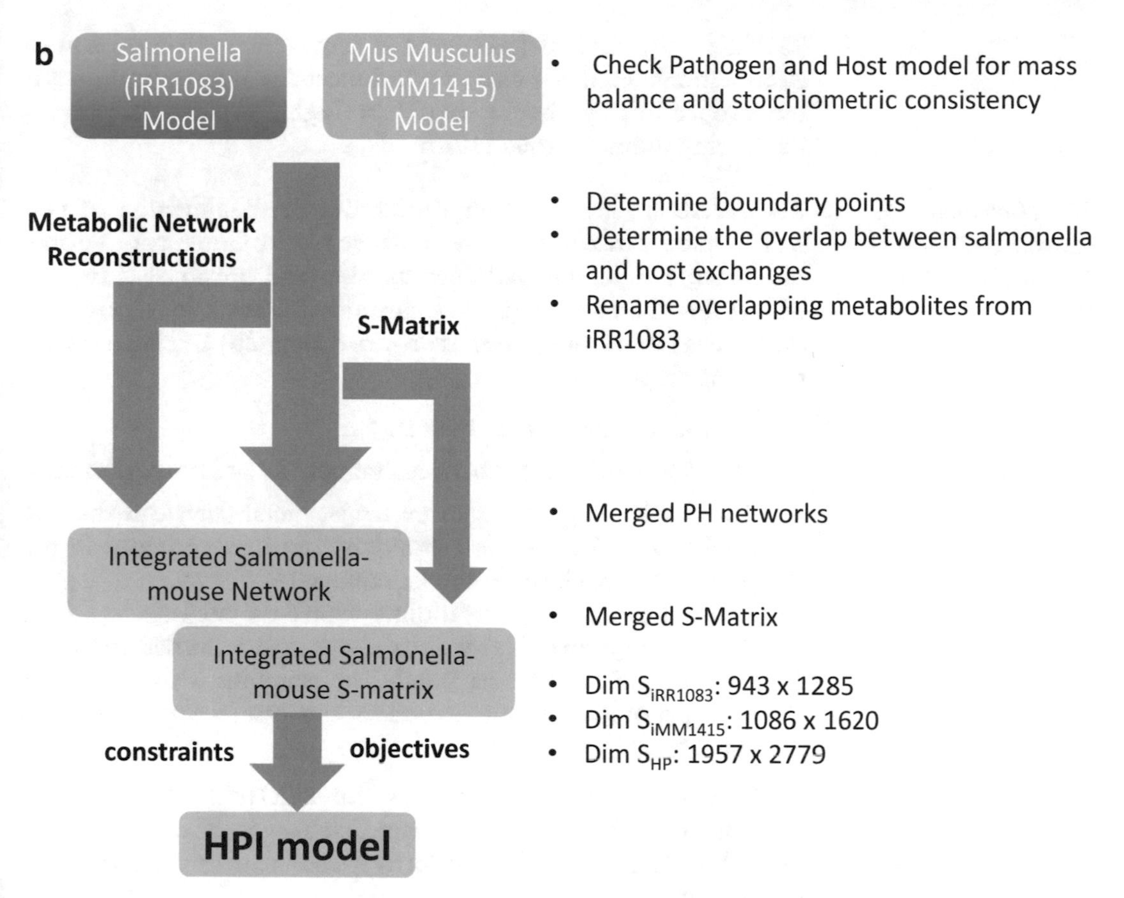

Fig. 2 (continued)

behavior of the model. These tests can range from simple point flux tests like growth (maximizing biomass) under certain media conditions, testing broad substrate ranges for growth capabilities all the way to single-gene deletion analysis that represent gene essentiality and virulence in the case of a pathogen. During pre-integration check, the network boundary points need to be identified.

In this example the pathogen is an intracellular microbe, so we determine the overlap between the boundary metabolites and compartment in the host which the microbe will invade/infect. If the two models were generated using the same reference databases and metabolite name abbreviations and annotations, then this becomes much easier. The use of unique chemical identifiers including ChEBI, SMILES, and InChI string identifiers greatly facilitates this process, particularly if the GEMs were constructed using different reference databases or with different abbreviation standards, then one must rely on the molecular formulas for comparison. For **HPI** models, specifically, the abbreviations of the boundary metabolites need to be compared and checked. Appropriate annotations via standardized notation DurmuCs 2015 (e.g., molecular formula,

SMILES [40], ChEBI [41], etc.) facilitate such a comparison of abbreviations. Hence, naive or blind mechanical integration without proper quality checks should be highly avoided to prevent developing uninformative HPI models.

3.2 Metabolic Network and Stoichiometric Matrix Merge

The metabolic network merge precedes the conjunction of two stoichiometric matrices. There are three types of interactions possible between host and pathogen as discussed previously [16, 21, 26]. A flowchart of the process currently followed for mouse and salmonella model integration is provided (Fig. 2b) and includes the following steps.

1. Identify boundary metabolites.
2. Determine the overlap between salmonella and host exchanges.

   The overlapping pathogen exchange metabolites can then be used to identify the corresponding exchange reactions in the pathogen model; these will be removed.
3. Rename overlapping metabolites from Salmonella.
   If the compartment(s) within the microbe overlap with the host, they should be renamed. The reactions within the host model should also be renamed (in our case, we appended the suffix "_sal").

The overlapping metabolites in iRR1083 thus directly exchange with iMM1415 cytoplasm. Exchanges for metabolites in iRR1083 not shared by host directly exchange with the extracellular environment.

The simplicity of formulation of the HPI models although trivial on the surface involves a critical step of identifying the overlapping set of metabolites and their corresponding abbreviation mappings between the host and pathogen metabolites. While this step may be facilitated by automated string comparison scripts, the final result needs to be manually evaluated to (1) remove any inappropriate annotations and (2) identify potentially missed overlapping annotations.

In order to produce a more biologically relevant host model, we used a transcriptomic expression data set in order to generate a context-specific "infected" mouse macrophage. There are a number of different approaches for generating context-specific models with this data; we used GIMME [33] with the biomass constrained to at least 85% of the maximum.

In the case of iRR1083 and iMM1415, there were 199 exchange metabolites in salmonella, 125 of which were in the cytoplasm compartment of the mouse cell, iMM1415. The overlapping (and non-overlapping) metabolites can be tracked and stored in a metabolite-mapping file (mapping compartment-specific metabolites between iRR1083 and iMM1415).

The functional integration of pathogen and host models to create a HPI model is a nontrivial process. Naive conjunction of two models can lead to computation of biologically infeasible steady states. For example, active flux through a reaction of the host/pathogen in the absence of biomass/energy production is obviously a violation of mass conservation laws. A simple illustration of the integration of the iRR1083 and iMM1415 models is shown (Fig. 2b). After the preparatory steps delineated in the above section, the stoichiometric matrices are merged. This integration is accomplished by using the metabolite-mapping file generated in the previous step. The matrices can then be merged in a single step, or sequentially. In the COBRA toolbox, the "addReaction" function can serve this purpose.

3.2.1 Calculating Matrix Properties

Elementary Matrix calculations including size, rank, etc. of the newly created HPI matrix allow dimensionality assessment and are critical in debugging and analysis of results.

Initially, the mouse and salmonella models had 1086 and 943 - compartment-specific metabolites, respectively, with 1620 and 1285 reactions, respectively. The merged model contains 1957 metabolites and 2779 reactions, as would be expected.

Mouse model null space size 1042 and left null space size 90. The merging of the models results in decreased dimensionality of the steady state flux space to 921 and interestingly increases in the left null space to 99.

The check to confirm if the host organelle compartments exchange any metabolite with the pathogen in the newly created stoichiometric matrix can be done by verifying that

$$m_{\text{mac-sal}} < m_{\text{mac}} + m_{\text{sal}}$$

(m representing the number of metabolites)

which holds true for the above integration.

3.3 Integrated Host-Pathogen Testing or Model Functionality Testing

Once the merged model has been constructed, detailed testing needs to be performed in order to ensure that the linkages between the two models were successful.

3.3.1 The Functionality Tests Suite

The first three steps are similar to pre-integration model check discussed earlier. These include checking for mass balances and stoichiometric inconsistencies, FVA analysis and revisiting bounds of some constraints and the Functionality Test Suite of the HPI model will also enable a basis for comparison and assist subsequent analyses.

Depending on the type and complexity of new constraints that are applied to the integrated host-pathogen model, there are situations that may introduce behavior that violates mass conservation. Since transport reactions in iRR1083 may potentially interact with transported metabolites in iMM1415 to produce non-physical results (i.e., violation of mass conservation), mass balance checks needs to be re-performed by closing off all exchange reactions and performing flux variability analysis to ensure there is no "free" production of metabolites. For intracellular pathogens like salmonella, the test needs to be applied to the isolated pathogen within the host as well.

The two optimizations that would cover the largest percentage of the network would be to carry out optimization of the host biomass and that of the pathogen biomass in the merged model (note that prior to merging the models, the uptake constraints for each of the models should be set at an appropriate level to allow uptake). In our example under consideration we expect the maximum of the mouse biomass to be identical to the pre-merge mouse model, but the maximum of salmonella may differ from the pre-merge salmonella growth optimum. If the model fails either of biomass tests, then there is likely a metabolite transport problem between the models, which then requires re-assessing the overlap and mapping between the metabolites.

3.3.2 The Interdependence Test

The interdependence test can be performed in multiple ways using similar corresponding reactions in the host and pathogen, complementary reactions, or directly opposing reactions [26]. This test requires identifying objective functions that are expected to influence or be influenced by the coupling between the host and pathogen. The pseudo-reaction that represents biomass composition is a very good candidate for such tests [21]; however, secretion/uptake of disease-specific metabolites, energy formation via ATP production, oxidative phosphorylation, may also be used. We consider assessment of the pareto-optimality front [42] for the biomass of mouse and salmonella (Fig. 3a) as well as maximum ATP production for the mouse and salmonella (Fig. 3b). In Fig. 3a we see that the growth of Salmonella is not significantly affected by the biomass demands of the murine macrophage. Similarly for low Salmonella growth rates the mouse macrophage biomass production is unaffected, until a precipice is reached at approximately 0.6 biomass arbitrary units (mass/time) of Salmonella. Figure 3b displays the pareto-optimality front for ATP production. Although the amount of ATP generated by the host and infecting microbe differs by orders of magnitude, we see that mouse ATP generation can significantly restrict the infecting microbe and vice versa. Again arbitrary flux units were employed and scaling of the biomass objective functions relative to one another was not performed, since these calculations are presented for illustrative purposes. A scientific systems investigation into the infection of the murine macrophage by salmonella would require adjusting biomass pseudo-reaction coefficients for each of the organisms as well as setting appropriate uptake constraints (in measured and appropriately scaled units).

In practice further evaluation of function would be performed through the functionality test suite, which is outside the scope of this tutorial description. After integration testing is complete, one can proceed to simulation testing, as previously outlined [26].

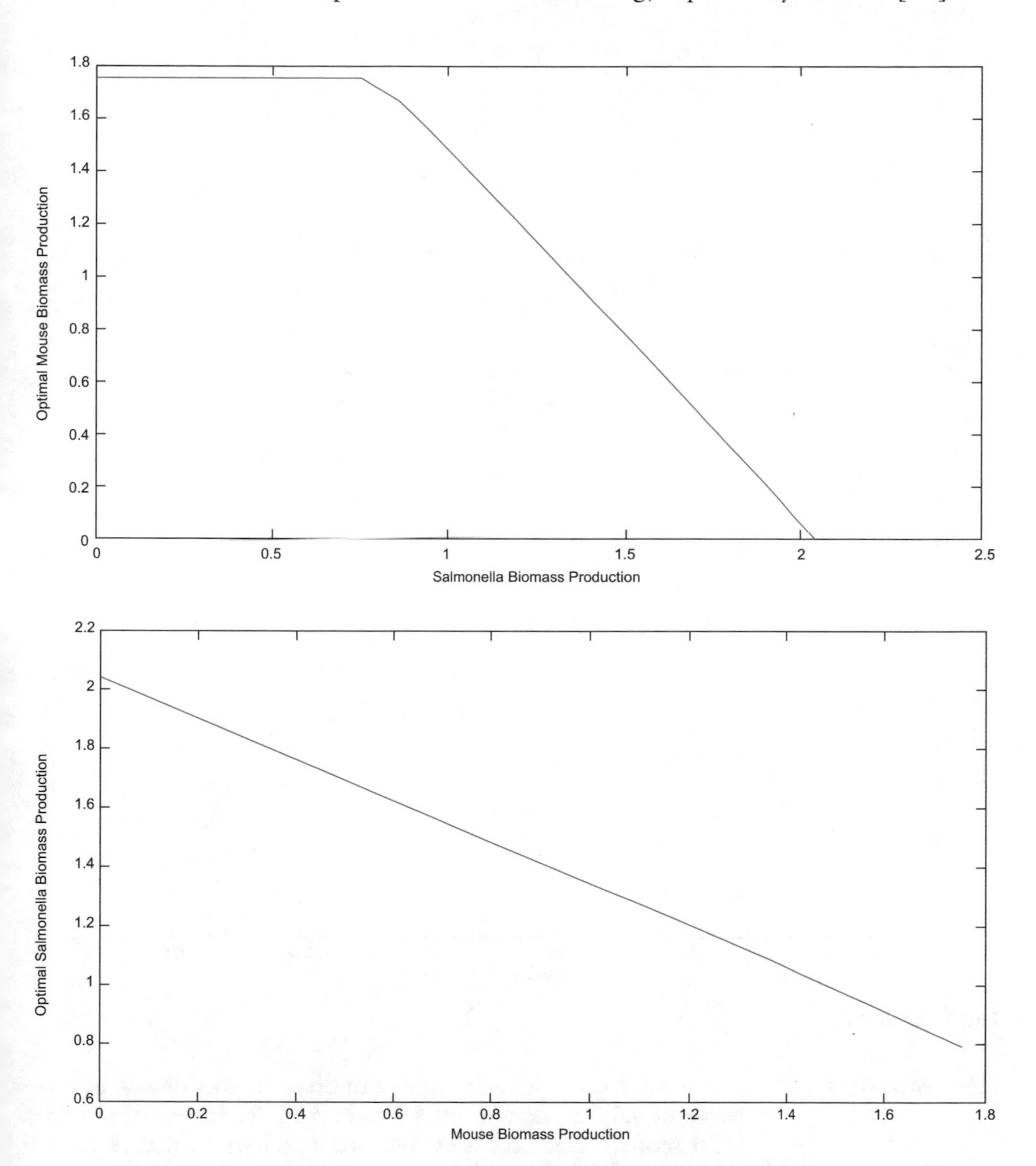

Fig. 3 (**a**) Pareto optimality front of mouse vs. salmonella biomass. Top panel, fixed salmonella biomass demand while maximizing mouse biomass production. Bottom panel fixed mouse biomass demand while maximizing salmonella biomass production. (**b**) Pareto optimality front of mouse vs. salmonella ATP demand. Top panel, fixed salmonella ATP demand while maximizing mouse ATP production. Bottom panel fixed mouse ATP demand while maximizing salmonella ATP production

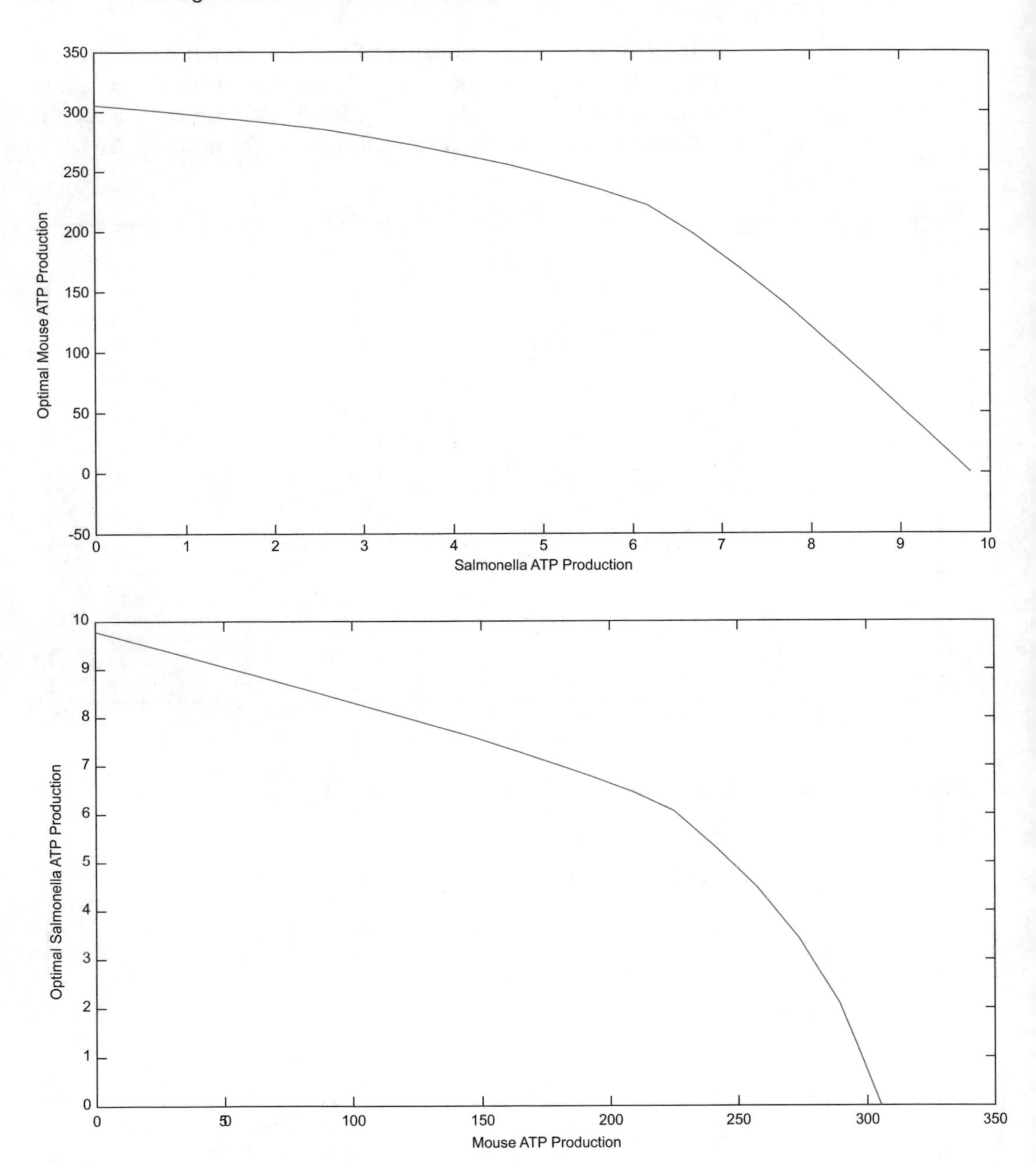

Fig. 3 (continued)

3.4 Simulations

There are many simulations one can do to predict disease phenotypes or validate experimental data in the infection scenario using HPI models. The protocols followed as similar to that discussed elsewhere for individually constructed pathogen or host models [14, 26]. A few representative examples are discussed below.

3.4.1 Characterizing Host-Related Metabolic Function

The knowledge of biochemical functionality of the tissue/organelle being modeled in the host-pathogen interaction would help to test for metabolic functions specific to the host. The HPI model can also be tested for the same functions; however, it may fare poorly in terms of abilities as compared to the global network.

3.4.2 Gene Essentiality

Gene deletion analysis of the HPI model can potentially predict virulent and avirulent genes for the pathogen [9, 42, 43] with better accuracy than the individual model. Gene essentiality simulations are set up as discussed previously in the literature [13, 17]. The identification of lethality is done by setting the upper bound and lower bound of an all reactions connected to a gene to zero. Flux balance analysis is then performed to calculate change in the growth/biomass function. One can set thresholds in comparison to the optimal biomass to assess virulence [21].

3.5 Constraints

Interfacial constraints between the host and the pathogen need to be set based on the literature in order to simulate any infectious states [21, 26]. This essentially allows context-specific modeling by setting the simulation environment to be more representative of the actual biological environment.

3.5.1 Metabolic Dependency Constraints

Constraints can be designed to reflect any potential metabolic dependencies between the host and pathogen [16, 26] that ultimately determines their interaction. Although the pareto-optimality evaluation in Fig. 3 revealed interdependence between the host and microbe, additional simulations are often performed to further characterize the directionality and strength of the interdependence. Figure 4 illustrates the tradeoff between Salmonella biomass

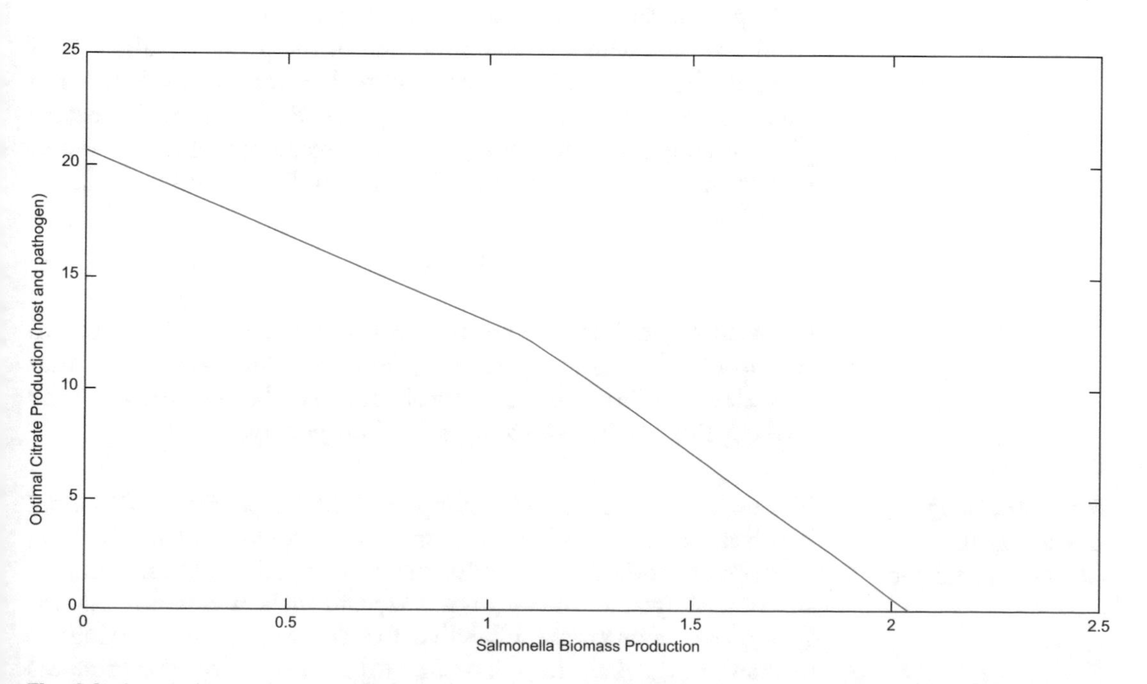

Fig. 4 Optimality tradeoff between Salmonella growth and net citrate production. Salmonella growth is fixed at different rates (arbitrary units, mass/time) while citrate production from the host-pathogen model is optimized

production with increased net citrate production for the host and pathogen. As expected, a negative slope is present with two regions; at lower salmonella biomass production, there can be higher net citrate secretion, but at higher salmonella growth rates, the slope becomes more steep with less citrate available for secretion. Evaluation of these dependencies under different conditions can in turn be used to inter-species dependencies under various conditions.

3.5.2 Nutrient Availability Constraints

These constraints are critical since they directly determine the capabilities of pathogen and host response. They literally delineate the availability of metabolites and micro-nutrients in the niche the pathogen occupies in the host and shape adaptation and evolution of the pathogen behavior based on its requirements for survival, proliferation, or persistence [21]. These constraints are also by far the most simple to implement as uptake rates (mmol/gDW-h). In some cases, scaling these uptake rates to the number of C-atoms allows comparison across limiting substrates of varying molecular sizes [16].

3.5.3 Coupling Constraints

Coupling constraints mathematically describe the issue of shared resources, allocation, and potential competition in the pathogen host interaction and explicitly link the host and pathogen models. These issues go beyond plain sharing of metabolites and address how cellular resource is allocated toward synthesis and relative activities of needed enzymes mimicking the role of metabolic regulation. For example, one can constrain two fluxes (e.g., a reaction in the host and a reaction in the microbe) to be linked by adding a linear relation between the two reactions. Further in conditions for which one is interested in the presence/absence of categorization, the relationships will then require formulation as a Mixed Integer Linear Programming (MILP) problem. For example, the coupling between flux through reaction i of the host and pathogen to achieve a certain objective can be related by a simple equation as:

$$\boldsymbol{\alpha} v_i^{\mathrm{h}} + \boldsymbol{\beta} v_j^{\mathrm{p}} = 1$$

where in v_i and v_j are the reaction fluxes for reaction i (in host) and reaction j (in pathogen). $\boldsymbol{\alpha}$ and $\boldsymbol{\beta}$ are non-unity coefficients that modulate the fluxes when the stoichiometric balance between the two reactions is known to be in fixed proportions [26].

3.6 Redefining Objectives and Objective Functions

The next step in creating the integration protocol of the HPI model involves redefining objective functions. Prerequisite to defining appropriate and novel objective functions is a detailed understanding of pathogen and host physiology and their interaction. Identification of appropriate objective functions is under continuous evaluation through in silico, in vitro, and in vivo experiments [42]. Generally, separated into two categories, single-objective and multi-objective problems can typically represent some of the complex biological phenotypes observed in vivo [44–46]. Integration of multiple objectives through weighted objectives and multi-level optimization allow for mathematical investigation of tradeoffs and represent better the host-pathogen interaction.

Bi-level optimization used typically in metabolic engineering processes [47] could be applied across the pathogen host boundary and is a powerful technique that allows one to describe competing objectives between the pathogen and host allowing simultaneous or opposing maximization of uptake of resource with maximizing growth.

Another example of multi-reaction linear programming is pareto-optimality analysis [48] that quantitatively accounts for the tradeoffs between reaction i and reaction j. Evaluating the flux range (minimum and maximum flux) through the ith reaction allows generation of pareto-optimal curves. The flux through reaction i is then fixed to span through the whole flux range in steps, while maximizing reaction j. The procedure is then repeated with i and j exchanged.

Many tools used to compute functional states rely on linear programming and are FBA based and characterize a set of optimal solutions. Constraint-based methods that characterize solution spaces like flux variability analysis [49], sampling [50], extreme pathway analysis rather than optimal solutions alone represent physiologically relevant objectives for host-pathogen interaction analysis. For example, pathogen objective function (biomass) can be tailored to better represent infection states being simulated. In the case of mycobacterium, the pathogen is latent and stops active division in the macrophage. Since the nutritional requirements change due to the harsh environment in the phagosome, the biomass composition changes. Randomized sampling for each of the biomass precursors in the biomass composition reaction allows characterization of the solution spaces. Combined with linear regression analysis, the iterative removal and addition of biomass precursors can result in redefining the best biomass composition under certain functional disease states [21].

3.7 Conclusions

Computational models of host-pathogen interactions can be harnessed to increase our understanding of complex and multifactorial disease states. There is a need to explore/develop intelligent methods for probing the solution spaces of HPI models to accomplish "differential diagnosis" that could lead to pinpointing strategies for treatment. Comparing healthy and diseased states could eventually help discover new mechanisms in the dynamics underlying many pathophysiological states and lead to hypothesis generation in novel therapeutic space.

4 Notes

1. Integration of two toy networks is depicted (Fig. 5) followed by the stoichiometric matrix merge (Fig. 6). The toy networks are visualized (Fig. 5) as schematics of a pathogen network (panel a), host network (panel b), and an integrated host-pathogen network (panel c). The corresponding stoichiometric matrices for each of the models are delineated in Fig. 6 (panels a, b, and c, respectively). S and Q are non-overlapping metabolites, within the host and hence R10, R15, and R16 will not be able to carry a flux. Intracellular organelles are not described in this toy example; however if the pathogen infects the host and resides in a specific organelle like a phagosome within the host, the procedure would be the same.

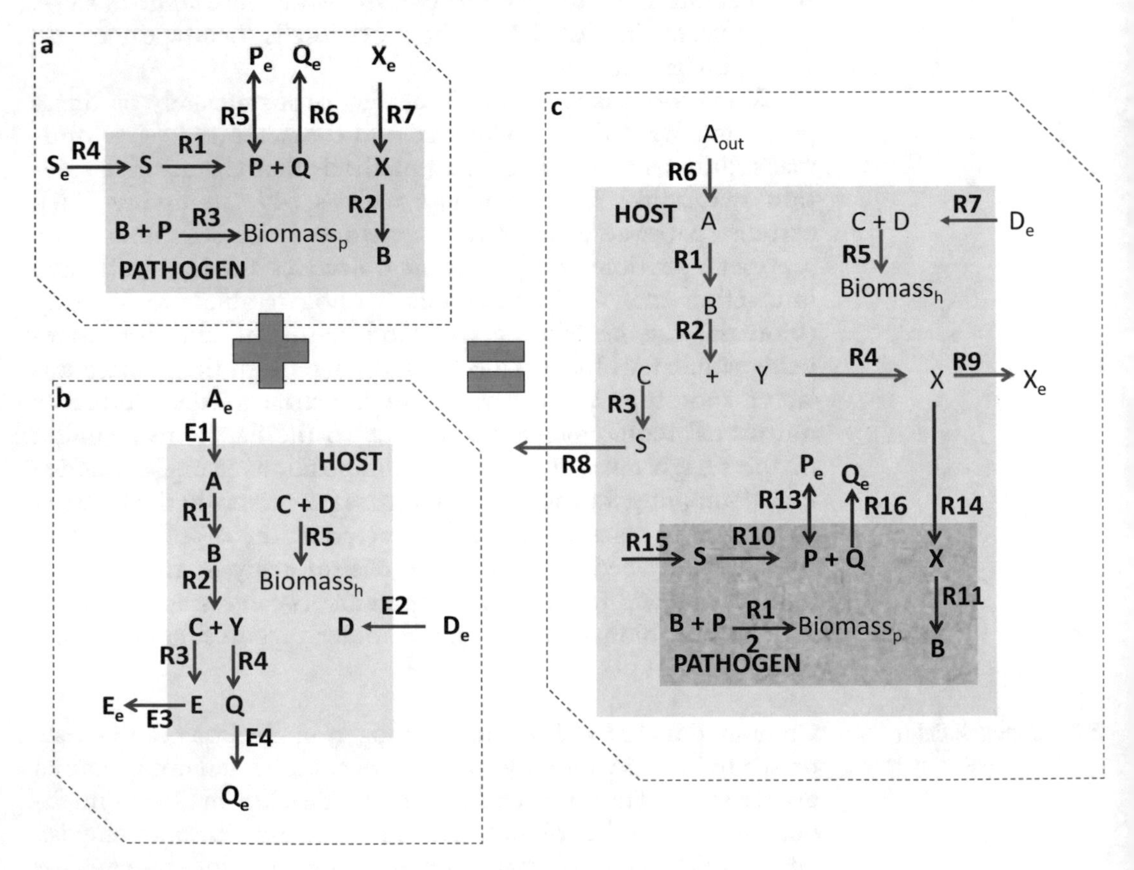

Fig. 5 Pathogen-host interaction metabolic network reconstruction. Toy networks are depicted for the pathogen (**a**), the host (**b**), and the pathogen-host interaction (**c**)

a

P	INTRACELLULAR			TRANSPORT				EXCHANGE			
	R1	**R2**	**R3**	**R4**	**R5**	**R6**	**R7**	**E1**	**E2**	**E3**	**E4**
S	-1	0	0	1	0	0	0	0	0	0	0
S_e	0	0	0	-1	0	0	0	-1	0	0	0
B	1	0	-1	0	1	0	0	0	0	0	0
B_e	0	0	0	0	-1	0	0	0	-1	0	0
Q	1	0	0	0	0	1	0	0	0	0	0
Q_e	0	0	0	0	0	-1	0	0	0	-1	0
X	0	-1	0	0	0	0	1	0	0	0	0
X_e	0	0	0	0	0	0	-1	0	0	0	-1
P	0	1	-1	0	0	0	0	0	0	0	0
LB	0	0	0	-∞	-∞	-∞	-∞	-10	-10	-10	-10
UB	∞	∞	+∞	∞	+∞	+∞	+∞	0	0	0	0

b

S =

H	INTRACELLULAR					TRANSPORT				EXCHANGE			
	R1	**R2**	**R3**	**R4**	**R5**	**R6**	**R7**	**R8**	**R9**	**E1**	**E2**	**E3**	**E4**
A	-1	0	0	0	0	1	0	0	0	0	0	0	0
A_e	0	0	0	0	0	-1	0	0	0	-1	0	0	0
B	1	-1	0	0	0	0	0	0	0	0	0	0	0
C	0	1	-1	0	-1	0	0	0	0	0	0	0	0
D	0	0	0	0	-1	0	1	0	0	0	0	0	0
D_e	0	0	0	0	0	0	-1	0	0	0	-1	0	0
E	0	0	1	0	0	0	0	1	0	0	0	0	0
E_e	0	0	0	0	0	0	0	-1	0	0	0	-1	0
Y	0	1	0	-1	0	0	0	0	0	0	0	0	0
Q	0	0	0	1	0	0	0	0	1	0	0	0	0
Q_e	0	0	0	0	0	0	0	0	-1	0	0	0	-1
LB	0	0	0	0	-∞	-∞	-∞	-∞	-∞	-10	-10	-10	-10
UB	∞	∞	∞	∞	∞	∞	∞	∞	∞	0	0	0	0

Fig. 6 Pathogen-host interaction stoichiometric matrix construction. The S-matrix of the toy networks drawn in Fig. 2 is depicted for the pathogen (**a**), the host (**b**), and the pathogen-host interaction (**c**)

c

	R1	R2	R3	R4	R5	R6	R7	R8	R9	R10	R11	R12	R13	R14	R15	R16	E1	E2	E3	E4	
A	-1	0	0	0	0	1	0	0	0	0	0	0	0	0	0	0	0	0	0	0	
A_e	0	0	0	0	0	-1	0	0	0	0	0	0	0	0	0	0	-1	0	0	0	
B	1	-1	0	0	0	0	0	0	0	0	0	0	1	0	0	0	0	0	0	0	
C	0	1	-1	0	-1	0	0	0	0	0	0	0	0	0	0	0	0	0	0	0	
D	0	0	0	0	-1	0	1	0	0	0	0	0	0	0	0	0	0	0	0	0	
D_e	0	0	0	0	0	0	-1	0	0	0	0	0	0	0	0	0	0	-1	0	0	
E	0	0	1	0	0	0	0	1	0	0	0	0	0	0	0	0	0	0	0	0	
E_e	0	0	0	0	0	0	0	-1	0	0	0	0	0	0	0	0	0	0	-1	0	
Y	0	1	0	-1	0	0	0	0	0	0	0	0	0	0	0	0	0	0	0	0	
X	0	0	0	1	0	0	0	0	1	0	0	0	0	1	0	0	0	0	0	0	
X_e	0	0	0	0	0	0	0	0	-1	0	0	0	0	0	0	0	0	0	0	-1	
Q	0	0	0	0	0	0	0	0	0	0	0	0	0	0	1	-1	0	0	0	0	
Q_e	0	0	0	0	0	0	0	0	0	0	0	0	0	0	0	1	0	0	0	0	
S_p	0	0	0	0	0	0	0	0	0	-1	0	0	0	0	0	0	0	0	0	0	
B_p	0	0	0	0	0	0	0	0	0	1	0	-1	-1	0	0	0	0	0	0	0	
Q_p	0	0	0	0	0	0	0	0	0	1	0	0	0	0	-1	0	0	0	0	0	
X_p	0	0	0	0	0	0	0	0	0	0	-1	0	0	-1	0	0	0	0	0	0	
P_p	0	0	0	0	0	0	0	0	0	0	1	-1	0	0	0	0	0	0	0	0	
LB	0	0	0	0	-∞	-∞	-∞	0	0	-∞	-∞	-∞	-∞	0	0	-∞	-∞	-20	-20	-20	-20
UB	∞	∞	∞	∞	∞	∞	∞	∞	∞	∞	∞	∞	∞	∞	∞	∞	∞	0	0	0	0

Fig. 6 (continued)

Acknowledgments

We thank Avinash V. Ghanate for helping us with GEO expression data sets used in the Salmonella-mouse model.

References

1. Burgner D, Jamieson SE, Blackwell JM (2006) Genetic susceptibility to infectious diseases: big is beautiful, but will bigger be even better? Lancet Infect Dis 6(10):653–663. https://doi.org/10.1016/S1473-3099(06)70601-6
2. Mardinoglu A, Nielsen J (2012) Systems medicine and metabolic modelling. J Intern Med 271(2):142–154. https://doi.org/10.1111/j.1365-2796.2011.02493.x
3. Chan IS, Ginsburg GS (2011) Personalized medicine: progress and promise. Annu Rev Genomics Hum Genet 12:217–244
4. Noble D (2008) Claude Bernard, the first systems biologist, and the future of physiology. Exp Physiol 93(1):16–26
5. Palsson B (2006) Systems biology: properties of reconstructed networks. Cambridge University Press, Cambridge

6. Price ND, Reed JL, Palsson BØ (2004) Genome-scale models of microbial cells: evaluating the consequences of constraints. Nat Rev Microbiol 2(11):886–897. https://doi.org/10.1038/nrmicro1023
7. Thiele I, Palsson BØ (2010) A protocol for generating a high-quality genome-scale metabolic reconstruction. Nat Protoc 5(1):93–121. https://doi.org/10.1038/nprot.2009.203
8. Feist AM, Palsson BØ (2008) The growing scope of applications of genome-scale metabolic reconstructions using Escherichia coli. Nat Biotechnol 26(6):659–667. https://doi.org/10.1038/nbt1401
9. Feist AM, Herrgård MJ, Thiele I, Reed JL, Palsson BØ (2009) Reconstruction of biochemical networks in microorganisms. Nat Rev Microbiol 7(2):129–143. https://doi.org/10.1038/nrmicro1949
10. Lewis NE, Nagarajan H, Palsson BO (2012) Constraining the metabolic genotype-phenotype relationship using a phylogeny of in silico methods. Nat Rev Microbiol 10 (4):291–305. https://doi.org/10.1038/nrmicro2737
11. Oberhardt MA, Gianchandani EP (2014) Genome-scale modeling and human disease: an overview. Front Physiol 5:527. https://doi.org/10.3389/fphys.2014.00527
12. Thorleifsson SG, Thiele I (2011) rBioNet: A COBRA toolbox extension for reconstructing high-quality biochemical networks. Bioinformatics 27(14):2009–2010. https://doi.org/10.1093/bioinformatics/btr308
13. Schellenberger J, Que R, Fleming RMT, Thiele I, Orth JD, Feist AM, Zielinski DC, Bordbar A, Lewis NE, Rahmanian S, Kang J, Hyduke DR, Palsson BØ (2011) Quantitative prediction of cellular metabolism with constraint-based models: the COBRA Toolbox v2.0. Nat Protoc 6(9):1290–1307. https://doi.org/10.1038/nprot.2011.308
14. Sadhukhan PP, Raghunathan A (2014) Investigating host–pathogen behavior and their interaction using genome-scale metabolic network models. Immunoinformatics 1184:523–562
15. Raghunathan A, Shin S, Daefler S (2010) Systems approach to investigating host-pathogen interactions in infections with the biothreat agent Francisella. Constraints-based model of Francisella tularensis. BMC Syst Biol 4:118. https://doi.org/10.1186/1752-0509-4-118
16. Thiele I, Heinken A, Fleming RMT (2013) A systems biology approach to studying the role of microbes in human health. Curr Opin Biotechnol 24(1):4–12. https://doi.org/10.1016/j.copbio.2012.10.001
17. Raghunathan A, Reed J, Shin S, Palsson B, Daefler S (2009) Constraint-based analysis of metabolic capacity of Salmonella typhimurium during host-pathogen interaction. BMC Syst Biol 3:38. https://doi.org/10.1186/1752-0509-3-38
18. Jamshidi N, Palsson BØ (2007) Investigating the metabolic capabilities of Mycobacterium tuberculosis H37Rv using the in silico strain iNJ661 and proposing alternative drug targets. BMC Syst Biol 1:26. https://doi.org/10.1186/1752-0509-1-26
19. Duarte NC, Becker SA, Jamshidi N, Thiele I, Mo ML, Vo TD, Srivas R, Palsson BØ (2007) Global reconstruction of the human metabolic network based on genomic and bibliomic data. Proc Natl Acad Sci U S A 104(6):1777–1782. https://doi.org/10.1073/pnas.0610772104
20. Sigurdsson MI, Jamshidi N, Steingrimsson E, Thiele I, Palsson BØ (2010) A detailed genome-wide reconstruction of mouse metabolism based on human Recon 1. BMC Syst Biol 4(1):140
21. Bordbar A, Lewis NE, Schellenberger J, Palsson BØ, Jamshidi N (2010) Insight into human alveolar macrophage and M. tuberculosis interactions via metabolic reconstructions. Mol Syst Biol 6(1):422
22. Bordbar A, Feist AM, Usaite-Black R, Woodcock J, Palsson BO, Famili I (2011) A multi-tissue type genome-scale metabolic network for analysis of whole-body systems physiology. BMC Syst Biol 5:180. https://doi.org/10.1186/1752-0509-5-180
23. Bordbar A, Palsson BO (2012) Using the reconstructed genome-scale human metabolic network to study physiology and pathology. J Intern Med 271(2):131–141
24. Huthmacher C, Hoppe A, Bulik S, Holzhütter H-G (2010) Antimalarial drug targets in Plasmodium falciparum predicted by stage-specific metabolic network analysis. BMC Syst Biol 4:120. https://doi.org/10.1186/1752-0509-4-120
25. Wallqvist A, Fang X, Tewari SG, Ye P, Reifman J (2016) Metabolic host responses to malarial infection during the intraerythrocytic developmental cycle. BMC Syst Biol 10(1):58. https://doi.org/10.1186/s12918-016-0291-2
26. Jamshidi N, Raghunathan A (2015) Cell scale host-pathogen modeling: another branch in the evolution of constraint-based methods. Front Microbiol 6:1032. https://doi.org/10.3389/fmicb.2015.01032

27. Devoid S, Overbeek R, DeJongh M, Vonstein V, Best AA, Henry C (2013) Automated genome annotation and metabolic model reconstruction in the SEED and Model SEED. Methods Mol Biol 985:17–45. https://doi.org/10.1007/978-1-62703-299-5_2
28. Wang Y, Eddy JA, Price ND (2012) Reconstruction of genome-scale metabolic models for 126 human tissues using mCADRE. BMC Syst Biol 6:153. https://doi.org/10.1186/1752-0509-6-153
29. Blazier AS, Papin JA (2012) Integration of expression data in genome-scale metabolic network reconstructions. Front Physiol 3:299. https://doi.org/10.3389/fphys.2012.00299
30. Machado D, Herrgård M (2014) Systematic evaluation of methods for integration of transcriptomic data into constraint-based models of metabolism. PLoS Comput Biol 10(4):e1003580. https://doi.org/10.1371/journal.pcbi.1003580
31. Bordbar A, Jamshidi N, Palsson BO (2011) iAB-RBC-283: a proteomically derived knowledge-base of erythrocyte metabolism that can be used to simulate its physiological and patho-physiological states. BMC Syst Biol 5:110. https://doi.org/10.1186/1752-0509-5-110
32. Radrich K, Tsuruoka Y, Dobson P, Gevorgyan A, Swainston N, Baart G, Schwartz J-M (2010) Integration of metabolic databases for the reconstruction of genome-scale metabolic networks. BMC Syst Biol 4(1):114
33. Becker SA, Palsson BO (2008) Context-specific metabolic networks are consistent with experiments. PLoS Comput Biol 4(5):e1000082. https://doi.org/10.1371/journal.pcbi.1000082
34. Lakshmanan M, Koh G, Chung BK, Lee D-Y (2012) Software applications for flux balance analysis. Briefings Bioinf 15(1):108–122. bbs069
35. Waltemath D, Karr JR, Bergmann FT, Chelliah V, Hucka M, Krantz M, Liebermeister W, Mendes P, Myers CJ, Pir P, Alaybeyoglu B, Aranganathan NK, Baghalian K, Bittig AT, Burke PEP, Cantarelli M, Chew YH, Costa RS, Cursons J, Czauderna T, Goldberg AP, Gomez HF, Hahn J, Hameri T, Gardiol DFH, Kazakiewicz D, Kiselev I, Knight-Schrijver V, Knupfer C, Konig M, Lee D, Lloret-Villas A, Mandrik N, Medley JK, Moreau B, Naderi-Meshkin H, Palaniappan SK, Priego-Espinosa-D, Scharm M, Sharma M, Smallbone K, Stanford NJ, Song J-H, Theile T, Tokic M, Tomar N, Toure V, Uhlendorf J, Varusai TM, Watanabe LH, Wendland F, Wolfien M, Yurkovich JT, Zhu Y, Zardilis A, Zhukova A, Schreiber F (2016) Toward community standards and software for whole-cell modeling. IEEE Trans Biomed Eng 63(10):2007–2014. https://doi.org/10.1109/TBME.2016.2560762
36. Ebrahim A, Almaas E, Bauer E, Bordbar A, Burgard AP, Chang RL, Dräger A, Famili I, Feist AM, Fleming RM (2015) Do genome-scale models need exact solvers or clearer standards? Mol Syst Biol 11(10):831
37. Dräger A, Palsson BØ (2014) Improving collaboration by standardization efforts in systems biology. Front Bioeng Biotechnol 2:61. https://doi.org/10.3389/fbioe.2014.00061
38. Bornstein BJ, Keating SM, Jouraku A, Hucka M (2008) LibSBML: an API library for SBML. Bioinformatics 24(6):880–881. https://doi.org/10.1093/bioinformatics/btn051
39. Godinez I, Haneda T, Raffatellu M, George MD, Paixão TA, Rolán HG, Santos RL, Dandekar S, Tsolis RM, Bäumler AJ (2008) T cells help to amplify inflammatory responses induced by Salmonella enterica serotype Typhimurium in the intestinal mucosa. Infect Immun 76(5):2008–2017. https://doi.org/10.1128/IAI.01691-07
40. Weininger D (1988) SMILES, a chemical language and information system. 1. Introduction to methodology and encoding rules. J Chem Inf Comput Sci 28(1):31–36
41. Degtyarenko K, De Matos P, Ennis M, Hastings J, Zbinden M, McNaught A, Alcántara R, Darsow M, Guedj M, Ashburner M (2008) ChEBI: a database and ontology for chemical entities of biological interest. Nucleic Acids Res 36(suppl 1):D344–D350
42. Schuetz R, Kuepfer L, Sauer U (2007) Systematic evaluation of objective functions for predicting intracellular fluxes in Escherichia coli. Mol Syst Biol 3:119. https://doi.org/10.1038/msb4100162
43. Oberhardt MA, Puchałka J, Martins dos Santos VAP, Papin JA (2011) Reconciliation of genome-scale metabolic reconstructions for comparative systems analysis. PLoS Comput Biol 7(3):e1001116. https://doi.org/10.1371/journal.pcbi.1001116
44. Thiele I, Jamshidi N, Fleming RMT, Palsson BØ (2009) Genome-scale reconstruction of Escherichia coli's transcriptional and translational machinery: a knowledge base, its mathematical formulation, and its functional characterization. PLoS Comput Biol 5(3):e1000312. https://doi.org/10.1371/journal.pcbi.1000312

45. Zomorrodi AR, Islam MM, Maranas CD (2014) d-OptCom: dynamic multi-level and multi-objective metabolic modeling of microbial communities. ACS Synth Biol 3(4):247–257. https://doi.org/10.1021/sb4001307
46. Nagrath D, Avila-Elchiver M, Berthiaume F, Tilles AW, Messac A, Yarmush ML (2007) Integrated energy and flux balance based multi-objective framework for large-scale metabolic networks. Ann Biomed Eng 35(6):863–885. https://doi.org/10.1007/s10439-007-9283-0
47. Burgard AP, Pharkya P, Maranas CD (2003) Optknock: a bilevel programming framework for identifying gene knockout strategies for microbial strain optimization. Biotechnol Bioeng 84(6):647–657. https://doi.org/10.1002/bit.10803
48. Harcombe WR, Riehl WJ, Dukovski I, Granger BR, Betts A, Lang AH, Bonilla G, Kar A, Leiby N, Mehta P, Marx CJ, Segrè D (2014) Metabolic resource allocation in individual microbes determines ecosystem interactions and spatial dynamics. Cell Rep 7(4):1104–1115. https://doi.org/10.1016/j.celrep.2014.03.070
49. Mahadevan R, Schilling CH (2003) The effects of alternate optimal solutions in constraint-based genome-scale metabolic models. Metab Eng 5(4):264–276
50. Schellenberger J, Palsson BØ (2009) Use of randomized sampling for analysis of metabolic networks. J Biol Chem 284(9):5457–5461. https://doi.org/10.1074/jbc.R800048200

Chapter 10

Metabolic Model Reconstruction and Analysis of an Artificial Microbial Ecosystem

Chao Ye, Nan Xu, Xiulai Chen, and Liming Liu

Abstract

Microbial communities are widespread in the environment, and to isolate and identify species or to determine relations among microorganisms, some 'omics methods like metagenomics, proteomics, and metabolomics have been used. When combined with various 'omics data, models known as artificial microbial ecosystems (AME) are powerful methods that can make functional predictions about microbial communities. Reconstruction of an AME model is the first step for model analysis. Many techniques have been applied to the construction of AME models, e.g., the compartmentalization approach, community objectives method, and dynamic analysis approach. Of these approaches, species compartmentalization is the most relevant to genetics. Besides, some algorithms have been developed for the analysis of AME models. In this chapter, we present a general protocol for the use of the species compartmentalization method to reconstruct a model of microbial communities. Then, the analysis of an AME is discussed.

Key words Microbial communities, Omics methods, Artificial microbial ecosystem models

1 Introduction

Microbiome community ecology (or environmental microbiology) is the science that studies the interrelations between microorganisms and their biotic and abiotic environments [1]. It deals with the three major domains of life—Eukaryota, Archaea, and Bacteria—as well as viruses. A microbial ecosystem, which consists of various microorganisms, plays a primary role in the regulation of biogeochemical systems in virtually all of our planet's environments, including some of the most extreme, from frozen environments and acidic lakes, to hydrothermal vents at the bottom of the deepest oceans, and some of the most familiar, such as the human small intestine [2].

The relations in microbial communities are mostly classified into competition, mutualism, and parasitism. Competition is widely present in both the intraspecific and interspecific relations. It happens when nutrition or space is limited. The results of

Marco Fondi (ed.), *Metabolic Network Reconstruction and Modeling: Methods and Protocols*, Methods in Molecular Biology, vol. 1716, https://doi.org/10.1007/978-1-4939-7528-0_10, © Springer Science+Business Media, LLC 2018

competition can lead to both growth inhibition or elimination of a certain organism. Mutualism means two or more organisms living together, supplying extra nutrition not present in the environment; this situation can create conditions more suitable for survival. Then, both species can grow better in coculture than in monoculture. Parasitism can be described as one species acquiring nutrition from another species. The former is called a parasite, and the latter is a host. The host provides not only nutrition but also the living place for the parasite. The result of parasitism is usually the death of the host. The relations among microorganisms in a microbial ecosystem have been widely used for environmental remediation, industrial production, and human health.

There are two key approaches to the research on microbial ecology. One of these is to identify and quantify the diversity of microorganisms, and the other is to reveal activity and the pathophysiological function of a microorganism in the community. Because the microbial ecosystem is complicated, most of these microorganisms are nonculturable. For example, in the gut environment, 40–80% of microorganisms cannot be cultured in vitro; this situation has limited the research into the diversity of gut microorganisms with traditional culture methods. Because there may be both mutualism and competition between two species, the exact function of a typical microorganism in a microbial community is difficult to identify. Besides, in a microorganism, detailed differences in proteins and metabolites between monoculture and coculture cannot be identified systemically by conventional detection methods.

To overcome these problems, many 'omics strategies, such as metagenomics, proteomics, and metabolomics, have been applied to exploration of microbial ecosystems. Metagenomics is the study of microbial metagenomes from environmental samples and is independent of the ability to cultivate microbes in the laboratory. Metagenomics not only overcomes the difficulty with culturing of microorganisms but also can be combined with bioinformatics methods. The research on metagenomics has become a hotspot and frontier of microbial research. It has been widely applied to the analysis of environmental biodiversity, climate change, extreme environments, the human intestinal tract, oil pollution remediation, and bio metallurgy [3–6]. Proteomics is an established discipline concerned with studying protein composition and function in various types of cells, tissues, or body fluids on a large-scale, high-throughput, and systematic level. The development of proteomic technologies has greatly stimulated the progress of proteomic research, and these technologies have been widely used in various research fields, such as the study of wastewater sludge microbial communities, which are utilized for enhanced biological removal of phosphorus [7]. Metabolomics is the scientific study of chemical processes involving metabolites. Specifically, metabolomics is the "systematic study of the unique chemical fingerprints that specific cellular processes leave behind": the study of their small-molecule metabolite profiles [8]. Ma et al. analyzed an artificial microbial community for two-step production of vitamin C combined with proteomics and metabolomics analyses [9]. Based on these 'omics technologies, it was proved that *Bacillus megaterium* can secrete essential metabolites for the growth of *Ketogulonicigenium vulgare* and for vitamin C production.

Although some new strategies have been applied to the research on microbial ecosystems, one of the challenges of systems biology and functional genomics is to integrate proteomic, transcriptomic, and metabolomic information to gain a better understanding of cellular biology. A genome scale metabolic model (GSMM) is a new systems biology tool, which integrates multi-omics data, including genomics, transcriptomics, proteomics, and metabolomics. The core task related to GSMMs is to build the relations gene-protein-reaction. Since the first *Haemophilus influenza* GSMM was constructed in 1999, GSMMs have been developed for more than 15 years. Over 100 GSMMs have been constructed for various organisms (Fig. 1). Some typical organisms, such as *Escherichia coli* and *Saccharomyces cerevisiae*, have more than one GSMM [10] (http://imgmd.jiangnan.edu.cn/database/).

GSMMs have commonly been used in metabolic engineering, model-driven discovery, prediction of cellular phenotypes, analysis of biological network properties, studies of evolutionary processes, and interspecies interactions [11–13].

1. Metabolic engineering: using the OptForce algorithm, Ranganathan et al. identified some common but mostly chain-specific genetic interventions pointing to the possibility of offline tuning of overproduction of specific fatty acids (in terms of chain length). In accordance with the OptForce prioritization of interventions, *fabZ* and acyl-ACP thioesterase were upregulated and *fadD* was deleted to create a strain that produces 1.70 g/L or 0.14 g of a fatty acid per gram of glucose (~39% maximum theoretical yield; C14–16 fatty acids) in the minimal M9 medium [14].
2. Model-driven discovery: discrepancies between GSMM predictions and empirical results can point to gaps in current knowledge or to areas with functional discrepancies. This approach in turn allows one to systematically formulate testable hypotheses. For example, incorrect predictions about *talAB* mutants grown on xylose led the authors to discover a novel pathway catalyzed by the gene products of *pfkA* and *fbaA* [15].
3. Prediction of cellular phenotypes: for simple organisms, the physiology of bacteria is remarkably complex. The diverse set of biochemical pathways in bacteria has a vast phenotypic

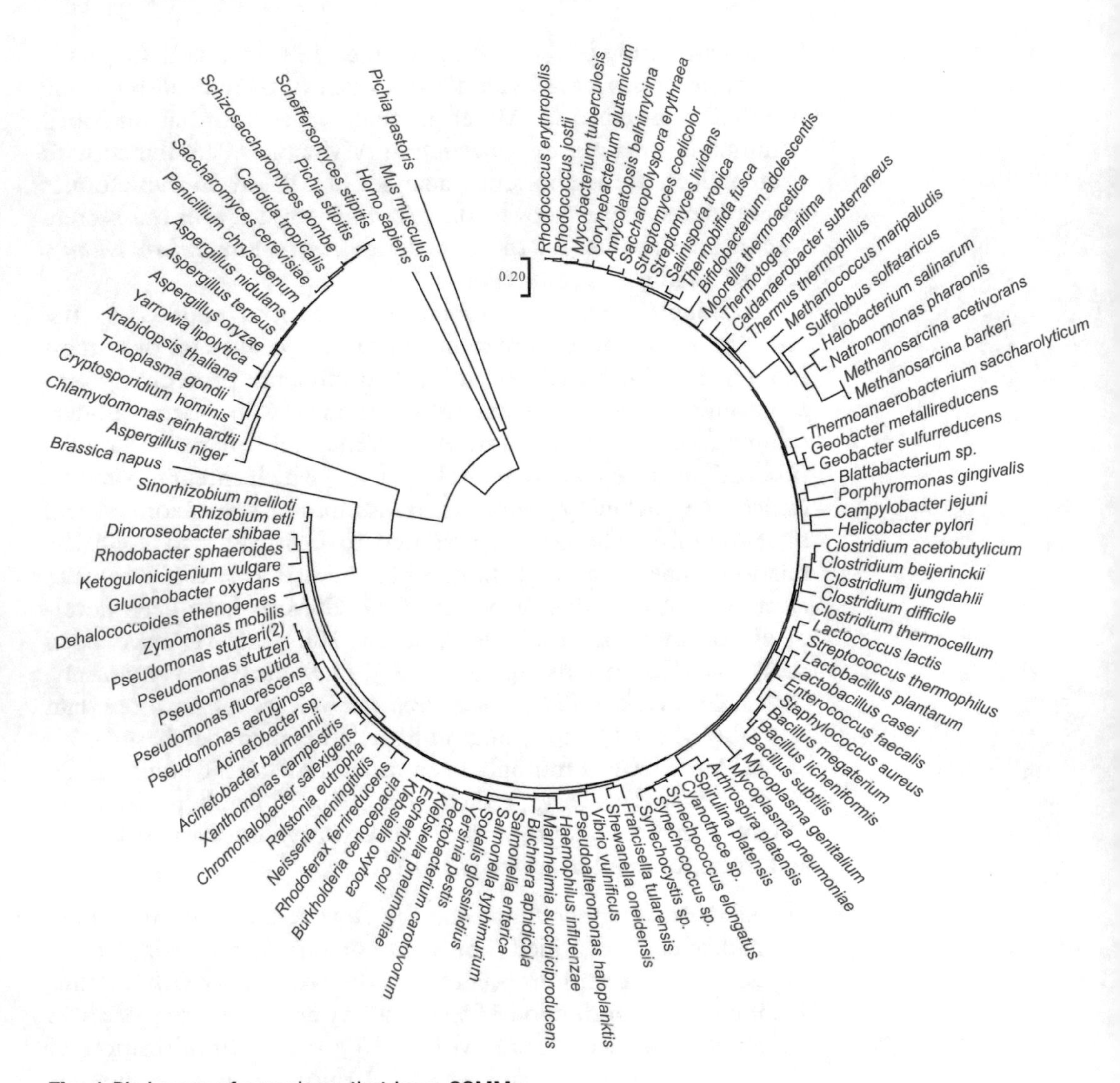

Fig. 1 Phylogeny of organisms that have GSMMs

potential that has enabled the bacteria to thrive in a plethora of environments ranging from volcanic vents on the bottom of the ocean, clouds, glaciers, and the human gut. To understand this phenotypic potential, researchers have turned to GSMMs to interpret and predict cellular phenotypes. Constraint-based modeling with GSMMs has allowed researchers to rapidly predict growth phenotypes in various genetic [16] and environmental conditions [17], to explore different objectives of microbial metabolism [18, 19], to examine the driving force behind cellular function, and to better understand the suboptimal behavior of cells after a perturbation [20] and latent pathway activation [21].

4. Analysis of biological network properties: In network analysis, biochemical reactions are transformed into a unipartite or bipartite graph, where the nodes and links take the form of metabolites and enzymatic reactions. Once formulated as a graph, the network can be sampled and explored using a variety of minimally biased mathematical and algorithmic methods to arrive at biologically insightful conclusions. For example, synthetic lethals—which are defined as two genes whose independent deletion is not lethal, but simultaneous elimination is lethal—are often a consequence of network redundancy or parallel pathways [22]. The converse of synthetic lethals, synthetic rescues, are defined as a gene pair where the deletion of one of the genes is lethal, but the simultaneous deletion of both genes is nonlethal. Synthetic rescues can be used to rescue a nonviable single-gene deletion phenotype by rewiring the network in such a way as to compensate for the deleterious effect of the initial genetic perturbation [23, 24].
5. Studies of evolutionary processes: Computational frameworks using GSMMs have been employed to simulate bacterial evolution through random gene deletions. These studies have shown that there appears to be a conserved reaction set that is similar among organisms with similar lifestyles [25]; this state of affairs reflects the common enzymatic machinery necessary to metabolize specific carbon sources. It has also been shown that although genes are lost at random, the order in which genes are lost follows a coordinated and consistent pattern, 40% of which can be accounted for by the metabolic model when compared with available phylogenic data [26]. The *E. coli* GSMM also provides a context in which phylogeny data can be understood. Comparative genomics in the context of constraint-based modeling by means of the *E. coli* GSMM has led investigators to assert that the dominant mechanism of bacterial evolution in *E. coli* is horizontal gene transfer. This transfer is highly dependent on the genomic content of the organism [27, 28] and involves genes that are mostly environment specific and located at the periphery of the metabolic network [29].
6. Interspecies interactions: An artificial microbial ecosystem (AME) consisting of *K. vulgare* and *B. megaterium* is currently used in a two-step fermentation process for vitamin C production. In order to gain a comprehensive understanding of the metabolic interactions between the two bacteria, a two-species stoichiometric metabolic model (*i*WZ-KV-663-BM-1055) consisting of 1718 genes, 1573 metabolites, and 1891 reactions (excluding exchange reactions) was constructed on the basis of separate GSMMs of *K. vulgare* and *B. megaterium* [30]. By means of this AME, the relation between *K. vulgare*

and *B. megaterium* was identified as both mutualism and competition.

Recently, based on a GSMM, many AMEs have been constructed. Since the first published AME in 2007 (a mutualistic microbial community) [31], the accumulating body of work has highlighted many unique challenges related to microbial community modeling. To date, over 25 AMEs have been built and applied to the analysis of microbial communities [32]. These AMEs have also been applied to research into the relations within microbial communities. For example, an AME consisting of *Geobacter sulfurreducens* and *Rhodoferax ferrireducens* was identified as competition in anoxic subsurface environments, and this relation can have an influence on the in situ bioremediation of uranium-contaminated groundwater [33]. In the two-step production of vitamin C, an AME consisting of *K. vulgare* and *B. megaterium* was constructed. The relations between *K. vulgare* and *B. megaterium* were predicted to be both mutualistic and competitive on the basis of the AME [30]. Besides, to understand the genotype–phenotype relations in *Mycoplasma tuberculosis* and human alveolar macrophages, a host–pathogen AME named *i*AB-AMØ-1410-Mt-661 was built [34].

Because the compartmentalization approach is widely applied to the construction of AMEs [32], this chapter is intended as a general protocol for the reconstruction of an AME (within the compartmentalization framework). In addition, we will describe the main tools used for AME analysis.

2 Materials

The compartmentalization framework is an intuitive and simple way to represent microbial interactions. This approach has been used more frequently than any others, and provides mechanistic insights into community metabolism, in good agreement with experiments [32]. In this section, we describe what is needed to start to reconstruct a metabolic model of an artificial microbial ecosystem based on compartmentalization.

2.1 Genome Sequence

To start the reconstruction of a model, you will need a FASTA file embedding a set of contigs or coding sequences of the genome you want to analyze. GenBank (USA, http://www.ncbi.nlm.nih.gov/genbank/), EMBL (Europe, https://www.ebi.ac.uk/), and DDBJ (Japan, http://www.ddbj.nig.ac.jp/) are three famous DNA databases that accumulate gene and genome sequence information. For example, GenBank contains 11453 organism genomes (8859 bacterial, 2010 eukaryotic, and 584 archaeal, as of September 18, 2016). Almost any original sequenced genomes can be downloaded from these databases.

Table 1
Selected characteristics of software platforms for reconstruction and simulation of metabolic networks

	Model SEED	RAVEN	COBRA toolbox	SuBliMinal	IMGMD
Input	Genome annotated in RAST	Annotated genome sequence	GSMM	Species name	Species genome sequence
Reference database	SEED	KEGG	N/A	KEGG, MetaCyc	IMGMD
Interface	Web	Matlab	Matlab	Command line	Web
License	Free	Free (requires a Matlab license)	Free (requires a Matlab license)	Free	Free
Output	SBML, Excel	SBML, Excel	SBML, Excel	SBML	Excel
Supports simulations	Yes	Yes	Yes	No	No

2.2 Tools for Auto-reconstruction of a Model

Because the process of model construction is complex, an interesting protocol was published in 2010 to address this issue. In this protocol, model construction is subdivided into four parts: creation of a draft reconstruction, manual reconstruction refinement, conversion from reconstruction to a mathematical model, and network evaluation. All these modules contain 96 steps (specific operations). The manual model construction process is time-consuming, which can take up to half a year or more. Some software or platforms for model auto-reconstruction have been developed (Table 1), including Model SEED [35], RAVEN [36], COBRA Toolbox [37], SuBliMinal [38], and IMGMD [10]. These tools have their advantages and disadvantages [39]. For instance, Model SEED is a web service that includes the RAST genome annotation tool. Based on the annotation results, a model for a specific organism can be reconstructed automatically. Given that the RAST service (http://rast.nmpdr.org/rast.cgi) can only annotate prokaryotes, Model SEED has limited applicability to eukaryotes. Besides, model construction by Model SEED will take a long time, according to the job numbers. IMGMD is also a web platform that serves for model construction. It is based on the results of genome homologous alignment. Users can upload a target organism's genome sequence and choose relevant parameters. After the submission of the job to IMGMD, results will be returned within 1 day. Nonetheless, a model constructed by IMGMD is a draft model. It still needs to be further processed to obtain a GSMM. The COBRA Toolbox is based on the Matlab platform, which is commonly used for model

Table 2
Data sources frequently used for metabolic reconstruction

Name	Link	References
Biochemical databases		
KEGG	http://www.genome.jp/kegg/	[40]
BRENDA	http://www.brenda-enzymes.info/	[41]
TCDB	http://www.tcdb.org/	[42]
MetaCyc	http://www.metacyc.org/	[43]
Organism-specific databases		
EcoCyc	http://ecocyc.org/	[44]
SGD	http://www.yeastgenome.org/	[45]
Protein localization databases		
PSORT	http://www.psort.org/psortb/	[46]
CELLO	http://cello.life.nctu.edu.tw/	[47]
COBRA simulation environments		
COBRA toolbox	http://opencobra.github.io/cobratoolbox/	[37]
libSBML	http://sbml.org/Software/libSBML	

construction. The COBRA Toolbox requires users to have basic Matlab knowledge and an advanced computer configuration for model analysis.

2.3 Databases

The construction of an AME requires many online databases and software packages. By the function during model construction, these databases can be classified into biochemical databases, organism-specific databases, protein localization databases, and COBRA simulation environments (Table 2) [48].

2.4 Modeling Resource Availability

As in the compartmentalization framework for AME construction, two separate GSMMs were first modified as different cell compartments in an AME. It is necessary to search for the constructed GSMMs in related model databases. There are some relevant databases, such as IMGMD [10] (http://imgmd.jiangnan.edu.cn/database/index.html) and BIGG Models [49] (http://bigg.ucsd.edu/), where a researcher can download standardized GSMMs. The IMGMD database has accumulated 169 published GSMMs of microorganisms since 1999. BIGG Models contain more than 75 high-quality manually curated genome scale metabolic models. If species constituting the microbial community are found in corresponding models in these databases, these models can be used as the foundation for AME construction. Otherwise, users should construct basic GSMMs de novo.

3 Methods

3.1 Genome Annotation

Genome annotation is the process of attaching biological information to sequences [50]. It involves three main steps: (1) identification of portions of the genome that do not code for proteins; (2) identification of elements in the genome, a process called *gene prediction*; and (3) attaching biological information to these elements. Generally, genome-wide annotation of gene structures is subdivided into two distinct phases. In the first phase, the "computation" phase, expressed sequence tags (ESTs), proteins, and other components are aligned to the genome and then *ab initio* and/or evidence-driven gene predictions are generated. In the second phase, the "annotation" phase, these data are synthesized into gene annotations [51]. To date, a variety of software tools have been developed to enable scientists to view and share genome annotations [51, 52]. Among these tools, KAAS [53] and RAST [35, 54] are often used for genome annotation for the construction of a model, automatically (Figs. 2 and 3).

3.2 Construction of a Draft Model of an Individual Species

The core of a GSMM is the association gene-protein-reaction. According to the genome annotation results, a gene and protein have been identified. By means of a metabolic database like KEGG and MetaCyc, reactions can be linked to genes and proteins, manually (Table 3). Besides, some tools, e.g., SEED [55], BIGG Models [49], and IMGMD [10], can automatically generate the gene-protein-reaction relation. Besides, when a model is constructed, there are several steps that still require manual curation to ensure quality and accuracy of the metabolic model [56].

3.3 Model Refinement

Once the draft model has been constructed, it still needs to be refined. According to the protocol of model construction, there are 32 steps related to the refinement of a reconstructed model [48]. Some key steps such as determining reaction directionality, determining the subcellular location, addition of extracellular and periplasmic transport reactions and a biomass reaction are described below.

1. Determining reaction directionality: because the reactions obtained from the KEGG database are all reversible (Fig. 2a), some reactions from other reference models may not be correct. These reactions should be confirmed by means of literature data and some metabolic databases such as MetaCyc [43] and Biopath [57].

```
▼ Carbohydrate metabolism
  ▼ 00010 Glycolysis / Gluconeogenesis [PATH:ko00010]
      Gglean000479.1; K00844  HK; hexokinase [EC:2.7.1.1]
      Gglean007239.1; K00844  HK; hexokinase [EC:2.7.1.1]
      Gglean008926.1; K00844  HK; hexokinase [EC:2.7.1.1]
      Gglean009355.1; K00844  HK; hexokinase [EC:2.7.1.1]
      Gglean009902.1; K01810  GPI, pgi; glucose-6-phosphate isomerase [EC:5.3.1.9]
      Gglean010018.1; K00850  pfkA, PFK; 6-phosphofructokinase 1 [EC:2.7.1.11]
      Gglean005205.1; K03841  FBP, fbp; fructose-1,6-bisphosphatase I [EC:3.1.3.11]
      Gglean013039.1; K01624  FBA, fbaA; fructose-bisphosphate aldolase, class II [EC:4.1.2.13]
      Gglean002382.1; K01624  FBA, fbaA; fructose-bisphosphate aldolase, class II [EC:4.1.2.13]
      Gglean005903.1; K01624  FBA, fbaA; fructose-bisphosphate aldolase, class II [EC:4.1.2.13]
      Gglean008940.1; K01803  TPI, tpiA; triosephosphate isomerase (TIM) [EC:5.3.1.1]
      Gglean010089.1; K01803  TPI, tpiA; triosephosphate isomerase (TIM) [EC:5.3.1.1]
      Gglean011318.1; K00134  GAPDH, gapA; glyceraldehyde 3-phosphate dehydrogenase [EC:1.2.1.12]
      Gglean007599.1; K00927  PGK, pgk; phosphoglycerate kinase [EC:2.7.2.3]
      Gglean009638.1; K15633  gpmI; 2,3-bisphosphoglycerate-independent phosphoglycerate mutase [EC:5.4.2.12]
      Gglean007418.1; K15634  gpmB; probable phosphoglycerate mutase [EC:5.4.2.12]
      Gglean006460.1; K01689  ENO, eno; enolase [EC:4.2.1.11]
      Gglean013006.1; K00873  PK, pyk; pyruvate kinase [EC:2.7.1.40]
      Gglean012210.1; K00161  PDHA, pdhA; pyruvate dehydrogenase E1 component alpha subunit [EC:1.2.4.1]
      Gglean007026.1; K00162  PDHB, pdhB; pyruvate dehydrogenase E1 component beta subunit [EC:1.2.4.1]
      Gglean011586.1; K00627  DLAT, aceF, pdhC; pyruvate dehydrogenase E2 component (dihydrolipoamide acetyltransferase) [EC:2.3.1.12]
      Gglean002292.1; K00382  DLD, lpd, pdhD; dihydrolipoamide dehydrogenase [EC:1.8.1.4]
      Gglean013235.1; K00016  LDH, ldh; L-lactate dehydrogenase [EC:1.1.1.27]
      Gglean006373.1; K00016  LDH, ldh; L-lactate dehydrogenase [EC:1.1.1.27]
      Gglean001791.1; K01568  PDC, pdc; pyruvate decarboxylase [EC:4.1.1.1]
      Gglean009676.1; K01568  PDC, pdc; pyruvate decarboxylase [EC:4.1.1.1]
      Gglean013336.1; K00121  frmA, ADH5, adhC; S-(hydroxymethyl)glutathione dehydrogenase / alcohol dehydrogenase [EC:1.1.1.284 1.1.1.1]
      Gglean013326.1; K13953  adhP; alcohol dehydrogenase, propanol-preferring [EC:1.1.1.1]
      Gglean002045.1; K13953  adhP; alcohol dehydrogenase, propanol-preferring [EC:1.1.1.1]
      Gglean002668.1; K13953  adhP; alcohol dehydrogenase, propanol-preferring [EC:1.1.1.1]
      Gglean002978.1; K13953  adhP; alcohol dehydrogenase, propanol-preferring [EC:1.1.1.1]
```

Fig. 2 KEGG Orthology (KO) assignment results of KAAS genome annotation

contig_id	feature_id	type	location	start	stop	strand	function	aliases	figfam	evidence_codes	nucleotide_sequence	aa_sequence
gi|6317992	fig|1051650.	peg	gi|63179929	1	1350	+	Chromosomal replication initiator		FIG0000	icw(2);DNA_repli	atgcccaatttagaggaactttg	MPNLEELWAYLNDK
gi|6317992	fig|1051650.	peg	gi|63179929	1523	2662	+	DNA polymerase III beta subunit (		FIG0006	icw(2);DNA_repli	atgaaatttacgatcacccgatc	MKFTITRSTFLKTLNI
gi|6317992	fig|1051650.	peg	gi|63179929	2779	2663	-	hypothetical protein				atgtggctgctttgcattttgatgc	MWLLCILMRTLSVHF
gi|6317992	fig|1051650.	peg	gi|63179929	3240	3452	+	FIG002958: hypothetical protein		FIG0000	icw(2);DNA_repli	atgacaacaatcgaaatcacg	MTTIEITTPFLTLGQF
gi|6317992	fig|1051650.	peg	gi|63179929	3449	4564	+	DNA recombination and repair pro		FIG0000	isu;DNA_repair,_	atgaaactggatcacttggtgct	MKLDHLVLKNYRNY
gi|6317992	fig|1051650.	peg	gi|63179929	4796	4641	-	Alpha-galactosidase (EC 3.2.1.22		FIG0000	idu(21);Galactos	atgcgttcactggcacagaaac	MRSLAQKPAHKDLE
gi|6317992	fig|1051650.	peg	gi|63179929	4817	6778	+	DNA gyrase subunit B (EC 5.99.1		FIG0002	isu;DNA_gyrase_	gtgacggacaagaaagaaac	MTDKKETAEEKKDE
gi|6317992	fig|1051650.	peg	gi|63179929	6840	9461	+	DNA gyrase subunit A (EC 5.99.1		FIG0000	icw(1);DNA_gyra	atggatgatcgccaagaaagc	MDDRQESRITNVNL
gi|6317992	fig|1051650.	peg	gi|63179929	10270	9566	-	Deoxyribose-phosphate aldolase (		FIG0000	idu(1);Deoxyribos	atgacagcttatactttagatcaa	MTAYTLDQFSRMIDH
gi|6317992	fig|1051650.	peg	gi|63179929	10546	10403	-	Alpha-galactosidase (EC 3.2.1.22		FIG0000	idu(21);Galactos	gtgcagcaacctgcacataag	MQQPAHKDLDRND
gi|6317992	fig|1051650.	peg	gi|63179929	10577	11008	+	Single-stranded DNA-binding prot		FIG0007	icw(1);DNA_repa	atgttgaatagtgtgagtctaac	MLNSVSLTGRLTKEF
gi|6317992	fig|1051650.	peg	gi|63179929	11136	11432	+	SSU ribosomal protein S6p		FIG0000	isu;Ribosome_S	atggctgaaaccaaatatgaa	MAETKYEVTYIIRPD
gi|6317992	fig|1051650.	peg	gi|63179929	11463	12059	+	Single-stranded DNA-binding prot		FIG0007	icw(1);DNA_repa	atgcttaacagtgttgcattgaca	MLNSVALTGRLTRD
gi|6317992	fig|1051650.	peg	gi|63179929	12146	12382	+	SSU ribosomal protein S18p @ S		FIG0000	isu;Staphylococc	atggcacaacaacgccgtggt	MAQQRRGGRRRRK
gi|6317992	fig|1051650.	peg	gi|63179929	12824	14302	+	Cytochrome d ubiquinol oxidase s		FIG0009	idu(1);Terminal_	atggggctcacgttagccacga	MGLTLATIPLAVTNL
gi|6317992	fig|1051650.	peg	gi|63179929	14295	14984	+	Cytochrome d ubiquinol oxidase s		FIG0000	icw(1);Bacterial_	atgcctagcctttcaaatttacaa	MPSLSNLQVIFLGIIS
gi|6317992	fig|1051650.	peg	gi|63179929	15899	15069	-	Mobile element protein		FIG0130	ff	atgattcaagatcaattgagccg	MIQDQLSRGHRITVI
gi|6317992	fig|1051650.	peg	gi|63179929	16204	15929	-	Mobile element protein		FIG0130	ff	atgaccaatacagctattcgcta	MTNTAIRYTPEFKQT
gi|6317992	fig|1051650.	peg	gi|63179929	16318	16704	+	Cytochrome d ubiquinol oxidase subunit II (			icw(1);Bacterial_	ttgtttgccatcttattgccgtttaa	MFAILLPFNTQFFTN
gi|6317992	fig|1051650.	peg	gi|63179929	16728	16925	+	FIG00743550: hypothetical protei		FIG0074	ff	atgatgaaaaagatctggatca	MMKKIWINIFWGLLV
gi|6317992	fig|1051650.	peg	gi|63179929	17591	16998	-	Predicted integral membrane prot		FIG0043	ff	atgaataaaggtcgagttgaag	MNKGRVEAFTDAVIA
gi|6317992	fig|1051650.	peg	gi|63179929	18383	17988	-	hypothetical protein				ttgcttgataaagttacttttaaaa	MLDKVTFKMYLRFSI
gi|6317992	fig|1051650.	peg	gi|63179929	18898	18596	-	FIG00749270: hypothetical protei		FIG0074	ff	atggcttatctcaattcaattttac	MAYLNSILQLPMALL
gi|6317992	fig|1051650.	peg	gi|63179929	20034	19177	-	CAAX amino terminal protease fa		FIG0135	ff	atggtcgataacagaactttgct	MVDNRTLLKEIIAVSL
gi|6317992	fig|1051650.	peg	gi|63179929	20296	20114	-	FIG007491: hypothetical protein Y		FIG0134	ff	atgacatcattaactatgttttcca	MTSLTMFSTLSLQD
gi|6317992	fig|1051650.	peg	gi|63179929	20295	20417	+	hypothetical protein				atggaatacctccgatttttcag	MEYLRFFQYLVATG
gi|6317992	fig|1051650.	peg	gi|63179929	20929	20753	-	hypothetical protein				atgtttgatcgcattccttgtgact	MFDRIPCDFFFALRK
gi|6317992	fig|1051650.	peg	gi|63179929	22250	21015	-	FIG00743123: hypothetical protei		FIG0074	ff	atgaaattcaataaagtcatga	MKFNKVMITLVAAVT
gi|6317992	fig|1051650.	peg	gi|63179929	25027	22454	-	FIG00906165: hypothetical protei		FIG0090	ff	atgaaaccactaactaaaaat	MKPLTKNLWRNIRD
gi|6317992	fig|1051650.	peg	gi|63179929	25741	25040	-	ABC transporter ATP-binding prot		FIG0000	idu(4);Bacitracin_	atggcatatatcgaagtcaaac	MAYIEVKHESKRYK
gi|6317992	fig|1051650.	peg	gi|63179929	26021	26572	+	Transcriptional regulator		FIG0131	ff	atgtcggggacgaaggataat	MSGTKDNRRVQYT
gi|6317992	fig|1051650.	peg	gi|63179929	26616	27422	+	Predicted hydrolase of the HAD su		FIG0062	ff	atgcagaatgtaaaattaattgc	MQNVKLIASDMDQT
gi|6317992	fig|1051650.	peg	gi|63179929	27427	27747	+	FIG00744246: hypothetical protei		FIG0074	ff	atggcagttatctcaatacgtgt	MAVISIRVAIGVYGFL
gi|6317992	fig|1051650.	peg	gi|63179929	29090	28587	-	hypothetical protein				ttggagttgttacaattaatgcttg	MELLQLMLGVLGFTV
gi|6317992	fig|1051650.	peg	gi|63179929	31260	29110	-	FIG00749963: hypothetical protei		FIG0074	ff	atggcagatgaagcagtacga	MADEAVRAVQKWLN
gi|6317992	fig|1051650.	peg	gi|63179929	32378	31605	-	Predicted permease		FIG0134	ff	atgaatttttcgatctttgcctttcta	MNFSIFAFLLIAGIGA
gi|6317992	fig|1051650.	peg	gi|63179929	32556	33161	+	Predicted integral membrane prot		FIG0043	ff	atgcacatgtttgaaaatagtaa	MHMFENSKSRLDAIS
gi|6317992	fig|1051650.	peg	gi|63179929	33501	33725	+	hypothetical protein				atgaaaaaagttaacatgcata	MKKVNMHNLGTSTIT

Fig. 3 The RAST genome annotation results

2. Determining the subcellular location: depending on the species used for model construction, there are different software tools that can be applied to the prediction of a subcellular location. For eukaryotic organisms, the WoLF PSORT [58] is often chosen, while for prokaryotic organisms (gram-positive, gram-negative, or archaea), the PSORTb software [46] has

Table 3
Gene-protein-reaction association in the draft model of *Ketogulonicigenium vulgare*

Local gene	Protein name	Reaction
KVU_0281	D-lactate dehydrogenase	NAD + (R)-lactic acid <=> H + NADH + Pyruvate
KVU_2458 KVU_PB0044	Malate dehydrogenase	NAD + Malate <=> H + NADH + Oxaloacetate
KVU_PB0117	Malate dehydrogenase (NADP+)	NADP + malate => Pyruvate + CO_2 + NADPH
KVU_1326	Isocitrate dehydrogenase	Isocitrate + NADP <=> CO_2 + 2-Oxoglutarate + NADPH

been used to predict protein subcellular localization. Besides, web server CELLO can be used for both eukaryotic and prokaryotic organisms [59] (Fig. 4).

3. Adding extracellular and periplasmic transport reactions: by means of the TransportDB and TCDB databases, transport proteins can be identified. Then, transport reactions can be added to the draft model, so that some gaps can be filled directly at this step. Besides, after addition of transport reactions between intracellular and extracellular components, the model can take up nutrition from the culture medium or secrete metabolites to the extracellular space.
4. Determining a biomass reaction: this reaction consists of the precursors of DNA, RNA, proteins, lipids, cell wall, vitamins, and cofactors, which are the main cell components. The detailed proportion of each fraction can be obtained by literature mining. For some statins, the lack of fraction data can be overcome by experimental measurement or by means of data from similar strains.
5. Filling gaps in the model: at this step, two tools can be used to identify gaps. GapAnalysis is an algorithm used to identify gaps in a model with the linear programming (LP) solver. In GapAnalysis results, "product" means that a source of the metabolite cannot be found. "Substrate" means that this metabolite cannot be consumed. Besides, GapFind is an algorithm with the mixed-integer linear programming (MILP) solver. Furthermore, "allGaps" means all gaps found by GapFind, "rootGap" means all gaps related to roots without production (or consumption), and "downstreamGaps" means all downstream gaps. It should be noted that not all gaps need to be filled: an investigator must act according to the literature or other biochemical databases.

a

SeqID: tr|F9Y3X6|F9Y3X6_KETVW Bifunctional protein GlmU OS=Ketogulonicigenium vulgare (strain WSH-001) GN=glmU PE=3 SV=1

Analysis Report:

CMSVM+	Unknown	[No details]
CWSVM+	Unknown	[No details]
CytoSVM+	Cytoplasmic	[No details]
ECSVM+	Unknown	[No details]
ModHMM+	Unknown	[No internal helices found]
Motif+	Unknown	[No motifs found]
Profile+	Unknown	[No matches to profiles found]
SCL-BLAST+	Cytoplasmic	[matched 81620441: Bifunctional protein glmU]
SCL-BLASTe+	Unknown	[No matches against database]
Signal+	Unknown	[No signal peptide detected]

Localization Scores:

Cytoplasmic	9.97
CytoplasmicMembrane	0.00
Cellwall	0.01
Extracellular	0.02

Final Prediction:

Cytoplasmic 9.97

----

b

k used for kNN is: 27
YOR014W nucl 21, cyto_nucl 14, cyto 3
YAL016W cyto 9, mito 8, nucl 5, cyto_pero 5
YGL190C nucl 20, cyto 4.5, cyto_mito 3
YJR149W cyto 12.5, cyto_nucl 12, nucl 8.5, mito 3, cysk 3
YJR025C cyto 19, cyto_nucl 12.333, cyto_pero 11.333, nucl 4.5
YIL107C nucl 20.5, cyto_nucl 13, cyto 4.5
YOL136C nucl 17, cyto 8
YHR183W cyto 19, cyto_nucl 11.5, pero 4, nucl 2
YGR256W cyto 22.5, cyto_nucl 12.5, pero 2
YJR155W mito 13, cyto 8.5, cyto_pero 6.5, pero 3.5
YOL165C cyto 14.5, mito 11, cyto_nucl 8
YNL331C cysk 12, mito 7, cyto 5.5, cyto_nucl 4.5
YFL057C mito 11, cysk 10, cyto 4
YCR107W mito 12.5, cyto_mito 10, nucl 7, cyto 6.5
YDL243C cysk 22, cyto 3
YFL056C cysk 16, cyto 4.5, cyto_pero 3.5, mito 3, nucl 2
YGL115W cyto_nucl 15.5, nucl 14, cyto 13
YHR047C cyto 13.5, cyto_nucl 12.5, nucl 10.5
YBL074C nucl 17.5, cyto_nucl 13.5, cyto 8.5
YLR027C cyto 10, cyto_nucl 8.5, pero 8, nucl 5, mito 4
YKL106W mito 17, mito_nucl 12.333, nucl 5.5, cyto_nucl 5.166, cyto 3.5

c

CELLO RESULTS

SeqID: tr|F9Y3X6|F9Y3X6_KETVW Bifunctional protein GlmU OS=Ketogulonicigenium vulgare (strain WSH-001) GN=glmU PE=3 SV=1

Analysis Report:

SVM	LOCALIZATION	RELIABILITY
Amino Acid Comp.	Cytoplasmic	0.909
N-peptide Comp.	Cytoplasmic	0.982
Partitioned seq. Comp.	Cytoplasmic	0.472
Physico-chemical Comp.	Cytoplasmic	0.885
Neighboring seq. Comp.	Membrane	0.597

CELLO Prediction:

	Cytoplasmic	3.568 *
	Membrane	1.186
	Extracellular	0.201
	CellWall	0.045

****

Fig. 4 The subcellular location results predicted by different software packages. (**a**) Results predicted by PSORTb; (**b**) results predicted by WoLF PSORT; (**c**) results predicted by CELLO

3.4 Model Verification

Once these key steps are done, a refined model has been constructed. The refined model should be evaluated. The verification can be categorized into qualitative and quantitative methods. In a qualitative method, by means of the flux balance analysis (FBA) algorithm, the biomass equation is derived to determine (in a simulation) whether the model can utilize particular carbon sources or nitrogen sources. These carbon or nitrogen sources are determined by literature mining or in experiments. In the quantitative method, the constraint conditions must agree with an experiment. Then, the FBA results on the maximal growth rate and product synthesis rate are compared with the results of the experiment. If the model has passed the verification, it can be used as a basic model for the construction of an AME.

3.5 Construction of an AME Based on GSMMs

After a researcher obtains the GSMMs constituting an AME, these two GSMMs need to be further refined so that they can be used for the simulation of an AME.

1. The formats of basic models obtained from other databases may vary, e.g., Systems Biology Markup Language (SBML) or EXCEL. For the convenient model modification, they should be unified in the EXCEL format. With the COBRA Toolbox, SBML models can be transformed into EXCEL using the function "writeCbModel(model, xls, filename)."
2. Once a model is obtained in the EXCEL format, it needs to be further standardized. This is because EXCEL models consist of two sheets: reactions and metabolites. Metabolites can be presented in various forms. For example, in *E. coli* model *i*AF1260, pyruvate is denoted as pyr. In *Saccharomyces cerevisiae* model Yeast 1.0 and *Yarrowia lipolytica* model *i*NL895, pyruvate is PYR and s_1277, respectively. According to IMGMD [10], BIGG Models [49], or MetaNetX [60], metabolites in these basic models can be standardized. For the list of reactions, a formula should be replaced with these unified metabolites. Besides, these reactions should also be in the same format.
3. After these two basic models are standardized, they should be viewed as two separate compartments in the AME. In the AME for vitamin C production [30], *K. vulgare* and *B. megaterium* were denoted as k and b, respectively. The space representing the community of *K. vulgare* and *B. megaterium* was defined as x. Another compartment (e) represented the environment, through which the metabolites can be exchanged in the medium and the community space (Fig. 5).
4. Finally, these two recompartmentalized models should be integrated into one model. Thus, an AME consisting of two species metabolic models is now fully reconstructed (Fig. 6).

a

```
>> changeCobraSolver('glpk','lp')

ans =

     1

>> [gaps] = gapAnalysis(model)

gaps =

    '4HLT[c]= substrate & product'
    '5FTHF[e]= substrate & product'
    '5OPRO[c]= product'
    'ABAL[c]= substrate & product'
    'ACACP[e]= substrate & product'
    'ADPR[c]= product'
    'AGMT[c]= substrate & product'
    'AICAD[c]= substrate & product'
    'APS[c]= substrate'
    'BA[c]= substrate'
    'BA[e]= product'
    'CDPGL[c]= substrate & product'
    'CDPGL[e]= substrate & product'
    'CPGI[c]= substrate'
    'DAD5[c]= substrate'
    'DHPT[c]= product'
    'DITP[c]= product'
    'DPRO[c]= substrate & product'
    'DTDPDM[c]= substrate & product'
    'ETHAM[c]= substrate & product'
    'FOFMET[c]= substrate'
    'G3PG[c]= substrate & product'
    'GABA[c]= substrate & product'
    'GAL[c]= substrate'
    'GDPFUC[c]= substrate'
    'GLUCARATE[c]= substrate & product'
    'GLYASNL[c]= substrate'
    'GLYASN[c]= substrate & product'
    'GLYCEROCHO[c]= substrate & product'
    'GLYCEROPE[c]= substrate & product'
```

b

```
>> changeCobraSolver('glpk','milp')

ans =

     1

>> [allGaps,rootGaps,downstreamGaps] = gapFind(model)

allGaps =

    'NAD[c]'
    'HISOL[c]'
    'H[c]'
    'NADH[c]'
    'HISTIDINAL[c]'
    'H2O[c]'
    'HIS[c]'
    'NADPH[c]'
    'DHSK[c]'
    'NADP[c]'
    'SME[c]'
    'ASPSA[c]'
    'HSER[c]'
    'CBHCAP[c]'
    'OICAP[c]'
    'ACLAC[c]'
    '33HMEOXOBUT[c]'
    'DH3MVA[c]'
    'ABUT[c]'
    'HMOP[c]'
    'DHMP[c]'
    '3PG[c]'
    'PHP[c]'
    'BASP[c]'
    'PI[c]'
    'ABAL[c]'
    'GABA[c]'
    'NAGLUP[c]'
    'NAGLUS[c]'
    'GLU5P[c]'
```

Fig. 5 Gaps in the draft model of *K. vulgare* according to different algorithms. (**a**) Gaps identified by GapAnalysis; (**b**) gaps identified by GapFind

3.6 Analysis of an Artificial Microbial Ecosystem

As in the analysis of GSMMs, an AME also involves these constraint-based methods. The COBRA Toolbox is necessary for AME analysis [37]. Besides, some algorithms like OptCom [61], d-OptCom [62], and community flux balance analysis (cFBA) [63] have been specifically developed for microbial communities. With these computational methods, the relations among a genotype, phenotype, and dynamic community structure of a microbial community can be identified.

3.6.1 Comparison of the Effects of Coculture and Monoculture

In the analysis of a microbial community, it is necessary to determine whether a strain can grow better when cultured with another microorganism. On the basis of a single objective function,

a

```
model =

                   mets: {649x1 cell}
               metNames: {649x1 cell}
            metFormulas: {649x1 cell}
                   rxns: {830x1 cell}
               rxnNames: {830x1 cell}
             subSystems: {830x1 cell}
                     lb: [830x1 double]
                     ub: [830x1 double]
                    rev: [830x1 double]
                      c: [830x1 double]
                      b: [649x1 double]
                      S: [649x830 double]
             rxnGeneMat: [830x663 double]
                  rules: {830x1 cell}
                grRules: {830x1 cell}
                  genes: {663x1 cell}
       confidenceScores: {830x1 cell}
          rxnECNumbers: {830x1 cell}
              rxnNotes: {830x1 cell}
         rxnReferences: {830x1 cell}
              proteins: {830x1 cell}
    metFormulasNeutral: {649x1 cell}
        metCompartment: {649x1 cell}
             metKEGGID: {649x1 cell}
        metInChIString: {649x1 cell}
             metSmiles: {649x1 cell}
             metCharge: [649x1 double]
           metPubChemID: {649x1 cell}
            metChEBIID: {649x1 cell}
```

b

```
model =

                   mets: {993x1 cell}
               metNames: {993x1 cell}
            metFormulas: {993x1 cell}
                   rxns: {1298x1 cell}
               rxnNames: {1298x1 cell}
             subSystems: {1298x1 cell}
                     lb: [1298x1 double]
                     ub: [1298x1 double]
                    rev: [1298x1 double]
                      c: [1298x1 double]
                      b: [993x1 double]
                      S: [993x1298 double]
             rxnGeneMat: [1298x1055 double]
                  rules: {1298x1 cell}
                grRules: {1298x1 cell}
                  genes: {1055x1 cell}
       confidenceScores: {1298x1 cell}
          rxnECNumbers: {1298x1 cell}
              rxnNotes: {1298x1 cell}
         rxnReferences: {1298x1 cell}
              proteins: {1298x1 cell}
    metFormulasNeutral: {993x1 cell}
        metCompartment: {993x1 cell}
             metKEGGID: {993x1 cell}
        metInChIString: {993x1 cell}
             metSmiles: {993x1 cell}
             metCharge: [993x1 double]
           metPubChemID: {993x1 cell}
            metChEBIID: {993x1 cell}
```

c

```
model =

                   mets: {1582x1 cell}
               metNames: {1582x1 cell}
            metFormulas: {1582x1 cell}
                   rxns: {2136x1 cell}
               rxnNames: {2136x1 cell}
             subSystems: {2136x1 cell}
                     lb: [2136x1 double]
                     ub: [2136x1 double]
                    rev: [2136x1 double]
                      c: [2136x1 double]
                      b: [1582x1 double]
                      S: [1582x2136 double]
             rxnGeneMat: [2136x1718 double]
                  rules: {2136x1 cell}
                grRules: {2136x1 cell}
                  genes: {1718x1 cell}
       confidenceScores: {2136x1 cell}
          rxnECNumbers: {2136x1 cell}
              rxnNotes: {2136x1 cell}
         rxnReferences: {2136x1 cell}
              proteins: {2136x1 cell}
    metFormulasNeutral: {1582x1 cell}
        metCompartment: {1582x1 cell}
             metKEGGID: {1582x1 cell}
        metInChIString: {1582x1 cell}
             metSmiles: {1582x1 cell}
             metCharge: [1582x1 double]
           metPubChemID: {1582x1 cell}
            metChEBIID: {1582x1 cell}
```

Fig. 6 Comparison of GSMMs and AME for vitamin C production. (**a**) *K. vulgare* model iWZ663, (**b**) *B. megaterium* model iMZ1055, (**c**) vitamin C production AME iWZ-KV-663-BM-1055

according to experimental data, one biomass function of a particular strain is fixed. The other strain's biomass is chosen as the objective function for simulation. With the culture medium simulated by setting the low boundary of related exchange reactions, FBA is used to calculate the optimized flux distribution for a target strain. In a comparison of the FBA results with monoculture, effects of coculture can be simulated. For example, in the AME for vitamin C production, in the l-sorbose-CSLP medium, when the *B. megaterium* biomass was fixed at 0.1/h, the growth of *K. vulgare* was 0.178 1/h, which represents a 1.1-fold increase over monoculture [30]. A similar result can also be obtained for the growth of *B. megaterium* (Table 4).

3.6.2 Uncovering the Interspecies Metabolite Transfers in the Community

In a microbial community, interspecies metabolite transfers play a key role in the research into the relation between two species. Metabolomics has been applied to elucidation of organismal responses to abiotic pressures and to investigation of the responses of organisms to another biota. Ma et al. analyzed an artificial microbial community for two-step production of vitamin C [9], and approximately 100 metabolites were identified. Of these metabolites, glutamic acid, 5-oxo-proline, L-sorbose, 2-keto-l-gulonic acid (2-KGA), 2,6-dipicolinic acid, and tyrosine served as potential biomarkers to distinguish different time-series samples. On the basis of the FBA results on the AME for *K. vulgare* and *B. megaterium*, 24 interspecies metabolites were identified

Table 4
Changes in the growth rate after comparison between monoculture and coculture

Strain	*K. vulgare* Monoculture(/h)	*K. vulgare* Coculture(/h)	*K. vulgare* Fold-change	*B. megaterium* Monoculture(/h)	*B. megaterium* Coculture(/h)	*B. megaterium* Fold-change
Growth rate	0.084	0.178	1.1	0.275	2.098	6.6

Table 5
Predicted exchanged metabolites between the GSMMs of *K. vulgare* and *B. megaterium*

Metabolites	$Mets_{B\text{-}K}$[a]	$Mets_{K\text{-}B}$
Amino acids	Lysine, arginine, asparagine, cysteine, histidine, tryptophan	Phenylalanine
Nucleic acids	Adenine, guanine, uracil	None
Vitamins and cofactors	Biotin, nicotinate, pantothenate, riboflavin, thiamin diphosphate, dihydrofolic acid	None
Organic acids	Chorismic acid, succinate, pyruvate	Fumarate, formate
Others	Glycerol, sulfite, ethanol	None

[a]$Mets_{B\text{-}K}$ represents metabolites transported from *B. megaterium* to *K. vulgare*, while $Mets_{K\text{-}B}$ represents the reverse transport direction

(Table 5). Both the metabolomics and simulation results showed that *B. megaterium* provides key elements necessary for *K. vulgare* to grow better and to produce more 2-KGA.

3.6.3 Identification of a Relation for Species in the Community

We know that the relations in a microbial community can be classified as competition, mutualism, and parasitism. Accordingly, we can identify the detailed interspecies relations with the help of the AME. Besides comparing the growth differences and interspecies metabolite transfers in monoculture and coculture, an algorithm named RobustnessAnalysis can be utilized for further identification. RobustnessAnalysis performs analysis of robustness for a reaction of interest and an objective of interest.

RobustnessAnalysis is implemented in the COBRA Toolbox, and to run it on your reconstructed AME in the COBRA Toolbox, use the function [controlFlux, objFlux] = robustnessAnalysis (model, controlRxn, nPoints, plotResFlag, objRxn, objType).

For inputs, where the model is the AME that was imported into COBRA, controlRxn is the reactions of interest whose value is to be controlled.

As for optional inputs, nPoints is the number of flux values per dimension (Default = 20), plotResFlag is Plot results

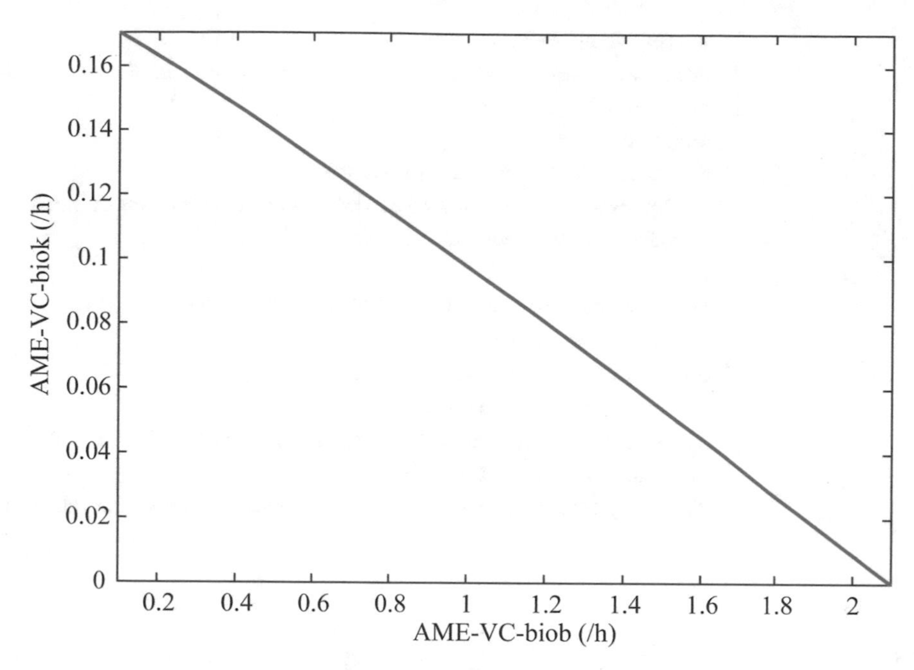

Fig. 7 Results on robustness between the growth rates of *B. megaterium* and *K. vulgare*. AME-VC-biob: growth rate of *B. megaterium*; AME-VC-biok: growth rate of *K. vulgare*

(Default = true), objRxn is Objective reaction to be maximized (Default = whatever is defined in the model), objType is maximize ("max") or minimize ("min") objective (Default = "max").

As for outputs, controlFlux is a flux value within the range of the maximum and minimum for a reaction of interest. objFlux is an optimal value of the objective reaction at each control reaction flux value.

To explore the exact interspecies relations in the vitamin C AME, RobustnessAnalysis was applied. The results showed that the higher the *B. megaterium* growth rate, the lower the *K. vulgare* growth rate will be (Fig. 7). This approach proved that there is also a competition relation between *B. megaterium* and *K. vulgare*. Considering the different growth rates in monoculture and coculture, competition and mutualism both exist between *B. megaterium* and *K. vulgare* [30].

4 Notes

1. An error was occurred when we tried to convert the AME model from Excel form to SBML form using "writeCbModel (model,'SBML',' *i*WZ-KV-663-BM-1055.xml')." And the error code was descripted as

```
Error using convertCobraToSBML (line 148)
Unknown compartment: b. Be sure that b is specified in
compSymbolList and
compNameList.
Error in writeCbModel (line 255)
sbmlModel = convertCobraToSBML(model,sbmlLevel,sbmlVersion,
compSymbolList,compNameList)
;
```

Because during the construction of vitamin C AME, we assumed the *K. vulgare* as compartment "k," and *B. megaterium* as "b." However, these two compartments were not recognized by convertCobraToSBML. To solve this problem, we should define the compartments "k" and "b" in the program of convertCobraToSBML.

References

1. Caumette P, Bertrand J-C, Normand P (2015) Some historical elements of microbial ecology. In: Bertrand J-C, Caumette P, Lebaron P, Matheron R, Normand P, Sime-Ngando T (eds) Environmental microbiology: fundamentals and applications. Springer, New York, pp 9–24
2. Bowler C, Karl DM, Colwell RR (2009) Microbial oceanography in a sea of opportunity. Nature 459(7244):180–184
3. Sun X, Gao Y, Yang Y (2013) Recent advancement in microbial environmental research using metagenomics tools. Biodivers Sci 21 (4):393–400
4. Yamada T (2011) Enterotypes of the human gut microbiome. Nature 473(7346):174–180
5. Chikere CB, Okpokwasili GC, Chikere BO (2011) Monitoring of microbial hydrocarbon remediation in the soil. 3 Biotech 1 (3):117–138
6. Singh BK, Bardgett RD, Smith P, Reay DS (2010) Microorganisms and climate change: terrestrial feedbacks and mitigation options. Nat Rev Microbiol 8(11):779–790
7. Wilmes P, Wexler M, Bond PL (2008) Metaproteomics provides functional insight into activated sludge wastewater treatment. PLoS One 3(3):e1778
8. Daviss B (2005) Growing pains for metabolomics. Scientist 19(8):25–28
9. Ma Q, Zhou J, Zhang WW, Meng XX, Sun JW, Yuan YJ (2011) Integrated proteomic and metabolomic analysis of an artificial microbial community for two-step production of vitamin C. PLoS One 6(10):e26108
10. Ye C, Xu N, Dong C, Ye Y, Zou X, Chen X, Guo F, Liu L (2017) IMGMD: a platform for the integration and standardisation of *In silico* Microbial Genome-scale metabolic models. Sci Rep-UK 7(1):727
11. McCloskey D, Palsson BØ, Feist AM (2013) Basic and applied uses of genome-scale metabolic network reconstructions of *Escherichia coli*. Mol Syst Biol 9(1):661
12. Oberhardt MA, Palsson BØ, Papin JA (2009) Applications of genome-scale metabolic reconstructions. Mol Syst Biol 5(1):320
13. Durot M, Bourguignon PY, Schachter V (2009) Genome-scale models of bacterial metabolism: reconstruction and applications. FEMS Microbiol Rev 33(1):164–190
14. Ranganathan S, Tee TW, Chowdhury A, Zomorrodi AR, Yoon JM, Fu Y, Shanks JV, Maranas CD (2012) An integrated computational and experimental study for overproducing fatty acids in *Escherichia coli*. Metab Eng 14(6):687–704
15. Nakahigashi K, Toya Y, Ishii N, Soga T, Hasegawa M, Watanabe H, Takai Y, Honma M, Mori H, Tomita M (2009) Systematic phenome analysis of *Escherichia coli* multiple-knockout mutants reveals hidden reactions in central carbon metabolism. Mol Syst Biol 5 (1):306
16. Fong SS, Palsson BO (2004) Metabolic gene-deletion strains of *Escherichia coli* evolve to

computationally predicted growth phenotypes. Nat Genet 36(10):1056–1058

17. Ibarra RU, Edwards JS, Palsson BO (2002) *Escherichia coli* K-12 undergoes adaptive evolution to achieve *in silico* predicted optimal growth. Nature 420(6912):186–189
18. Schuetz R, Kuepfer L, Sauer U (2007) Systematic evaluation of objective functions for predicting intracellular fluxes in *Escherichia coli*. Mol Syst Biol 3:119
19. Selvarasu S, Ow DSW, Lee SY, Lee MM, Oh SKW, Karimi IA, Lee DY (2009) Characterizing *Escherichia coli* DH5 alpha Growth and Metabolism in a complex medium using genome- scale flux analysis. Biotechnol Bioeng 102(3):923–934
20. Segre D, Vitkup D, Church GM (2002) Analysis of optimality in natural and perturbed metabolic networks. Proc Natl Acad Sci U S A 99 (23):15112–15117
21. Nishikawa T, Gulbahce N, Motter AE (2008) Spontaneous reaction silencing in metabolic optimization. PLoS Comput Biol 4(12): e1000236
22. Ghim CM, Goh KI, Kahng B (2005) Lethality and synthetic lethality in the genome-wide metabolic network of *Escherichia coli*. J Theor Biol 237(4):401–411
23. Motter AE, Gulbahce N, Almaas E, Barabasi AL (2008) Predicting synthetic rescues in metabolic networks. Mol Syst Biol 4:168
24. Kim DH, Motter AE (2009) Slave nodes and the controllability of metabolic networks. New J Phys 11:113047
25. Pal C, Papp B, Lercher MJ, Csermely P, Oliver SG, Hurst LD (2006) Chance and necessity in the evolution of minimal metabolic networks. Nature 440(7084):667–670
26. Yizhak K, Tuller T, Papp B, Ruppin E (2011) Metabolic modeling of endosymbiont genome reduction on a temporal scale. Mol Syst Biol 7:479
27. Notebaart RA, Kensche PR, Huynen MA, Dutilh BE (2009) Asymmetric relationships between proteins shape genome evolution. Genome Biol 10(2):R19
28. Pal C, Papp B, Lercher MJ (2005) Adaptive evolution of bacterial metabolic networks by horizontal gene transfer. Nat Genet 37 (12):1372–1375
29. Pal C, Papp B, Lercher MJ (2005) Horizontal gene transfer depends on gene content of the host. Bioinformatics 21:222–223
30. Ye C, Zou W, Xu N, Liu L (2014) Metabolic model reconstruction and analysis of an artificial microbial ecosystem for vitamin C production. J Biotechnol 182–183:61–67
31. Stolyar S, Van Dien S, Hillesland KL, Pinel N, Lie TJ, Leigh JA, Stahl DA (2007) Metabolic modeling of a mutualistic microbial community. Mol Syst Biol 3:92
32. Biggs MB, Medlock GL, Kolling GL, Papin JA (2015) Metabolic network modeling of microbial communities. Wiley Interdiscip Rev Syst Biol Med 7(5):317–334
33. Zhuang K, Izallalen M, Mouser P, Richter H, Risso C, Mahadevan R, Lovley DR (2010) Genome-scale dynamic modeling of the competition between *Rhodoferax* and *Geobacter* in anoxic subsurface environments. ISME J 5 (2):305–316
34. Bordbar A, Lewis NE, Schellenberger J, Palsson BØ, Jamshidi N (2010) Insight into human alveolar macrophage and *M. tuberculosis* interactions via metabolic reconstructions. Mol Syst Biol 6(1):422
35. Overbeek R, Olson R, Pusch GD, Olsen GJ, Davis JJ, Disz T, Edwards RA, Gerdes S, Parrello B, Shukla M (2014) The SEED and the rapid annotation of microbial genomes using subsystems technology (RAST). Nucleic Acids Res 42(D1):D206–D214
36. Agren R, Liu LM, Shoaie S, Vongsangnak W, Nookaew I, Nielsen J (2013) The RAVEN toolbox and its use for generating a genome-scale metabolic model for *Penicillium chrysogenum*. PLoS Comput Biol 9(3):e1002980
37. Schellenberger J, Que R, Fleming RM, Thiele I, Orth JD, Feist AM, Zielinski DC, Bordbar A, Lewis NE, Rahmanian S (2011) Quantitative prediction of cellular metabolism with constraint- based models: the COBRA Toolbox v2. 0. Nat Protoc 6(9):1290–1307
38. Swainston N, Smallbone K, Mendes P, Kell D, Paton N (2011) The SuBliMinaL toolbox: automating steps in the reconstruction of metabolic networks. J Integr Bioinform 8 (2):186–202
39. Ravikrishnan A, Raman K (2015) Critical assessment of genome-scale metabolic networks: the need for a unified standard. Brief Bioinform 16(6):1057–1068
40. Kanehisa M, Sato Y, Kawashima M, Furumichi M, Tanabe M (2016) KEGG as a reference resource for gene and protein annotation. Nucleic Acids Res 44(D1):D457–D462
41. Chang A, Schomburg I, Placzek S, Jeske L, Ulbrich M, Xiao M, Sensen CW, Schomburg D (2014) BRENDA in 2015: exciting developments in its 25th year of existence. Nucleic Acids Res 43(D1):D439–D446
42. Saier MH, Reddy VS, Tsu BV, Ahmed MS, Li C, Moreno-Hagelsieb G (2015) The transporter classification database (TCDB): recent

advances. Nucleic Acids Res 44(D1): D372–D379

43. Caspi R, Billington R, Ferrer L, Foerster H, Fulcher CA, Keseler IM, Kothari A, Krummenacker M, Latendresse M, Mueller LA (2016) The MetaCyc database of metabolic pathways and enzymes and the BioCyc collection of pathway/genome databases. Nucleic Acids Res 44(D1):D471–D480
44. Keseler IM, Mackie A, Peralta-Gil M, Santos-Zavaleta A, Gama-Castro S, Bonavides-Martínez C, Fulcher C, Huerta AM, Kothari A, Krummenacker M (2013) EcoCyc: fusing model organism databases with systems biology. Nucleic Acids Res 41(D1):D605–D612
45. Sheppard TK, Hitz BC, Engel SR, Song G, Balakrishnan R, Binkley G, Costanzo MC, Dalusag KS, Demeter J, Hellerstedt ST (2015) The *Saccharomyces* genome database variant viewer. Nucleic Acids Res 44(D1): D698–D702
46. Nancy YY, Wagner JR, Laird MR, Melli G, Rey S, Lo R, Dao P, Sahinalp SC, Ester M, Foster LJ (2010) PSORTb 3.0: improved protein subcellular localization prediction with refined localization subcategories and predictive capabilities for all prokaryotes. Bioinformatics 26 (13):1608–1615
47. Yu CS, Chen YC, Lu CH, Hwang JK (2006) Prediction of protein subcellular localization. Proteins 64(3):643–651
48. Thiele I, Palsson BØ (2010) A protocol for generating a high-quality genome-scale metabolic reconstruction. Nat Protoc 5(1):93–121
49. King ZA, Lu J, Drager A, Miller P, Federowicz S, Lerman JA, Ebrahim A, Palsson BO, Lewis NE (2016) BiGG Models: a platform for integrating, standardizing and sharing genome-scale models. Nucleic Acids Res 44(D1): D515–D522
50. Stein L (2001) Genome annotation: from sequence to biology. Nat Rev Genet 2 (7):493–503
51. Yandell M, Ence D (2012) A beginner's guide to eukaryotic genome annotation. Nat Rev Genet 13(5):329–342
52. Stothard P, Wishart DS (2006) Automated bacterial genome analysis and annotation. Curr Opin Microbiol 9(5):505–510
53. Moriya Y, Itoh M, Okuda S, Yoshizawa AC, Kanehisa M (2007) KAAS: an automatic genome annotation and pathway reconstruction server. Nucleic Acids Res 35(Web Server):W182–W185
54. Aziz RK, Bartels D, Best AA, DeJongh M, Disz T, Edwards RA, Formsma K, Gerdes S, Glass EM, Kubal M (2008) The RAST Server: rapid annotations using subsystems technology. BMC Genomics 9(1):75–89
55. Devoid S, Overbeek R, DeJongh M, Vonstein V, Best AA, Henry C (2013) Automated genome annotation and metabolic model reconstruction in the SEED and Model SEED. Methods Mol Biol 985:17–45
56. Kim TY, Sohn SB, Kim YB, Kim WJ, Lee SY (2012) Recent advances in reconstruction and applications of genome-scale metabolic models. Curr Opin Biotechnol 23(4):617–623
57. Sacher O, Reitz M, Gasteiger J (2009) Investigations of enzyme-catalyzed reactions based on physicochemical descriptors applied to hydrolases. J Chem Inf Model 49(6):1525–1534
58. Horton P, Park K-J, Obayashi T, Fujita N, Harada H, Adams-Collier C, Nakai K (2007) WoLF PSORT: protein localization predictor. Nucleic Acids Res 35(suppl 2):W585–W587
59. CS Y, Lin CJ, Hwang JK (2004) Predicting subcellular localization of proteins for Gram-negative bacteria by support vector machines based on n-peptide compositions. Protein Sci 13(5):1402–1406
60. Moretti S, Martin O, Tran TV, Bridge A, Morgat A, Pagni M (2016) MetaNetX/MNXref - reconciliation of metabolites and biochemical reactions to bring together genome-scale metabolic networks. Nucleic Acids Res 44(D1): D523–D526
61. Zomorrodi AR, Maranas CD (2012) OptCom: a multi-level optimization framework for the metabolic modeling and analysis of microbial communities. PLoS Comput Biol 8(2): e1002363
62. Zomorrodi AR, Islam MM, Maranas CD (2014) d-OptCom: dynamic multi-level and multi- objective metabolic modeling of microbial communities. ACS Synth Biol 3 (4):247–257
63. Khandelwal RA, Olivier BG, Roling WFM, Teusink B, Bruggeman FJ (2013) Community flux balance analysis for microbial consortia at balanced growth. PLoS One 8(5):e64567

Chapter 11

RNA Sequencing and Analysis in Microorganisms for Metabolic Network Reconstruction

Eva Pinatel and Clelia Peano

Abstract

There is a strict interplay between metabolic networks and transcriptional regulation in bacteria; indeed, the transcriptome regulation, affecting the expression of large gene sets, can be used to predict the likely "on" or "off" state of metabolic genes as a function of environmental factors. Up to date, many bacterial transcriptomes have been studied by RNAseq, hundreds of experiments have been performed, and Giga bases of sequences have been produced. All this transcriptional information could potentially be integrated into metabolic networks in order to obtain a more comprehensive view of their regulation and to increase their prediction power.

To get high-quality transcriptomic data, to be integrated into metabolic networks, it is paramount to clearly know how to produce highly informative RNA sequencing libraries and how to manage RNA sequencing data.

In this chapter, we will get across the main steps of an RNAseq experiment: from removal of ribosomal RNAs, to strand-specific library preparation, till data analysis and integration. We will try to share our experience and know-how, to give you a precise protocol to follow, and some useful recommendations or tips and tricks to adopt in order to go straightforward toward a successful RNAseq experiment.

Key words RNAseq, Gene expression analysis, Data integration

1 Introduction

The availability of next-generation sequencing (NGS) technologies and the advent of functional genomics have dramatically changed the approach to studying gene expression. RNA-sequencing has revolutionized the analysis of transcriptomes in both prokaryotes and eukaryotes and recently it has emerged as the best approach to study bacterial transcriptomes. Actually, advances in RNA sequencing (RNAseq) technology have innovated the study of bacterial transcriptomes [1] and high-resolution RNAseq data allowed transcriptional features to be pinpointed and quantified. The true power of RNAseq resides in its potential as an analytical tool for quantifying differential expression of transcripts, including operons, transcription units within operons, noncoding and antisense

Marco Fondi (ed.), *Metabolic Network Reconstruction and Modeling: Methods and Protocols*, Methods in Molecular Biology, vol. 1716, https://doi.org/10.1007/978-1-4939-7528-0_11,

RNAs, promoter activity, and terminator efficiency [2]. RNAseq generates digital information that allows transcriptional features to be located with single-nucleotide precision in a strand-specific manner, thus RNAseq facilitates quantitative computational analysis of any selected region of the transcriptome [3].

The necessary prerequisite for an efficient RNAseq study is that the data generated should have the potential to provide a deeper understanding of the biological problem of interest. To this aim, it is necessary to adopt the best experimental procedure taking into account all the steps of the workflow, from total RNA processing to library preparation and sequencing, up to the final step of data analysis.

To obtain reliable information, it is crucial that RNA sequencing achieves a sufficient "coverage" to detect even rare or low expressed transcripts. In bacterial cells, one of the main disadvantages of the transcriptomic approach is the high amount of ribosomal RNA (rRNAs), which accounts for more than 95% of total cellular RNA, greatly reducing the obtainment of enough transcript coverage. Thus, efficient removal of rRNA is the first critical step to deal with when programming a successful RNAseq experiment in bacteria. Unlike for eukaryotic mRNA, polyadenylation of bacterial mRNAs is limited and is mostly involved in targeting mRNA for degradation by PNPase [4]; hence, bacterial mRNA cannot be readily isolated from other RNA sources by hybridization to immobilized poly-T or enriched through reverse transcription with poly-T primers. Therefore, a major challenge in RNAseq applications in bacterial cells is the enrichment for all transcript species other than rRNA.

Papers describing the use of high-throughput sequencing for transcriptomics in bacteria have used mRNA enrichment methods usually based on depletion of rRNA utilizing three alternative approaches [5–7]: (1) hybridization capture of rRNAs by antisense oligonucleotides followed by pulldown through binding to magnetic beads, (2) degradation of processed RNA such as mature rRNA and tRNA by a 5′–3′ exonuclease that specifically digests RNA species with a 5′-monophosphate end, (3) utilization of a random primer mix selectively designed to enrich the mRNA portion of bacterial total RNA.

Among the rRNA capture approaches the MICROBExpress Kit (Ambion) has widely been applied in RNA-seq studies [6–9], including metatranscriptomics [10, 11]. On the other hand, the use of RiboMinus Bacteria Transcriptome Isolation Kit (Invitrogen), based on a similar capture method, has only been reported in one study [12]. The rRNA capture approach is the only method suitable for precise quantitative analyses. However, as it is based on 16S and 23S rRNA-specific capture probes, depletion efficiency of these methods varies between bacterial species.

In the work from Giannoukos et al. a very extensive comparative analysis of five different rRNA removal methods was performed [13], and for the first time the authors emphasized the efficiency of the Ribo-Zero rRNA removal kit, developed by Epicentre and recently commercialized by Illumina. They demonstrated that the Ribo-Zero kit, also based on rRNA capture, is very efficient in removing bacterial rRNAs, both on pure cultures and on RNAs extracted from fecal samples, while preserving mRNAs relative abundances. In the last 4 years the Ribo-Zero method has been widely used and has become the gold standard for ribosomal RNA removal both in Eukaryotes and in Prokaryotes [14, 15].

Alternatively to the capture-based methods, rRNA removal can also be achieved through its degradation by specific enzymes. An example is the mRNA-ONLY Prokaryotic mRNA Isolation Kit (Epicentre Biotechnologies), based on selective degradation of processed RNAs by the enzyme terminator 5′-phosphate-dependent exonuclease (TEX). This enzyme exclusively degrades RNA molecules carrying a 5′-monophosphate (i.e., processed RNA such as rRNA and tRNA), while mRNAs, carrying a 5′-triphosphate group, are not affected [16]. An additional advantage of rRNA degradation with TEX consists in the enrichment of primary transcripts with 5′-triphosphate ends, thus allowing identification of transcription start sites [17, 18]. This methodology, termed differential RNAseq (dRNAseq), is extremely informative for promoter mapping and identification of small RNAs [19–21]. In some cases, methods based on selective degradation of rRNAs have been used in combination with subtractive hybridization to optimize rRNA removal [22, 23].

Another enzymatic method was developed for mRNA enrichment: it was based on the use of a duplex-specific nuclease (DSN) to remove rRNA and it was applied with good results, both in terms of mRNA coverage and robustness, in transcriptome profiling of *Escherichia coli* grown in four different conditions [24].

Unlike the approaches described so far, which are based on rRNA removal or degradation, the Ovation Prokaryotic RNA-Seq System kit developed by NuGEN (NuGEN Technologies, San Carlos, CA, USA) was based on the synthesis of first and second strand cDNA using a random primer mix selectively designed to enrich the mRNA portion of bacterial total RNA. The selective random primers are designed against a sequence database composed of 50 bacterial and archaeal species representing all of the major phylogenetic subgroups. This kit, alone or in combination with the capture-based MICROBExpress kit, was tested by Peano et al. [25] proving to be particularly efficient in removing rRNAs from GC-rich bacteria, and subsequently it was applied for the analysis of *Burkholderia thailandensis*, *Planobispora rosea*, and *Acinetobacter Baumanii* transcriptomes in different growing conditions [26–28].

After choosing the most efficient rRNA removal method it is very important to select the right protocol for library preparation. To get highly informative and complete results from an RNAseq experiment, it is advisable to produce strand-specific data that give more detailed and correct results than non-stranded data both in estimating expression values, in assembling transcripts, and in improving genome annotation and its manual refinement.

The transcriptome has long been studied by reverse transcribing single-stranded RNA into double-stranded cDNA, but by assessing gene expression through dscDNA the strand information of the RNA is lost. Indeed, in a standard RNAseq library preparation protocol, cDNAs of a desired size, generated from retro transcription of fragmented totalRNAs with random hexamer primers, or from fragmented full-length cDNAs, are ligated to DNA adapters before amplification and sequencing [29]. While simple, this approach loses the information about which DNA strand corresponds to the sense strand of RNA. Lack of strand specificity would make it difficult to identify antisense and novel RNA species and cause inaccurate measurement of annotated RNAs expression.

With the advent of many strand-specific RNA library preparation protocols, high numbers of RNA sequencing experiments have been performed and huge amounts of stranded RNA sequencing data generated [30–32]. Several methods have been developed to capture the directionality of RNA in cDNA libraries [33]. One of the first approaches was based on attaching different adapters directly to the 5′ and 3′ ends of the RNA molecule. Originally designed for small RNA-Seq [34], this method begins with the removal of 3′phosphate group from fragmented RNA and addition of a 5′ phosphate group. This is followed by sequential ligations of a 5′ adenylated 3′ adapter using a truncated RNA ligase II and a 5′ adapter ligation using RNA ligase I. The sequence difference between 5′ and 3′ adapters preserves RNA strandedness. This approach suffers from substantial biases due to the influence of both 5′ and 3′ end sequences on the ligation steps. This issue, however, has been mitigated significantly by using random nucleotides at the ligation end of each adapter [35].

The strand-specific library protocol published by Parkhomchuk et al. [30] has become the most popular among such protocols being both relatively simple and effective. The protocol is called "dUTP second strand marking method" and consists of using dUTPs instead of dTTPs during the synthesis of the second strand in the cDNA synthesis step during sample preparation. Then prior to PCR amplification the uracil in the second strand is degraded using Uracil-*N*-Glycosylase (UNG), as the second strand is partly degraded, only the first strand undergoes amplification in the subsequent PCR. This particular strand-specific protocol was evaluated as superior in terms of simplicity and data quality in a comparative study of strand-specific protocols [36]; however, since it requires extra enzymatic and purification steps that can cause material loss, it is not suitable for low-input samples.

There have been several attempts to develop new methods based on adding adapters to fragmented RNA. One of these methods, Peregrine, uses template-switching after priming with a random hexamer that contains a small tag [37]. Another method, BrAD-Seq, takes advantages of temporary strand separation in double-stranded DNA to introduce a tag sequence [38]. One more example is sequential ligation of tags to RNA that has been fragmented to 200 nt, preserving even small RNAs [39, 40].

In 2012, Jim Pease and Roy Sooknanan from Epicentre published a Nature Methods Application Note on the combination of RiboZero™ rRNA removal with the ScriptSeq™ v2 protocol that enabled researchers to go from total RNA to cluster-ready strand-specific RNAseq libraries in less than 1 day [41]. The RNAseq libraries produced with this protocol are almost completely free of contaminating ribosomal RNA (rRNA) and provide for directional paired-end and multiplex sequencing on Illumina platforms. The ScriptSeq v2 RNA-Seq Library Preparation Kit uses a patented terminal-tagging process (Fig. 1) to generate directional RNA-seq

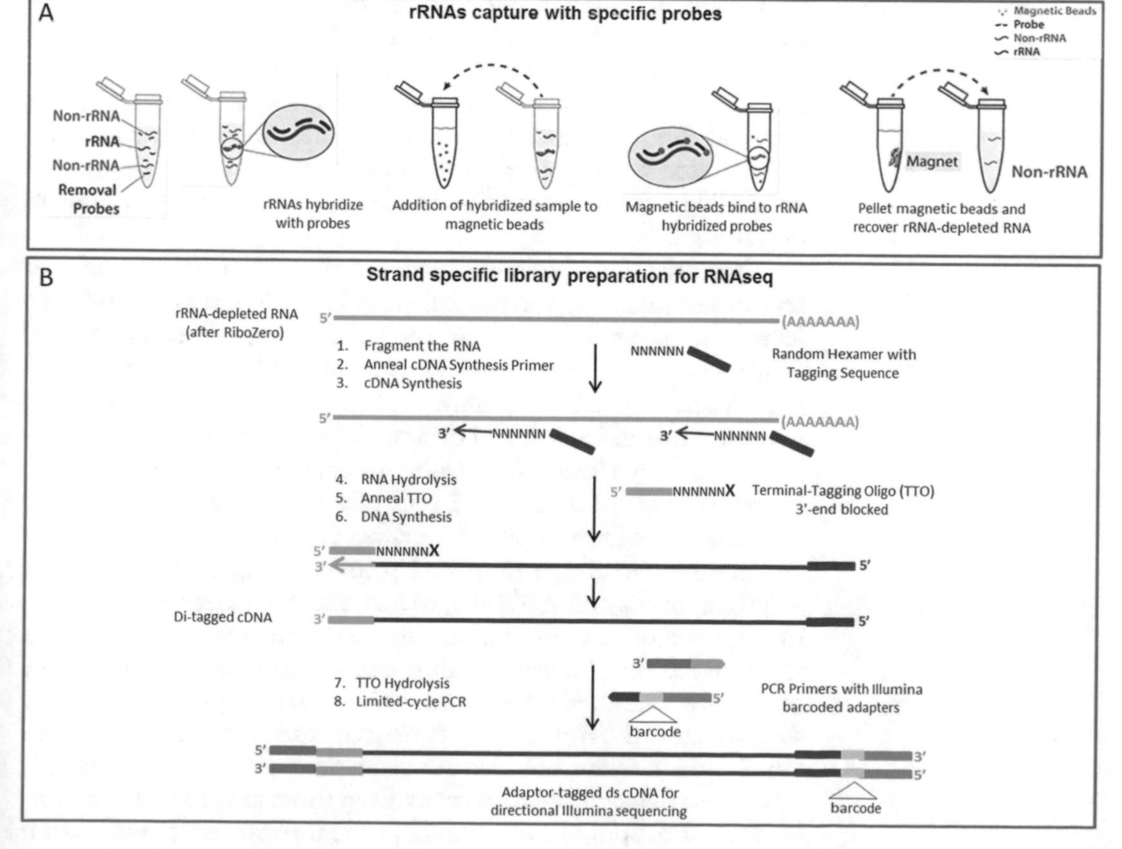

Fig. 1 Ribosomal RNA removal (**a**) and strand-specific library preparation for RNAseq (**b**)

libraries. Briefly, 500 pg to 50 ng of Ribo-Zero-treated RNA is fragmented and reverse transcribed using random primers containing a 5′-tagging sequence. The 5′-tagged cDNA is then tagged again at its 3′ end by a terminal-tagging reaction. Following purification, the di-tagged cDNA is amplified by limited-cycle PCR, which completes the addition of the Illumina adaptor sequences, amplifies the library for subsequent cluster generation, and adds an optional Illumina Index or user-defined barcode. The amplified RNAseq library is purified and is ready for cluster generation and sequencing (Fig. 1).

This combined protocol has become commercially available and recently, after the acquisition of Epicentre by Illumina, it has become the gold standard for strand-specific RNAseq (strand-spec-RNAseq).

Thanks to the availability of different RiboZero kits it is possible to manage total RNA from many different sources (i.e., Human/Mouse/Rat/Bacteria), in particular this kit is strongly emerging in the field of Bacteria transcriptome analysis because of its high efficiency and sensitivity in removing ribosomal RNAs both in $Gram^+$ and $Gram^-$ bacteria.

Even more recently this combined kit has been proposed for strand-spec-RNAseq library preparation for metatranscriptomic analysis; the ScriptSeq Complete Gold Epidemiology Kit removes ribosomal RNA from intact and partially degraded mixtures of Human/Mouse/Rat and Bacteria in a single step, it creates stranded Illumina sequencing libraries and allows the production of complete transcriptome sequencing data of both coding and noncoding transcripts thus being ideal for host/pathogen interaction studies.

Once the most suitable protocol to remove rRNAs and to produce sequencing libraries has been identified, it is necessary to focus the attention on the sequencing experimental design. Sequencing can produce single-end (SE) or paired-end (PE) reads, the first are mainly used for quantitative analysis (i.e., transcript quantification and differential expression) while the second are usually chosen for qualitative analysis which aim to refine the transcriptional landscape. In the case of eukaryotes PE reads are required for isoform expression analysis and transcript discovery. In the case of microorganisms read pairs can help to have a better definition of transcript 3′ ends, to increase the transcripts coverage, to improve operon definition, and to accurately map reads on repeated regions if dealing with bacteria whose genome sequences are known. Another relevant factor is sequencing depth or library size, which is the number of sequenced reads required for a given sample [42]. When libraries are sequenced to a deeper level it is possible to detect more transcripts, even those expressed at very low levels or noncoding RNAs, and to get a more precise quantification of gene expression [43]. Finally, an important factor is the choice of the replicates number, which should be defined considering both the biological and the technical variability of the experiment which are correlated to the stability of the system under study and to the reproducibility of the sample preparation procedures, as well as to the desired statistical power [42].

The choice of proper analytical tools for RNAseq data analysis is not a trivial task, in particular for the analysis of bacterial transcriptomes. As a matter of fact, only few tools have been developed specifically for the analysis of prokaryotic transcriptomes, the majority have been developed for eukaryotic ones and then shifted to the analysis of bacterial RNAseq data, whose features frequently challenge some of the basic assumptions of the algorithms that underlie these data analysis tools.

Pipelines for RNAseq data analysis can have different starting points, due to the presence or absence of the reference genome for the bacterial species under study. The lack of a reference genome requires an initial step of de novo transcriptome assembly, which is followed by the two main steps of the standard data analysis pipeline. The first step consists in the alignment of the reads to the reference genome or to the assembled transcriptome, which is succeeded by reads assignment to annotated genes in order to determine the number of reads mapping inside each gene (gene counts); the second step includes gene counts normalization and differential expression analysis.

Usually, bacterial RNAseq datasets consist of millions of single-end reads 35–75 bp in length. To align these raw sequences in the absence of a reference genome de novo assemblers like Rockhopper [44], Trinity [45] and SOAPdenovo2 [46] have been already tested in bacteria and benchmarked [47]. In the presence of a reference genome or once a de novo assembled one is available, several tools have been developed and tested in bacteria to align the raw reads, but up to date there is not a golden standard. Among them we can cite Rockhopper [44], Bowtie2 [48], SHRiMP [49], and BWA [50] for ss-RNAseq, while segemehl [51] is by far the most adopted in dRNAseq alignments. Some of these alignment algorithms have been recently benchmarked on a huge set of prokaryotic RNAseq data by McClure and colleagues [44].

There are two tricky points that should be taken into consideration handling alignment algorithms to map bacterial RNAseq reads: the first is the fine-tuning of mapping stringency level according to the specific experimental characteristics; usually high stringency is preferred for stable and well-annotated genomes and species with low genomic variability, but the thresholds need to be relaxed in the case of hyper-variable strains or poorly defined genomes. The second is the management of reads aligning to multiple regions of the genome; the standard procedure eliminates those reads because their mapping quality is low, but when their evaluation is relevant they can be considered for gene coverage determination purposes. A "simple" solution for metabolic models data integration is to force the aligner to assign the reads randomly only to one of the repeated loci, during gene coverage evaluation the duplicated genes can be substituted with a "centroid-element" accounting for the sum of the reads mapping on all the duplicates. This arrangement is feasible because for metabolic models the locus of transcription is not relevant but they only take into account the proteins encoded by the duplicated genes.

The coverage of each gene is measured in terms of number of reads overlapping it. Those measures, commonly defined gene counts, can be obtained using the functions embedded in common sequence managing tools (i.e., BedTools [52]) or on purpose packages as HTSeq count [53]. The major point to consider when dealing with prokaryotic RNAseq data is due to the organization of genes into operons, and thus many reads could map across two contiguous genes. In these cases, to avoid read multiplication, it is possible to assign the read to the gene having the major sequence overlap.

After obtaining gene counts the data analysis pipeline involves two consecutive stages: normalization and differential expression analysis, which are usually embedded in all-in-one tools. Widely used for the analysis of prokaryotic gene expression are Rockhopper, which adopts upper-quartile normalization, DeSEQ2 [54], whose normalization relies on size factors (estimated through the median-of-ratios method [55]) and edgeR [56], using the trimmed mean of M-values (TMM) for count normalization. All of them are based on differential expression testing using a negative binomial model as the reference distribution and, up to date, no benchmarking of their performances in the analysis of bacterial transcriptomes has been carried out.

Some peculiarities of prokaryotic gene expression can challenge the basic assumptions necessary to safely apply the previously listed algorithms, especially referring only to the default options:

1. The number of genes that constitute bacterial genomes can be too low to precisely estimate the model parameters.
2. The number of differentially expressed genes can vary a lot. In some cases, it can be more than 50% of the total representing a too broad variation to obtain a correct normalization, because the number of genes not differentially expressed is too low. In other cases it can be really small, just 1% of the total number of genes.
3. Sometimes, the most of the reads (from 55 to 98% of the total) map only into very few highly expressed genes (top 5% of total genes, i.e., housekeeping genes), thus affecting the subsequent normalization step.

For all these reasons the analysis tools herein considered should be properly evaluated case by case when managing bacterial RNAseq data.

Recently, some all-inclusive pipelines have been developed specifically to analyze prokaryotic RNA-seq data, among them Rockhopper [44], READemption [57], and SPARTA [58] should be mentioned. The first is the more versatile, designed to perform the analysis both from pre-assembled and not assembled bacterial genomes and to perform both qualitative (operon transcripts and ncRNA prediction from sequencing tracks) and quantitative analysis (DEGs calculation). It is easy to use because it is programmed through javascript and it does not require programming skills. Otherwise, READemption and SPARTA perform only quantitative analysis and were developed in python requiring shell usage.

Up to date, there is no optimal workflow for RNAseq analysis due to the wide variety of different applications of this technique. Different analytical strategies can be adopted depending on the organism being studied, for example if its genome sequence is available or not could change a lot in the analysis workflow. Each experiment could potentially have its own optimal analytical strategy requiring the choice of different methods for alignment, transcript quantification, normalization, and differential expression analysis. Furthermore, quality control checks should be properly applied at each step of the analysis to ensure the reproducibility and the reliability of the results. Indeed one of the key points in an RNAseq data analysis pipeline, whose importance is usually underestimated, is the set up of proper quality control checks in order to evaluate sequencing performances, library quality (in terms of genome alignment levels, adapter presence, rRNA removal, and strand specificity), global gene coverage levels, and sample reproducibility. These checks enable estimating the level of reliability of the obtained results and, as a consequence, defining if a library needs to be prepared or sequenced again to be fairly compared to the other samples.

In the following sections, we will explain in detail our reference protocol to perform RNAseq experiments and data analysis in microorganisms with known genome sequences. Data analysis description, and the example of RNAseq analysis pipeline reported, has the particular aim to explain how to produce the different transcriptomic data inputs required for integration into metabolic networks.

2 Materials and Tools

2.1 Best RNAseq Protocol to Feed the Model and Recommendations from Our Hands-On Experience

In the last 4 years we acquired great experience in comparing and testing different protocols for RNA sequencing and analysis in different microorganisms having different levels of genome complexity, GC content, and annotation [25–28]. This training course has led us to try to find a protocol always more efficient in rRNA removal, increasingly powerful in producing strand-specific libraries and more effective in generating sequencing data of good quality perfectly suitable to be integrated in metabolic network reconstruction.

Currently, to prepare strand-specific libraries for RNAseq in microorganisms, we use the protocol developed by Jim Pease and Roy Sooknanan in 2012 [41] and recently commercialized as a kit by Illumina under the name of ScriptSeq Complete kit (Bacteria). This kit merges the Ribo-Zero rRNA removal step and the preparation of a strand-specific tagged library ready for Illumina sequencing; in Fig. 1 the details of the different steps of the protocol are schematically represented.

Warning:

- The total RNA must be treated with DNAseI before the Ribo-Zero removal step in order to avoid any genomic DNA contamination in the following steps of the protocol.

Recommendation:

- Before starting with the rRNA removal step the quality and quantity of total RNA should be checked by Agilent Bioanalyzer to get a RIN value (RNA Integrity Number). We suggest proceeding only with not degraded total RNAs having a RIN>7.

The Ribo-Zero removal step is optimized for the removal of all sizes of rRNA both from intact or degraded total RNA samples (starting from 50 ng up to 5 μg). Total RNAs are mixed with the rRNA Removal Reagents in the solution for 10 min at 68 °C. The mixture is then added to Ribo-Zero magnetic beads and incubated for 5 min at 50 °C followed by the removal of the beads bringing the rRNAs captured with a magnetic stand (2 min). The rRNA-depleted RNA is recovered either by ethanol precipitation or column-purification method (Fig. 1a).

Recommendations:

- Even if the protocol suggests washing the magnetic beads in batch, determining the amount required for the total number of reaction and washing them all together in a single 1.5 ml Eppendorf tube (sufficient for 6 reactions), we strongly suggest following the procedure for individual washing. The individual washing ensures that the correct amount of magnetic beads necessary for efficient rRNA removal is available in each tube at the end of the washing step (i.e., 65 μl).
- Even if the protocol considers the addition of RiboGuard RNase Inhibitor optional we suggest always adding it immediately after the washing procedure and mixing well by vortexing.

- After adding the RNAs mixture to the Ribo-Zero magnetic beads, we strongly recommend performing the 5 min incubation at 50 °C in a thermal block with slow shaking (about 300 rpm).
- Even if the protocol suggests for the purification of the rRNA-depleted sample three possible methods: ethanol precipitation, AMPure beads, or spin columns, we strongly recommend purifying it with ethanol precipitation considering some modifications of the standard precipitation protocol: (1) place the tubes for at least 2 h at −80 °C for efficient RNA precipitation; (2) perform the final washing steps with Ethanol 75% instead of 70% to better recover the smallRNAs.
- The rRNA-depleted sample should be resuspended, after air-drying of the RNA pellet, in a small volume of RNase-Free water (max 15 μl), dissolved in water for 10 min at room temperature and for 10 min in ice, then immediately stored at −80 °C. To get a correct quantification of the rRNA-depleted sample, we suggest keeping the samples overnight at −80 °C and quantifying them the day after.
- Before starting with the library preparation step check the rRNA removal by Agilent Bioanalyzer, the peaks corresponding to the ribosomal RNAs should have been almost totally disappeared and the software should calculate an RIN<1.

The ScriptSeq library preparation step employs random-primed, first-strand cDNA synthesis from rRNA-depleted mRNA that incorporates a platform-specific 3′-sequencing tag (Fig. 1b). The RNA and excess oligonucleotides are then enzymatically hydrolyzed and a mixture comprising a terminal-tagging oligonucleotide (TTO) and a DNA synthesis reagent is added. The TTO contains a known 5′-sequence tag, a 3′-random sequence, and a terminally blocked 3′ end to prevent priming of DNA synthesis. The 3′ ends of the cDNA molecules are extended, incorporating a complement to the sequence tag, and forming cDNA molecules with known sequence tags at their 5′ and 3′ ends for directionality. Excess TTO is enzymatically degraded and the di-tagged cDNA molecules are purified. The complete di-tagged cDNA synthesis process is performed in a single-reaction tube. Next, platform-specific capture sequences, which can include a barcode, are added to the di-tagged cDNA molecules by limited-cycle PCR, and the products are purified. The adaptor-tagged library is now ready for cluster generation and Illumina sequencing (Fig. 1b).

Recommendations:

- To purify the cDNA after the synthesis and terminal tagging steps, we recommend using the MiniElute PCR Purification kit (Qiagen) and not the AMPure XP system (Agencour).

- To purify the RNAseq library we strongly suggest using always the AMPure XP beads in order to efficiently remove the adapters dimers and short fragments that can affect the good quality of sequencing.
- The quality of the strand-specific RNAseq library obtained should be checked by Agilent Bioanalyzer to verify the adapter dimer removal and to evaluate the length range of fragments, and quantified by Real Time PCR using the KAPA Library Quantification Kit (Roche) to correctly estimate the library concentration before sequencing.

2.2 Files and Programs Required for the RNAseq Analysis

To perform an RNAseq data analysis the genomic sequence of the bacteria under study, its annotation in tabular format, and the files containing the sequenced reads are required. Genomic sequences of microorganisms and their annotations are available in NCBI at the following link: https://www.ncbi.nlm.nih.gov/genomes/MICROBES/microbial_taxtree.html. Both the RefSeq and GeneBank versions of these files are usually available, but you should pay attention because for some species the number of annotated features can vary between the two versions.

Table 1 lists all the programs and packages adopted to develop the RNAseq analysis pipeline.

3 RNAseq Analysis Pipeline

Here below we will describe the RNAseq data analysis pipeline that we apply for bacterial transcriptomes. The pipeline is based on four main steps: (a) reads alignment and quality filtering; (b) alignment visualization; (c) gene counts production; and (d) gene counts normalization and differential expression analysis. We embedded in the analysis workflow five checkpoints regarding: (1) the evaluation of sequencing performances; (2) the calculation of genome alignment levels; (3) the estimation of the quality of the library (adapter presence, rRNA removal, and strand specificity); (4) the assessment of global gene coverage; and (5) the control of sample reproducibility (Fig. 2). Please note that the entire workflow refers to the analysis of single-end strand-specific reads obtained from Illumina sequencers.

3.1 Reads Alignment and Quality Filters

Sequencing outputs are usually files in .fastq (or .fq) format, eventually compressed. Raw-data registered on public archives like SRA (Sequence Read Archive) instead need to be converted to fastq format following the link instructions https://www.ncbi.nlm.nih.gov/books/NBK158900/.

Fastq structure includes for each sequenced read four lines of information:

Table 1
List of programs and packages adopted to develop the pipeline

Name	Version	Webpage for download	Manual
Trimmomatic	0.36	http://www.usadellab.org/cms/index.php?page=trimmomatic	http://www.usadellab.org/cms/uploads/supplementary/Trimmomatic/TrimmomaticManual_V0.32.pdf
FastQC	0.11.4	http://www.bioinformatics.babraham.ac.uk/projects/fastqc/	https://biof-edu.colorado.edu/videos/dowell-short-read-class/day-4/fastqc-manual
BEDTools	2.26.0	http://bedtools.readthedocs.io/en/latest/content/installation.html	http://bedtools.readthedocs.io/en/latest/content/bedtools-suite.html
SAMtools	0.1.19	http://www.htslib.org/download/	http://www.htslib.org/doc/samtools.html
BOWTIE2	2.2.9	https://sourceforge.net/projects/bowtie-bio/files/bowtie2/2.2.9/	http://bowtie-bio.sourceforge.net/bowtie2/manual.shtml
igv	2.3.8	http://software.broadinstitute.org/software/igv/download	http://software.broadinstitute.org/software/igv/UserGuide
R	3.1.2	https://cran.rstudio.com/	https://cran.r-project.org/manuals.html
DESeq2 (R library) + dependencies	1.6.3	https://bioconductor.org/packages/release/bioc/html/DESeq2.html	https://bioconductor.org/packages/release/bioc/vignettes/DESeq2/inst/doc/DESeq2.pdf
ggplot2 (R library) + dependencies	2.1	https://cran.r-project.org/web/packages/ggplot2/index.html	http://tutorials.iq.harvard.edu/R/Rgraphics/Rgraphics.html

```
@M02520:110:000000000-AH9U0:1:1101:17136:1358 1:N:0:1
CGACGCCAGCGCGCTGCTGCTGAACAGCGGCTTCACGGTCTCCATCTTCTCTCCCTTCCTCTTC
+
8--8-@C7+C+@+@F+@E,CF,,,;,,;++8+;,,,,:8,99,,:,,9,9,9,9696,9999C@
```

The first line contains the read identifiers (i.e., sequencing machine code and read position on the flow cell), the second line shows the nucleotide sequence, the third line is simply a plus, and the last reports the quality score, which estimates the reliability of the called nucleotide for each base of the sequence. Illumina MiSeq sequencer expresses the quality score using the Phred+33 scoring system, but the scoring system could be different for other Illumina sequencers.

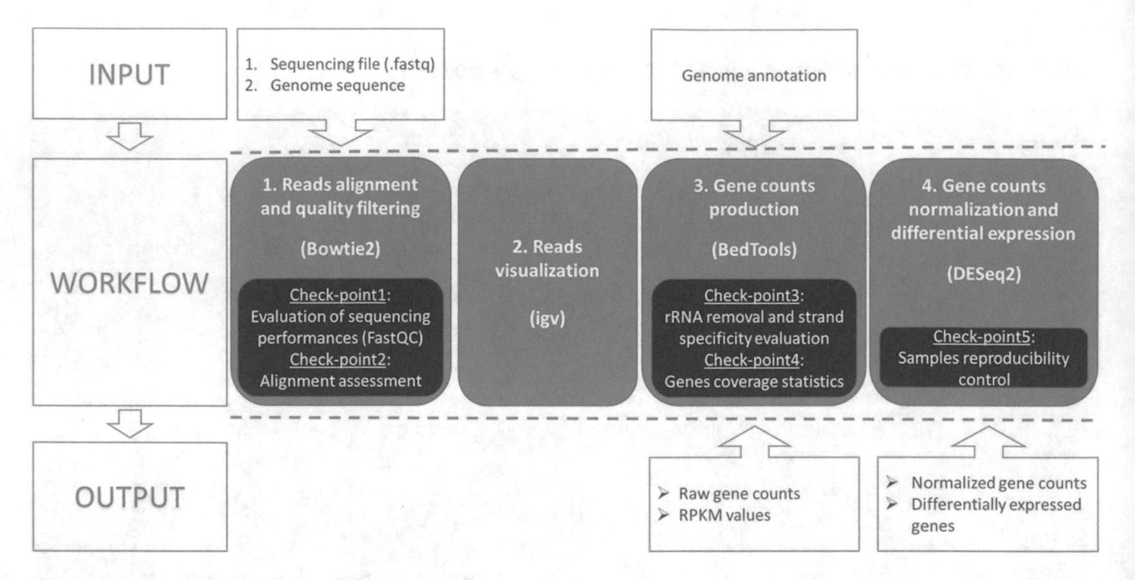

Fig. 2 RNAseq data analysis workflow

3.1.1 Check-Point 1: Evaluation of Sequencing Performances and Library Preparation

The first quality control check is necessary to evaluate sequencing performance, to this aim we adopt the program FastQC [59]. It examines the sequence quality and at the same time it evaluates some libraries characteristics, providing warnings if the sequencing efficiency and quality differs from expected parameters.

You can run FastQC simply opening the interactive window (typing fastqc on the terminal) and following the instructions. Once you get the FastQC output you should focus on two different scores named "per base sequence quality" and "per sequence quality." If for these points warnings appear it could mean that sequencing quality was not so good, and probably several bases have been wrongly identified. A low sequencing quality could affect the subsequent step of reads alignment. The alignment step could be affected also by the presence of adapter sequences inside the reads, this phenomenon is usually observed when sequencing libraries are made of short insert fragments. The output generated by the last FastQC release quantifies the level of adapters contamination in a specific field named "adapters."

To improve alignment performances when warnings appear in the cited FastQC output fields, we suggest, if possible, shortening (trim) the read length to keep only the portion of the read with good quality scores and to eliminate the adapters.

For Illumina reads you can do it in a single step using Trimmomatic [60]

```
java -jar trimmomatic-0.36.jar SE -phred33 input.fq.gz output.fq.gz
ILLUMINACLIP:<full path to the selected adapter files within trimmomatic
directory>:2:30:10 CROP:X MINLEN:Z
```

This command performs the following:

1. *Remove Illumina TruSeq adapters (custom adapter file can be provided).*
2. *Shorten the read to X read length (set X according to the FastQC results, in correspondence of the quality drop length).*
3. *Eliminate reads shorter than Z (usually reads too short to be correctly aligned).*

Alternatively, you can cut the reads only when the quality falls below a certain threshold. The SLIDINGWINDOW parameter cuts each read when in a window of A base pairs, the mean quality falls below the threshold B.

```
java -jar trimmomatic-0.36.jar SE -phred33 input.fq.gz output.fq.gz
ILLUMINACLIP:<full path to the selected adapter files within trimmomatic
directory>:2:30:10 MINLEN:Z SLIDINGWINDOW:A:B
```

Warning: always control the quality score system of your fastq file, and change the –phred33 parameter included in the script if your scoring system is different.

Tips&Tricks: don't mind about the other warnings signaled by FastQC as GC% can be different from the expected in GC-rich or AT-rich bacteria. Sequence length distribution can be bi-modal as some sequencing procedures already trim the adapters during fastq production, while over-represented sequence and duplication warnings are frequent for bacteria having small genomes sequenced at high depth.

The first step in the pipeline consists in reads alignment or "mapping." The alignment algorithms map the reads onto the reference genome and establish their best mapping position and the strandedness. To perform the alignment the genome sequence of the bacterial strain under study, the fastq file containing the reads, and the alignment algorithm are needed. In the following example, the genome sequence file, in fasta format, is named genomeX.fna; the fastq file is indicated as SampleA.fastq, and the alignment algorithm is Bowtie2.

Bowtie2 requires a genome index to perform the alignment, the index is already provided on bowtie web-site for the organisms included in Illumina iGenome collection; otherwise, it is necessary to run the following command:

```
bowtie2-build genomeX.fna <path to output directory / index basename>
```

then it is possible to align the reads:

```
bowtie2 --end-to-end -q --phred33 --very-sensitive -x <path to genome index directory / index basename> -U <SampleA.fastq or comma separated list of read files> -S SampleA.sam 2>
SampleA.bowtie.log
```

This command line instructs bowtie2 to align to the genome indexed in -x folder the single-end (-U) fastq files displaying (-q) –phred33 qualities and to collect the results into SampleA.sam. We use end-to-end mapping and very-sensitive quality because, in our hands, it is the most solid procedure for consolidated genomes and good quality reads. If you want to permit the reads soft-clipping or to relax alignment parameters refer to Bowtie2 manual.

Tips&Tricks:sometimes bacterial genomes show big duplications, which can involve also genes relevant for the model. Usually, reads aligned to multiple genomic loci (multi-mapping) are reported only once in the final file. The program can be forced to randomly assign the multi-mapping-reads to a single genomic locus by using the –non-deterministic option; otherwise, to get out multiple alignment loci, -k parameter should be specified.

The alignment output consists in a .sam format file (https://samtools.github.io/hts-specs/SAMv1.pdf) which, in the case of Bowtie2, contains several fields, listed here below, that can be useful for quality filtering:

FLAG: a bitwise flag indicating the properties of the read as, for example, if it is aligned or not and on which strand it maps.

MAPQ: the read mapping quality, which is a measure of the wrong mapping probability.

CIGAR: an alphanumeric string reporting the number of sequence matches (which can be matches or mismatches), insertion, deletion, and soft clipping present in the alignment.

AS:i: the read alignment score accounting for the number of mismatches and indels present.

XS:i: indicating, when present, multi-mapping reads.

To select high-quality (HQ) reads we use SAMtools [61] standard functions and awk [62].

```
samtools view -S -h -F4 -q30 SampleA.sam | awk 'BEGIN
{OFS="\t";FS="\t"}{split($12,a,":"); if (a[3]>-15) print
$0}' - | samtools view -S -h -b - >SampleA.HQ.unsorted.bam
samtools sort SampleA.HQ.unsorted.bam
SampleA.HQ rm SampleA.HQ.unsorted.bam
```

This command eliminates unmapped reads (-F4) and reads with MAPQ score lower than 30, then it selects reads with alignment score (AS.:i:) greater than −15 which generally correspond to a maximum of 3 mismatches or 1 indel plus 1 or 2 mismatches. Pay attention to the fact that zero is the maximum score for end-to-end alignment but scoring system changes for local alignment option. In the final step, the command orders and compresses the file to binary format, producing an .HQ.bam file.

Tips&Tricks:if you need to consider also duplicated genes you can include also multi-mapping reads in the HQ.bam as follows:

```
cat <(samtools view -S -H SampleA.sam) <( samtools view -S -F4
-q30 SampleA.sam | grep -v "XS:i" - | awk 'BEGIN{OFS="\t";
FS="\t"}{split($12,a,":"); if (a[3]>-15) print $0}' -) <(sam-
tools view -S SampleA.sam |grep "XS:i" - |awk 'BEGIN{OFS="\t";
FS="\t"}{split($12,a,":"); if (a[3]>=-15) print $0}' -) |
samtools view - S -h -b - >SampleA.duplunsorted.HQ.bam
samtools sort SampleA.duplunsorted.HQ.bam SampleA.dupl.HQ
rm SampleA.duplunsorted.HQ.bam
```

3.1.2 Check-Point 2: Alignment Assessment

To evaluate alignment performances the information of bowtie log file (SampleA.bowtie.log) and quality filtering should be combined. The bowtie log file reports the number of uniquely aligned, multi-aligned, and not aligned reads. The number of multi-mapping reads depends on the amount of duplicated regions and genes present in the analyzed genome and also on their level of expression. As a consequence, uniquely mapping reads can vary in a range between 65 and 95% because of the peculiarities of the analyzed transcriptome but also because there can be a strong bias introduced by the different growing conditions under study that can influence the library preparation step.

Usually, more than 90% of mapping reads should be of High Quality (HQ-mapping-reads), if this percentage drops too much take into consideration to revise the HQ selection parameters. To calculate the number of HQ-mapping-reads run samtools flagstats and store these results into SampleA_quality_stats file as follows:

```
samtools flagstat SampleA.HQ.bam > SampleA.HQ.bam.stats
cat SampleA.bowtie.log >SampleA_quality_stats
awk 'BEGIN{OFS="\t"}{if (NR==3) print "Number HQ reads",$1 }'
SampleA.HQ.bam.stats >>SampleA_quality_stats
```

3.2 Aligned Reads Visualization

To visualize the aligned reads open source programs like IGV [63] can be used if .bam file index is provided. Visual inspection can help in the identification of regions with low alignment levels and, to inspect the effects of quality filtering, it is possible to change program settings in order to visualize also low-quality mapping reads, in this way the original file (including all the reads) and the high-quality one can be compared.

```
samtools view -S -h -b -F4 SampleA.sam >SampleA.unsorted.bam
samtools sort SampleA.unsorted.bam SampleA
samtools index SampleA.bam
samtools index SampleA.HQ.bam
```

3.3 Gene Counts Production

This step of the pipeline aims to obtain genes coverage evaluation by assigning an annotation to the regions where reads are mapped. Gene coverages, expressed in terms of strand-specific read counts, are necessary to perform differential expression analysis.

To allow gene coverage evaluation, the annotation file (whose suffix is feature_table.txt.gz) should be properly split and parsed as follows:

```
awk 'BEGIN{FS="\t";OFS="\t"}{if ($2=="protein_coding" &&
$1=="gene") print $7,$8,$9,$17,$2,$10}'
genomeX_feature_table.txt >genomeX.mRNA.bed
awk 'BEGIN{FS="\t";OFS="\t"}{if ($2=="tRNA" && $1=="gene")
print $7,$8,$9,$17,$2,$10}' genomeX_feature_table.txt
>genomeX.tRNA.bed
awk 'BEGIN{FS="\t";OFS="\t"}{if ($2=="rRNA" && $1=="gene")
print $7,$8,$9,$17,$2,$10}' genomeX_feature_table.txt
>genomeX.rRNA.bed
awk 'BEGIN{FS="\t";OFS="\t"}{if (NR!=1 && $2!="rRNA" &&
$2!="tRNA" && $2!="protein_coding" && $2!="" && $1=="gene")
print $7,$8,$9,$17,$2,$10}' genomeX_feature_table.txt
>genomeX.otherRNA.bed
```

Applying these commands four files containing respectively mRNAs, tRNAs, rRNAs, and otherRNAs in bed6 format will be generated. Bed six file format is a six-column file with the following structure: Column.1 genomic_accession; Column.2 start; Column.3 end; Column.4 Locus_tag; Column.5 Class; Column.6 strand.

3.3.1 Check Point 3: rRNA Removal and Strand Specificity Evaluation

To measure the efficacy of the rRNAs removal step it is necessary to take into consideration the full set of mapping reads, without removing the multi-mapping ones, because rRNA genes are usually present in many identical copies in bacterial genomes.

To attribute each read to a feature (annotated transcript) we require at least 50% of read overlap; in this way, it is possible to limit read count duplication if a read maps across two contiguous genes. This choice increases the number of reads mapping to unannotated regions, and in particular to UTR that are poorly annotated in bacteria. It is advisable to evaluate the pros and cons, when choosing minimal overlap parameter, because they are strongly related to gene density of the bacterial genome under investigation. A valid alternative could be to eliminate the minimal overlap requirement and then discard all the reads mapping across multiple features. For all these calculations we take advantage of BEDTools [50].

```
bedtools bamtobed -i SampleA.bam > SampleA.bed
TRNA_READS_COVERAGE=`bedtools intersect -f 0.5 -b
genomeX_tRNA.bed -a SampleA.bed | cut -f 4 | sort | uniq | wc -l`
RRNA_READS_COVERAGE=`bedtools intersect -f 0.5 -b
genomeX_rRNA.bed -a SampleA.bed | cut -f 4 | sort | uniq | wc -l`
MRNA_READS_COVERAGE=`bedtools intersect -f 0.5 -b
genomeX_mRNA.bed -a SampleA.bed | cut -f 4 | sort | uniq | wc -l`
samtools flagstat SampleA.bam > SampleA.bam.stats
BAM_READS=`awk '{if (NR==3) print $1 }' SampleA.bam.stats`
PERC_RRNA_READS_COVERAGE=`echo "$RRNA_READS_COVERAGE
$BAM_READS" |awk 'BEGIN{OFS="\t"}{print "rRNA relative percentage",
$1/$2*100}'`
PERC_MRNA_READS_COVERAGE=`echo "$MRNA_READS_COVERAGE
$BAM_READS" |awk 'BEGIN{OFS="\t"}{print "mRNA relative percentage",
$1/$2*100}'`
PERC_TRNA_READS_COVERAGE=`echo "$TRNA_READS_COVERAGE
$BAM_READS" |awk 'BEGIN{OFS="\t"}{print "tRNA relative percentage",
$1/$2*100}'`
```

Afterward the overall level of strand-specific coverage for each transcript category is evaluated as follows:

```
cat genomeX_mRNA.bed genomeX_rRNA.bed genomeX_tRNA.bed
genomeX_otherRNA.bed >genomeX_AllTranscripts.bed
ALL_SS_COVERAGE=`bedtools intersect -f 0.5 -a SampleA.bed -b
genomeX_AllTranscripts.bed -wb | awk 'BEGIN{FS="\t";OFS="\t";
SS=0;OS=0;tot=0}{if ($6==$12){SS+=1}else{OS+=1};tot=tot+1}
END{print "ALL:Strand Specific =\t"(SS/tot)*100"\nALL:Oppo-
site Strand =\t"(OS/tot)*100"\nALL:Tot Reads on TX =\t"tot}'`
MRNA_SS_COVERAGE=`bedtools intersect -f 0.5 -a SampleA.bed -b
genomeX_mRNA.bed -wb | awk 'BEGIN{FS="\t";OFS="\t";SS=0;OS=0;
tot=0}{if ($6==$12){SS+=1}else{OS+=1};tot=tot+1}END{print
```

```
"mRNA:Strand Specific =\t"(SS/tot)*100"\nmRNA:Opposite Strand
=\t"(OS/tot)*100"\nmRNA:Tot Reads on TX =\t"tot}' '
RRNA_SS_COVERAGE='bedtools intersect -f 0.5 -a SampleA.bed -b
genomeX_rRNA.bed -wb | awk 'BEGIN{FS="\t";OFS="\t";SS=0;OS=0;
tot=0}{if ($6==$12){SS+=1}else{OS+=1};tot=tot+1}END{print
"rRNA:Strand Specific =\t"(SS/tot)*100"\nrRNA:Opposite Strand
=\t"(OS/tot)*100"\nrRNA:Tot Reads on TX =\t"tot}' '
TRNA_SS_COVERAGE='bedtools intersect -f 0.5 -a SampleA.bed -b
genomeX_tRNA.bed -wb | awk 'BEGIN{FS="\t";OFS="\t";SS=0;OS=0;
tot=0}{if ($6==$12){SS+=1}else{OS+=1};tot=tot+1}END{print
"tRNA:Strand Specific =\t"(SS/tot)*100"\ntRNA:Opposite Strand
=\t"(OS/tot)*100"\ntRNA:Tot Reads on TX =\t"tot}' '
```

When the final aim of the RNAseq data analysis is the obtainment of transcriptomic data that should be integrated into metabolic models, the gene counts production step should take into consideration only genes annotated as CDS both for strand-specific gene coverage evaluation and for differential expression analysis. However, it is possible to extend the list of annotated genes used for gene counts determination, by including also tRNAs and pseudogenes that can be recovered from their corresponding bed6 files.

Here below the commands for extracting strand-specific reads from the .bam file:

```
cat genomeX_mRNA.bed >Interesting_genes.bed
bedtools intersect -abam SampleA.HQ.bam
-b Interesting_genes.bed -s > SampleA.HQ_SS.bam
```

Then it is possible to estimate the number of strand-specific reads covering each of the interesting genes as follows:

```
bedtools bamtobed -i SampleA.HQ_SS.bam > SampleA.HQ_SS.bed
bedtools intersect -b Interesting_genes.bed
-a SampleA.HQ_SS.bed -f 0.5 | bedtools coverage -b stdin
-a Interesting_genes.bed > SampleA.genes_coverage.bed
cut -f 4,7 SampleA.genes_coverage.bed > SampleA.read_count
```

These commands allow the transformation of the .bam file into a .bed format; thus, it is possible to compare it with the .bed file containing the list of interesting genes and to assign to each gene only the reads mapping on it with an overlap of at least 50%. Please check that the SampleA.genes_coverage.bed contains one row for each gene of interest and that it is structured as follows: chromosome name, gene_start, gene_end, gene_name, gene_type, strand, reads mapping on the gene, nucleotides of the gene covered by reads, gene length, and gene coverage percentage. This is a key step to obtain correct read counts.

Several tools used for transcriptomic data integration into metabolic models require as input gene expression measurements relative to a single experimental condition, which will be used to rank or subdivide the genes according to their expression levels. To this aim raw count values need to be normalized on gene length adopting a within-sample normalization; otherwise, the longer genes will have more chance to be ranked in the top positions than the shorter ones. RPKM (reads per kilobase per million of mapped reads) estimates are often used to cope with this problem and can be easily elaborated from raw counts with the following commands:

```
MAPPED_READS=`wc -l SampleA.HQ_SS.bed| awk '{print $1}' -`
cat <(awk -v mapped="$MAPPED_READS" 'BEGIN{OFS="\t"}{if
($7>0){print $4,$7*1000000000/($9*mapped)}else{print $4,0}}'
SampleA.genes_coverage.bed | sort ) >
SampleA.genes_coverage.RPKM
```

3.3.2 Check Point 4: Genes Coverage Statistics

To better evaluate the strand-specific coverage for each sample on the list of interesting genes, some parameters can be estimated to gain a genome-wide evaluation of gene coverage: the percentage of expressed genes, the percentage of genes covered by at least five reads, the number of reads covering the 90% of the genes, and the percentage of genes covered for at least 50% of their length.

These parameters are useful specially to compare the final gene coverage obtained for different samples and to decide if further sequencing is required.

```
ONE_COV=`awk 'BEGIN{FS="\t";OFS="\t";c=0}{if ($2>0)c=c+1}END
{print "Percentage of expressed genes",c/NR*100}' SampleA.
read_count`
DHT_COV=`awk -v dth=4 'BEGIN{FS="\t";OFS="\t";c=0}{if
($2>dth)c=c+1}END{print "Percentage of genes covered at least
by five reads",c/NR*100}' SampleA.read_count`
NUMBER_OF_MRNA=`wc -l SampleA.read_count | cut -f1 -d " "`
PERCENTILE_10=`echo $NUMBER_OF_MRNA-\($NUMBER_OF_MRNA/10\) |
bc`
PERCENTILE_10_COV=`sort -k2,2rn SampleA.read_count | awk -v
pos=$PERCENTILE_10 'BEGIN{FS="\t";OFS="\t"}{if (NR==pos)
print "Reads covering 90 percent of the genes",$2}'`
COV_50=`mawk 'BEGIN{FS="\t";OFS="\t";c=0}{if ($10>0.5)
c=c+1}END{print "Percentage of genes covered for at half of
their lenght",c/NR*100}' SampleA.genes_coverage.bed`
printf "$PERC_MRNA_READS_COVERAGE\n$PERC_RRNA_READS_COVERAGE
\n$PERC_TRNA_READS_COVERAGE\n$ALL_SS_COVERAGE\n
$MRNA_SS_COVERAGE\n$RRNA_SS_COVERAGE\n$TRNA_SS_COVERAGE\n
$ONE_COV\n$DHT_COV\n$PERCENTILE_10_COV\n$COV_50"
>>SampleA_quality_stats
```

After running all the check points code you will end up with a file containing the quality statistics for each of the analyzed samples enabling an easy comparison of the alignment level, library preparation, and sequencing depth among them.

3.4 Gene Counts Normalization and Differential Expression Analysis

Once the gene counts for all the samples have been obtained, it is possible to proceed to the last part of the analysis which is performed within R environment [64]. To do data normalization and differential expression analysis we suggest using DESeq2 tool. In this example, we propose the analysis of a simple two conditions and two replicas dataset, but this tool can be applied also to complex experimental designs, you can refer to DESeq2 manual for further information.

Start the environment just typing

```
R
```

and load the libraries needed for the analysis and all the gene count files.

```
library("DESeq2")
library("ggplot2")
Treated_rep_A <- read.table("/path to read/Sample1.read/
count", row.names=1)
Treated_rep_B <-read.table("/path to read/Sample2.read/
count", row.names=1)
Control_rep_A <-read.table("/path to read/Sample3.read/
count", row.names=1)
Control_rep_B <-read.table("/path to read/Sample4.read/
count", row.names=1)
```

Then combine the gene count data into a single matrix and build the dataframe necessary to decipher the experimental design and to generate a DESeq dataset.

```
My_experiment<-cbind(Treated_rep_A,Treated_rep_B,
Control_rep_A,Control_rep_B)
colnames(My_experiment)<-c("Treated_rep_A","Treated_rep_B",
"Control_rep_A","Control_rep_B")
colData_My_experiment = data.frame(row.names =colnames
(My_experiment), replica=c("A","B","A","B"),
condition=c("Treated","Treated","Untreated","Untreated"))
dds <- DESeqDataSetFromMatrix(countData = My_experiment,
colData= colData_My_experiment, design = ~ condition)
```

3.4.1 Check Point 5: Samples Reproducibility Control

This check point is very important and should be performed before proceeding with the differential expression analysis. DESeq2 implements a lot of graphical outputs to check sample reproducibility and to control the effects of data normalization. You can refer to the manual for extensive information.

```
rld <- rlogTransformation(dds, blind=TRUE)
data <- plotPCA(rld, intgroup=c("condition", "replica"),
returnData=TRUE)
percentVar <- round(100 * attr(data,"percentVar"))
pdf("PCA_samples")
ggplot(data, aes(PC1, PC2, color=condition, shape=replica)) +
geom_point(size=3) + xlab(paste0("PC1:",percentVar[1],
"%variance"))+ylab(paste0("PC2:",percentVar[2],"%variance"))
dev.off()
```

Few simple steps are required for differential expression calculation: first you have to define the control condition. DESeq2 automatically classifies as control the first condition in alphabetical order, so be careful and properly set up the name of the control condition; second you should run the calculation and third you can get the results.

```
dds$condition <- relevel(dds$condition, ref="untreated")
dds2 <- DESeq(dds)
res <- results(dds2)
```

Finally, write the results to a file that can be used outside of R environment.

```
write.csv(as.data.frame(res),
file="condition_treated_VS_untreated_results.csv")
```

The results file is organized into six columns, the baseMean column contains the average of the normalized count values, the log2FoldChanges column contains the fold changes expressed in log 2 scale, and the standard error column contains the error values associated with log 2 fold change estimate. The following three columns contain respectively the statistical value, the relative p-value, and the p-value adjusted for multiple testing. The statistical analysis applied by DESeq2 is based on the testing of the null hypothesis for each gene to determine if its log2 fold change can be significantly related to the treatment or not.

Please note that log2FoldChanges values obtained with DESeq2 are not exactly the ones calculated from normalized counts, but the result of a shrinkage step performed during the model elaboration which aims to limit high fold change values deriving from low count genes.

Between sample normalized counts can be stored as follows:

```
write.csv(counts(dds,normalized=TRUE),
file="normalized_counts.csv")
```

as well as the log2 fold changes values obtained using standard calculation (this option is only available in the last DESeq2 version)

```
resMLE <- results(dds, addMLE=TRUE)
write.csv(as.data.frame(resMLE),
file="condition_treated_VS_untreated_resultswhithMLE.csv")
```

Note: the command-lines reported in this chapter have been splitted for graphical reasons; you can find the original script here: https://github.com/epinatel/Bacterial_RNAseq/blob/master/README.md

4 Integration of Transcriptomic Data into Metabolic Models

The algorithms used to integrate transcriptomic data into metabolic models can be divided among those handling the expression data relative to a single time-point or condition [65–68] and those handling data derived from multiple conditions or Time-Course experiments [69–73]. In the first case, the transcriptomic data are needed to define not expressed genes [67], or highly, moderate, and poorly expressed ones [65] on the bases of expression value distribution or static thresholds. In other cases these tools can also directly use continuous gene expression values [68], but, all of them, require as input within-sample normalized counts. The tools integrating multiple datasets instead require two types of input. In some cases these tools adopt between-sample normalized counts, which are then binarized according to user-defined thresholds or more complex elaborations [69, 71, 73]. While, in other cases, they require the output of a differential expression analysis because they evaluate fold changes and p values to define the reactions activity state in the analyzed conditions [70, 72].

References

1. Croucher NJ, Thomson NR (2010) Studying bacterial transcriptomes using RNA-seq. Curr Opin Microbiol 13:619–624. https://doi.org/10.1016/j.mib.2010.09.009
2. Conway T, Creecy JP, Maddox SM et al (2014) Unprecedented high-resolution view of bacterial operon architecture revealed by RNA sequencing. MBio 5(4):e01442–e01414. https://doi.org/10.1128/mBio.01442-14
3. Creecy JP, Conway T (2015) Quantitative bacterial transcriptomics with RNA-seq. Curr Opin Microbiol 23:133–140. https://doi.org/10.1016/j.mib.2014.11.011
4. Arraiano CM, Andrade JM, Domingues S et al (2010) The critical role of RNA processing and degradation in the control of gene expression. FEMS Microbiol Rev 34:883–923

5. Passalacqua KD, Varadarajan A, Ondov BD et al (2009) Structure and complexity of a bacterial transcriptome. J Bacteriol 191:3203–3211
6. Perkins TT, Kingsley RA, Fookes MC et al (2009) A strand-specific RNA-Seq analysis of the transcriptome of the typhoid bacillus Salmonella typhi. PLoS Genet 5:e1000569
7. Yoder-Himes DR, Chain PS, Zhu Y et al (2009) Mapping the Burkholderia cenocepacia niche response via high-throughput sequencing. Proc Natl Acad Sci U S A 106:3976–3981
8. Passalacqua KD, Varadarajan A, Byrd B, Bergman NH (2009) Comparative transcriptional profiling of Bacillus cereus sensu lato strains during growth in CO2-bicarbonate and aerobic atmospheres. PLoS One 4:e4904
9. Filiatrault MJ, Stodghill PV, Bronstein PA et al (2010) Transcriptome analysis of Pseudomonas syringae identifies new genes, noncoding RNAs, and antisense activity. J Bacteriol 192:2359–2372
10. Gilbert JA, Field D, Huang Y et al (2008) Detection of large numbers of novel sequences in the metatranscriptomes of complex marine microbial communities. PLoS One 3:e3042
11. Poretsky RS, Hewson I, Sun S et al (2009) Comparative day/night metatranscriptomic analysis of microbial communities in the North Pacific subtropical gyre. Environ Microbiol 11:1358–1375
12. Dodd D, Moon YH, Swaminathan K et al (2010) Transcriptomic analyses of xylan degradation by Prevotella bryantii and insights into energy acquisition by xylanolytic bacteroidetes. J Biol Chem 285:30261–30273
13. Giannoukos G, Ciulla DM, Huang K et al (2012) Efficient and robust RNA-seq process for cultured bacteria and complex community transcriptomes. Genome Biol 13:r23
14. Sigurgeirsson B, Emanuelsson O, Lundeberg J (2014) Analysis of stranded information using an automated procedure for strand specific RNA sequencing. BMC Genomics 15:631. https://doi.org/10.1186/1471-2164-15-631
15. Sultan M, Amstislavskiy V, Risch T et al (2014) Influence of RNA extraction methods and library selection schemes on RNA-seq data. BMC Genomics 15:675. https://doi.org/10.1186/1471-2164-15-675
16. Sharma CM, Hoffmann S, Darfeuille F et al (2010) The primary transcriptome of the major human pathogen Helicobacter pylori. Nature 464:250–255
17. Jager D, Sharma CM, Thomsen J et al (2009) Deep sequencing analysis of the Methanosarcina mazei Go1 transcriptome in response to nitrogen availability. Proc Natl Acad Sci U S A 106:21878–21882
18. Irnov I, Sharma CM, Vogel J, Winkler WC (2010) Identification of regulatory RNAs in Bacillus subtilis. Nucleic Acids Res 38:6637–6651
19. Bischler T, Tan HS, Nieselt K, Sharma CM (2015) Differential RNA-seq (dRNA-seq) for annotation of transcriptional start sites and small RNAs in Helicobacter pylori. Methods 86:89–101. https://doi.org/10.1016/j.ymeth.2015.06.012
20. Thomason MK, Bischler T, Eisenbart SK et al (2015) Global transcriptional start site mapping using differential RNA sequencing reveals novel antisense RNAs in Escherichia coli. J Bacteriol 197:18–28. https://doi.org/10.1128/JB.02096-14
21. Sharma CM, Vogel J (2014) Differential RNA-seq: the approach behind and the biological insight gained. Curr Opin Microbiol 19:97–105. https://doi.org/10.1016/j.mib.2014.06.010
22. Wurtzel O, Sapra R, Chen F et al (2010) A single-base resolution map of an archaeal transcriptome. Genome Res 20:133–141
23. Vivancos AP, Guell M, Dohm JC et al (2010) Strand-specific deep sequencing of the transcriptome. Genome Res 20:989–999
24. Yi H, Cho YJ, Won S et al (2011) Duplex-specific nuclease efficiently removes rRNA for prokaryotic RNA-seq. Nucleic Acids Res 39 (20):e140
25. Peano C, Pietrelli A, Consolandi C et al (2013) An efficient rRNA removal method for RNA sequencing in GC-rich bacteria. Microb Inform Exp 3(1):1
26. Rossi E, Longo F, Barbagallo M, Peano C, Consolandi C, Pietrelli A, Jaillon S, Garlanda CLP, Landini P (2016) Glucose availability enhances lipopolysaccharide production and immunogenicity in the opportunistic pathogen Acinetobacter baumannii. Future Microbiol 11 (3):335–349. https://doi.org/10.2217/fmb.15.153
27. Tocchetti A, Bordoni R, Gallo G et al (2015) A genomic, transcriptomic and proteomic look at the GE2270 producer Planobispora rosea, an uncommon actinomycete. PLoS One 10(7): e0133705. https://doi.org/10.1371/journal.pone.0133705
28. Peano C, Chiaramonte F, Motta S et al (2014) Gene and protein expression in response to different growth temperatures and oxygen availability in *Burkholderia thailandensis*. PLoS ONE 9(3): e93009. https://doi.org/10.1371/journal.pone.0093009

29. Hrdlickova R, Toloue M, Tian B (2016) RNA-Seq methods for transcriptome analysis. Wiley Interdiscip Rev RNA 8(1). https://doi.org/10.1002/wrna.1364
30. Parkhomchuk D, Borodina T, Amstislavskiy V et al (2009) Transcriptome analysis by strand-specific sequencing of complementary DNA. Nucleic Acids Res 37(18):e123. https://doi.org/10.1093/nar/gkp596
31. Zhong S, Joung JG, Zheng Y et al (2011) High-throughput illumina strand-specific RNA sequencing library preparation. Cold Spring Harb Protoc 6:940–949. https://doi.org/10.1101/pdb.prot5652
32. Weissenmayer BA, Prendergast JGD, Lohan AJ, Loftus BJ (2011) Sequencing illustrates the transcriptional response of Legionella pneumophila during infection and identifies seventy novel small non-coding RNAs. PLoS One 6(3):e17570. https://doi.org/10.1371/journal.pone.0017570
33. Borodina T, Adjaye J, Sultan M (2011) A strand-specific library preparation protocol for RNA sequencing. Methods Enzymol 500:79–98. https://doi.org/10.1016/B978-0-12-385118-5.00005-0
34. Hafner M, Landgraf P, Ludwig J et al (2008) Identification of microRNAs and other small regulatory RNAs using cDNA library sequencing. Methods 44:3–12. https://doi.org/10.1016/j.ymeth.2007.09.009
35. Jayaprakash AD, Jabado O, Brown BD, Sachidanandam R (2011) Identification and remediation of biases in the activity of RNA ligases in small-RNA deep sequencing. Nucleic Acids Res 39(21):e141. https://doi.org/10.1093/nar/gkr693
36. Levin JZ, Yassour M, Adiconis X et al (2010) Comprehensive comparative analysis of strand-specific RNA sequencing methods. Nat Methods 7:709–715. https://doi.org/10.1038/nmeth.1491
37. Langevin SA, Bent ZW, Solberg OD et al (2013) Peregrine: a rapid and unbiased method to produce strand-specific RNA-Seq libraries from small quantities of starting material. RNA Biol 10:502–515. https://doi.org/10.4161/rna.24284
38. Townsley BT, Covington MF, Ichihashi Y et al (2015) BrAD-seq: breath adapter directional sequencing: a streamlined, ultra-simple and fast library preparation protocol for strand specific mRNA library construction. Front Plant Sci 6:366. https://doi.org/10.3389/fpls.2015.00366
39. Miller DFB, Yan PS, Buechlein A et al (2013) A new method for stranded whole transcriptome RNA-seq. Methods 63:126–134. https://doi.org/10.1016/j.ymeth.2013.03.023
40. Miller DFB, Yan PX, Fang F et al (2015) Stranded whole transcriptome RNA-Seq for all RNA types. Curr Protoc Hum Genet 84:11.14.1–11.14.23. https://doi.org/10.1002/0471142905.hg1114s84
41. Pease J, Sooknanan R (2012) A rapid, directional RNA-seq library preparation workflow for Illumina[reg] sequencing. Nat Methods 9: i–ii. https://doi.org/10.1038/nmeth.f.355
42. Conesa A, Madrigal P, Tarazona S et al (2016) A survey of best practices for RNA-seq data analysis. Genome Biol 17:13. https://doi.org/10.1186/s13059-016-0881-8
43. Haas BJ, Chin M, Nusbaum C et al (2012) How deep is deep enough for RNA-Seq profiling of bacterial transcriptomes? BMC Genomics 13:734. https://doi.org/10.1186/1471-2164-13-734
44. McClure R, Balasubramanian D, Sun Y et al (2013) Computational analysis of bacterial RNA-Seq data. Nucleic Acids Res 41:e140. https://doi.org/10.1093/nar/gkt444
45. Haas BJ, Papanicolaou A, Yassour M et al (2013) De novo transcript sequence reconstruction from RNA-seq using the Trinity platform for reference generation and analysis. Nat Protoc 8:1494–1512. https://doi.org/10.1038/nprot.2013.084
46. Luo R, Liu B, Xie Y et al (2012) SOAPdenovo2: an empirically improved memory-efficient short-read de novo assembler. Gigascience 1:18. https://doi.org/10.1186/2047-217X-1-18
47. Tjaden B (2015) De novo assembly of bacterial transcriptomes from RNA-seq data. Genome Biol 16(1). https://doi.org/10.1186/s13059-014-0572-2
48. Langmead B, Salzberg SL (2012) Fast gapped-read alignment with Bowtie 2. Nat Methods 9:357–359. https://doi.org/10.1038/nmeth.1923
49. Rumble SM, Lacroute P, Dalca AV et al (2009) SHRiMP: accurate mapping of short color-space reads. PLoS Comput Biol 5:e1000386. https://doi.org/10.1371/journal.pcbi.1000386
50. Li H, Durbin R (2009) Fast and accurate short read alignment with Burrows-Wheeler transform. Bioinformatics 25:1754–1760
51. Hoffmann S, Otto C, Kurtz S et al (2009) Fast mapping of short sequences with mismatches, insertions and deletions using index structures. PLoS Comput Biol 5:e1000502. https://doi.org/10.1371/journal.pcbi.1000502
52. Quinlan AR, Hall IM (2010) BEDTools: a flexible suite of utilities for comparing genomic features. Bioinformatics 26:841–842

53. Anders S, Pyl PT, Huber W (2015) HTSeq–a Python framework to work with high-throughput sequencing data. Bioinformatics 31:166–169. https://doi.org/10.1093/bioinformatics/btu638
54. Love MI, Huber W, Anders S (2014) Moderated estimation of fold change and dispersion for RNA-seq data with DESeq2. Genome Biol 15:550. https://doi.org/10.1186/s13059-014-0550-8
55. Anders S, Huber W (2010) Differential expression analysis for sequence count data. Genome Biol 11:R106
56. Robinson MD, McCarthy DJ, Smyth GK (2010) edgeR: a bioconductor package for differential expression analysis of digital gene expression data. Bioinformatics 26:139–140. https://doi.org/10.1093/bioinformatics/btp616
57. Förstner KU, Vogel J, Sharma CM (2014) READemption-a tool for the computational analysis of deep-sequencing-based transcriptome data. Bioinformatics 30:3421–3423. https://doi.org/10.1093/bioinformatics/btu533
58. Johnson BK, Scholz MB, Teal TK, Abramovitch RB (2016) SPARTA: simple program for automated reference-based bacterial RNA-seq transcriptome analysis. BMC Bioinformatics 17:66. https://doi.org/10.1186/s12859-016-0923-y
59. Andrews S. (2010). FastQC: a quality control tool for high throughput sequence data. Available online at: http://www.bioinformatics.babraham.ac.uk/projects/fastqc
60. Bolger AM, Lohse M, Usadel B (2014) Trimmomatic: a flexible trimmer for Illumina sequence data. Bioinformatics 30:2114–2120. https://doi.org/10.1093/bioinformatics/btu170
61. Li H, Handsaker B, Wysoker A et al (2009) The sequence alignment/map format and SAMtools. Bioinformatics 25:2078–2079. https://doi.org/10.1093/bioinformatics/btp352
62. Aho A V, Kernighan BW, Weinberger PJ (1987) The Awk Programming Language, Editor: Prentice Hall, ISBN-13: 978-0201079814
63. Thorvaldsdóttir H, Robinson JT, Mesirov JP (2013) Integrative Genomics Viewer (IGV): High-performance genomics data visualization and exploration. Brief Bioinform 14:178–192. https://doi.org/10.1093/bib/bbs017
64. Team RDC (2012) R: A language and environment for statistical computing. the R Foundation for Statistical Computing, Vienna, Austria
65. Zur H, Ruppin E, Shlomi T (2010) iMAT: an integrative metabolic analysis tool. Bioinformatics 26:3140–3142. https://doi.org/10.1093/bioinformatics/btq602
66. Becker SA, Palsson BO (2008) Context-specific metabolic networks are consistent with experiments. PLoS Comput Biol 4(5): e1000082. https://doi.org/10.1371/journal.pcbi.1000082
67. Åkesson M, Förster J, Nielsen J (2004) Integration of gene expression data into genome-scale metabolic models. Metab Eng 6:285–293. https://doi.org/10.1016/j.ymben.2003.12.002
68. Colijn C, Brandes A, Zucker J et al (2009) Interpreting expression data with metabolic flux models: predicting mycobacterium tuberculosis mycolic acid production. PLoS Comput Biol 5(8):e1000489. https://doi.org/10.1371/journal.pcbi.1000489
69. Collins SB, Reznik E, Segrè D (2012) Temporal expression-based analysis of metabolism. PLoS Comput Biol 8(11):e1002781. https://doi.org/10.1371/journal.pcbi.1002781
70. Navid A, Almaas E (2012) Genome-level transcription data of Yersinia pestis analyzed with a new metabolic constraint-based approach. BMC Syst Biol 6:150. https://doi.org/10.1186/1752-0509-6-150
71. Chandrasekaran S, Price ND (2010) Probabilistic integrative modeling of genome-scale metabolic and regulatory networks in Escherichia coli and Mycobacterium tuberculosis. Proc Natl Acad Sci 107:17845–17850. https://doi.org/10.1073/pnas.1005139107
72. Töpfer N, Jozefczuk S, Nikoloski Z (2012) Integration of time-resolved transcriptomics data with flux-based methods reveals stress-induced metabolic adaptation in Escherichia coli. BMC Syst Biol 6:148. https://doi.org/10.1186/1752-0509-6-148
73. Jensen PA, Papin JA (2011) Functional integration of a metabolic network model and expression data without arbitrary thresholding. Bioinformatics 27:541–547. https://doi.org/10.1093/bioinformatics/btq702

Chapter 12

Differential Proteomics Based on 2D-Difference In-Gel Electrophoresis and Tandem Mass Spectrometry for the Elucidation of Biological Processes in Antibiotic-Producer Bacterial Strains

Giuseppe Gallo and Andrea Scaloni

Abstract

Proteomics based on 2D-Difference In Gel Electrophoresis (2D-DIGE) coupled with mass spectrometry (MS) procedures can be considered a "gold standard" to determine quantitatively and comparatively protein abundances in cell extracts from different biological sources/conditions according to a gel-based approach. In particular, 2D-DIGE is used for protein specie separation, detection, and relative quantification, whenever tandem MS is used to obtain peptide sequence information that is managed according to bioinformatic procedures to identify the differentially represented protein species. The proteomic results consist of a dynamic portray of over- and down-represented protein species that, with the integration of gene ontology resources, allow obtaining a comprehensive understanding of the complex network of molecular signaling, regulatory circuits, and biochemical reactions occurring in cellular contexts. For this reason, proteomics has been widely used for studying molecular physiology of Gram-positive bacterial strains producing bioactive metabolites and belonging to actinomycete family. This highlighted the complex relationships linking overall regulatory processes and metabolic pathways to the biosynthesis of interesting bioactive molecules. In this chapter, we provide a detailed description of the procedures adopted to perform a differential proteomic analysis of the actinomycete *Microbispora* ATCC-PTA-5024, producing the promising NAI-107 lantibiotic. Although each experimental proteomic procedure has to be optimized to face the specific molecular characteristics of the organism under investigation, the protocols here described have also been used with minor modifications for proteomic studies on other bacterial strains, including the actinomycetes *Streptomyces coelicolor*, *S. ambofaciens*, *Amycolatopsis balhimycina*, and the Gram-negative proteobacteria *Klebsiella oxytoca* and *Pseudoalteromonas haloplanktis*.

Key words 2D-DIGE, Protein separation, Relative quantification, NanoLC-ESI-LIT-MS/MS, Protein identification, Bioinformatics

1 Introduction

In the last decade, 2D-Difference In-Gel Electrophoresis (2D-DIGE) coupled with mass spectrometry (MS) procedures has been widely used to analyze quantitatively and comparatively

Marco Fondi (ed.), *Metabolic Network Reconstruction and Modeling: Methods and Protocols*, Methods in Molecular Biology, vol. 1716, https://doi.org/10.1007/978-1-4939-7528-0_12,

protein abundances in cell extracts from different biological conditions, according to a holistic perspective [1]. In 2D-DIGE, proteins are resolved according to their isoelectric point and molecular weight values, similarly to classical 2D-gel electrophoresis (2D-GE), by performing a *first dimension* isoelectric focusing (IEF) separation, followed from a *second dimension* separation through SDS-PolyAcrylamide Gel Electrophoresis (PAGE). Thus, resolved proteins are present on a 2D gel array, in which each component (in principle) occurs as a specific gel spot. Few thousands of protein species are generally separated in a single experiment.

Whereas in 2D-GE different samples are run separately on individual gels and, after staining, are compared for their images by dedicated software, 2D-DIGE allows more than one sample to be separated simultaneously on the same 2D-gel, including an internal standard made by pooling all protein samples. This is achieved by a sub-stoichiometrical labeling of each sample with a specific CyDye fluorescent dye (Fig. 1), which is followed by acquisition of 2D-DIGE images by specific scanner instrumentations, and final analysis by dedicated software. In the whole, 2D-DIGE allows: (1) reducing gel-to-gel variations, simplifying gel matching; (2) improving protein spot quantification permitting the measurement of normalized spot volumes; (3) detecting and defining spot boundaries; (4) calculating mean normalized abundances of protein spots in different 2D-gels belonging to the same class (e.g., 2D-gels from the same condition); (5) defining a ratio for each protein in the different classes, with a statistical evaluation of the reliability of results, making significant/acceptable protein abundance variations as low as 20% [2] (Fig. 2). For the reasons mentioned above, 2D-DIGE represents a technological advancement with respect to 2D-GE.

After further staining with methods compatible with MS analysis, e.g., silver nitrate- or SyproRuby-based procedures, spots of interest are visualized on 2D-gels, picked manually or robotically for further identification of gel-contained proteins [3, 4]. The latter goal is routinely achieved by MS procedures, which measure the mass values of the peptides generated by *in-gel* treatment of spot pieces with trypsin, or provide the above-mentioned information together with peptide fragmentation results. In the first case, a peptide mass fingerprinting information is generated, based on recorded mass values. In the second case, information on mass value and sequence of resulting peptides is obtained [5, 6]; the latter is generally achieved by nanoLiquid Chromatography-ElectroSpray Ionization-Tandem MS instruments, such as the nanoLiquid Chromatography-ElectroSpray-Linear Ion Trap -Tandem MS (nanoLC-ESI-LIT-MS/MS) spectrometer described in this chapter. In both the cases, raw data generated are then searched by dedicated software engines against updated nonredundant

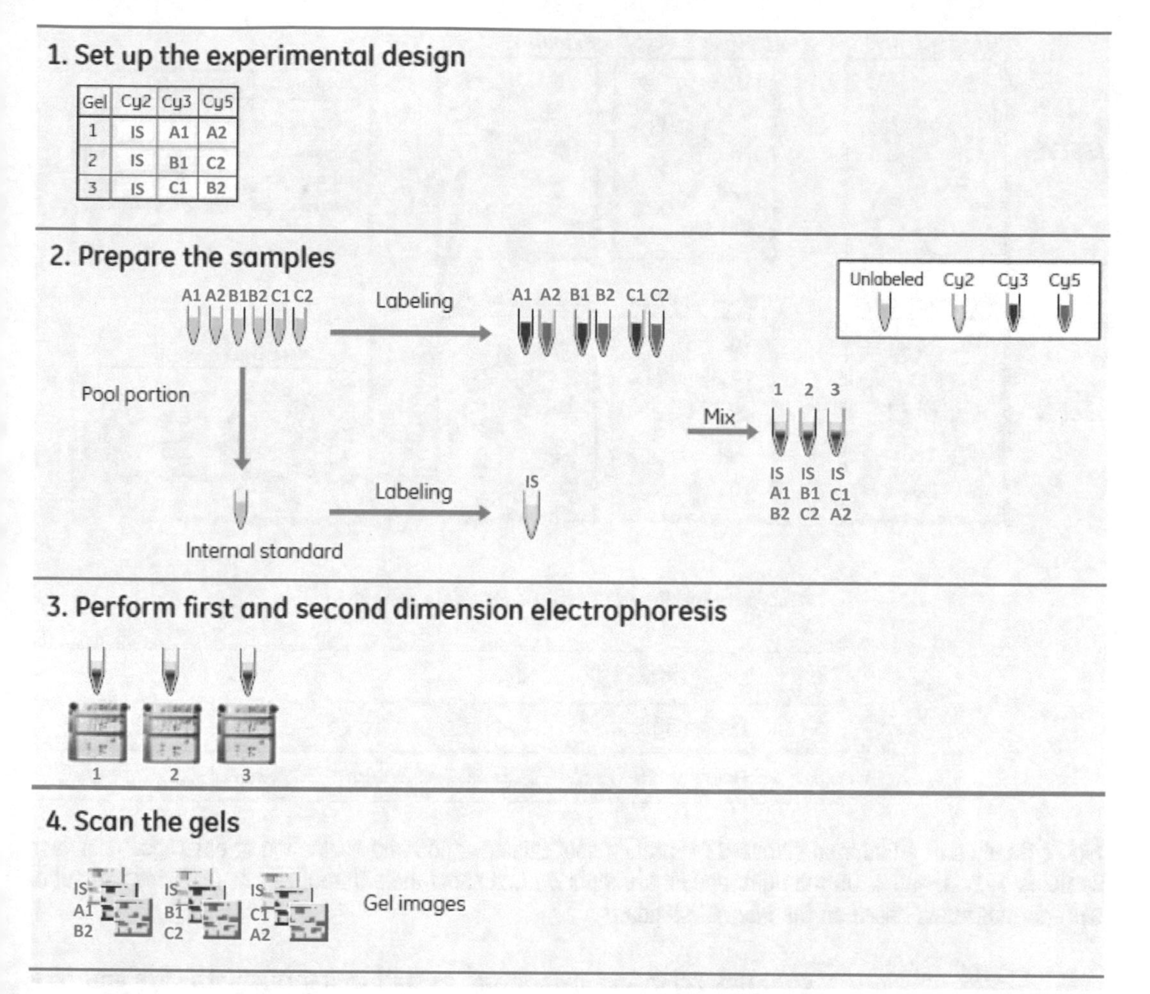

Fig. 1 Workflow of 2D-DIGE experiments. Swapping of CyDye fluor labeling and sample matching randomization are performed to minimize any biases. A, B and C represent different conditions; numbering refers to corresponding biological replicas. IS: internal standard made of by pooled sample protein mix

database, with the aim of identifying proteins present in each gel spot. Database searching is performed by using information on preventive protein/spot treatment, instrument-dependent mass tolerance, proteolytic enzyme used, and others.

On this basis, quantitative proteomics reveals dynamic portrays of over- and down-represented protein abundances that, coupled with gene ontology resources, allow obtaining a comprehensive understanding of complex network of molecular signaling, regulatory circuits, and biochemical reactions occurring in cellular contexts. For this reason, quantitative proteomics has been widely used for studying molecular physiology of bacterial strains belonging to the actinomycete family [7–15]. In fact, actinomycetes, filamentous Gram-positive bacteria, are the most prolific source of natural bioactive molecules, like antibiotics [16, 17]; understanding of

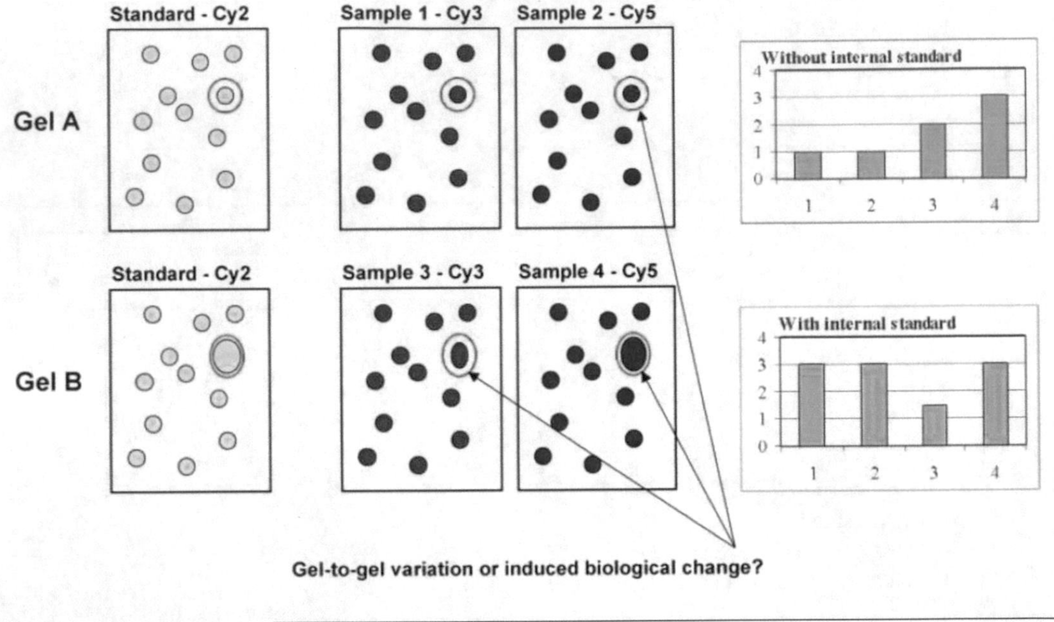

Gel	Cy2 (Standard)	Cy3	Cy5
A	Pool samples 1–4	Sample 1 - untreated	Sample 2 - treated
B	Pool samples 1–4	Sample 3 - treated	Sample 4 - untreated

Fig. 2 Benefits of an internal standard for protein spot quantification and evaluation of differences between samples 1, 2, 3, and 4. On the right, panels are reported that show the different results achieved without or with normalization based on the internal standard

complex relationships among overall metabolic pathways and biosynthesis of interesting bioactive molecules may be of interest in the development of robust and cost-effective production processes based on actinomycete strain fermentation [18, 19].

In this chapter, we provide a detailed description of the proteomic procedures used to obtain results in differential studies on actinomycetes based on 2D-DIGE and MS analyses, such as that described in Gallo et al. [7]. The methodologies here reported comprise different and complex experimental procedures and technologies, which were used to cover the whole proteomic pipeline in actinomycetes, i.e., from protein extraction to protein identification and data interpretation. Due to the high complexity of the procedures and technologies used in proteomics, a further dedicated reading on these topics is highly recommended. Worth mentioning is also the fact that biological sample extraction/manipulation has to be optimized every time according to the biological source under investigation or further different aspects related to organism biology. Introducing or not minor modifications, these procedures were used successfully for proteomic analyses on different bacterial strains including actinomycetes—like *Microbispora* ATCC-PTA-5024 [7], *Amycolatopsis balhimycina* [8, 9], *Streptomyces coelicolor* [20–22], *S. lividans* [23], *S. ambofaciens* (manuscript in preparation)—and Gram-negative proteobacteria like *Klebsiella oxytoca* [24] and *Pseudoalteromonas haloplanktis* (manuscript in preparation) strains.

2 Materials

Prepare all the solutions using ultrapure water (prepared by purifying deionized water to attain a sensitivity of 18 MΩ cm at 25 °C) and analytical grade reagents (unless indicated otherwise). All the unsterile plastic disposals—like safe-lock tubes and pipette tips—have to be sterilized by autoclave treatment. Prepare and store all reagents at room temperature (unless indicated otherwise). All the chemicals should be considered potentially hazardous. Therefore, they have to be handled in accordance with the principles of good laboratory practice (wear laboratory overalls and safety gloves, work in a cleaned fume hood). Diligently follow all waste disposal regulations when disposing waste materials.

2.1 Solutions for Total Protein Extraction

1. Washing solution (WS): 10 mM Tris–HCl pH 7.5, 5 mM EDTA, 1 mM DTT, 0.5 mM PMSF, 4 mg/mL leucopeptin, 0.7 mg/mL pepstatin, and 5 mg/mL benzamidin (*see* **Note 1**). WS was prepared in a 50 mL sterile screw cap plastic tube (Table 1).
2. Sonication solution (SS): SS was prepared just before use by adding in a sterile screw cap plastic tube 30 μL of 10% w/v SDS per 1 mL of WS.

Table 1
Amounts of constituents to prepare 50 mL of WS

Compound	Stock solution	Volume (mL)
Dithiothreitol (DTT)	1 M	0.05
Phenylmethylsulfonyl fluoride (PMSF)	0.1 M dissolved in isopropanol	0.250
Tris-HCl	1 M, pH 7.5	0.5
Ethylenediaminetetraacetic acid (EDTA)	0.5 M	0.5
Leupeptin	5 mg/mL	0.05
Pepstain	1 mg/mL in ethanol	0.350
Benzamidine	4 mg/mL	0.05

Make up to 50 mL with autoclaved ultrapure water

Table 2
Amounts of constituents to prepare 50 mL of SB

Reagent	Quantity
Urea	21 g
Thiourea	7.61 g
Tris (1 M)	1.5 mL
CHAPS	2 g

Make up to 50 mL with autoclaved ultrapure water

2.2 Sample Preparation for 2D-DIGE Analysis

1. Sample buffer (SB): 30 mM Tris, 7 M urea, 2 M thiourea, 4% w/v CHAPS, pH 8.5 (Table 2). SB was prepared in a sterile screw cap plastic tube (*see* **Note 2**). Small aliquots were stored at −20 °C for up to 3 months.
2. Dialysis membrane tubing: membrane tubing (14.3 mm diameter, 12 kDa Mw cut off) were cut to 8–10 cm length; they were immersed in a 2 L glass beaker containing 1 L solution of 2% w/v sodium bicarbonate, 1 mM EDTA (pH 8.0). Then, they were boiled for 10 min; they were decanted of the solution, and were rinsed thoroughly with ultrapure water. Membrane tubing was boiled in 1 L of 1 mM EDTA (pH 8.0), for 10 min, allowing them to cool. Finally, they were stored at 4 °C; before use, membrane tubing was washed inside and out with ultrapure water.
3. Anhydrous dimethylformamide (DMF) (*see* **Note 3**).
4. CyDye fluorophores: specifically designed fluorescent dyes for protein minimal labeling to detect protein abundance differences in 2D-DIGE experiments. In this experiment, 5 nmol CyDye™ DIGE minimal labeling kit (GE Healthcare, Sweden) was used. For reconstitution of CyDyes, the dye tubes were allowed to warm at room temperature for 5 min, and then were added with 5 μL of 99.8% v/v DMF to prepare 1 mM the stock solutions; the latter were mixed by vortexing and then spun briefly by a microfuge to collect dye solution at the bottom of the tube. They were stored at −20 °C for not more than 3 months after reconstitution.
5. Immobilized pH gradient (IPG) gel strips: 18 cm Immobiline DryStrip gels with pH 3–10 nonlinear (NL) were used (GE Healthcare, Sweden).
6. IPG buffer: ampholyte-containing buffer, specifically formulated for use with IPG gel strips on the basis of pH range. For the experiments here reported, pH 3–10 NL IPG buffer was used (GE Healthcare, Sweden).

7. IEF buffer: commercial DeStreak Rehydration Solution (GE Healthcare, Sweden) was used. Before use, the solution was thawed at room temperature; then it was added with 20 mM DTT (*see* **Note 4**). The required IPG buffer was added just before use (*see* **Note 5**).
8. IEF apparatus for first dimension separation: Ettan IPGphor III apparatus with 18 cm IPG strip holders (GE Healthcare, Sweden).
9. Mineral oil: DryStrip cover fluid (GE Healthcare, Sweden).
10. SDS-PAGE system for second dimension separation: Ettan DALT*six* (GE Healthcare, Sweden).
11. Bromophenol blue solution: 1% w/v bromophenol blue, 50 mM Tris-base. 0.1 g of bromophenol blue and 0.06 g of Tris were dissolved in about 8 mL of ultrapure water, contained in a 15 mL sterile screw cap plastic tube. They were mixed by inversion, and then they were added with ultrapure water to reach a final volume of 10 mL.
12. Equilibration buffer A (SolA): 6 M urea, 30% v/v glycerol, 2% w/v SDS, 0.05 M Tris–HCl (from 1 M, pH 6.8), 2% w/v DTE (*see* **Notes 2** and **6**). 5 mL of SolA were prepared for each IPG strip in a sterile screw cap plastic tube. They were prepared just before use.
13. Equilibration buffer B (SolB): 6 M urea, 30% v/v glycerol, 2% w/v SDS, and 0.05 M Tris–HCl (from 1 M, pH 6.8), 2.5% w/v iodoacetamide, complete by adding few trace of bromophenol blue (*see* **Notes 2** and **6**). 5 mL of SolB were prepared for each IPG strip in a sterile screw cap plastic tube. They were prepared just before use.
14. SDS electrophoresis buffer: 25 mM Tris, 192 M glycine, 0.1% w/v SDS. The buffer volume required to carry out the SDS-PAGE was prepared by diluting a pre-prepared 10× stock solution (*see* **Note 7**).
15. Agarose overlay solution: 0.5% w/v agarose, 0.002% w/v bromophenol blue in SDS electrophoresis buffer (*see* **Note 8**).
16. Ammonium persulfate: 10% w/v solution in ultrapure water (*see* **Note 9**).
17. *N,N,N,N'*-Tetramethyl-ethylenediamine: store at 4 °C.

2.3 Solution for Silver Staining

1. Fixing solution: 50% v/v methanol, 10% v/v acetic acid, ultrapure water up to 1 L. It was stored at room temperature.
2. Sensitizing solution: 0.08% w/v sodium thiosufate pentahydrate ($Na_2S_2O_3 \cdot 5 \cdot H_2O$). To prepare 1 L of sensitizing solution, 0.8 g of sodium thiosulfate pentahydrate were solved in

600 mL of ultrapure water; they were added of additional ultrapure water to reach the final volume.

3. Silver solution: 0.4% w/v silver nitrate ($AgNO_3$), prepared just before use. To prepare 1 L of silver solution, 4 g of silver nitrate were solved in 600 mL of ultrapure water; they were added of further ultrapure water to reach the final volume. Silver solution was precooled at 4 °C. Each time, it was prepared fresh before use.
4. Developing solution: 2% w/v sodium carbonate, 0.04% w/v formaldehyde. To prepare 1 L of developing solution, 20 g of sodium carbonate were solved in about 800 mL of ultrapure water, and added with further ultrapure water to reach the final volume. Just before use, the sodium carbonate solution was added with 1080 μL of 37% w/v formaldehyde. Developing solution was kept at room temperature.
5. Stop solution: 1% v/v acetic acid. To prepare 1 L of stop solution, 10 mL of glacial acetic acid were added to 990 mL of deionized water. The stop solution was kept at room temperature.

2.4 Solution Used for Protein Reduction, Alkylation and Digestion

Water, acetonitrile, and formic acid used for the treatment of gel spots were LC-MS grade products.

Digestion solution: 50 mM NH_4HCO_3, 5 mM $CaCl_2$, trypsin (12.5 ng/mL). A bovine trypsin stock solution was prepared by weighing 25 μg of sequencing grade enzyme (Boehringer, Germany), which was solved in 250 μL of 1 mM HCl in order to avoid protein autoproteolysis. Just before use, 45 μL of trypsin stock solution were mixed with 150 μL of water, 150 μL of 100 mM NH_4HCO_3, and 15 μL of 120 mM $CaCl_2$.

2.5 Solutions Used for Mass Spectrometry-Based Protein Identification

Water, acetonitrile, and formic acid used for mass spectrometric identification of proteins within gel spots were LC-MS grade products. Water was degassed before use.

3 Methods

3.1 Biomass Sampling and Protein Extraction

1. Biomass samples of *Microbispora* ATCC-PTA-5024 strains were collected at different incubation times as described by Gallo et al. [6]. In particular, 6 mL of culture samples were collected from cultivations performed in triplicate replicas. In order to take and store samples, 10 mL sterile plastic pipettes and 10 mL screw cap plastic tubes were used, respectively. The tubes containing the culture aliquots were placed on ice until all the sampling procedure was accomplished. To separate spent growth medium from biomass, the samples were centrifuged

(3000 × *g* in a swing rotor, 10 min, at 4 °C). The supernatant was collected in a new 10 mL screw cap plastic tube. Both the spent medium and the biomass pellet samples were stored at −80 °C until use.

2. To remove any trace of spent medium, frozen biomass samples were suspended in 2 mL of WS and centrifuged as above for three times (*see* **Note 9**). After a resuspension in 2 mL of SS (*see* **Note 9**), the mycelium was disrupted by sonication (output control 4, 4 × 15 s pulses, Vibra Cell, USA) (*see* **Note 10**). The mycelium lysate was immediately put and kept on ice. Tubes containing mycelium lysates were then placed in a boiling water bath (5 min) to allow protein denaturation, and then they were cooled down on ice (15 min) to avoid protein refolding. To remove DNA and RNA, a mix of DNase (100 μg/mL) and RNase (50 μg/mL) was added to cellular lysates and incubated on ice for 20 min (*see* **Note 11**). In order to separate cell debris and unbroken cells from the protein extracts, cell lysates were dispensed equally into 2 mL microfuge tubes for each sample, before performing centrifugation (15,000 × *g*, for 15 min at 4 °C). After centrifugation, the supernatants containing protein extracts from the same cell lysate were collected into the same tube (*see* **Notc 12**).
3. Protein extracts were placed into dialysis membrane tubing, clipped to form a bag, and then dialyzed against ultrapure water (3 h at 4 °C) (*see* **Note 13**). Then, one end of the membrane bag was cut using clean scissors and the dialyzed protein extracts were collected with a 5 mL sterile pipette. Each protein sample was aliquoted into 0.5 mL volumes and placed into 2 mL tubes for precipitation with three volumes of acetone (overnight at −20 °C). After centrifugation (13,000 × *g*, 20 min, 4 °C), the supernatant was discharged and the protein pellet was subjected to vacuum concentrator centrifugation (15 min) to remove acetone (*see* **Note 14**). Finally, each protein pellet was solubilized in 50–100 μL SB. The SB-dissolved proteins extracted from the same biomass sample were collected together in a 1 mL microfuge tube. The pH of protein samples was evaluated by spotting 2–3 μL in a pH indicator strip and pH adjusted to 8.5 when required use (*see* **Note 15**). Then, the protein samples were stored at −80 °C until use (*see* **Note 16**).

3.2 CyDye Fluorescent Protein Labeling

1. The protein concentration of each sample was diluted to 5 μg/μL, which is the optimal concentration for the fluorescent minimal labeling protocol (*see* **Note 17**).
2. From these preparations and according to the experimental design (Tables 3 and 4), one or two aliquots of 50 μg of protein

Table 3
Example of experimental design for 2D-DIGE comparison including two conditions (namely A and B) and three biological replicas (1–3)

Gel	Cy5	Cy3	Cy2
1	A1	B1	Standard
2	A2	B2	Standard
3	A3	B3	Standard

Table 4
Example of experimental design for 2D-DIGE comparison including three conditions (namely A, B, and C) and four biological replicas (1–4)

Gel	Cy5	Cy3	Cy2
1	A1	B1	Standard
2	A2	C1	Standard
3	C3	B3	Standard
4	B4	C4	Standard
5	C2	A3	Standard
6	B2	A4	Standard

samples (10 μL) were transferred in new 0.5 mL microfuge tubes, one or two for each sample (*see* **Note 18**).

3. The pooled protein standard was prepared by mixing equal amounts of protein sample. A total of 300 μg is necessary to obtain six 50 μg-aliquots of pooled proteins; however, pooled protein standard was prepared in an excess (*see* **Note 19**) and, then, the required 50 μg (10 μL) aliquots were dispensed into six clean 0.5 mL microfuge tubes.
4. CyDye working solutions (400 pmol/μL) were prepared for each dye by mixing 2.1 μL of DMF and 1.4 μL of a CyDye flour in a 0.5 mL microfuge tube.
5. The 50 μg aliquots of samples were then labeled by adding 0.5 μL (200 pmol) of Cy(3 or 5)Dye, according to the experimental design (Tables 3 and 4); 50 μg aliquots of pooled protein standard were labeled by adding 0.5 μL (200 pmol) of Cy2 dye (*see* **Note 20**). Protein labeling was performed on ice, for 30 min, in the dark.
6. Labeling reaction was stopped by the addition of 0.5 μL of 10 mM lysine (10 min, on ice in the dark) (*see* **Note 20**).

Table 5
IEF program. The maximal current for strip was set to 50 μA and the temperature was maintained at 20 °C

Step	Voltage amplitude (V)	Voltage type	Time
"Active" rehydration	0	Not applicable	1 h
I	30	Step	10 h
II	200	Step	1 h
III	300	Step	30 min
IV	3500	Gradient	3 h
V	3500	Step	1 h
VI	8000	Gradient	30 min
VII	8000	Step	8 h

3.3 First Dimension Separation: IEF

1. The 50 μg-labeled protein samples were mixed according to the experimental design (Tables 3 and 4) in a clean microfuge tube.
2. The IEF buffer containing 20 mM DTT and the pH 3–10 NL IPG buffer were added to each mix up to 340 μL (*see* **Note 5**).
3. Each sample mix was loaded directly into the corresponding IPG strip holder (*see* **Note 21**). The sample mixes were distributed over the entire length of the strip holder, between the anode and cathode plates.
4. After the plastic cover foil having been removed carefully, IPG strips were placed gel-side down in the holder channel (*see* **Notes 22** and **23**) and covered using a suitable mineral oil to prevent evaporation (*see* **Note 24**).
5. An IEF program was performed in the dark, as described in Table 5, including a 10 h active (30 V) rehydration step (*see* **Note 25**). After IEF run, the IPG strips were stored at −80 °C until use (*see* **Note 26**).

3.4 Second Dimension Separation: SDS-PAGE

1. A day before using a Ettan DALT gel caster (GE, Healthcare), six 0.1% w/v SDS-12% T polyacrylamide gels (1 mm thick) were prepared (*see* **Note 27**). Low florescence glasses were used as cassette for gel polymerization (*see* **Note 28**). A 450 mL polyacrylamide gel solution was prepared by mixing the components described in Table 6 in a 1 L beaker, prior to be poured into the caster filling channel (*see* **Note 29**). Before use, SDS-containing polyacrylamide gels were immersed into the SDS-containing electrophoresis buffer placed in the Ettan DALT-*six* electrophoresis chamber, for 2–3 h, and cooled at 10 °C.
2. Each IPG strip was placed gel side up in one channel of the reswelling tray (GE Healthcare, Sweden) and incubated with

Table 6
Recipe to prepare 450 mL of 0.1% w/v SDS-12% w/v acrylamide solution to be poured in Ettan DALT gel caster for six 1 mm thick gels

Components	Volume required (mL)
30% w/w acrylamide/bis-acrilamide stock solution (29:1)	188
1.5 M Tris-Cl, pH 8.8	113
Water	140
10% w/v SDS	4.5
10% w/v APS	4.5
TEMED	0.062

5 mL of SolA (10 min, under gentle agitation). Then, SolA was poured off and the IPG strips were incubated with 5 mL of SolB (10 min, under gentle agitation); then, SolB was poured off (*see* **Note 30**).

3. The IPG strips were briefly rinsed by submerging it in the SDS-containing electrophoresis running buffer, which was placed in a 500 mL graduate cylinder (*see* **Note 22**).
4. One by one, the IPG strips were placed in the between of the two glass plates of each gel cassette. A thin plastic spacer was used to push gently on the plastic support of the IPG strip (*see* **Note 31**), sliding until it came in contact with the SDS-polyacrylamide gel upper edge (*see* **Note 32**).
5. The IPG strips were sealed using a melted agarose overlay solution, after waiting at least 1 min for agarose solidification (*see* **Note 33**).
6. Cassettes containing the IPG strip loaded on SDS-polyacrylamide gels were placed in the Ettan DALT*six*, following the instructions of the Ettan DALT user manual.
7. Electrophoresis was performed in the dark, at 10 °C, with 110 V and maximum setting of 40 μA per gel (*see* **Note 34**).

3.5 Image Digitalization

1. After electrophoresis, the gel cassettes were rinsed with ultrapure water. Then, they were scanned (*see* **Note 35**) using an Ettan DIGE imager (GE Healthcare, Sweden) to detect CyDye-labeled proteins (*see* **Note 36**). The images were acquired in a 2D-DIGE modality manner, selecting "DIGE name file format" and Cy2 as standard, with the following settings: pixel size 100 μm, and 0.8, 0.3, 0.5 s/pixel as exposition time for Cy2, Cy3, Cy5, respectively.

2. After image acquisition, the gels were carefully removed from the cassettes (*see* **Note 37**) and put in fitting (380 × 250 × 40 mm) glass baking tray for silver staining procedure.

3.6 Silver Staining Procedure

A protocol with minor modification from Shevchenko et al. [3] was adopted for protein staining.

1. The gels were briefly rinsed with about 250 mL of ultrapure water (*see* **Note 38**).
2. Each gel was incubated with 200 mL of fixing solution, with gentle agitation in an orbital shaker for about 60 min.
3. After having removed appropriately fix solution, four ultrapure water (250 mL) wash steps were performed, 1 h each with gentle agitation in an orbital shaker, but the last step was performed overnight at 4 °C.
4. The gels were treated with sensitizing solution (200 mL), for 5 min, and then rinsed with two washes (250 mL) of ultrapure water, 1 min each (*see* **Note 39**). All these steps were performed with gentle agitation in an orbital shaker.
5. Then the gels were incubated in 0.4% w/v silver solution for 1 h, at 4 °C, in the dark (*see* **Note 40**).
6. The gels were rinsed two times with ultrapure water, with gentle agitation in an orbital shaker, 1 min each (*see* **Note 41**).
7. Then, 200 mL of developing solution were added and, when it turned yellow, immediately replaced with other fresh 200 mL of developing solution (*see* **Note 40**).
8. Protein spot developing was stopped by adding 100 mL of stop solution (*see* **Note 42**).
9. After discarding the initial stop solution, the gels were added with additional 200 mL of fresh stop solution. The gels were stored at 4 °C for 2–3 months.

3.7 Image Analysis

A more detailed description can be found in the user manual of Image Master 2D platinum 7.0 DIGE-enabled.

1. Protein spots were automatically detected on the gel image above digitalized by a dedicated scanner.
2. Protein spots were automatically matched. To obtain the best automatic gel match, it was necessary to identify 30–50 landmarks in each gel. The landmarks were chosen as corresponding to the spots present in each gel of pooled protein standard images.
3. Individual spot abundance was automatically calculated from replicate 2D-gels as the mean spot volume (i.e., integration of

optical density over spot area); it was normalized to the corresponding protein spot of Cy2-labeled internal standard.

4. Prior to performing the differential quantitative analysis, the classes were created as comprising gels from the same condition.
5. By comparing classes and therefore the gels within those classes, it was calculated the protein abundance variations between different biological samples. Protein spots showing more than 1.3-fold change in normalized spot volume, with a $P < 0.05$ (ANOVA test), were considered differentially abundant.

3.8 Protein Reduction, Alkylation, and Digestion

1. To identify proteins from a spot of interest, a manual spot picking was performed from gels stained with silver staining method. To this aim, all spots of interest were excised with a clean scalpel, cut into cubes (1 × 1 mm), and transferred into microfuge tubes (0.5 mL) containing 1% v/v acetic acid solution.
2. For protein reduction and alkylation, spot gel particles were washed with water (100–150 μL) for 5 min, centrifuged and removed of the excess liquid [3]. Then, they were added with acetonitrile (100 μL) for 10–15 min until gel particles shrunk, become white, and stick together; then they centrifuged, removed of the liquid, and dried in a vacuum centrifuge. Gel pieces were swelled in 10 mM DTT in 0.1 M NH_4HCO_3 (100 μL) and incubated for 30 min, at 56 °C. Then, they were centrifuged, removed of the excess liquid and shrunk with acetonitrile (100 μL), as mentioned above. Gel particles were then added with 55 mM iodoacetamide in 0.1 M NH_4HCO_3 (100 μL) for 20 min, at room temperature, in the dark. Finally, they were removed of the iodoacetamide solution, washed with 0.1 M NH_4HCO_3 (200 μL) for 15 min, at 37 °C, centrifuged, removed of the excess liquid, and finally shrunk with acetonitrile (100 μL), as mentioned above. After centrifugation, the gel particles were removed of acetonitrile, and dried in a vacuum centrifuge.
3. For protein digestion, the dehydrated gel particles were swelled with digestion solution (50 mM NH_4HCO_3, 5 mM $CaCl_2$, trypsin—12.5 ng/mL) for 30–45 min, at 4 °C (ice bucket) [3]. After 15 min, they were check out for the absence of digestion buffer and eventually added of it. After that, they were removed of the remaining supernatant, added with 20 μL of 50 mM NH_4HCO_3, 5 mM $CaCl_2$ to cover gel pieces and to keep it wet during digestion, and left at 37 °C, overnight.
4. The gel particles were extracted for resulting peptides by adding 15 μL of 25 mM NH_4HCO_3, and left for 15 min, at 37 °C, under shaking. Then, they were centrifuged, added with an

equal volume of acetonitrile, and left for 15 min, at 37 °C, under shaking. After centrifugation, the supernatant was collected and stored in a dedicated vial containing the peptide extract. The remaining gel particles were added with 50 μL of 5% v/v formic acid, and left for 15 min, at 37 °C, under shaking. Then, they were centrifuged, added again with an equal volume of acetonitrile, and left for 15 min, at 37 °C, under shaking. After centrifugation, the supernatant was removed and added to the one present in the dedicated vial. Finally, pooled supernatants containing the peptide extract were dried in a vacuum centrifuge.

3.9 Characterization of Peptide Digests by nanoLC-ESI-LIT-MS/MS

1. Resulting peptide mixtures were desalted by μZip-TipC_{18} using 50% v/v acetonitrile, 5% v/v formic acid as an eluent. In particular, protein digests were loaded in 5% v/v formic acid (20 μL), washed with 5% v/v formic acid (3 × 20 μL), and eluted with 50% v/v acetonitrile, 5% v/v formic acid (20 μL).
2. Recovered peptides were dried in vacuum centrifuge, and then analyzed for protein identification by nLC-ESI-LIT-MS/MS, using an LTQ XL mass spectrometer (Thermo Fisher Scientific, USA) equipped with a Proxeon nanospray source connected to an Easy-nanoLC (Thermo Fisher Scientific) (*see* **Note 43**).
3. Peptides were resolved on an Easy C18 column (100 mm × 0.075 mm, 3 μm) (Thermo Fisher Scientific). Mobile phases were 0.1% v/v formic acid (solvent A) and 0.1% v/v formic acid in acetonitrile (solvent B), running at a total flow rate of 300 nL/min. Linear gradient was initiated 20 min after sample loading; solvent B ramped from 5 to 35% over 45 min, from 35 to 60% over 10 min, and from 60 to 95% over 20 min. Spectra were acquired in the range m/z 400–2000 (*see* **Note 44**). Each peptide mixture was analyzed under a collision-induced dissociation (CID)-MS/MS data-dependent product ion scanning procedure, enabling dynamic exclusion (repeat count 1 and exclusion duration 60 s), over the three most abundant ions (*see* **Note 45**). Mass isolation window and collision energy were set to m/z 3 and 35%, respectively.

3.10 Bioinformatic Analysis for Protein Identification

1. Raw data from nLC-ESI-LIT-MS/MS analysis were searched by MASCOT search engine (version 2.2.06, Matrix Science, UK) against an updated NCBI nonredundant database also containing *Microbispora* ATCC-PTA-5024 ORF product database based on the corresponding genome sequence (GenBank accession AWEV00000000) with the aim of identifying proteins from gel spots (*see* **Note 46**). Database searching was performed by using Cys carbamidomethylation and Met oxidation as fixed and variable protein modifications, respectively, a mass tolerance value of 1.8 Da for precursor ion and 0.8 Da

for MS/MS fragments, trypsin as proteolytic enzyme, and a missed cleavage maximum value of 2. Other MASCOT parameters were kept as default.

2. Protein candidates assigned according to at least two sequenced peptides with an individual peptide expectation value <0.05 (corresponding to a confidence level for peptide identification >95%) were considered confidently identified.
3. Definitive peptide assignment was always associated with manual spectra visualization and verification.

4 Notes

1. Stock solutions can be prepared at convenience and stored at −20 °C until use.
2. Urea solubilization needs heating. However, all the solutions containing urea must not be heated above 37 °C to prevent ammonium cyanate formation and, consequently, protein carbamylation. In addition, the SB solution should not contain DTT or any primary amines, such as ampholytes, as they will compete with proteins for *N*-Hydroxysuccinimide (NHS)-based reaction of CyDye fluors, thus compromising labeling efficiency.
3. DMF quality is crucial for successful in protein labeling. DMF has to be not contaminated by water; after opening, DMF will degrade producing amine compounds, which compete with proteins for reacting with NHS ester of CyDye fluors. Since DMF is quite cheap and CyDye fluors are very expensive, we suggest using a new DMF bottle every time for avoiding to waste fluorescent dyes and the efforts of a labor-intensive procedure.
4. Add 60 μL of a 1 M DTT stock solution directly into the vials containing 3 mL DeStreak Rehydration Solution.
5. An amount of 2.3 μL of IPG buffer was added to each mix (340 μL final volume) containing IEF buffer (304.7 μL) and the labeled protein samples (33 μL).
6. For a 2D-DIGE experiment comprising a total of 6 IPG strips to be equilibrated, 30 mL of SolA and 30 mL of SolB have to be prepared. For this purpose, prepare two 50 mL sterile screw cap plastic tubes by adding the following constituents in the order: 1.5 mL of 1 M Tris–HCl pH 6.8, 7.5 mL of ultrapure water, 10.8 g of urea, and 0.6 g of SDS. Then, close the tubes and mix to resuspend the powder into the aqueous phase. After that, place the tubes in a 37 °C water bath for approximately 15 min, mixing by inversion several times gently in order to avoid excessive foaming. When urea and SDS are almost

completely dissolved, add 9 mL of glycerol and, after having mixed by inversion several times gently, add ultrapure water up to 30 mL for each tube. To obtain SolA and SolB, add 0.6 g of DTE and 0.75 g of iodoacetamide, respectively. SolB can be completed by adding 240 μL of 1% w/v bromophenol blue solution, which is used as tracking dye during SDS-PAGE.

7. SDS electrophoresis buffer can be prepared as a 10× stock solution containing 250 mM Tris, 1920 M glycine, 1% w/v SDS. For this purpose, dissolve 60.5 g of Tris, 288 g of glycine, and 200 mL of 10% w/v SDS in a 2 L cylinder containing about 1600 mL of ultrapure water. A small magnetic stir bar can be introduced into the cylinder for using a magnetic mixer. After having removed the magnetic stir bar, add ultrapure water up to 2 L. The cylinder may be carefully closed using parafilm or food wrap and solution stored for months at room temperature.
8. Mix 0.5 g of agarose and 0.2 mL of 1% w/v bromophenol blue into 100 mL of SDS containing-electrophoresis buffer using a 250 mL conical flask. Then, heat it in the microwave using a low power setting until agarose melts. Allow the solution to cool slightly before use. Store at room temperature. Do not keep for more than 1 month.
9. The volumes of WS and SS to be used are based on empirical experience. Usually, the WS or SS volumes have to be at least twofold greater than that of the packed wet biomass to be resuspended. Thus, 2 mL of WS or SS are suitable if 0.5–1.0 mL of packed wet mycelium volumes have to be treated. This is necessary for washing biomass samples properly and for avoiding protein trapping by cell debris, which severely compromises protein extraction yield. Packed wet biomass has to be resuspended vigorously by inverting and vortexing.
10. Before use, the sonicator tip has to be carefully rinsed first with 70% v/v ethanol and then with ultrapure water; a single-pulse sonication cycle (output control 4, 15 s, Vibra Cell, USA) can be performed using ultrapure water. Exceeding water drops can be dried gently from sonicator tip by using a laboratory 3-MM filter paper or sterile gauzes (do not rub the tip). Rinse accurately the sonicator tip with ultrapure water at the end of each cycle (i.e., before placing a new sample tube) to avoid sample contamination. Tube containing samples have to be fastened using a laboratory clamp; their bottom has to be immerged in an iced bath in transparent glass beaker during the sonication cycle. During sonication pulse, the sonicator tip must not touch the tube wall; thus, it has to be placed almost in the middle of the suspension. Wait 10 s for each pulse of a sonication cycle to let the solution cool. At the end of the

sonication procedure, repeat sonicator tip rinsing first with 70% v/v ethanol and then with ultrapure water.

11. Prepare DNase and RNase mix 100× stock solution by dissolving 5 mg of DNase and 2.5 mg of RNAse in 0.5 mL of WS.
12. The obtained protein extracts can be either processed further or stored at −80 °C until use. To avoid protein sample freeze-thaw stress, it is recommended to proceed with dialysis step.
13. Dialysis step is not mandatory; however, it can be successful in removing those contaminants (like charged metabolites interfering with protein isoelectric focusing) that are not removed by the subsequent acetone precipitation step. After having closed one end of a membrane by folding over and tying it, put the protein extract in the formed membrane bag and then fold over, and tie the other open end. Paste one end of the dialysis tubing on the edge of 2 L beaker (containing a little bit less than 2 L ultrapure water) using a laboratory labeling adhesive tape, where the sample name has been recorded. Mix gently using a magnetic stir bar.
14. Acetone has to be removed completely since it severely compromises protein solubilization. To remove the residual volume of acetone (i.e., a drop in the bottom of the tube), perform a 15 min evaporation in a vacuum concentrator. However, since acetone evaporation time can also depend on protein amount, carefully control sample by visual inspection. It is also important to avoid prolonged treatment, since highly dried proteins are insoluble as well.
15. Usually, the pH value of SB-solubilized proteins is lower than pH 8.5; the latter value is the optimal value for CyDye fluorescent labeling. Thus, the pH has to be adjusted by adding drop by drop a pre-prepared pH 9.5 SB until a pH 8.5 is obtained.
16. The protein extracts have to be quantitatively and qualitatively analyzed. For this reason, it is convenient to set aside small aliquots of protein samples (2 or 3 × 0.5-mL microfuge tubes containing 10–20 μL) to perform test analyses. For quantitative evaluation, it could be convenient the use of commercial kits that are compatible with CHAPS, urea, and thiourea in order to determine accurately protein concentration values. For qualitative analysis, a 0.1% w/v SDS—12% T PAGE preliminary experiment should be performed in order to estimate the relative abundance of the protein species having high or low Mw values, the latter likely to be the results of polypeptide fragmentation.
17. Based on the described protocol for protein extraction, the concentration of SB-solubilized proteins is usually about 10 μg/μL or higher. Thus, dilution with SB has to be performed to bring the final concentration of proteins to 5 μg/μL.

However, if concentration values lower than 5 μg/μL are obtained, all the samples have to be set with the lowest concentration, taking into account that up to 1–2 μg/μL protein samples can be generally labeled.

18. Experimental design was set to include CyDye fluors swap for protein sample labeling to eliminate any dye-specific biases. Experimental design of 2D-DIGE experiment carried out using an Ettan DALT*six* (GE Healthcare) apparatus comprises up to six gels, where 12 samples (six Cy3- and six Cy5-labeled) and six protein mix aliquots from pooled standard (Cy2-labeled) can be combined. In the case of two different conditions (namely, A and B) and three biological replicas (1–3), the total number of protein samples is six (A1, A2, A3 and B1, B2, and B3). Whenever two technical replicas are used, although not strictly necessary, the total number of protein samples is 12. In the case of three different conditions (namely A, B, and C) and four biological replicas (1–4), the total number of protein samples to be analyzed is 12 (A1–4, B1–4, and C1–4).
19. In the above-mentioned case of two different conditions (namely A and B) and three biological replicas (1–3), for a total of six samples (A1–3 and B1–3), 75 μg of proteins (15 μL) were collected from each sample and mixed in a new 1.5 mL microfuge tube for a total of 450 μg of pooled protein mix. In the above-mentioned case of three different conditions (namely A, B, and C) and four biological replicas (1–4), for a total of 12 samples (A1–4, B1–4, C1–4), 50 μg of proteins (10 μL) were collected from each sample and mixed in a new 1.5 mL microfuge tube, obtaining a total of 600 μg of pooled proteins. In both the cases, the concentration of proteins remained 5 μg/μL; six 50 μg (10 μL) aliquots were thus dispensed into six clean 0.5 mL microfuge tubes for Cy2 dye labeling.
20. Mix by pipetting up and down and, then, spin briefly in a microfuge.
21. Every IPG strip holder has an identification number placed in its side. This number may help in associating the sample mix to each strip holder.
22. Hold IPG strips by the cathodic end plastic support using clean forceps.
23. Take care to distribute the protein mix solution evenly under the strip, avoiding to trap air bubbles.
24. Avoid oil overload since it can pour under the anode or cathode plates affecting electric current.

25. The IEF program was set empirically based on the protocol suggested by IPG strip supplier. In any case, it is important that for this program the final Volt-hours (Vhs) is greater than 70 kVhs. If the final Vhs is below 70 kVhs, then the last step can be prolonged consequently. In addition, the quality of IEF can be assessed by the appearance of the bromophenol blue tracking dye, which has to move completely to the anode as a yellow band, when the IEF protocol is finished.
26. Place the IPG strips individually on an aluminum foil used like a folder. Take care to avoid pressing the IPG strips. Write with a permanent pen the sample identification code. Place the foil in a box and put it safely at −80 °C.
27. Each couple of glass plates of a single-gel cassette consists of a short and a long one, with long protruding from the upper side of the cassette. Introduce a small piece (about $15 \times 10\ mm^2$) of 3MM laboratory paper marked with a number (from 1 to 6) into the between of each couple of glasses delimiting a single-gel cassette. Place them always in the same lower corner (right or left) so that it can also indicate the acid or basic side of IPG strips, accordingly.
28. Low fluorescent glass plates must be used to reduce background signal and improve image quality.
29. For a detailed description of components, for the preparation of Ettan DALT gels, for detailed protocols for loading gels into the caster, and for casting Ettan DALT gels into the Ettan DALT*six* system, please consult the corresponding user manual.
30. Do not over-equilibrate since proteins can diffuse during this step.
31. Each couple of glass plates of a single-gel cassette consists of a short and a long one, with long protruding from the upper side of the cassette. In order to put in contact the IPG strip and the upper side of the SDS-containing polyacrylamide gel, place gel side up the IPG strip on the protruding part of the long glass plate and carefully draw it to hold one end until it is centered by using forceps. Then, make the IPG strip sliding to the SDS-containing polyacrylamide gel by pressing on the IPG strip back, where plastic support is placed, using a plastic spacer or a flat ruler; take care of not damaging the IPG gel.
32. Fill the free room between the couple of glasses from the gel upper edge with SDS-containing electrophoretic buffer to facilitate IPG strip sliding; take care to not trap air bubbles at the interface between IPG strip and SDS-containing polyacrylamide gel.

33. Melt by microwave using medium power and then allow it to cool slightly (until it can be held by hand); pipette slowly the solution into the between of glass plate, starting from one side to the other one, taking care not to trap air bubbles.
34. Following the bromophenol blue as a tracking dye, the electrophoretic run will end approximately after 17 h; for this reason, it is reasonable to start SDS-PAGE at late afternoon.
35. Wear powder-free gloves since residual powder on the glass plates can fluoresce and scatter light, thus affecting image quality.
36. During a single-gel scanning, the other gels may still stay in the Ettan DALT*six* immerged in the SDS-containing electrophoresis buffer, in the dark, at 10 °C.
37. Place the cassette on the bench and, using a clean plastic spacer or a flat ruler, lift the short glass plate, and then remove gently the IPG strip. For transferring it into the baking tray, let the gel slide on the glass plate over a water stream created by rinsing thoroughly ultrapure water with a wash bottle.
38. Seal the backing tray using a common food wrap to protect the gel from dust. Do not touch the gel with the bare hands or metal objects; use instead plastic bars or glass bars or pipettes to handle the gel.
39. To obtain high sensitivity and low background, it is very important to follow closely the incubation time of the following steps. To this aim, it is suggested to work with not more than two gels by time from sensitizing to silver solution step.
40. Add the solution from the corner of the backing tray. Do not pour solutions directly on the gel, as it may result in unequal background.
41. It is suggested to work with not more than one gel by time from this step to the end.
42. The reaction can be stopped as soon as the desired intensity of the bands is obtained.
43. Using the mass spectrometric system reported here is not mandatory. Other mass spectrometers still based on a nanoelectrospray ionization source, but having different analyzer (quadrupole, ion trap, Orbitrap, time of flight, etc.) systems can be considered suitable for proteomic analysis. Mandatory is their capability to perform tandem mass spectrometry experiments to provide more stringent identification results. Since resolution and accuracy performances of the mass spectrometer have important reflections on raw data searching by bioinformatic approaches, instrument nature and its reflection on proteomic experiments have to be strictly considered.

44. Using the chromatographic system reported here is not mandatory. Other nano-chromatographers interfaced with various C_{18} columns can be considered suitable for proteomic analysis. Based on column and chromatographer (i.e., injector, pumping, and tubing system) nature, corresponding void volumes and elution performances can vary; accordingly, experimental parameters here reported have to be intended as referred to this specific chromatographic system.
45. The specific acquisition parameters reported here are dependent on the nature and the performances of the mass spectrometer. Depending on the instrument nature, other experimental setups can be considered to optimize final identification results.
46. Other search engines for bioinformatics database searching of acquired, mass spectrometric raw data can be used. In this case, no significant changes in the search parameters here reported have to be introduced, when considering raw data produced with a mass spectrometer having resolution and accuracy performances similar to the one reported in this chapter. When this is not the case (for instrument having better resolution and accuracy performances), more stringent search parameters can be used to improve final identification results.

References

1. Coombs KM (2011) Quantitative proteomics of complex mixtures. Expert Rev Proteomics 8 (5):659–677
2. Wu WW, Wang G, Baek SJ, Shen RF (2006) Comparative study of three proteomic quantitative methods, DIGE, cICAT, and iTRAQ, using 2D gel- or LC-MALDI TOF/TOF. J Proteome Res 5(3):651–658
3. Shevchenko A, Wilm M, Vorm O, Mann M (1996) Mass spectrometric sequencing of proteins silver-stained polyacrylamide gels. Anal Chem 68(5):850–858
4. Lopez MF, Berggren K, Chernokalskaya E et al (2000) A comparison of silver stain and SYPRO ruby protein gel stain with respect to protein detection in two-dimensional gels and identification by peptide mass profiling. Electrophoresis 21(17):3673–3683
5. Shevchenko A, Jensen ON, Podtelejnikov AV et al (1996) Linking genome and proteome by mass spectrometry: large-scale identification of yeast proteins from two dimensional gels. Proc Natl Acad Sci U S A 93(25):14440–14445
6. Shevchenko A, Tomas H, Havlis J et al (2006) In-gel digestion for mass spectrometric characterization of proteins and proteomes. Nat Protoc 1(6):2856–2860
7. Gallo G, Renzone G, Palazzotto E et al (2016) Elucidating the molecular physiology of lantibiotic NAI-107 production in *Microbispora* ATCC-PTA-5024. BMC Genomics 17:42
8. Licona-Cassani C, Lim S, Marcellin E, Nielsen LK (2014) Temporal dynamics of the *Saccharopolyspora erythraea* phosphoproteome. Mol Cell Proteomics 13(5):1219–1230
9. Yang Q, Ding X, Liu X et al (2014) Differential proteomic profiling reveals regulatory proteins and novel links between primary metabolism and spinosad production in *Saccharopolyspora spinosa*. Microb Cell Fact 13(1):27
10. Ye C, Ng IS, Jing K, Lu Y (2014) Direct proteomic mapping of *Streptomyces roseosporus* NRRL 11379 with precursor and insights into daptomycin biosynthesis. J Biosci Bioeng 117 (5):591–597
11. Chaudhary AK, Dhakal D, Sohng JK (2013) An insight into the "-omics" based engineering of streptomycetes for secondary metabolite overproduction. Biomed Res Int 2013:968518
12. Yin P, Li YY, Zhou J (2013) Direct proteomic mapping of *Streptomyces avermitilis* wild and industrial strain and insights into avermectin production. J Proteomics 79:1–12

13. Song E, Malla S, Yang YH (2011) Proteomic approach to enhance doxorubicin production in *panK*-integrated *Streptomyces peucetius* ATCC 27952. J Ind Microbiol Biotechnol 38 (9):1245–1253
14. Gallo G, Alduina R, Renzone G et al (2010) Differential proteomic analysis highlights metabolic strategies associated with balhimycin production in *Amycolatopsis balhimycina* chemostat cultivations. Microb Cell Fact 9:95
15. Gallo G, Renzone G, Alduina R et al (2010) Differential proteomic analysis reveals novel links between primary metabolism and antibiotic production in *Amycolatopsis balhimycina*. Proteomics 10(7):1336–1358
16. Monciardini P, Iorio M, Maffioli S et al (2014) Discovering new bioactive molecules from microbial sources. Microb Biotechnol 7 (3):209–220
17. Donadio S, Maffioli S, Monciardini P et al (2010) Sources of novel antibiotics–aside the common roads. Appl Microbiol Biotechnol 88 (6):1261–1267
18. Bibb MJ (2005) Regulation of secondary metabolism in streptomycetes. Curr Opin Microbiol 8(2):208–215
19. van Wezel GP, McDowall KJ (2011) The regulation of the secondary metabolism of *Streptomyces*: new links and experimental advances. Nat Prod Rep 28(7):1311–1333
20. Palazzotto E, Gallo G, Renzone G (2016) TrpM, a small protein modulating tryptophan biosynthesis and morpho-physiological differentiation in *Streptomyces coelicolor* A3(2). PLoS One 11(9):e0163422
21. Palazzotto E, Renzone G, Fontana P et al (2015) Tryptophan promotes morphological and physiological differentiation in *Streptomyces coelicolor*. Appl Microbiol Biotechnol 99 (23):10177–10189
22. Gallo G, Lo Piccolo L, Renzone G et al (2012) Differential proteomic analysis of an engineered *Streptomyces coelicolor* strain reveals metabolic pathways supporting growth on n-hexadecane. Appl Microbiol Biotechnol 94 (5):1289–1301
23. Alduina R, Giardina A, Gallo G et al (2005) Expression in *Streptomyces lividans* of *Nonomuraea* genes cloned in an artificial chromosome. Appl Microbiol Biotechnol 68 (5):656–662
24. Gallo G, Baldi F, Renzone G et al (2012) Adaptative biochemical pathways and regulatory networks in *Klebsiella oxytoca* BAS-10 producing a biotechnologically relevant exopolysaccharide during Fe(III)-citrate fermentation. Microb Cell Fact 11:152

Chapter 13

Techniques for Large-Scale Bacterial Genome Manipulation and Characterization of the Mutants with Respect to In Silico Metabolic Reconstructions

George C. diCenzo and Turlough M. Finan

Abstract

The rate at which all genes within a bacterial genome can be identified far exceeds the ability to characterize these genes. To assist in associating genes with cellular functions, a large-scale bacterial genome deletion approach can be employed to rapidly screen tens to thousands of genes for desired phenotypes. Here, we provide a detailed protocol for the generation of deletions of large segments of bacterial genomes that relies on the activity of a site-specific recombinase. In this procedure, two recombinase recognition target sequences are introduced into known positions of a bacterial genome through single cross-over plasmid integration. Subsequent expression of the site-specific recombinase mediates recombination between the two target sequences, resulting in the excision of the intervening region and its loss from the genome. We further illustrate how this deletion system can be readily adapted to function as a large-scale in vivo cloning procedure, in which the region excised from the genome is captured as a replicative plasmid. We next provide a procedure for the metabolic analysis of bacterial large-scale genome deletion mutants using the Biolog Phenotype MicroArray™ system. Finally, a pipeline is described, and a sample Matlab script is provided, for the integration of the obtained data with a draft metabolic reconstruction for the refinement of the reactions and gene-protein-reaction relationships in a metabolic reconstruction.

Key words Genome deletion, Site-specific recombinase, In vivo cloning, Phenotype microarray, Metabolic modeling, Functional genomics

1 Introduction

The explosion in the availability of complete bacterial genomes over the last decade has resulted in novel genes being identified at a rate that far exceeds our ability to characterize them. Approximately a third of the genes available through the Entrez Gene database are associated with the term hypothetical [1]. Highlighting this lack of understanding was the discovery that even a third of the essential *Mycoplasma mycoides* proteins are of unknown function [2]. The percentage of genes without a well-defined molecular function rises further when considering those with only general functional

Marco Fondi (ed.), *Metabolic Network Reconstruction and Modeling: Methods and Protocols*, Methods in Molecular Biology, vol. 1716, https://doi.org/10.1007/978-1-4939-7528-0_13, © Springer Science+Business Media, LLC 2018

annotations, e.g., epimerase, without a known substrate. At the same time, high-throughput phenotypic analyses often lead to the identification of cellular capabilities while the underlying genetic determinants remain unclear [3, 4].

One approach to bridging the gap and associating functions to uncharacterized genes is to employ a large-scale genome deletion approach. Such a technique allows for the simultaneous analysis of tens to hundreds of genes, rapidly narrowing down the search for the genes underlying a particular phenotype. Our group has been using large-scale genome deletion and cloning approaches to study the genomics of *Sinorhizobium meliloti* since the late 1980s [5–7]. We and others have employed this approach in the study of both simple and complex biological traits including: identification of carbon transport and metabolic loci [4, 5, 8–10], ion transport systems [11], toxin-antitoxin modules [6], and essential genes [12], as well as in the study of regulatory networks [13], osmotic tolerance [14], soil colonization [15], magnetotaxis [16], cytochrome *c* respiration [10], and the rhizobium—legume symbiosis [10, 17]. The combination of large-scale genome reduction studies with in silico genome-scale metabolic reconstructions can be a particularly powerful approach. Metabolic characterization of the deletion mutants can facilitate the refinement of the gene-protein-reaction relationships in a metabolic reconstruction and assist in the annotation of novel gene functions [4].

In this chapter, we detail a procedure for the construction of bacterial strains with large-scale genome deletions in which the exact endpoints of each deletion are known. This procedure involves introduction of the recombinase recognition target sequences into the genome at a pre-selected position. Here, this is accomplished through single-plasmid cross-over recombination, although alternative methods of introducing the target sequences can be employed [18]. Subsequent expression of the corresponding site-specific recombinase results in recombination between the two target sequences and deletion of the intervening genomic region. A protocol for adapting this deletion method to function as an in vivo cloning tool is also described. Finally, metabolic characterization of these mutants with the Phenotype MicroArray™ system [19] and the application of the resulting data during the manual curation stage of an in silico genome-scale metabolic network reconstruction is described.

2 Materials

2.1 General Materials

1. Supplies and equipment required for general DNA manipulations, including PCR, restriction digest, DNA fragment purification, and cloning.
2. General laboratory supplies and equipment, such as centrifuge tubes, petri dishes, test tubes, and incubators.

3. Liquid and solid media permissive to growth of all bacteria used in this procedure. A common option is LB medium (10 g/L tryptone, 5 g/L yeast extract, 5 g/L NaCl, and 15 g/L agar for solid medium).
4. Filter sterilized stock solutions of the appropriate selective agents.
5. Sterile 0.85% NaCl saline solution in water (if preferred, PBS can be used throughout the protocol in place of the 0.85% saline solution).
6. Autoclaved glass beads, or an L-shaped glass rod and 95% ethanol, or disposable sterile plastic rods for spreading of liquid culture on solid media.
7. An *E. coli* strain with a "helper plasmid" that expresses the RK2 *tra* genes (e.g., pRK600 [20]).
8. If performing in vivo cloning as described in Subheading 3.5 then an *E. coli* strain with a selectable marker (e.g., rifampicin resistant DH5α) is required.
9. NEBcutter (http://nc2.neb.com/NEBcutter2/) [21] for in silico determination of restriction enzyme digestion patterns.

2.2 A Site-Specific Recombinase System

The procedure for the construction of large, defined deletion mutants is based on a system involving a recombinase protein that detects and mediates recombination between two specific DNA sequences. The three main recombinase systems employed in bacterial genome editing are the Flp/*FRT*, Cre/*lox*, and phage ϕC31/*att* systems, and the key characteristics of these systems are described in Table 1. The choice of recombinase will influence the downstream applications (*see* Table 1), although all the systems can be employed for the construction of a basic deletion mutant. Our group employs the Flp/*FRT* system for construction of bacterial genome deletion mutants, and all our materials are freely available upon request.

2.3 A Vector Expressing the Site-Specific Recombinase

The minimum requirements for this vector are that:

1. The plasmid be replicative in the species of interest.
2. The vector is transferable via conjugation when the RK2 *tra* genes are expressed *in trans.*
3. The plasmid contains a selective marker.
4. The site-specific recombinase is expressed from a promoter active in the species of interest. In addition to these minimal requirements, two additional features are strongly recommended.
5. The vector expressing the recombinase is unstable in the organism of interest, such as pRK7813 [28], so that the vector can be readily lost from the population following the construction of the deletion (*see* **Note 1**); and
6. The site-specific recombinase is not constitutively expressed and is instead expressed from an inducible promoter (*see* **Note 2**).

Table 1
Site-specific recombinases employed in bacterial genome editing

Recombinase: Flp
Recognition site: *FRT* (5′-GAAGTTCCTATACTTTCTAGAGAATAGGAACTTC)
Notes: Following deletion formation, one *FRT* site remains in the genome that can be used in a second round of recombination to enlarge the deletion. This system cannot be used to delete two independent regions of the genome in the same strain; however, two existing deletions can be recombined into one strain. Additionally, once the two deletions are within the same strain, the region in between the two deletions can be removed by once again expressing the Flp recombinase
Examples: [6, 10, 22, 23]
Recombinase: Cre
Recognition site: *loxP* (5′-ATAACTTCGTATAGCATACATTATACGAAGTTAT) and variations [18, 24]
Notes: Following deletion formation, one *loxP* site remains in the genome that can be used in a second round of recombination to enlarge the deletion. Mutant lox sites have been reported that are recognized by the Cre recombinase but for which the *lox* site that is left over following recombination is a poor Cre substrate. Thus, using these mutant *lox* sites would allow for the construction of two or more independent deletions within the same strain
Examples: [16, 18, 24–26]
Recombinase: ϕC31
Recognitions site: *attB* (5′-GGGTGCCAGGGCGTGCCCTTGGGCTCCCCGGGCGCGTA) and *attP* (5′-CCCCAACTGGGGTAACCTTTGAGTTCTCTCAGTTGGGG)
Notes: Following deletion formation, the *attB* and *attP* sites are no longer present and instead an *attL* or *attR* site remains in the genome, which are not recognized by the ϕC31 recombinase. This system therefore cannot be used in a second round of recombination to enlarge the deletion, but can be readily used for the construction of two or more independent deletions within the same bacterial strain
Examples: We are unaware of studies that use this system for bacterial genome deletions, but it has been used to integrate novel DNA into bacterial genomes [12, 17, 27]

2.4 Vectors for the Integration of the Recognition Targets into the Genome

Two vectors for the integration of the recombinase recognition target sequences into the genome must be designed. The minimum requirements for these vectors are that:

1. These vectors have a narrow host range origin of replication (e.g., ColE1, p15A) and function as a suicide vector in the species of interest.
2. They each contain a selective marker. The selective markers on the two plasmids must be different from each other and different from the selective marker on the recombinase expression vector.
3. They each contain a copy of the recombinase recognition target sequence (Table 1). If using the ϕC31 recombinase system, one vector will contain the *attB* sequence, and the other will contain the *attP* sequence.

4. A multiple cloning site for the introduction of a region homologous to the genomic region where the recognition target is to be integrated; and
5. The vectors must share limited homology to avoid recombination between the two vectors.

Careful consideration of the design of these vectors is necessary. The design of the vector influences the number of selective markers that remain once the deletion is made, and an improper design will result in an inversion, not a deletion, of the region of interest (Fig. 1). Additionally, if designed appropriately, the recombinase

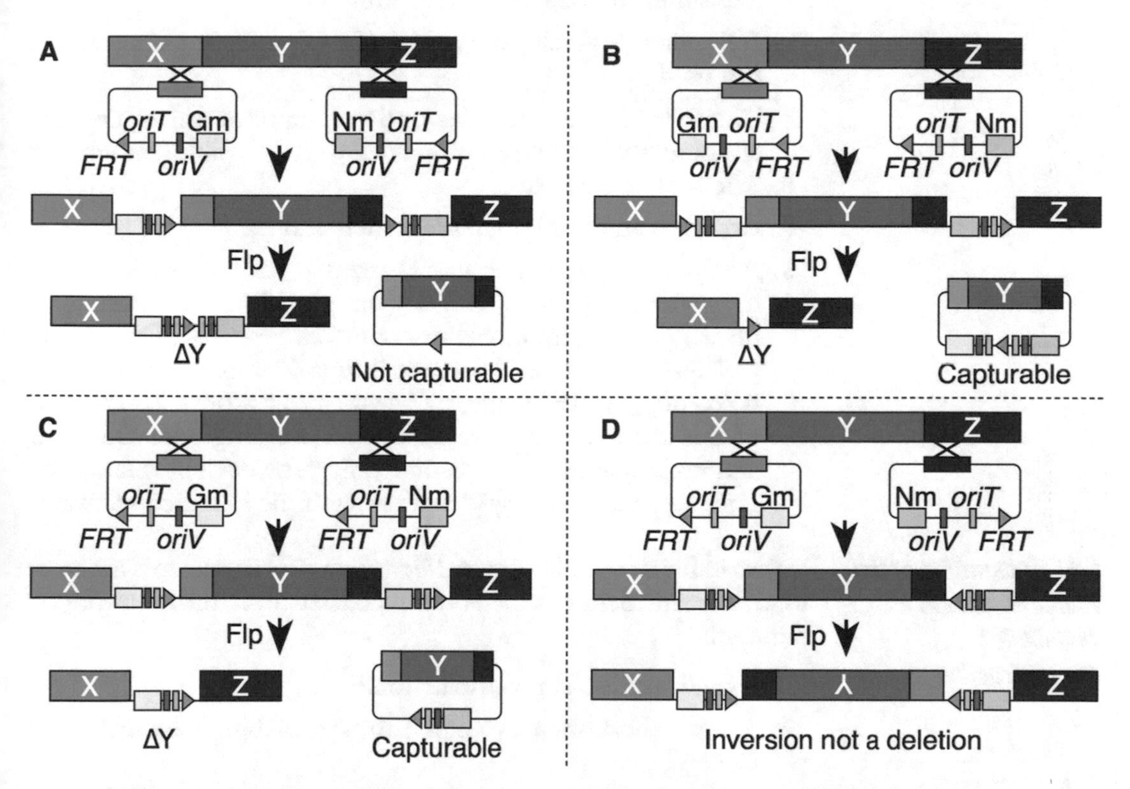

Fig. 1 Influence of vector design on the recombination end-product. A schematic illustrating how design of the vectors for the integration of the recombinase recognition target sequences into the genome will influence the structure of the resulting deletion. The Flp/*FRT* system is used as an example, but the schematic applies equally to all recombinase systems. Depending on how the integration vectors are organized, the final deletion may be marked by either (**a**) two selective markers, (**b**) zero selective markers, or (**c**) one selective marker. If the integration vectors are organized such that following recombination into the genome there are at least one *oriV*, *oriT*, and selective marker situated between the two FRT sites (**b**, **c**), the deleted region may be captured as a replicative vector in *E. coli*. Note that in all the cases, the region cloned into the vectors for recombination into the genome is not deleted following recombination between the *FRT* sites. In (**d**) an example is shown of how improper vector design will result in an inversion of the region of interest instead of a deletion. In this example, the *FRT* sites are introduced into the genome in the opposite direction of each, as opposed to direct orientation as desired, leading to the inversion

system can be used for both the construction of large-scale deletions and in vivo cloning of large genomic fragments (*see* **Note 3** and Fig. 1) [17, 29–31].

2.5 Biolog Phenotype MicroArray™ Analysis

If an analysis of the metabolic capabilities of the deletion strains with the Biolog Phenotype MicroArray™ technology is desired, the following are required:

1. At least one of Biolog plates for testing metabolic capacity (http://www.biolog.com). The primary plates for this purpose are the PM1 (carbon metabolism), PM2A (carbon metabolism), PM3 (nitrogen metabolism), and PM4 (phosphorus and sulfur metabolism) plates; and
2. A standard 96-well plate reader, if an OmniLog plate reader is not used.

   The following list includes optional supplies and equipment, although their use is recommended where possible (*see* **Note 4**):
3. The Biolog redox dye.
4. Access to an OmniLog plate reader; and
5. Analysis software, such as "DuctApe" [32] which is written in python (https://combogenomics.github.io/DuctApe/, https://www.python.org) or "opm" and associated extensions [33, 34], which is an R package (https://r-forge.r-project.org/R/?group_id=1573, https://www.r-project.org). The authors of this chapter only have experience working with DuctApe, and consequently, this chapter only provides information on Phenotype MicroArray™ analysis with the DuctApe software.

2.6 Integration of the Biolog Data with a Metabolic Reconstruction

If the results of the Phenotype MicroArray™ experiments are to be used for refinement of a metabolic reconstruction, the following are required:

1. A draft metabolic reconstruction.
2. A text editor such as gedit (https://wiki.gnome.org/Apps/Gedit).
3. A modeling framework, such as the CobraToolbox [35] in Matlab (http://opencobra.github.io, http://www.mathworks.com/products/matlab/) or Cobrapy [36] in python (http://opencobra.github.io, https://www.python.org).
4. A list of all genes removed in each deletion, as well as a list of genes removed in each deletion that are also present in the draft metabolic reconstruction; and
5. A script to replicate the Phenotype MicroArray™ experiment in silico with the metabolic reconstruction (*see* **Note 5** for a sample script).

3 Methods

Subheadings 3.1–3.4 provide procedures for the construction and confirmation of deletions based on our experience in generating Flp/*FRT*-mediated deletions in the *S. meliloti* genome [6]. In Subheading 3.5 we detail how this procedure can be easily adapted for the in vivo cloning of large genome regions, with an upper size limit of at least 70 kb [17]. The final two Subheadings 3.6 and 3.7 provide protocols that have been used in the metabolic characterization of *S. meliloti* deletion mutants using Phenotype MicroArrays™, and the use of this data in refining a genome-scale metabolic reconstruction [4].

3.1 Flanking a Genome Region with Recombinase Recognition Target Sequences

1. Decide upon the nucleotide (nt) positions where the deletion will start and end (e.g., start at nt position 100,000 and end at nt position 200,000).
2. PCR amplify and purify fragments between ~500 and 1000 nt in length that directly border the region to be deleted (e.g., 99,000–99,999 and 200,001–201,000). Ensure that the primers have the necessary 5′ sequences for cloning into the vectors that contain the recognition targets (*see* **Note 6**).
3. Purify the two vectors that contain the recombinase recognition target sequences (e.g., *FRT* sites, *see* Subheading 2.4) and digest with an appropriate restriction enzyme(s) for the introduction of the PCR products produced in **step 2**.
4. Clone the PCR products from **step 2** into the vectors digested in **step 3** and transform into *E. coli*. One PCR product should be cloned into one of the vectors, and the second PCR product into the second vector (*see* **Note 6**).
5. Inoculate overnight cultures of the organism of interest in which the deletion is to be made, an *E. coli* strain carrying one of the vectors produced in **step 4**, and an *E. coli* strain carrying the conjugation helper plasmid (e.g., pRK600). All the strains can be inoculated in complex medium such as LB, and the appropriate selective agents should be included for growth of the two *E. coli* strains. Grow cultures at the appropriate temperature (37 °C for *E. coli*) overnight with shaking.
6. The next day, subculture the *E. coli* strains to a 1:50 dilution in a fresh complex medium with the selective agent and grow for a further 4–5 h at 37 °C with shaking (*see* **Note 7**).
7. Perform conjugations with the *E. coli* cultures of **step 6** and the overnight culture of the recipient as detailed in Subheading 3.4 (*see* **Note 8**).
8. Once transconjugant colonies have been obtained, purify a colony by dilution streaking on a fresh selective medium petri dish, and incubate.

9. Once the transconjugant colonies from the above purification (**step 8**) have grown, purify further by again dilution streaking one colony from the petri dish of **step 8** and incubating.
10. Repeat **steps 5–9**, starting with overnight cultures of the following strains: the strain purified in **step 9**, an *E. coli* strain carrying the second of the vectors produced in **step 4**, and an *E. coli* strain carrying the helper plasmid (e.g., pRK600). The final strain contains the genome region of interest flanked by the recombinase recognition target sequences.

3.2 Deletion of the Region Flanked with Recombinase Recognition Target Sequences

1. Inoculate overnight cultures of the strain produced in Subheading 3.1, the *E. coli* strain carrying the vector encoding the site-specific recombinase, and an *E. coli* strain carrying the helper plasmid (e.g., pRK600). All the strains can be inoculated in a complex medium such as LB, and the appropriate selective agents should be included for the growth of the two *E. coli* strains (*see* **Note 9**). Grow cultures overnight at the appropriate temperature with shaking.
2. The next day, subculture the *E. coli* strains to a 1:50 dilution in a fresh complex medium with the selective agent and grow for a further 4–5 h at 37 °C with shaking (*see* **Note 7**).
3. Perform conjugations with the *E. coli* cultures of **step 2** and the overnight culture of the recipient, as detailed in Subheading 3.4 but with the following change. If using a vector containing an inducible recombinase gene (recommended), and if you wish to immediately generate the desired deletion, plate the 10^0–10^{-4} dilutions (Subheading 3.4, **step 11**) on both selective medium with and without the inducer compound present. Alternatively, if you wish to simply transfer the recombinase vector to the recipient without inducing deletion formation, plate the 10^0–10^{-4} dilutions only on medium lacking the inducer (*see* **Notes 10** and **11**).
4. Once transconjugant colonies have been obtained, determine the frequency of conjugation separately for transconjugant cells recovered in the presence or in the absence of the inducing agent (*see* **Note 10**).
5. Purify at least four colonies from the selective medium containing the inducing compound by dilution streaking on medium selective for the recipient cells containing the recombinase vector (*see* **Note 12**). The inducer compound does not need to be included in these plates. Incubate the petri dishes at the appropriate temperature for the necessary length of time.
6. Confirm the structure of the deletion as described in Subheading 3.3.
7. If the recombinase plasmid is stably inherited, streak purify one of the confirmed deletion mutants from **step 5** by dilution

streaking a second time. In this case, this strain is the final deletion mutant. Otherwise, if the recombinase plasmid is an unstable plasmid, and you wish to lose the plasmid from the cell, proceed to **step 8**.

8. Inoculate a culture of a confirmed deletion mutant. Use a complex medium without selective agents, and incubate overnight at the appropriate temperature.
9. The next day, serial dilute the overnight culture in saline as described in **step 10** of Subheading 3.4, treating the overnight cultures as the 10^0 dilution.
10. Spread plate (*see* **Note 13**) the 10^{-4} to 10^{-7} dilutions on a complex medium that does not contain compounds selective for the recombinase vector, and incubate the petri dishes at the appropriate temperature until colonies form.
11. Replica plate (e.g., patching with sterile toothpicks) ~100 colonies onto complex medium selective for cells with the recombinase vector and onto complex medium lacking selective agents specific for the recombinase vector. Include a control strain that is expected to grow in the presence of the selective agent, and a control strain expected not to grow. Incubate the petri dishes until growth is observed.
12. Streak purify a strain that does not grow in the presence of the selective agent. Streak the colony onto medium both with and without the selective agent to confirm the genotype. Assuming the strain fails to grow in the presence of the selective agent, this strain is the final deletion mutant and lacks the recombinase vector.

3.3 Confirmation of Deletion Formation

It is possible that some or even all of the transconjugants recovered in Subheading 3.2 do not contain the desired deletion (*see* **Note 14**). It is therefore necessary to confirm that the transconjugants contain the correct deletion structure through testing of the selective marker resistance/sensitivity profile and by PCR, as well as optionally through whole genome sequencing such as with Illumina sequencing.

1. If deletion formation is associated with the removal of at least one selective marker (Fig. 1b, c), screen the transconjugants for the presence/absence of the selective marker(s). If the correct selective marker profile is observed, or if deletion formation is not associated with loss of a selective marker, proceed to **step 2** (*see* **Note 15**). Otherwise, *see* **Note 16**.
2. Use PCR to confirm the correct structure of the deletion by designing primers that amplify genome segments directly outside of both borders of the deletion and directly inside both borders of the deletion (*see* Fig. 2 for an example). If the PCR

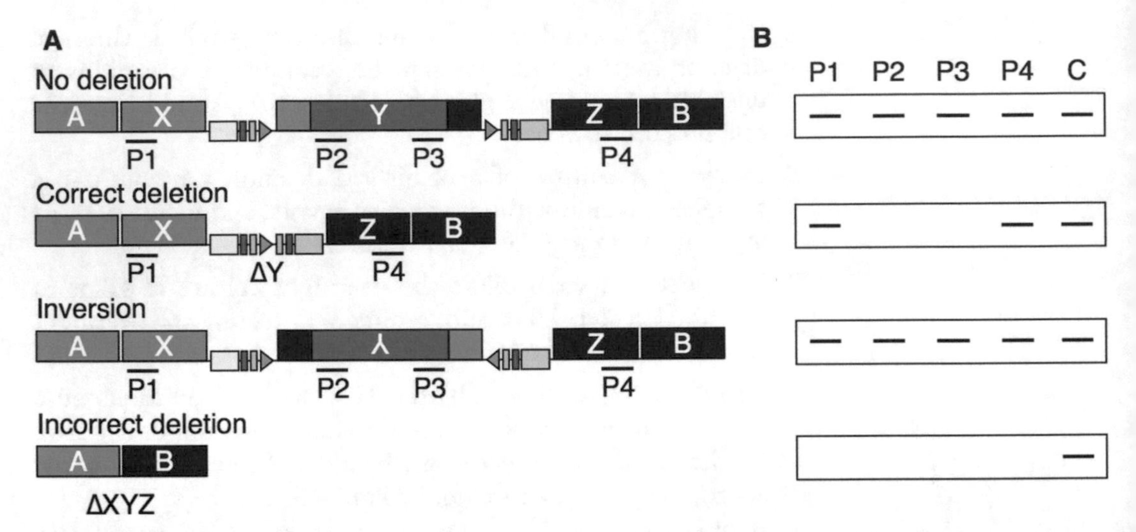

Fig. 2 Confirmation of deletion structure by PCR. (**a**) Schematic representations of sample genome structures following correct or incorrect deletion formation. The recommended positions of PCR products (P1, P2, P3, P4) for confirmation of the correct deletion structure are shown in all schematics. Ideally, the PCR products used for confirmation of the deletion structure do not overlap the fragments that were PCR amplified in **step 2** of Subheading 3.1. (**b**) A schematic representation of the expected banding pattern of the confirmation PCRs for each of the samples shown in (**a**). The PCR product "C" refers to a control PCR, which amplifies a genome segment situated far from the region to be deleted and is expected to always give a product

product profile is as expected, the correct deletion has been obtained. If the PCR product profile is not as expected, *see* **Note 16**.

3. Optionally, whole genome sequencing with the Illumina MiSeq or HiSeq platforms can be used to confirm the genome structure of the deletion mutants. Genomic DNA should be isolated and sequenced according to established procedures. An average genome coverage of at least 20× should be sufficient. The raw reads can be directly mapped to the reference genome of the organism using available software, such as Geneious (www.geneious.com), and the location of the deletion identified as a genome region with an absence of mapped reads. Compared to PCR, whole genome sequencing has the added benefit of allowing identification of whether a spontaneous second-site deletion, in addition to the desired deletion, is present in the recovered transconjugants.

3.4 Performing Conjugations and Isolating Transconjugants

1. Transfer 1 mL of each of the three cultures (recipient, donor, and helper) into separate 1.5 mL centrifuge tubes. Centrifuge the cells at 16,000 × *g* for 1 min, or as required for your organism of interest.
2. Remove the supernatant from all the three tubes, and resuspend the pellet in 0.9 mL of saline.

3. Centrifuge the cell suspensions as in **step 1**. Discard the supernatant and resuspend the pellet in 1 mL of saline.
4. In a fresh 1.5 mL tube, mix 100 μL of both *E. coli* cell suspensions.
5. In a fresh 1.5 mL tube, mix 50 μL of the cell suspension from **step 4** with 25 μL of the recipient organism.
6. On a complex medium agar petri dish (e.g., LB agar), pipette into one spot all 75 μL of the mixture from **step 5**, separately pipette into one spot 50 μL of the mixture from **step 4**, and separately pipette into one spot 25 μL of the recipient organism (*see* **Note 17**). Each of the three solutions should be present as separate spots/pools and should not be spread on the plate. All three cell suspensions can be plated either on the same petri dish or each on a separate petri dish. Ensure no bubbles are created during pipetting of the cell suspensions as splashing will occur when the bubbles pop.
7. Incubate the petri dish(es) from **step 6** overnight at the optimal temperature for your organism of interest. Incubate the petri dish lid-up, and be sure that when transferring the petri dish to the incubator not to spill the liquid spots.
8. The next day, add 1 mL of saline to each of three 1.5 mL centrifuge tubes, and label each to correspond to each of the three mating mixtures from **step 6**. Prepare an additional six 1.5 mL centrifuge tubes, containing 0.9 mL saline and label each corresponding to the 10^{-1}–10^{-6} dilutions of the 75 μL mating spot from **step 6**.
9. Use a sterile wooden stick or sterile metal loop to individually collect each of the mating mixtures from **step 6** and transfer to the corresponding centrifuge tubes containing 1 mL of saline from **step 8**. Vortex, or use a pipette, to fully resuspend the cells in each of the centrifuge tubes. These tubes are the 10^{0} dilutions.
10. Serial dilute the sample containing the 75 μL mating spot. Transfer 100 μL of the 10^{0} dilution to the 10^{-1} centrifuge tube that contains 900 μL of saline, and mix the solution. Repeat for the 10^{-1} dilution to create the 10^{-2} dilution, and continue until 10^{-6}. It is not necessary to serial dilute the samples for the 50 μL or 25 μL mating spots.
11. Spread plate (*see* **Note 13**) the 10^{0}–10^{-4} dilutions for the 75 μL mating mixtures on medium selective for the recipient cells that received the appropriate vector. Also plate the 10^{0} dilution for the 50 μL and 25 μL mating mixtures on the same medium.
12. Spread plate (*see* **Note 13**) the 10^{-4} to 10^{-6} dilutions for the 75 μL mating mixtures on medium selective for all recipient cells regardless of whether or not they received any vector.

13. Incubate all the plates at the appropriate temperature for your organism of interest (*see* **Notes 18** and **19**). Once colonies have formed, count the number of colonies and determine the frequency of conjugation (CFU/mL of transconjugants in the 10^0 dilution/CFU/mL of total recipients in the 10^0 dilution).

3.5 Adaptation of the Deletion Method for In Vivo Cloning of Large Genome Fragments

The large-scale deletion approach described in the previous sections can be readily adapted for the cloning of large DNA fragments. Thus far, we have used this method to clone fragments as large as 70 kb [17], and expect it is possible to clone larger regions as well [7]. Note, however, that this procedure only works if the strain carrying the *FRT* sites is constructed appropriately (*see* **Note 3** and Fig. 1b, c).

1. Use the methods of Subheadings 3.1 and 3.2 to construct a strain that contains two *FRT* sites flanking the genome region to be cloned, and contains the recombinase expression vector. At this point, do not induce expression of the recombinase.
2. Inoculate overnight cultures of the strain produced in **step 1**, an *E. coli* strain carrying the helper plasmid (e.g., pRK600), and an *E. coli* recipient strain with a selectable marker (e.g., rifampicin-resistant DH5α). All the strains can be inoculated in a complex medium such as LB, and the appropriate selective agents should be included. For the strain from **step 1**, selective agents to ensure maintenance of the recombinase vector and maintenance of the genome region to be cloned (e.g., using the selective markers associated with the recognition target sites) must be included.
3. The next day, subculture the *E. coli* helper strain to a 1:50 dilution in a fresh complex medium with the selective agent and grow for a further 4–5 h at 37 °C with shaking (*see* **Note 7**).
4. Using the strains from **steps 2** and **3**, set up conjugations as described in Subheading 3.4, with the following changes.
   (a) In Subheading 3.4, **step 6**, include the inducer compound for induction of recombinase expression in the complex medium agar petri dish.
   (b) When spread plating (*see* **Note 13**) for transconjugants in Subheading 3.4, **step 11**, only plate the 10^{-0} through 10^{-2} dilutions as the conjugation frequency is expected to be low and on the order of 10^{-7}. Transconjugants are to be isolated on medium selective for the *E. coli* recipient that gained at least one of the selective markers associated with the recombinase recognition target sequences.
   (c) In Subheading 3.4, **step 13**, incubate the petri dishes at 37 °C for 1 or 2 days for colonies to form.
5. Once transconjugant colonies have been obtained, purify at least four colonies by dilution streaking on a fresh selective medium petri dish, and incubate.

6. Once the transconjugant colonies from the above purification (**step 5**) have grown, purify each further by again dilution streaking one colony from each of the petri dishes of **step 8** and incubating.
7. Purify the plasmids from each of the four strains of **step 6**. Do so using standard molecular biology techniques, such as with alkaline lysis [37] or commercial plasmid purification kits.
8. Digest the vector with appropriate restriction enzymes and separate the resulting fragments on an agarose gel according to standard procedures. Compare the digest pattern with the expected digest pattern of the plasmid as determined with available software, such as NEBcutter [21]. If the digestion pattern matches the predicted pattern, then the region of interest was successfully cloned and isolated on a plasmid. If the correct digestion pattern is not observed, *see* **Note 20**.

3.6 Biolog Phenotype MicroArray™ Analysis

The deletions produced using the method described here can be characterized similarly to any single-gene deletion mutant. Characterization of their metabolic capabilities can provide invaluable information with respect to metabolic reconstructions. Many methods, including non-targeted metabolomics and Phenotype MicroArrays™, can be employed for this purpose [4, 38]. Here, we focus on the metabolic characterization of the large-scale deletion mutants using the Biolog Phenotype MicroArray™ technology [19] as the ease at which it can be performed with limited specialized equipment, analyzed, and integrated into the metabolic reconstruction makes it particularly useful. Many procedures for Phenotype MicroArray™ analysis can be found in the literature [3, 19, 39], and a protocol based on those procedures is described below.

1. Grow the wild type and all deletion strains of interest on any complex medium agar plates. Often it is possible to start the Phenotype MicroArray™ experiment from colonies grown on a solid medium, but in some cases it may be necessary to start from a liquid culture (*see* **Note 21**).
2. Use a cotton swab to resuspend the cells in 10 mL of a 0.85% NaCl solution to a final reading of 81% turbidity as determined with a Biolog turbidimeter, or to an OD_{600} ~ 0.1 in a 1 cm pathlength.
3. For each strain, combine 2 mL of each cell suspension from **step 2** with 22 mL of the desired minimal medium for the experiment. Ensure that the minimal medium used lacks the nutrient source to be tested (e.g., if using the carbon Biolog plates, the minimal medium must include all nutrients required for growth except for a carbon source).
4. Add 240 μL of Biolog redox dyeA into each 24 mL solution from **step 3** (*see* **Note 4**).

5. For each solution in **step 4**, add 100 μL to each well of the Biolog plates being employed.
6. Incubate all the plates in an OmniLog plate reader where available, otherwise a standard microtiter plate reader with temperature control may be employed (*see* **Note 4**). Incubate the plates at the appropriate growth temperature, and monitor reduction of the dye for 120 h, or however long is necessary for your specific organism and mutants.
7. Analyze the output from the OmniLog system using freely available software such as DuctApe [32] or opm [33, 34] (Fig. 3a, b; *see* **Note 22**).

A			activity			
plate_id	well_id	chemical	Wild type	Mutant 1	Mutant 2	Mutant 3
PM01	A01	Negative Control	0	0	0	0
PM01	A02	L-Arabinose	9	3	9	9
PM01	A03	N-Acetyl-D-Glucosamine	9	9	9	9
PM01	A04	D-Saccharic acid	0	0	0	0

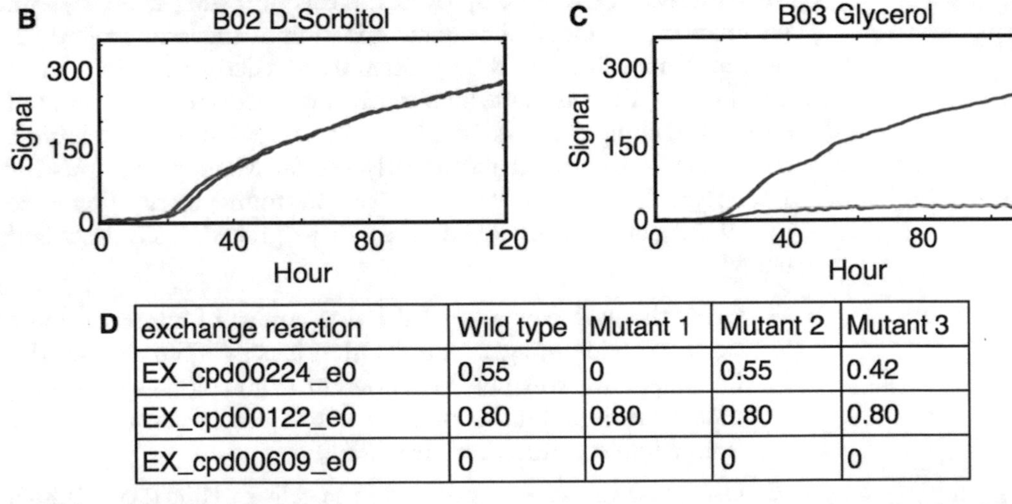

exchange reaction	Wild type	Mutant 1	Mutant 2	Mutant 3
EX_cpd00224_e0	0.55	0	0.55	0.42
EX_cpd00122_e0	0.80	0.80	0.80	0.80
EX_cpd00609_e0	0	0	0	0

Fig. 3 Sample output of the experimental and in silico Phenotype MicroArray™ experiments. (**a**–**c**) Sample output from the DuctApe analysis software. (**a**) A subset of an output table that provides a summary statistic summarizing how good each strain grew with each compound present in the tested Biolog plates. The higher the value, the better the growth. (**b**, **c**) Sample output curves illustrating the metabolic activity of wild type and one mutant strain in two different wells of the Biolog plate. (**d**) A subset of the table resulting from the script provided in **Note 5**. The objective function solution, which here is a specific growth rate, of each strain with each compound is shown. The first row was not printed by the script but added separately. The exchange reactions in column one can be switched with the corresponding compound name with simple scripts, or even in Excel, to allow direct comparison of the DuctApe output software. Sample data for this figure is taken from published work [4]

3.7 Incorporation of Phenotype MicroArray™ Data with a Draft Metabolic Reconstruction

The first step in using the experimental Phenotype MicroArray™ data to refine a draft metabolic reconstruction is to replicate the experiment in silico. Below, a procedure to do so and a sample Matlab script (*see* **Note 5**) are provided based on published work [4].

1. Create a list of all genes deleted in each of the deletion mutants examined.
2. Obtain a list of all genes present in the draft metabolic reconstruction, such as through the modelName.genes function in the cobraToolbox for Matlab.
3. Compare the list from **step 1** with the list from **step 2** in order to produce a comma separated list for each deletion mutant that indicates what genes are both removed in the mutant and also present in the draft metabolic reconstruction.
4. Use the data from **step 3** and a script such as that in **Note 5** to examine the in silico predicted phenotypes for each of the deletions studied with the Phenotype MicroArray™ analysis. In essence the script must perform the following: iteratively for the wild type and each mutant defined, the script must first delete all genes, and dependent reactions, that are removed in the deletion mutant and then iteratively provide a different nutrient source and test the ability of the model to produce biomass. The script provided in **Note 5** is a Matlab script making use of the cobraToolbox. Once the necessary user-defined variables are defined, this script can directly be used to replicate a Phenotype MicroArray™ experiment in silico (Fig. 3c).
5. Modify the draft metabolic reconstruction as necessary to reconcile the in silico predictions with the experimentally determined Phenotype MicroArray™ results. This can include modifying the gene-protein-reaction relationships present in the reconstruction, and/or addition or deletion of reactions.
6. Repeat **steps 4** and **5** as necessary until satisfied with the agreement between the in silico and experimental derived results. If during the curation new genes are added to the model, or genes are eliminated from the model, also repeat **steps 1–3**.

4 Notes

1. Usage of an unstable vector is not absolutely necessary, but is advantageous and can be required for certain applications. If it is desired to make a small deletion and enlarge it through successive rounds of deletion, the recombinase plasmid must be lost or eliminated from the cell following construction of the

initial deletion and prior to transfer of the next recombinase recognition target sequence containing plasmid. Similarly, if two unlinked deletions are to be recombined into the same strain, it is necessary to first remove the recombinase plasmid. Finally, the presence of the vector may impact the growth rate of the strain, it prevents re-usage of the associated selective marker, and other plasmids with the same origin cannot be co-maintained.

2. The use of an inducible promoter has two applications. First, it can assist in troubleshooting when the desired deletion cannot be recovered. If transconjugants carrying the recombinase plasmid are recovered less frequently when plated on media with the inducer compound than without the inducer compound, this indicates that the deletion is lethal. Otherwise, if transconjugants are recovered at the same frequency in both the cases, but none of the transconjugants carry the expected deletion, there is another issue (e.g., recognition targets are incorrectly integrated, the recombinase is not expressed). Second, it allows for recovery of strains carrying both recognition targets and the uninduced recombinase, which can subsequently be induced during specific environmental conditions and/or to study temporal effects [25]. For example, recombinase expression can be induced in a liquid culture to examine the effect on cell viability [6], or deletion formation can be induced under varying conditions to examine the effect of different factors on the phenotype of the deletion.

3. When designed appropriately, it is possible to not only delete a genome region but also capture the deleted region as a plasmid. This can be done by ensuring that an *oriT*, a narrow host range *oriV*, and a selective marker are present between the two integrated recognition targets (Fig. 1b, c). Induction of expression of the recombinase will result in the excision of the region intervening the two recognition targets, which will be present as a non-replicative vector that is lost from the cell. However, if the transfer genes are provided *in trans*, this non-replicative vector can be conjugated to a permissive host such as *Escherichia coli*, where it can be recovered as a replicative plasmid as described elsewhere [17].

4. The OmniLog plate reader can be replaced with a standard 96-well plate reader, with redox dye reduction measured at 590 nm. If desired, the ability of a deletion mutant to grow with the different compounds can be directly measured, such as with OD_{600}, instead of using a redox dye as an indicator of active metabolism. The use of published analysis software is not necessary and can be replaced with manual analysis, for example in Excel or with custom scripts. However, it is recommended to make use of the OmniLog plate reader, the redox dye, and

existing analysis software as this allows for a simple experimental and computational workflow that will provide meaningful data for use in metabolic reconstructions without an excessive effort.

5. A sample Matlab script for replicating a Phenotype MicroArray™ experiment in silico with a metabolic reconstruction is provided below. It is based on previously used code [4].

```
%% Load the model
model = readCbModel('modelName.xml'); % import the model
%% Define the strains to be studied
wildType = {}; % an empty string for the wild type
deletion1 = {'gene1','gene2','gene3'}; % list all genes re-
moved in deletion 1 and that are present in the reconstruction
deletion2 = {'gene4','gene5'}; % list all genes removed in
deletion 2 and that are present in the reconstruction. Repeat
for all deletions
strains = {wildType,deletion1,deletion2}; % list all the
strains defined above
%% Define the base medium composition
exchangeList = {'exchange1';'exchange2'}; % list all the
exchange reactions in the model
model = changeRxnBounds(model, exchangeList, 0, 'l'); % set
the lower bound of all exchange reactions to 0
baseMedium = {'nitrogen','sulfur','phosphorus','etc'}; % list
all exchange reactions that define the base medium. Should
contain all essential nutrients except for the nutrient class
being tested (e.g. except for a carbon source)
model = changeRxnBounds(model, baseMedium, -100, 'l'); % set
the lower bound of the exchange reactions of the base medium to
be non-growth rate limiting
%% Define the compounds to be studied
testCompounds = {'carbon1','carbon2','carbon3'}; % list the
exchange reactions for all compounds to be tested
%% Define the output variables
testCompounds2 = {'blank';'blank';'carbon1';'carbon2';'car-
bon3'}; % The same as the 'testCompounds' string, except it
begins with two fields containing 'blank' and compounds are
separated by ';' not ','
outputBiolog = {}; % the variable that will contain the final
data
outputBiolog = testCompounds2; % add the list of tested
compounds to the output variable
temp = {}; % a temporary variable
%% Perform the analysis
for strains = {wildType,deletion1,deletion2} % start of a for
loop to iteratively test each of the strains defined earlier
```

```
    strain = strains; % set the 'strain' variable equal to the current strain being tested
    genes = strain{1}; % set the 'genes' variable equal to all the genes present in the current strain
    temp = {'blank';'blank'}; % add two blank cells to the temp variable and clear existing data
      [modelMutant,hasEffect,constrRxnNames,deletedGenes] = deleteModelGenes(model,genes); % delete all genes in the 'genes' variable from the model. Save the model as modelMutant, and all reactions that should no longer exist are saved in the constrRxnNames variable
    modelRxnsRemoved = removeRxns(modelMutant,constrRxnNames); % delete all reactions from the model that should no longer exist
      for testCompounds = {'carbon1','carbon2','carbon3'} % start a second for loop to iteratively add one test compounds and test biomass production
        compoud = testCompounds; % set the 'compound' variable equal to the current compound
         model2 = changeRxnBounds(modelRxnsRemoved, compound, -5, 'l'); % add the tested compound to the growth medium at a growth rate limiting concentration
        solution = optimizeCbModel(model2, 'max'); % optimize the model
           temp(end+1) = {solution.f}; % add the objective function flux solution to the temp variable
      end % finish the for loop for testing of the various compounds
    outputBiolog = horzcat(outputBiolog,temp); % add the data in the temporary file to the output file
end % finish the for loop for testing the various strains
clearvars -except output; % clear all intermediate variables leaving just the output. The first column of the output file will contain the different exchange reactions, and each following column will contain the output for each of the strains in the order that they were listed in the 'strains' string (a value of 0 means the strain failed to grow with the corresponding compound listed in the first column)
```

6. Any method can be used for cloning of the PCR products into the vectors containing the recombinase recognition targets. Possible methods include classical ligation [37], Gibson cloning [40], SLiCE [41], or SLIC [42], among others. We work primarily with SLIC (*s*equence- and *l*igation-*i*ndependent *c*loning) for cloning of PCR products and we recommend the use of a ligation-independent cloning method as this precludes the need to digest the PCR product and therefore eliminates issues with identifying compatible restriction enzymes that can be

used to digest both the vector and PCR product without cleaving the PCR product or vector in half.

7. The use of fresh cultures increases conjugation efficiency, and thus this subculture step is to achieve a higher plasmid transfer rate. However, this step can normally be skipped if necessary due to time constraints.
8. Assuming a plasmid transfer frequency of ~10^{-2} to 10^{-3} per recipient cell, and assuming the plasmid recombines into the recipient genome at a frequency ~10^{-3} to 10^{-4} per recipient cell that received plasmid, the overall conjugation frequency is expected to be in the range of ~10^{-5} and 10^{-7}.
9. Selective agents can be included for growth of the strain produced in Subheading 3.1, although in most cases this is not necessary as the integrated vectors should not be easily lost.
10. Comparing the amount of transconjugants recovered on medium with the inducing agent versus without the inducing agent provides insight into the viability of the deletion. Assuming a plasmid transfer frequency of ~10^{-2} to 10^{-3} per recipient cell, the frequency of conjugation is expected to be ~10^{-2} to 10^{-3} per recipient regardless of whether the inducer compound is included in the selective medium. A large decrease (>tenfold different) in the recovery of transconjugants in the presence of the inducing compound compared to without the inducing agent indicates that the majority of cells in which the deletion was made failed to grow and form colonies. In such cases, it may not be possible to recover colonies containing the deletion.
11. Transferring of the recombinase vector to the recipient without the inducer compound present allows for the isolation of transconjugants carrying the uninduced recombinase without forming the deletion. Such a strain can be purified, and then grown with the inducer compound present at a later time to induce the deletion, which can be useful for studies as discussed in **Note 2**.
12. If attempting to isolate transconjugants carrying the recombinase vector but without forming the deletion, it is best to continuously select for a selective marker that is lost once the deletion is formed. Even without actively inducing the expression of the recombinase, we have found that background expression of the recombinase can be sufficient to facilitate the formation of the deletion in a subset of the population. Thus, it is necessary to continually select for strains not containing the deletion to ensure the genotype of the cell population is as desired.
13. To spread plate the cell suspensions, transfer by pipette 100 μL of the desired suspension to the petri dish. Spread the culture

either by placing sterile glass beads on the petri dish, shaking, and then discarding the glass beads, or by using a sterile "hockey stick" glass rod to spread the culture (sterilize the glass rod in between uses by dipping it in 95% ethanol and then burning the ethanol by very briefly bringing the glass rod in contact with the flame of the Bunsen burner—leaving the glass rod in the flame for an extended period will cause the rod to break).

14. There are several reasons for why a recovered transconjugant may not contain any deletion. The primary reason is that it is possible for either of the vectors integrated into the genome through single cross-over plasmid integration to recombine back out of the genome at a frequency ~10^{-3} to 10^{-4}, leaving only one recombinase recognition target sequence in the genome. Additionally, at a frequency not greater than 10^{-7} there may be a spontaneous mutation within the recombinase gene or either recognition sequence that would inactivate the system. Another possibility is that the deletion strategy was incorrectly designed, and the recognition targets are not in the same orientation, and so an inversion is produced instead of a deletion (Fig. 1d).

15. A correct selective marker profile confirms that a deletion has been made. However, our own experience tells us that this should not be considered sufficient evidence that the correct deletion has been made. It is possible the plasmids were targeted to the incorrect genome location, or a spontaneous genome deletion event removed the selective marker(s) while making an incorrect deletion. Therefore, PCR confirmation is strongly recommended even if the correct selective marker profile is detected.

16. If the structure of the deletion is not confirmed, then more transconjugant colonies can be screened through replica patching ~100 of the transconjugant colonies obtained in Subheading 3.2, **step 4**. If the correct deletion is still not detected, then conjugation of the recombinase expression vector can be repeated. If the correct deletion is still not obtained, and especially if there is a large difference in the number of transconjugants obtained on media with and without the inducing agent, deletion of the region under the tested conditions may be lethal (*see* **Note 10**). However, if there is no difference in the number of transconjugants obtained on media with and without the inducing agent, then it may be that the deletion construct is incorrectly designed (Fig. 1d) and should be re-examined.

17. The 50 and 25 μL mating spots serve as negative controls. When eventually plated on medium selective for recipient

cells that had gained the vector containing the recombinase recognition target, no colonies should form on these plates. If colonies do appear, there was a source of contamination at some point of the experiment, and the experiment should be repeated.

18. It may be necessary to incubate the petri dishes a day or two longer than normal for colonies to grow due to the large number of cells plated on the petri dish.
19. This note is applicable only to conjugations involving a vector that functions as a suicide vector in the recipient cell. In these cases, it is common for there to be greater background growth on the petri dishes on which the 75 μL spot was streaked compared to the petri dishes on which the 50 μL and 25 μL spots were streaked. This is due to slight growth of recipient cells that gain the vector containing the recombinase recognition target but in which the vector did not recombine into the recipient's genome.
20. It is necessary to confirm that the recovered plasmid is correct as there are two main reasons for why the incorrect plasmid might be obtained. It is possible for the vectors containing the recombinase recognition target sequence to recombine back out of the genome at a frequency ~10^{-3} to 10^{-4}, and for just one or both of these vectors to be transferred to the *E. coli* recipient. Additionally, if the recognition target sequences were introduced into the incorrect genome location, the wrong genome region would be cloned.
21. In some cases, a deletion mutant might experience an extra long lag phase compared to the wild type upon inoculation into the liquid minimal medium. This can negatively impact the experiment and analysis. In such cases, it is recommended to not start the Phenotype MicroArrayTM experiment from solid agar, but to instead use liquid cultures as the inoculum. To do so, grow all appropriate mutants and wild type in a minimal medium containing all nutrients required to grow until the cultures have reached a sufficiently high density. Pellet and wash the cells from each culture twice with 0.85% saline, and resuspend the washed pellets in 1 mL of 0.85% saline. Use this cell resuspension to inoculate 10 mL of a 0.85% NaCl solution to 81% turbidity as determined with a Biolog turbidimeter, or to an OD600 ~0.1 in a 1 cm pathlength, and proceed to **step 3** of Subheading 3.6. In some cases, significant background growth in the negative control may be observed with this method, indicating that the cells may need to be starved of the nutrient type (e.g., nitrogen) being tested. If this occurs, after washing and resuspending the cultures, dilute to an OD600 ~0.1 in medium lacking the nutrient class to be tested

(e.g., no nitrogen source) and incubate overnight, or longer as necessary. Then wash the cells with saline as before, and proceed to preparing the 10 mL saline cell resuspension.

22. The DuctApe analysis software allows automated prediction of genes whose deletion in a mutant may be responsible for the observed metabolic phenotype. In order to make use of this function, it is necessary to produce a fasta proteome file (.faa) for the wild type and for each mutant. The wild-type proteome file should contain the amino acid sequences for all proteins encoded by the wild-type genome. The mutant proteome files should contain the amino acid sequences for all proteins that are encoded by the wild type but not by the mutant (i.e., the proteins deleted in the mutant).

References

1. diCenzo GC, Finan TM (2015) Genetic redundancy is prevalent within the 6.7 Mb *Sinorhizobium meliloti* genome. Mol Genet Genomics 290:1345–1356. https://doi.org/10.1007/s00438-015-0998-6
2. Hutchison CA, Chuang R-Y, Noskov VN et al (2016) Design and synthesis of a minimal bacterial genome. Science 351:aad6253–aad6253. https://doi.org/10.1126/science.aad6253
3. Biondi EG, Tatti E, Comparini D et al (2009) Metabolic capacity of *Sinorhizobium* (*Ensifer*) *meliloti* strains as determined by Phenotype MicroArray analysis. Appl Environ Microbiol 75:5396–5404. https://doi.org/10.1128/AEM.00196-09
4. diCenzo GC, Checcucci A, Bazzicalupo M et al (2016) Metabolic modelling reveals the specialization of secondary replicons for niche adaptation in *Sinorhizobium meliloti*. Nat Commun 7:12219. https://doi.org/10.1038/ncomms12219
5. Charles TC, Finan TM (1991) Analysis of a 1600-kilobase *Rhizobium meliloti* megaplasmid using defined deletions generated in vivo. Genetics 127:5–20
6. Milunovic B, diCenzo GC, Morton RA, Finan TM (2014) Cell growth inhibition upon deletion of four toxin-antitoxin loci from the megaplasmids of *Sinorhizobium meliloti*. J Bacteriol 196:811–824. https://doi.org/10.1128/JB.01104-13
7. Chain PS, Hernández-Lucas I, Golding B, Finan TM (2000) *oriT*-directed cloning of defined large regions from bacterial genomes: identification of the *Sinorhizobium meliloti* pExo megaplasmid replicator region. J Bacteriol 182:5486–5494. https://doi.org/10.1128/JB.182.19.5486-5494.2000
8. MacLean AM, White CE, Fowler JE, Finan TM (2009) Identification of a hydroxyproline transport system in the legume endosymbiont *Sinorhizobium meliloti*. Mol Plant Microbe Interact 22:1116–1127. https://doi.org/10.1094/MPMI-22-9-1116
9. MacLean AM, MacPherson G, Aneja P, Finan TM (2006) Characterization of the β-ketoadipate pathway in *Sinorhizobium meliloti*. Appl Environ Microbiol 72:5403–5413. https://doi.org/10.1128/AEM.00580-06
10. Yurgel SN, Mortimer MW, Rice JT et al (2013) Directed construction and analysis of a *Sinorhizobium meliloti* pSymA deletion mutant library. Appl Environ Microbiol 79:2081–2087. https://doi.org/10.1128/AEM.02974-12
11. Cheng J, Poduska B, Morton RA, Finan TM (2011) An ABC-type cobalt transport system is essential for growth of *Sinorhizobium meliloti* at trace metal concentrations. J Bacteriol 193:4405–4416. https://doi.org/10.1128/JB.05045-11
12. diCenzo G, Milunovic B, Cheng J, Finan TM (2013) The tRNA^{arg} gene and *engA* are essential genes on the 1.7-Mb pSymB megaplasmid of *Sinorhizobium meliloti* and were translocated together from the chromosome in an ancestral strain. J Bacteriol 195:202–212. https://doi.org/10.1128/JB.01758-12
13. Ying B-W, Seno S, Kaneko F et al (2013) Multilevel comparative analysis of the contributions of genome reduction and heat shock to the *Escherichia coli* transcriptome. BMC Genomics 14:25. https://doi.org/10.1186/1471-2164-14-25
14. Dominguez-Ferreras A, Perez-Arnedo R, Becker A et al (2006) Transcriptome profiling reveals the importance of plasmid pSymB for

osmoadaptation of *Sinorhizobium meliloti*. J Bacteriol 188:7617–7625. https://doi.org/10.1128/JB.00719-06

15. diCenzo GC, MacLean AM, Milunovic B et al (2014) Examination of prokaryotic multipartite genome evolution through experimental genome reduction. PLoS Genet 10: e1004742. https://doi.org/10.1371/journal.pgen.1004742
16. Ullrich S, Schüler D (2010) Cre-*lox*-based method for generation of large deletions within the genomic magnetosome island of *Magnetospirillum gryphiswaldense*. Appl Environ Microbiol 76:2439–2444. https://doi.org/10.1128/AEM.02805-09
17. diCenzo GC, Zamani M, Milunovic B, Finan TM (2016) Genomic resources for identification of the minimal N_2-fixing symbiotic genome. Environ Microbiol 18:2534–2547. https://doi.org/10.1111/1462-2920.13221
18. Döhlemann J, Brennecke M, Becker A (2016) Cloning-free genome engineering in *Sinorhizobium meliloti* advances applications of Cre/-*loxP* site-specific recombination. J Biotechnol 233:160–170. https://doi.org/10.1016/j.jbiotec.2016.06.033
19. Bochner BR (2001) Phenotype microarrays for high-throughput phenotypic testing and assay of gene function. Genome Res 11:1246–1255. https://doi.org/10.1101/gr.186501
20. Finan TM, Kunkel B, De Vos GF, Signer ER (1986) Second symbiotic megaplasmid in *Rhizobium meliloti* carrying exopolysaccharide and thiamine synthesis genes. J Bacteriol 167:66–72
21. Vincze T, Pósfai J, Roberts RJ (2003) NEBcutter: a program to cleave DNA with restriction enzymes. Nucleic Acids Res 31:3688–3691
22. Leprince A, de Lorenzo V, Völler P et al (2012) Random and cyclical deletion of large DNA segments in the genome of *Pseudomonas putida*. Environ Microbiol 14:1444–1453. https://doi.org/10.1111/j.1462-2920.2012.02730.x
23. Wang Y, Wang Z, Cao J et al (2014) FLP-FRT-based method to obtain unmarked deletions of *CHU_3237* (*porU*) and large genomic fragments of *Cytophaga hutchinsonii*. Appl Environ Microbiol 80:6037–6045. https://doi.org/10.1128/AEM.01785-14
24. Lambert JM, Bongers RS, Kleerebezem M (2007) Cre-*lox*-based system for multiple gene deletions and selectable-marker removal in *Lactobacillus plantarum*. Appl Environ Microbiol 73:1126–1135. https://doi.org/10.1128/AEM.01473-06
25. Harrison CL, Crook MB, Peco G et al (2011) Employing site-specific recombination for conditional genetic analysis in *Sinorhizobium meliloti*. Appl Environ Microbiol 77:3916–3922. https://doi.org/10.1128/AEM.00544-11
26. Yu BJ, Sung BH, Koob MD et al (2002) Minimization of the *Escherichia coli* genome using a Tn5-targeted Cre/*loxP* excision system. Nat Biotechnol 20:1018–1023. https://doi.org/10.1038/nbt740
27. Heil JR, Cheng J, Charles TC (2012) Site-specific bacterial chromosome engineering: ΦC31 integrase mediated cassette exchange (IMCE). J Vis Exp 61:e3698. https://doi.org/10.3791/3698
28. Jones JD, Gutterson N (1987) An efficient mobilizable cosmid vector, pRK7813, and its use in a rapid method for marker exchange in *Pseudomonas fluorescens* strain HV37a. Gene 61:299–306
29. Posfai G, Koob MD, Kirkpatrick HA, Blattner FR (1997) Versatile insertion plasmids for targeted genome manipulations in bacteria: isolation, deletion, and rescue of the pathogenicity island LEE of the *Escherichia coli* O157:H7 genome. J Bacteriol 179:4426–4428
30. Posfai G, Koob M, Hradecná Z et al (1994) In vivo excision and amplification of large segments of the *Escherichia coli* genome. Nucleic Acids Res 22:2392–2398
31. Wilson JW, Figurski DH, Nickerson CA (2004) VEX-capture: a new technique that allows in vivo excision, cloning, and broad-host-range transfer of large bacterial genomic DNA segments. J Microbiol Methods 57:297–308. https://doi.org/10.1016/j.mimet.2004.01.007
32. Galardini M, Mengoni A, Biondi EG et al (2014) DuctApe: a suite for the analysis and correlation of genomic and OmniLog™ Phenotype Microarray data. Genomics 103:1–10. https://doi.org/10.1016/j.ygeno.2013.11.005
33. Vaas LAI, Sikorski J, Michael V et al (2012) Visualization and curve-parameter estimation strategies for efficient exploration of phenotype microarray kinetics. PLoS One 7:e34846. https://doi.org/10.1371/journal.pone.0034846
34. Vehkala M, Shubin M, Connor TR et al (2015) Novel R pipeline for analyzing biolog phenotypic microarray data. PLoS One 10: e0118392. https://doi.org/10.1371/journal.pone.0118392
35. Schellenberger J, Que R, Fleming RMT et al (2011) Quantitative prediction of cellular metabolism with constraint-based models: the

COBRA Toolbox v2.0. Nat Protoc 6:1290–1307. https://doi.org/10.1038/nprot.2011.308

36. Ebrahim A, Lerman JA, Palsson BØ, Hyduke DR (2013) COBRApy: COnstraints-based reconstruction and analysis for python. BMC Syst Biol 7:74. https://doi.org/10.1186/1752-0509-7-74
37. Sambrook J, Fritsch EF, Maniatis T (1989) Molecular cloning: a laboratory manual. Cold Spring Harbor Laboratory Press, New York
38. Fei F, diCenzo GC, Bowdish DME et al (2016) Effects of synthetic large-scale genome reduction on metabolism and metabolic preferences in a nutritionally complex environment. Metabolomics 12:23. https://doi.org/10.1007/s11306-015-0928-y
39. Spini G, Decorosi F, Cerboneschi M et al (2015) Effect of the plant flavonoid luteolin on *Ensifer meliloti* 3001 phenotypic responses. Plant Soil 399:159–178. https://doi.org/10.1007/s11104-015-2659-2
40. Gibson DG, Young L, Chuang R-Y et al (2009) Enzymatic assembly of DNA molecules up to several hundred kilobases. Nat Methods 6:343–345. https://doi.org/10.1038/nmeth.1318
41. Zhang Y, Werling U, Edelmann W (2014) Seamless ligation cloning extract (SLiCE) cloning method. Methods Mol Biol 1116:235–244. https://doi.org/10.1007/978-1-62703-764-8_16
42. Jeong J-Y, Yim H-S, Ryu J-Y et al (2012) One-step sequence- and ligation-independent cloning as a rapid and versatile cloning method for functional genomics studies. Appl Environ Microbiol 78:5440–5443. https://doi.org/10.1128/AEM.00844-12

Chapter 14

Computational Prediction of Synthetic Lethals in Genome-Scale Metabolic Models Using Fast-SL

Karthik Raman, Aditya Pratapa, Omkar Mohite, and Shankar Balachandran

Abstract

In this chapter, we describe Fast-SL, an in silico approach to predict synthetic lethals in genome-scale metabolic models. Synthetic lethals are sets of genes or reactions where only the simultaneous removal of all genes or reactions in the set abolishes growth of an organism. In silico approaches to predict synthetic lethals are based on Flux Balance Analysis (FBA), a popular constraint-based analysis method based on linear programming. FBA has been shown to accurately predict the viability of various genome-scale metabolic models. Fast-SL builds on the framework of FBA and enables the prediction of synthetic lethal reactions or genes in different organisms, under various environmental conditions. Predicting synthetic lethals in metabolic network models allows us to generate hypotheses on possible novel genetic interactions and potential candidates for combinatorial therapy, in case of pathogenic organisms. We here summarize the Fast-SL approach for analyzing metabolic networks and detail the procedure to predict synthetic lethals in any given metabolic model. We illustrate the approach by predicting synthetic lethals in *Escherichia coli*. The Fast-SL implementation for MATLAB is available from https://github.com/RamanLab/FastSL/.

Key words Flux balance analysis, Genome-scale metabolic networks, Synthetic lethality, Constraint-based analysis, Systems biology, Gene essentiality

1 Introduction

Genome-scale metabolic networks are in silico reconstructions of the metabolism of organisms, which contain information about known metabolic reactions and their associated enzymes. Genome-scale metabolic models (GEMs) constructed from these networks coupled with additional constraints arising from stoichiometry and thermodynamics have been widely used to predict metabolic capabilities of organisms using computational techniques such as Flux Balance Analysis (FBA) [1–8]. GEMs have also been used to identify potential drug targets by identifying essential and synthetic lethal genes in pathogenic organisms using FBA [9, 10], and targets for metabolic engineering [11].

Marco Fondi (ed.), *Metabolic Network Reconstruction and Modeling: Methods and Protocols*, Methods in Molecular Biology, vol. 1716, https://doi.org/10.1007/978-1-4939-7528-0_14,

Synthetic lethal (SL) gene sets are sets of genes where only the simultaneous removal of all genes in the set abolishes growth of an organism [12]. Of the general gene knockout sets, synthetic lethal sets are of particular interest while identifying combinatorial drug targets [13, 14] that make it difficult for organism to resist. Synthetic lethal sets also reveal complex interactions in metabolic networks and have been analyzed in the past for predicting novel genetic interactions and estimating the extent of robustness of biological networks [15]. In silico, synthetic lethal sets can be identified by simulating the effect of removal of gene sets from the reconstructed GEM of an organism.

Computational approaches to identify synthetic lethal reactions in GEMs have built on the framework of FBA. The earliest approaches extended FBA to exhaustively analyze all possible combinations of genes [16, 17]. While exhaustively enumerating multiple knockouts, the number of gene combinations to be evaluated increases exponentially with size, because of the combinatorial explosion of possible knockouts. Therefore, many previous studies have used a computer cluster to parallelize the knockout simulations [16, 17]. Other approaches involving a bi-level Mixed Integer Programming (MILP) formulation [12, 18] have been suggested to be effective in pruning the search space for knockout behaviors. However, the convergence of these methods can be slower in case of large networks. Furthermore, the MILP-based approaches do not have the inherent parallelism of the exhaustive enumeration method. Recently, we proposed a new algorithm, Fast-SL [19], which retains the parallelism of exhaustive enumeration, and additionally minimizes the L_1-norm of the flux distribution predicted by FBA. Fast-SL offers a framework to identify both synthetic lethal reaction and gene sets, while both reducing the search space and maintaining the inherent parallelism of combinatorial analysis. Fast-SL is currently available on GitHub and is compatible with the COnstraint-Based Reconstruction and Analysis (COBRA) toolbox [20, 21], which has emerged as a *de facto* standard for (constraint-based) metabolic network analysis.

In this chapter, we present a detailed protocol for in silico prediction of synthetic lethality based on FBA using the Fast-SL framework, and illustrate it by applying it to predict synthetic lethal genes and reactions in *Escherichia coli*. The detailed algorithm as well as comparisons with other methods can be found in the original Fast-SL paper [19].

2 Materials

In this section, we describe the essential tools for performing in silico analyses using metabolic models.

2.1 Software Tools

The most popular toolbox for analyzing metabolic networks is the COBRA Toolbox, which is available for both MATLAB as "COBRA Toolbox for MATLAB" [21] and for Python, as "COBRA for Python" [22] as a part of the openCOBRA project (opencobra.github.io). In this chapter, we focus only on using COBRA Toolbox for MATLAB (www.mathworks.com/products/matlab/) for in silico synthetic lethality analysis. Installation instructions for MATLAB are available as part of MathWorks® documentation (www.mathworks.com/help/install/index.html). COBRA toolbox is available from github.com/opencobra/cobratoolbox/zipball/master and the installation instructions are available online (opencobra.github.io/cobratoolbox/docs/index.html). Following the installation of the COBRA toolbox, it is important to set up at least one linear programming (LP) solver, for solving LP problems such those that arise in FBA.

2.1.1 Solvers

As mentioned earlier, FBA involves solving an LP problem for large metabolic models with thousands of reactions and metabolites. Further, for combinatorial analyses such as identifying synthetic lethal sets, we solve many such FBA LP problems, one for each combinatorial reaction or gene deletion. Therefore, it is essential to have an efficient solver to solve these large LP problems with thousands of variables and constraints. The COBRA toolbox is by default equipped with the GLPK (GNU Linear Programming Kit) solver and also supports MATLAB's `lpsolve` routine. Additionally, it supports other more efficient solvers listed in Table 1. Most of these solvers are free for academic use. A comparison of running time of few of these solvers on a metabolic model can be found in **Note 1**. For Fast-SL, we recommend the use of IBM CPLEX, which offers better running times. Following the installation of the solver, make sure that the COBRA toolbox is able to access the solvers, by using the following initialization command:

Table 1
List of solvers supported by COBRA Toolbox

Solver	Website	Pricing
GLPK	http://www.gnu.org/software/glpk/	Free
Gurobi	http://www.gurobi.com	Free (Academic)
CPLEX	http://www.ibm.com/developerworks/downloads/ws/ilogcplex/	Free (Academic)
Mosek	http://www.mosek.com/products/mosek	Free (Academic)
LP Solve	http://lpsolve.sourceforge.net/5.5/	Free
TOMLAB CPLEX	http://tomopt.com/tomlab/products/cplex/	Paid
Lindo	http://www.lindo.com/	Paid

```
>> initCobraToolbox
```

If the installation has been successful, an output such as the following must be observed (the installation below uses Gurobi 6):

```
>> initCobraToolbox
Define LP solver...
LP solver: gurobi6

Define MILP solver...
MILP solver: gurobi6

Define QP solver...
QP solver: gurobi6

Define MIQP solver...
MIQP solver: gurobi6

Define CB map output...
CB map output: svg

TranslateSBML worked with the test .xml file: Ecoli_core_ECO-
SAL.xml
```

2.1.2 Fast-SL

Installation. Fast-SL is available for free and can be downloaded from GitHub (https://github.com/RamanLab/FastSL/archive/master.zip) and unzipped to a folder FastSL-master. Fast-SL is built for MATLAB, and can therefore run on all platforms with a MATLAB installation (and a suitable LP solver, alongside the COBRA toolbox). All the required functions in the downloaded folder "FastSL-master" can be added in MATLAB (i.e., including the functions in MATLAB's search path) using following command:

```
>> addpath(genpath('FastSL-master/'))
```

2.2 Genome-Scale Metabolic Models (GEMs)

GEMs contains all known metabolic reactions in an organism and their associated enzymes [1, 2, 8]. The available genomic and proteomic data are used to identify the metabolic reactome of the organism. A metabolic *reactome* ("reaction" + "omics") is a collection of all known biochemical metabolic reactions and pathways that occur in an organism. GEMs have been developed for many prokaryotes, and are available for several eukaryotic organisms including yeast, green algae, and mammals [23]. Table 2 lists some example resources from which GEMs may be obtained for simulation.

Table 2
List of example resources for obtaining GEMs. Statistics are up-to-date as of 1st November 2016

Resource	Website	Remarks	Reference(s)
BiGG	http://bigg.ucsd.edu/	Hosts 79 models	[24]
MetaNetX	http://www.metanetx.org/	Hosts 163 models from ModelSeed, Path2Models, conforming to a uniform namespace	[25]
UCSD Organisms Page	http://systemsbiology.ucsd.edu/InSilicoOrganisms/OtherOrganisms	Links to 160 existing published models	[23]
ModelSeed	http://modelseed.org/	Large number of models derived via KEGG/MetaCyc/SEED	[26]
Path2Models	https://www.ebi.ac.uk/biomodels-main/path2models	Can automatically convert from KEGG to Systems Biology Markup Language (SBML) that is compatible with the COBRA toolbox	[27]
MEMOSys	https://memosys.i-med.ac.at/MEMOSys/home.seam	Hosts 20 models of 16 different organisms	[28]
BioMet Toolbox	http://biomet-toolbox.org/index.php?page=models	Hosts eight fungal and six bacterial models online, along with tools for analyses	[29]
KEGG	http://www.genome.jp/kegg/	Classic resource on metabolic reactions and pathways	[30]
Pathway Tools	http://pathwaytools.org/	Classic resource for metabolic reconstruction and construction of organism-specific databases	[31]

Below, we detail the important components of GEMs [32]:

- *Reactions*: In a GEM, reactions are identified from the enzymatic information obtained from the annotated genome sequence and other proteomic data of target organism. These reactions include those happening within the cell (internal reactions), transport reactions (that shuttle metabolites between different cellular compartments), extracellular reactions such as uptake reactions (e.g., uptake of different media components), and exchange reactions (metabolite exchange with the extracellular environment).
- *Metabolites*: This is an exhaustive list of metabolites participating in the metabolic reactions mentioned above.
- *Stoichiometric Matrix*: The information on all the reactions present in the model is represented in the form of a stoichiometric matrix, usually denoted as $S_{m \times r}$, where r and

m represent the number of reactions and the number of metabolites participating in these reactions respectively. The individual values, $s_{ij} \in \mathbf{S}$ are the stoichiometric coefficients of a particular metabolite i, occurring in a reaction j; these coefficients are negative for reactants and positive for products.

- *Bounds*: The reaction bounds, i.e., lower and upper bounds, suggest the reaction directionality and also specify the media conditions via uptake reactions. Typically, for internal reactions, the bounds are set to a large positive value (e.g., +1000) as an upper bound, and a large negative value (e.g., −1000) or zero as the lower bound, depending on whether the reaction is reversible or irreversible, respectively. For exchange and transport reactions, the bounds are indicative of whether the metabolite is taken up from the environment, secreted out or can go both ways.
- *Biomass reaction*: The biomass composition of an organism is incorporated in a GEM as a "biomass reaction." This reaction is set up in such a way that the flux through this reaction equals the exponential phase growth-rate of an organism [7]. The biomass reaction accounts for known biomass constituents of an organism, with stoichiometry indicative of contributions to overall cellular biomass. The reaction typically includes macromolecules such as nucleic acids, amino acids, lipids, and carbohydrates that are essential for an organism's survival.
- *Gene–Protein–Reaction Associations*: Perturbations to a model are commonly made via reaction deletions. This involves constraining the flux through a particular reaction to zero. For gene deletions, a set of reactions are constrained to zero, based on the associations between genes and reactions. These associations, known as gene–protein–reaction (GPR) associations, encode the relationship(s) between genes, proteins and how they catalyze reactions, in the form of Boolean rules. This is evidently a many-to-many mapping. Although GPR associations are not needed for predicting synthetic lethal reaction sets (Subheading 3.5), they are essential for predicting gene essentiality or synthetic lethal gene sets (Subheading 3.6).
- *ATP Maintenance*: Many FBA models contain additional constraints for non-growth-associated maintenance (NGAM) and growth-associated maintenance (GAM), which essentially check if the model can additionally support the synthesis of a certain amount of ATP for non-metabolic activities. This experimentally determined value, when fit to the GEMs, often improves the growth rate predictions. For example, the *E. coli* model *i*AF1260 uses an NGAM value of 8.39 mmol ATP $g\text{DW}^{-1}\ \text{h}^{-1}$ and a GAM value of 59.81 mmol ATP $g\text{DW}^{-1}$, corresponding to the best fit with experimental data [33].

- *Other information*: A GEM may additionally include other information such as reaction subsystems, standard enzyme and metabolite names, charges on metabolites and compartments in the cell.

For a more extended account of GEMs, we refer the reader elsewhere [23, 34].

To perform synthetic lethality analysis, the model must accurately describe at least the following: (a) the stoichiometry of each reaction, (b) the appropriate medium for growth, in terms of the uptake reactions, (c) upper and lower bounds for each reaction, (d) a biomass objective function, and (e) the GPR relationships (not necessary to infer synthetic reaction lethals).

3 Methods

In this section, we explain how one can use Fast-SL to predict synthetic lethals in a given metabolic model. As an exemplar, we use the popular *E. coli* model *i*AF1260 [33]. At this point, it is expected that the installation of MATLAB, the COBRA toolbox, a suitable LP solver and Fast-SL are all complete. We then begin with a COBRA toolbox-compatible SBML model (Subheading 3.1), verify the model's usability by performing a simple FBA (Subheading 3.2), and identify reactions that must not be deleted from the model at any point (Subheading 3.3). Following this, we explain the Fast-SL methodology (Subheading 3.4) and illustrate the application of Fast-SL on the *E. coli* *i*AF1260 model for predicting synthetic lethal reaction sets (Subheading 3.5) and synthetic lethal gene sets (Subheading 3.6).

3.1 Loading the Metabolic Model

GEMs are commonly exchanged in SBML format [35]. The information about various contents of the GEM (Subheading 2.1.1) is encoded differently depending on the software/toolbox requirements. Many reconstructions comply with the specifications laid down by the COBRA toolbox, which can be described as follows.

The SBML model of a metabolic network can be divided into three major sections. The first part contains the general "header" information of about the reconstructed model such as SBML version, organism name, the unit definitions, and the model compartments.

```
<?xml version="1.0" encoding="UTF-8"?>
<sbml xmlns="http://www.sbml.org/sbml/level2" level="2"
version="1" xmlns:html="http://www.w3.org/1999/xhtml">
    <model id="MODELID_3307911" name="E. coli iAF1260">
        <listOfUnitDefinitions>
            <unitDefinition id="mmol_per_gDW_per_hr">
```

```
                    <listOfUnits>
                     <unit kind="mole" scale="-3" multiplier="1"
offset="0" />
                                   <unit kind="gram" exponent="-1"
multiplier="1" offset="0" />
                                 <unit kind="second" exponent="-1"
multiplier="0.00027777" offset="0" />
          </listOfUnits>
        </unitDefinition>
      </listOfUnitDefinitions>
      <listOfCompartments>
        <compartment id="C_e" name="Extraorganism" />
        <compartment id="C_c" name="Cytosol" />
        <compartment id="C_p" name="Periplasm" />
      </listOfCompartments>
```

The second section has the information about all the chemical species that model can recognize such as namc, species ID, formula, charge, compartment, etc.

```
<listOfSpecies>
      <species id="M_13dpg_c" name="_3-Phospho-D-glyceroyl
phosphate" compartment="C_c" charge="-4"
boundaryCondition="false">
         <notes>
                    <html xmlns="http://www.w3.org/1999/
xhtml">FORMULA: C3H4O10P2
              </html>
         </notes>
     </species>
       <species id="M_3pg_c" name="_3-Phospho-D-glycerate"
compartment="C_c" charge="-3" boundaryCondition="false">
         <notes>
                    <html xmlns="http://www.w3.org/1999/
xhtml">FORMULA: C3H4O7P
              </html>
         </notes>
     </species>
      <species id="M_6pgc_c" name="_6-Phospho-D-gluconate"
compartment="C_c" charge="-3" boundaryCondition="false">
         <notes>
                    <html xmlns="http://www.w3.org/1999/
xhtml">FORMULA: C6H10O10P
              </html>
         </notes>
     </species>
```

The third section is the list of all reactions encoded in the model with details of name, id, reactants, products, gene associations, subsystem, parameters, lower bound, upper bound, objective coefficient, etc.

```
<reaction id="R_ACALD" name="acetaldehyde dehydrogenase
(acetylating)" reversible="true">
    <notes>
        <html:p>Abbreviation: R_ACALD</html:p>
        <html:p>Synonyms: _0</html:p>
        <html:p>EC Number: 1.2.1.10</html:p>
        <html:p>SUBSYSTEM: Pyruvate Metabolism</html:p>
        <html:p>Equation: [c] : acald + coa + nad &lt;==&gt;
accoa + h + nadh</html:p>
        <html:p>Confidence Level: 0</html:p>
        <html:p>GENE ASSOCIATION: (b1241) or (b0351)</html:
p>
    </notes>
    <listOfReactants>
        <speciesReferencespecies="M_acald_c"stoichiometry="1"
/>
        <speciesReference species="M coa_c" stoichiometry="1" />
        <speciesReference species="M_nad_c" stoichiometry="1" />
    </listOfReactants>
    <listOfProducts>
        <speciesReference species="M_accoa_c" stoichiometry="1" />
        <speciesReference species="M_h_c" stoichiometry="1" />
        <speciesReference species="M_nadh_c" stoichiometry="1" />
    </listOfProducts>
    <kineticLaw>
        <mathxmlns="http://www.w3.org/1998/Math/MathML">
        <ci>FLUX_VALUE</ci>
        </math>
    <listOfParameters>
        <parameter id="LOWER_BOUND" value="-999999" uni-
ts="mmol_per_gDW_per_hr" />
        <parameter id="UPPER_BOUND" value="999999" uni-
ts="mmol_per_gDW_per_hr" />
        <parameter id="OBJECTIVE_COEFFICIENT" value="0" />
        <parameter id="FLUX_VALUE" value="0" units="mmol_-
per_gDW_per_hr" />
    </listOfParameters>
  </kineticLaw>
</reaction>
```

Although this format was followed by several genome-scale models, many discrepancies such as inconsistencies in the nomenclature, absence of critical information were observed. In order to streamline the reconstructions, Flux Balance Constraints (FBC) package [36] was introduced, which prescribes several guidelines to represent the contents of GEMs. More details about the FBC package can be found at SBML.org (http://sbml.org/Documents/Specifications/SBML_Level_3/Packages/Flux_Balance_Constraints_(flux)).

The SBML model compatible with COBRA Toolbox is loaded using the following command:

```
>> model=readCbModel('ecoli_iAF1260.xml')
```

Note that the latest version of COBRA Toolbox (for MATLAB) supports reading SBML-FBC v2 files too (command is the same as above). Once the model is loaded we can see the following output on the screen:

```
model =

   modelVersion: [1x1 struct]
                rxns: {2382x1 cell}
                mets: {1668x1 cell}
                   S: [1668x2382 double]
                 rev: [2382x1 double]
                   c: [2382x1 double]
            metNames: {1668x1 cell}
                  lb: [2382x1 double]
                  ub: [2382x1 double]
         metFormulas: {1668x1 cell}
           metCharge: [1668x1 double]
               rules: {2382x1 cell}
               genes: {1261x1 cell}
          rxnGeneMat: [2382x1261 double]
             grRules: {2382x1 cell}
            rxnNames: {2382x1 cell}
          metCHEBIID: {1668x1 cell}
           metKEGGID: {1668x1 cell}
        metPubChemID: {1668x1 cell}
      metInChIString: {1668x1 cell}
                   b: [1668x1 double]
         description: 'iAF1260.xml'
```

A detailed description of all the above fields is available in the documentation of the COBRA toolbox, but it is easy to identify most of the common "components" of the GEM here, namely reaction IDs (`rxns`), metabolite IDs (`mets`), stoichiometric matrix (`S`), reversibility (`rev`), the objective function (`c`), lower bounds of reactions (`lb`), and upper bounds of reactions (`ub`). Note that if some of these key components of the model are not correctly readable from the SBML file (*see* **Note 2**), it would not be possible to run further simulations.

3.2 Flux Balance Analysis

Reaction stoichiometry and other physico-chemical constraints limit the various states an organism's phenotype can take. FBA uses a constraint-based approach to predict the phenotypic behavior of an organism [5, 6], and is based on LP. FBA predicts viability of a phenotype by measuring the flow of metabolites through the metabolic network [7]. FBA has also been proven to efficiently predict the effects of single and multiple gene knockouts in an organism with reliable accuracy [37, 38]. A more detailed explanation of FBA can be found elsewhere [6, 7, 33, 34]. **Note 3** presents a brief discussion of the validity of FBA predictions, and pointers to better understand model mis-predictions.

It is important to note that the gene essentiality and synthetic lethality are medium-specific, i.e., a gene may not be essential or lethal under certain growth conditions. Therefore, it is important to first identify the growth medium in which we wish to predict essential and synthetic lethal sets and set the upper and lower bounds for reactions appropriately. The solution for the FBA problem can be obtained by using the COBRA toolbox [20]. Before performing an FBA, it is important to initialize the COBRA toolbox, as described in Subheading 2.1.1, using the `initCobraToolbox` command. If the toolbox is successfully initialized, with appropriate solvers, we can perform an FBA using the following command:

```
>> FBAsolution = optimizeCbModel(model,'max','one')

FBAsolution =
           x: [2382x1 double]
           f: 0.7367
           y: []
           w: []
        stat: 1
    origStat: 'OPTIMAL'
      solver: 'gurobi6'
        time: 0.1233
```

The important components of the "FBAsolution" returned above are *x*, the optimal flux distribution for the given model, *f*, the predicted growth rate. Other components give information on the status of the simulation (feasible/infeasible/etc.). An *f* value of zero at this point is an indicator that the model is not being simulated correctly; it is therefore necessary to carefully examine the model for incorrect bounds, errors in metabolites/stoichiometry, or an incorrect biomass function. Also, *see* **Note 2**.

3.3 Identifying Uptake, Exchange, and Dead-End Reactions

Fast-SL is primarily designed to speed up the process of identifying synthetic lethal sets. Identifying and eliminating reactions that may not necessarily belong to a set of synthetic lethal sets, will help reduce the initial search space. To identify the set of dead-end reactions, exchange, and uptake reactions, we can use the removeDeadEnds and findExcRxns functions from the COBRA Toolbox:

```
>> [model,removedMets,removedRxns] = removeDeadEnds(model)
```

The above command removes all the dead-end metabolites or reactions, which either participate in only one reaction or can only be produced or only consumed. Following output occurs on the console along with model:

```
removedMets =
    '14glucan[c]'
    '14glucan[e]'
     '14glucan[p]'

      ...
removedRxns =
        '14GLUCANabcpp'
        '14GLUCANtexi'
        '2DGLCNRx'
      ...
```

Exchange and nutrient uptake reactions can be found by the use of the command,

```
>> [selExc,selUpt] = findExcRxns(model)
```

The output consists of two Boolean vectors indicating whether each reaction in model is exchange or nutrient uptake respectively.

Another reaction crucial for maintaining the overall energy balance in a metabolic model is the non-growth-associated "ATP maintenance" reaction, often referred to as "ATPm." The ATPm reaction is an essential reaction in terms of the organism's survival, and for accurate predictions of growth, as noted earlier (Subheading 2.1.1). In order to account for this, ATPm is added to the list of excluded/eliminated reactions (along with uptake, exchange, and dead-end reactions) from the list of reactions we wish to find synthetic lethal sets. For the sake of convenience, we will refer to this as "*eliList*," i.e., the list of reactions eliminated during identification of synthetic lethal sets, for the rest of this chapter.

3.4 Predicting Synthetic Lethals In Silico: The Fast-SL Algorithm

As discussed in Subheading 3.2, the in silico simulation of a metabolic model through FBA involves solving an LP problem. To simulate the effects of gene deletions on the growth of an organism, we first identify the reactions that are affected by this gene deletion, using the GPR associations encoded in the model. We then remove the reactions that are affected by the gene deletion from the model and re-solve the FBA problem to predict a new growth rate of the organism under the gene deletion. By comparing the wild-type growth rate and the mutant growth rate after gene deletion, we can predict the phenotype of the organism when the corresponding gene is deleted. Often, studies use either a 1% or 5% of wild-type growth rate as a cutoff for the mutant growth rate [12, 33, 39], i.e., if FBA predicts a maximum growth rate less than this cut-off value, then this gene is considered essential for organism's survival ("lethal phenotype"). A similar procedure is incorporated to study synthetic lethality using FBA, i.e., we remove all the reactions associated with two or more genes, observe the maximum growth-rate using FBA and compare it with the cut-off value. If deleting the genes individually does not result in a lethal phenotype, but deleting both the genes simultaneously results in a significant reduction of maximum growth rate predicted, we consider those genes to be synthetic lethals.

In order to enumerate all synthetic lethal genes, many studies suggested an exhaustive enumeration approach [4, 16, 17]. In the exhaustive enumeration approach, each gene is explicitly removed from the metabolic model, followed by computing the growth-rate and consequently, identifying the essential genes. This process is repeated for all two-gene combinations to identify synthetic lethal gene pairs and all three-gene combinations to identify synthetic lethal triplets and so on. The main drawback of this approach is that it becomes expensive in terms of time and computation for models with a large number of genes due to combination explosion. However, this method can be easily parallelized, as the simulations for two different sets of knockouts can be performed independently (and simultaneously). Still, the number of simulations to be performed can be prohibitively large: for the *i*AF1260 model with 1260 genes, the number of triple gene deletions that need to be performed exceeds 200 million [19].

Alternative approaches to identify synthetic lethals in metabolic models use a bi-level mixed-integer linear programming (MILP) formulation of FBA problem [12]. While the MILP-based methods avoid the combinatorial explosion of exhaustive enumeration, solving MILPs for large models can be difficult due to their NP-hard nature. In addition, due to the nature of the MILP formulation, these methods are not parallelizable and we can only identify one synthetic lethal set at a time. Therefore, in this chapter, we focus on Fast-SL [19], a method that follows the idea of exhaustive enumeration, but dramatically reduces the number of combinations to be evaluated for predicting essentiality and synthetic lethality.

Fast-SL, while maintaining the inherent parallelism of the exhaustive enumeration approach, identifies and eliminates non-lethal sets without the need for reformulating the FBA problem as a bi-level MILP. GEMs contain a large number of reactions, not all of which may be active under all conditions. Fast-SL identifies a sparse solution to the FBA LP; any reaction that does not carry a flux in this solution cannot be lethal to the organism. This is because if there is a flux distribution ("sparse solution" mentioned above) that supports growth with some reactions carrying no flux, deleting those reactions from the model will not affect growth. Fast-SL successively computes the following sets of reactions:

1. J_{sl}, the set of single lethal reactions,
2. $J_{dl} \subset J^2$, the set of synthetic lethal reaction pairs, and
3. $J_{tl} \subset J^3$, the set of synthetic lethal reaction triplets

At the core of computing these sets is the sparse solution to FBA, which involves computing a flux distribution corresponding to maximum growth rate, while minimizing the sum of absolute values of the fluxes, i.e., the L_1-norm of the flux vector. In other words, we compute the "minimum norm" solution to the FBA LP problem. This is in turn easily reformulated as an LP. Figure 1 further intuits the nature of search space elimination performed by Fast-SL.

Fast-SL offers significant reduction in the search space for identifying essential and synthetic lethal reactions/genes (≈4000-fold in the case of *E. coli*, for synthetic lethal triplets). Further, our results also exactly match with exhaustive simulations performed on

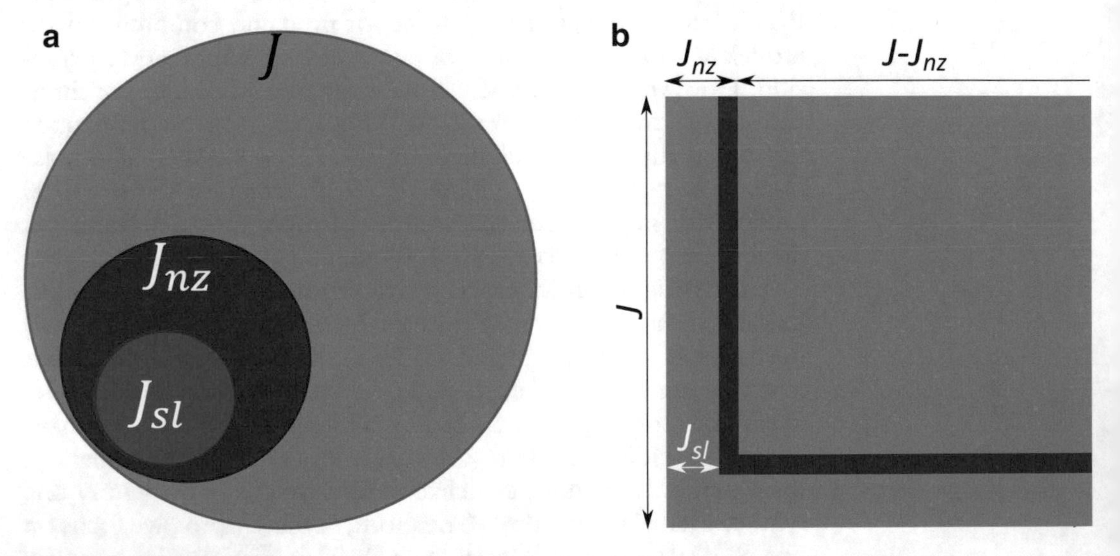

Fig. 1 A representation of the search space eliminations performed by Fast-SL. The Venn diagram in (**a**) indicates the set of all reactions, J, the set of reactions carrying a nonzero flux in the sparse solution (J_{nz}, nz standing for non-zero) and J_{sl}, the set of single lethal reactions. Note that J_{sl} is fully contained in J_{nz}—this is the basis of extending the approach to double lethals, as shown in (**b**). Panel (**b**) represents a matrix of size $|J| \times |J|$. Two reactions both drawn from outside J_{nz} cannot together constitute a synthetic double lethal (by extending the argument for single lethals). All synthetic lethal pairs lie in the narrow "red region" (roughly $\{J_{nz} - J_{sl}\} \times \{J - J_{sl}\}$), drawn to scale for *E. coli*. Even in the narrow red region, further gains are made by re-applying the idea of the minimum norm solution (*see* ref. 19). Thus Fast-SL ends up solving a very small fraction of the LPs solved during exhaustive enumeration, to obtain the same result, by a careful elimination of many non-lethal sets

a computer cluster [19]. Similar to the exhaustive enumeration approach, Fast-SL can also be parallelized (available as `parallelSL` and `parallelSL_Gene` in MATLAB). Fast-SL can identify both synthetic lethal reaction and gene sets. A detailed explanation of the Fast-SL algorithm can be found in the original research paper [19]. The following sections detail the procedure for using Fast-SL to identify synthetic lethal reaction and gene sets, using *E. coli* *i*AF1260 as an example GEM.

3.5 Procedure for Using Fast-SL to Identify Synthetic Lethal Reaction Sets

Fast-SL function that identifies all possible synthetic lethal reaction sets can be called as follows:

```
>> fastSL(model,cutoff,order,eliList,atpm)
```

Note that it is important to initialize the COBRA toolbox (*see* Subheading 2.1.1) before running the above command, failing which Fast-SL will report an error using solveCobraLP "No solver found. call changeCobraSolver(solverName)."

Inputs:

1. *COBRA Metabolic Model structure*: The COBRA toolbox representation of a GEM (a MATLAB structure) is required as an input for Fast-SL reaction formulation. This is most easily obtained by parsing a compatible SBML file corresponding to the model available from databases such as those listed in Table 2 or published literature. Following fields in the model are required for simulations; other fields are not essential.

   S—Stoichiometric matrix

   b—Right hand side

   c—Objective coefficients

   lb—Lower bounds

   ub—Upper bounds

   rxns—Reaction names

   description—Model name

Optional Inputs:

1. *cut-off*: Cutoff specifies the percentage value to determine the lethal phenotype. Generally, a phenotype is considered lethal if the maximum growth rate obtained after FBA is less than specified cutoff. Default value is set to 1% of the in silico maximum wild-type growth rate, as used in previous studies [39].

2. *order*: Order specifies the required size of the synthetic lethal sets. Default order is two, corresponding to double lethal sets.
3. *eliList*: eliList is the list of reactions that can/must be ignored for lethality analysis (as described in Subheading 3.3).
4. *atpm*: This variable represents the reaction ID of the ATP maintenance reaction in SBML representation of the GEM. It can either be supplied with the eliminated reactions list (eliList) or separately, and is not considered for synthetic lethality analysis. Generally, it is referred to as "ATPm" in most of the SBML models and is automatically identified, if the name matches.

Output:

After we supply the necessary inputs and invoke the Fast-SL function, the progress bars appear (as shown in Fig. 2), indicating the progress of identifying single and multiple lethal sets.

At the end of the simulation, the following message appears on the screen:

```
 Saving Single, Double and Triple Lethal Reactions List...
Done.
```

The output from Fast-SL is saved in a MATLAB file named `<model.description>_Rxn_lethals.mat`, which contains all the synthetic lethal reaction sets. Here, model.description stands for the entry under the "description" field under the model loaded into the MATLAB workspace. Usually, it is the model name or filename. For the *i*AF1260 model, with the description

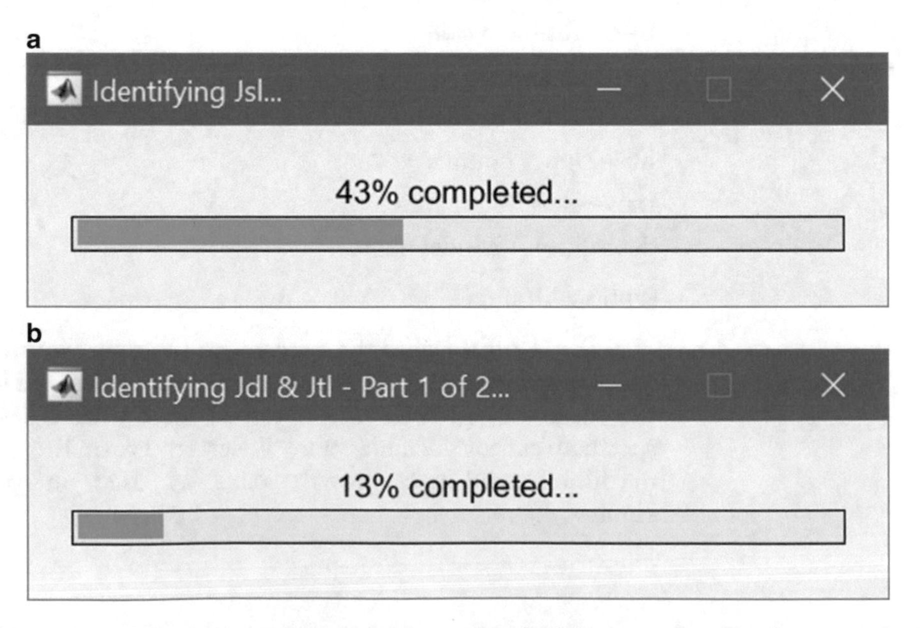

Fig. 2 Screenshots of the progress bars that appear when Fast-SL is running. The figure panels show an example of the progress bars displayed in MATLAB when Fast-SL is computing (**a**) J_{sl} and (**b**) J_{dl} and J_{tl}, respectively

"iAF1260.xml," the results are stored in `iAF1260.xml_Rxn_lethals.mat`, in the current working directory of MATLAB. The names of the cell arrays in the .mat file with the synthetic lethal reaction sets are `Jsl` (single lethal deletions), `Jdl` (double lethal sets), and `Jtl` (triple lethal sets). J_{sl}, J_{dl}, and J_{tl} have also been described in Subheading 3.4.

```
>> Jdl
Jdl =
    'ACKr'          'ACt2rpp'
    'ACKr'          'ACtex'
...
    'TKT2'          'TALA'
    'TKT2'          'TKT1'
>> Jtl
Jtl =
    '5DGLCNR'       'GAPD'          'TPI'
    '5DGLCNR'       'PGK'           'TPI'
...
    'RPE'           'TALA'          'TDPDRR'
    'RPE'           'TALA'          'TDPGDH'
```

The IDs ACKr, 5DGLCNR etc. correspond to the reaction IDs used in the input SBML file/COBRA model structure.

3.6 Procedure for Using Fast-SL to Identify Synthetic Lethal Gene Sets

Synthetic lethal gene deletions can be identified by Fast-SL using the following command, which additionally considers the GPR associations in model:

```
>> fastSLgenes(model, cutoff, order, flag)
```

Inputs:

1. *COBRA Metabolic Model structure*: The COBRA toolbox representation of a GEM (a MATLAB structure) is again required as an input for "Fast-SL gene" formulation. This is most easily obtained by parsing the SBML file corresponding to a model available from databases such as those listed in Table 2 or published literature. Following fields in the model are required for simulations; other fields are not essential.

   S—Stoichiometric matrix

   b—Right hand side

   c—Objective coefficients

   lb—Lower bounds

   ub—Upper bounds

rxns—Reaction names
genes—Gene Names
rules—Gene-Reaction Rules
rxnGenemat—Reactions-gene matrix
description—Model name

Optional Inputs:

1. *cutoff*: specifies the percentage value to determine the lethal phenotype. Default value is set to 1% of the in silico maximum wild-type growth rate.
2. *order*: specifies the required size of the synthetic lethal sets. Default order is two, corresponding to double lethal sets.
3. *flag*: 1 for a more extensive search (to consider unusual GPRs, which are seldom encountered in most GEMs), default is 0.

Output:
Again, progress bars similar to those discussed in Subheading 3.5 and the following message appear at the end of the simulation.

```
Initializing...
Done...
Mapping lethal reactions to genes ...
Done...
Saving Single, Double and Triple Lethal Genes List...
Done.
```

Similar to the output in Subheading 3.5, the output is saved in a MATLAB file named `<model.description>_Gene_lethals.mat`, which contains all synthetic lethal gene sets. The names of the cell arrays in the `.mat` file with the gene deletions are `sgd` (single gene deletions), `dgd` (double gene deletions), `tgd` (triple gene deletions).

```
>> dgd
dgd =
    'b0002'    'b3940'
    'b0048'    'b1606'
...
    'b2905'    'b4388'
    'b3829'    'b4019'
>> tgd
tgd =
    'b0114'    'b1779'    'b3731'
    'b0114'    'b1779'    'b3732'
...
    'b2987'    'b3493'    'b3728'
    'b3089'    'b3653'    'b4067'
```

The IDs b0002 etc. correspond to the gene names used in the input SBML file/COBRA model structure.

4 Notes

1. *Relative performance of various solvers.* Although the results of simulations do not vary with different solvers, the time performance may vary very much. Therefore, it may be important to choose a faster solver to work on larger metabolic networks. Table 3 shows the time taken by different solvers to find the FBA solution for the *E. coli i*AF1260 model, on a desktop computer with a Core-i7 processor, 16 GB RAM, 1 TB Hard Disk and running MATLAB R2015a on Kubuntu 15.04 (computed by averaging the time taken for a single FBA 10^4 times). We observe that the CPLEX solver is faster compared to GLPK or Gurobi.
2. *Quality of the models.* Although 100+ genome-scale reconstructions are available currently, it is not uncommon to find models with missing annotation on metabolites and genes [40]. Further, some models may not be exactly compatible with the COBRA toolbox [41]. As discussed in [41], however, it is often easy to fix the models, by a close examination of the potential sources of error. The most common source of error would likely be a mis-specification of lower and upper bounds, or a lack of precise definition of the medium of growth. This typically would result in a zero growth rate for the model while performing FBA (Subheading 3.3). In such cases, it is necessary to "debug" the model, before proceeding with further analyses, in terms of synthetic lethality prediction etc. Also *see* ref. 42.
3. *Validation of model predictions.* Although FBA has been demonstrated to predict growth rates and phenotypes with high degrees of accuracy (>80% for well-studied organisms such as *Saccharomyces cerevisiae* [43] and *E. coli* [44]), it is not uncommon to find predictions that do not agree with experimental observations [34]. The same paper [34] also discusses in detail the likely causes of model mis-prediction. The model predictions are nevertheless very useful as a starting point for prioritizing further experimental investigations.

Table 3
Comparison of average running time (in seconds) of a single FBA problem for various solvers

Solver	Time to solve FBA (s)
GLPK 2.7	0.1499
Gurobi 6.5.1	0.0410
IBM ILOG CPLEX 12.1.0	0.0045

Acknowledgments

The authors thank Aarthi Ravikrishnan for useful discussions. The use of the high-performance computing facility HPCE at IIT Madras is gratefully acknowledged. This work was supported by the grant BT/PR4949/BRB/10/1048/2012 from the Department of Biotechnology, Government of India, and the Initiative for Biological Systems Engineering (IBSE) at IIT Madras.

References

1. Duarte NC, Becker SA, Jamshidi N et al (2007) Global reconstruction of the human metabolic network based on genomic and bibliomic data. Proc Natl Acad Sci U S A 104:1777–1782. https://doi.org/10.1073/pnas.0610772104
2. Kim HU, Kim SY, Jeong H et al (2011) Integrative genome-scale metabolic analysis of Vibrio vulnificus for drug targeting and discovery. Mol Syst Biol. https://doi.org/10.1038/msb.2010.115
3. Thiele I, Vo TD, Price ND, Palsson BØ (2005) Expanded metabolic reconstruction of Helicobacter pylori (iIT341 GSM/GPR): an in silico genome-scale characterization of single- and double-deletion mutants. J Bacteriol 187:5818–5830. https://doi.org/10.1128/JB.187.16.5818-5830.2005
4. Thiele I, Hyduke DR, Steeb B et al (2011) A community effort towards a knowledge-base and mathematical model of the human pathogen Salmonella Typhimurium LT2. BMC Syst Biol. https://doi.org/10.1186/1752-0509-5-8
5. Varma A, Palsson BØ (1994) Stoichiometric flux balance models quantitatively predict growth and metabolic by-product secretion in wild-type Escherichia coli W3110. Appl Environ Microbiol 60:3724–3731
6. Kauffman KJ, Prakash P, Edwards JS (2003) Advances in flux balance analysis. Curr Opin Biotechnol 14:491–496. https://doi.org/10.1016/j.copbio.2003.08.001
7. Orth JD, Thiele I, Palsson BØ (2010) What is flux balance analysis? Nat Biotechnol 28:245–248. https://doi.org/10.1038/nbt.1614
8. Edwards JS, Palsson BØ (2000) The Escherichia coli MG1655 in silico metabolic genotype: its definition, characteristics, and capabilities. Proc Natl Acad Sci U S A 97:5528–5533
9. Raman K, Rajagopalan P, Chandra N (2005) Flux balance analysis of mycolic acid pathway: targets for anti-tubercular drugs. PLoS Comput Biol 1:e46. https://doi.org/10.1371/journal.pcbi.0010046
10. Hartman HB, Fell DA, Rossell S et al (2014) Identification of potential drug targets in Salmonella enterica sv. Typhimurium using metabolic modelling and experimental validation. Microbiology 160:1252–1266. https://doi.org/10.1099/mic.0.076091-0
11. Alper H, Jin Y-S, Moxley JF, Stephanopoulos G (2005) Identifying gene targets for the metabolic engineering of lycopene biosynthesis in Escherichia coli. Metab Eng 7:155–164. https://doi.org/10.1016/j.ymben.2004.12.003
12. Suthers PF, Zomorrodi A, Maranas CD (2009) Genome-scale gene/reaction essentiality and synthetic lethality analysis. Mol Syst Biol 5:301. https://doi.org/10.1038/msb.2009.56
13. Lee D-S, Burd H, Liu J et al (2009) Comparative genome-scale metabolic reconstruction and flux balance analysis of multiple Staphylococcus aureus genomes identify novel antimicrobial drug targets. J Bacteriol 191:4015–4024. https://doi.org/10.1128/JB.01743-08
14. Sigurdsson G, Fleming RMT, Heinken A, Thiele I (2012) A systems biology approach to drug targets in Pseudomonas aeruginosa biofilm. PLoS One 7:e34337+. https://doi.org/10.1371/journal.pone.0034337
15. Raghunathan A, Reed J, Shin S et al (2009) Constraint-based analysis of metabolic capacity of Salmonella typhimurium during host-pathogen interaction. BMC Syst Biol 3:38. https://doi.org/10.1186/1752-0509-3-38
16. Deutscher D, Meilijson I, Kupiec M, Ruppin E (2006) Multiple knockout analysis of genetic robustness in the yeast metabolic network. Nat Genet 38:993–998. https://doi.org/10.1038/ng1856
17. Henry CS, Xia F, Stevens R (2009) Application of high-performance computing to the

reconstruction, analysis, and optimization of genome-scale metabolic models. J Phys Conf Ser 180:12025. https://doi.org/10.1088/1742-6596/180/1/012025

18. von Kamp A, Klamt S (2014) Enumeration of smallest intervention strategies in genome-scale metabolic networks. PLoS Comput Biol 10:e1003378. https://doi.org/10.1371/journal.pcbi.1003378
19. Pratapa A, Balachandran S, Raman K (2015) Fast-SL: an efficient algorithm to identify synthetic lethal sets in metabolic networks. Bioinformatics 31:3299–3305. https://doi.org/10.1093/bioinformatics/btv352
20. Becker SA, Feist AM, Mo ML et al (2007) Quantitative prediction of cellular metabolism with constraint-based models: the COBRA Toolbox. Nat Protoc 2:727–738. https://doi.org/10.1038/nprot.2007.99
21. Schellenberger J, Que R, Fleming RMT et al (2011) Quantitative prediction of cellular metabolism with constraint-based models: the COBRA Toolbox v2.0. Nat Protoc 6:1290–1307. https://doi.org/10.1038/nprot.2011.308
22. Ebrahim A, Lerman JA, Palsson BØ, Hyduke DR (2013) COBRApy: COnstraints-Based Reconstruction and Analysis for Python. BMC Syst Biol 7:74. https://doi.org/10.1186/1752-0509-7-74
23. Feist AM, Herrgård MJ, Thiele I et al (2009) Reconstruction of biochemical networks in microorganisms. Nat Rev Microbiol 7:129–143. https://doi.org/10.1038/nrmicro1949
24. Schellenberger J, Park J, Conrad T, Palsson B (2010) BiGG: a Biochemical Genetic and Genomic knowledgebase of large scale metabolic reconstructions. BMC Bioinformatics. https://doi.org/10.1186/1471-2105-11-213
25. Ganter M, Bernard T, Moretti S et al (2013) MetaNetX.org: a website and repository for accessing, analysing and manipulating metabolic networks. Bioinformatics 29:815–816. https://doi.org/10.1093/bioinformatics/btt036
26. Henry CS, DeJongh M, Best AA et al (2010) High-throughput generation, optimization and analysis of genome-scale metabolic models. Nat Biotechnol 28:977–982. https://doi.org/10.1038/nbt.1672
27. Büchel F, Rodriguez N, Swainston N et al (2013) Path2Models: large-scale generation of computational models from biochemical pathway maps. BMC Syst Biol 7:116. https://doi.org/10.1186/1752-0509-7-116
28. Pabinger S, Rader R, Agren R et al (2011) MEMOSys: bioinformatics platform for genome-scale metabolic models. BMC Syst Biol. https://doi.org/10.1186/1752-0509-5-20
29. Garcia-Albornoz M, Thankaswamy-Kosalai S, Nilsson A et al (2014) BioMet Toolbox 2.0: genome-wide analysis of metabolism and omics data. Nucleic Acids Res 42:W175–W181. https://doi.org/10.1093/nar/gku371
30. Kanehisa M, Goto S (2000) Kyoto encyclopedia of genes and genomes. Nucleic Acids Res 28:27–30. https://doi.org/10.1093/nar/28.1.27
31. Karp PD, Paley SM, Krummenacker M et al (2010) Pathway Tools version 13.0: integrated software for pathway/genome informatics and systems biology. Brief Bioinform 11:40–79. https://doi.org/10.1093/bib/bbp043
32. Thiele I, Palsson BØ (2010) A protocol for generating a high-quality genome-scale metabolic reconstruction. Nat Protoc 5:93–121. https://doi.org/10.1038/nprot.2009.203
33. Feist AM, Henry CS, Reed JL et al (2007) A genome-scale metabolic reconstruction for *Escherichia coli* K-12 MG1655 that accounts for 1260 ORFs and thermodynamic information. Mol Syst Biol 3:121. https://doi.org/10.1038/msb4100155
34. Joyce AR, Palsson BØ (2008) Predicting gene essentiality using genome-scale in silico models. Methods Mol Biol 416:433–457. https://doi.org/10.1007/978-1-59745-321-9_30
35. Finney A, Doyle JC, Kitano H et al (2004) Evolving a lingua franca and associated software infrastructure for computational systems biology: the Systems Biology Markup Language (SBML) project. Syst Biol (Stevenage) 1:41–53. https://doi.org/10.1049/sb:20045008
36. Olivier BG, Bergmann FT (2015) Flux balance constraints, Version 2 Release 1
37. Edwards JS, Palsson BØ (2000) Metabolic flux balance analysis and the in silico analysis of Escherichia coli K-12 gene deletions. BMC Bioinformatics 1:1
38. Harrison R, Papp B, Pal C et al (2007) Plasticity of genetic interactions in metabolic networks of yeast. Proc Natl Acad Sci U S A 104:2307–2312. https://doi.org/10.1073/pnas.0607153104
39. Deutscher D, Meilijson I, Schuster S, Ruppin E (2008) Can single knockouts accurately single out gene functions? BMC Syst Biol 2:50. https://doi.org/10.1186/1752-0509-2-50
40. Ravikrishnan A, Raman K (2015) Critical assessment of genome-scale metabolic

networks: the need for a unified standard. Brief Bioinform 16:1057–1068. https://doi.org/10.1093/bib/bbv003

41. Ebrahim A, Almaas E, Bauer E et al (2015) Do genome-scale models need exact solvers or clearer standards? Mol Syst Biol 11:831. 10.15252/msb.20156157
42. Basler G (2015) Computational prediction of essential metabolic genes using constraint-based approaches. 1279 pp 183–204
43. Duarte NC, Herrgård MJ, Palsson BØ (2004) Reconstruction and validation of Saccharomyces cerevisiae iND750, a fully compartmentalized genome-scale metabolic model. Genome Res 14:1298–1309. https://doi.org/10.1101/gr.2250904
44. Covert MW, Palsson BØ (2002) Transcriptional regulation in constraints-based metabolic models of Escherichia coli. J Biol Chem 277:28058–28064. https://doi.org/10.1074/jbc.M201691200

Chapter 15

Coupling Fluxes, Enzymes, and Regulation in Genome-Scale Metabolic Models

Paul A. Jensen

Abstract

Genome-scale models have expanded beyond their metabolic origins. Multiple modeling frameworks are required to combine metabolism with enzymatic networks, transcription, translation, and regulation. Mathematical programming offers a powerful set of tools for tackling these "multi-modality" models, although special attention must be paid to the connections between modeling types. This chapter reviews common methods for combining metabolic and discrete logical models into a single mathematical programming framework. Best practices, caveats, and recommendations are presented for the most commonly used software packages. Methods for troubleshooting large sets of logical rules are also discussed.

Key words Metabolic modeling, Flux balance analysis, Mixed-integer programming, MILP, Transcriptional regulation, Boolean networks, Constraint-based modeling

1 Introduction

Constraint-based reconstruction and analysis (COBRA) models were developed to compute and analyze the fluxes through metabolic reactions [1]. COBRA models are typically solved using linear programming (LP), a well-developed and robust form of mathematical programming [1]. LP models assume variables (reaction fluxes) are continuous, but enzymes and transcription factors that regulate metabolism are often treated as discrete—either "on" or "off." Early studies on COBRA models solved the continuous and discrete networks separately [2]. Boolean rules describing regulatory interactions and enzyme availability were evaluated to determine which reactions could carry flux. Only these reactions were used in a subsequent LP to calculate a flux distribution by optimizing a metabolic objective. Separating the continuous and discrete networks simplifies the model solution process; however, it also restricts the types of questions that can be asked about a COBRA model. Questions like "what transcription factors should be activated to optimize growth?" or "are any three gene deletions

Marco Fondi (ed.), *Metabolic Network Reconstruction and Modeling: Methods and Protocols*, Methods in Molecular Biology, vol. 1716, https://doi.org/10.1007/978-1-4939-7528-0_15,

synthetically lethal?" require the reaction fluxes and gene states be solved simultaneously [3, 4]. By converting COBRA models from linear programs to mixed integer linear programs (MILPs), both continuous and discrete cellular networks can be interrogated as a single optimization problem [5].

Simultaneously analyzing continuous and discrete networks with MILPs significantly complicates the model solution process [6]. Translating logical rules into linear constraints requires many choices that affect the efficiency and stability of the MILP. The interface between variables for (continuous) reaction fluxes and (discrete) enzymes also requires careful design. Fortunately, software tools are available to automate most of the conversion, but researchers must understand the underlying algorithms that create and solve their models. This chapter explains in detail how to couple continuous and discrete network models.

2 Materials

2.1 COBRA Toolbox

The COBRA toolbox [7] is a collection of Matlab functions for manipulating and simulating with COBRA models. The toolbox is maintained by a community of developers and includes implementations of several published algorithms. A Python port of the COBRA toolbox is available [8], although the Python version includes only a subset of the functionality of the original Matlab toolbox.

Models in the COBRA toolbox are stored as a Matlab structure ("struct") with fields containing the stoichiometric matrix, metabolic objective, flux bounds, and names of the metabolites and reactions. The structure also includes a list of the model's genes and the gene-protein-reaction (GPR) relationships. The GPRs are stored as character strings in two ways: first as the original rule from the model file (the rules field), and second as a binary expression that can be evaluated by Matlab (the grRules field). For example, in the metabolic model iJB1441 of *Burkholdheria multivorans* [9], reaction rJB00022 (glucose 1-dehyrogenase) is catalyzed by the products of either of two genes, Bmul_0907 or Bmul_1003. The corresponding entry in rules is

```
(Bmul_0907) or (Bmul_1003)
```

while the entry in grRules is

```
(x(258)) | (x(277))
```

When a COBRA model is loaded, a list of unique genes is extracted from the GPRs and stored (in alphabetical order) in the field genes. To create the expressions in grRules, two transformations are applied to the strings in rules. First, the words "and" and "or" are substituted with the equivalent Matlab operators "&" and "|." Second, the gene names are replaced with the string "x(i)," where *i* is the index of the gene name in the model's gene list (the field genes). In the example above, gene Bmul_0907 is gene 258 in the iJB1441 model, and gene Bmul_1003 is gene 277.

To simulate a gene deletion with the COBRA toolbox, the deleteModelGenes function first creates a logical vector named *x*. The number of entries in *x* is equal to the number of genes in the model, and all entries are initially set to one. The entries for the genes to be deleted are then set to zero, and each string in grRules is evaluated. The resulting values indicate if a reaction should be allowed to carry flux in the model. If the GPR is false, the reaction bounds are set to zero.

It is important to note that the COBRA toolbox's approach to gene deletions does not allow multiple genes to be deleted sequentially. The COBRA model does not "remember" which genes have been previously deleted, and every invocation of deleteModelGenes evaluates the GPRs using only the genes listed in the present call. To illustrate this caveat, consider the above GPR for *B. multivorans*. If both Bmul_0907 and Bmul_1003 are deleted, reaction rJB00022 should not be able to carry flux. Indeed, if deleteModelGenes is called with both genes, the GPR evaluates to false since the entries in *x* for both Bmul_0907 and Bmul_1003 are set to zero. If instead the user tried to delete Bmul_0907 first and then delete Bmul_1003 with a subsequent call to deleteModelGenes, reaction rJB00022 would remain active. In the first call to delete Bmul_0907, the GPR for rJB00022 evaluates to one, since the reaction requires only one of the two associated enzymes. In the second call with Bmul_1003, the GPR still evaluates to one, since there is no memory in the COBRA model that Bmul_0907 was previously deleted. When deleting multiple genes with the COBRA toolbox, all genes must be deleted in a single call to deleteModelGenes. If the user wishes to delete subsequent genes, they must store the list of previously deleted genes and call deleteModelGenes with all of the genes that should be missing from the model.

The COBRA toolbox makes a clear distinction between genes and reactions. Genetic manipulations are separate from model simulations. Deleting model genes uses GPR strings to identify reactions that should not carry flux. The bounds for these reactions are set to zero to enforce the effects of the gene deletion on subsequent simulations. The user is responsible for tracking the gene deletions that have been applied to the model to ensure correct results when deleting new genes.

2.2 TIGER Toolbox

The TIGER toolbox [10] is a Matlab package that extends the features of the COBRA toolbox, particularly the handling of GPRs, logical constraints, and transcriptional regulation. In TIGER models, both reactions and genes appear as variables, and additional variables can be added to represent transcription factors, signaling molecules, and other non-metabolic processes. These values of reaction fluxes and genes are solved simultaneously in TIGER models, allowing algorithms to interrogate both the genetic and metabolic states.

Similar to COBRA toolbox models, TIGER models are stored as Matlab "structs." Unlike COBRA toolbox models, TIGER models are, by default, stored as mixed-integer linear programs (MILPs), rather than as linear programs (LPs). The discrete variables in the MILP are used to encode the Boolean logic of the GPRs and other regulatory constraints.

TIGER models are created from COBRA toolbox models using the cobra_to_tiger function. The stoichiometric matrix, reaction bounds, objective function, and reaction and metabolite names are stored unchanged in the TIGER model structure. TIGER adds other fields to the model structure to support logical constraints. TIGER parses the GPRs into a tree-like intermediate structure; these trees are used to construct MILP constraints that enforce the GPR logic and connect genes to reactions (*see* Subheading 3.1). The rules and grRules fields containing the original GPR strings are copied into the TIGER structure, but they are unused after the model is built.

TIGER models contain more variables than the original COBRA toolbox model. A variable is created for each reaction and gene. A number of intermediate variables are also created. The intermediate variables do not correspond to any biological entity, but are necessary to build linear constraints for the GPRs. All variable names are stored in a field varnames, with the reaction names appearing first (in the order specified in the COBRA rxns field), followed by the genes (in the order specified in the COBRA genes field), and finally the intermediate variables. Reactions are represented by continuous variables, while genes are, by default, binary.

To simulate genetic perturbations with TIGER, users set the bounds of the genes directly. (Users can set bounds by either modifying the lb and ub vectors or by using the set_vars function—the latter is recommended.) To simulate a gene knockout, users set the corresponding gene to zero. If the deletion prevents a reaction from carrying flux (as determined by the GPRs), the constraints in the TIGER model will automatically force the corresponding flux to zero. Unlike the COBRA toolbox, the effects of gene deletions are cumulative. The effect of missing genes is not enforced until the model is solved, so genes can be deleted sequentially.

It is important to remember that COBRA toolbox models are strict subsets of TIGER models. There is no information lost when converting a COBRA toolbox model into a TIGER model, but some information is lost when extracting a COBRA toolbox model from a TIGER model. TIGER allows constraints beyond the Boolean logic of the GPR, and these advanced features cannot be represented in a COBRA toolbox model. TIGER implements only those algorithms that cannot be programmed using COBRA toolbox models. Many other analyses can be performed using the COBRA toolbox before converting to a TIGER model.

Representing both reaction fluxes and genes with a TIGER models allows algorithms to analyze connections between the metabolic and genetic networks [11–13]. However, the flexibility of TIGER models comes with a cost. MILP models usually take longer to solve, and logical constraints can create numerical difficulties [6]. The following section explains how TIGER models are built and how they can be efficiently solved.

2.3 Solvers

Both TIGER and the COBRA toolbox rely on external mathematical programming packages to solve LPs and MILPs. Fortunately, both TIGER and the COBRA toolbox provide a single interface for accessing multiple solvers, so users can easily switch solvers based on performance and availability.

The open source GLPK [14] is a free LP/MILP solver and part of the GNU software project. GLPK is available on all major operating systems and can be used by both TIGER and the COBRA toolbox. While GLPK is a viable option for LPs, it may struggle with large or complex MILPs (*see* **Note 1**).

Two widely used commercial MILP solvers are CPLEX [15] and Gurobi [16]. Both are freely available for academic use after registration. CPLEX and Gurobi are used to solver large MILPs across a variety of disciplines and industries, and both are compatible with TIGER and the COBRA toolbox. Performance of either solver depends on the particular model being solved. A common set of MILP benchmark programs (http://plato.asu.edu/bench.html) includes several models where one solver significantly outperforms the other. When encountering a difficult to solve model, switching between Gurobi and CPLEX may offer a performance boost; it is worth considering both options.

3 Methods

3.1 Boolean Logic

To encode GRPs into MILPs, TIGER uses a variation of the SR-FBA formulism [5]. This section details how general Boolean logic can be encoded by linear constraints of binary variables. Optimizations specific to GPRs are also discussed.

Boolean statements are built using logical conjunctions (x AND y) and logical disjunctions (x OR y). These expressions can be used to create Boolean rules describing relationships between Boolean variables. Rules are constructed using one of two operators: $\Rightarrow$ (implies) and $\Leftrightarrow$ (if and only if). The difference between these operators is subtle but important. Consider the rule $x \Rightarrow z$. If x is true, then z must be true. However, if x is false, z can be either true or false. In the rule $x \Leftrightarrow z$, z is true only when x is true, and false only when x is false.

We begin with the simplest rule, $x \Rightarrow z$, and assume all variables are binary unless stated otherwise. The rule $x \Rightarrow z$ is equivalent to the MILP constraint $z \geq x$, since $x = 1$ forces $z = 1$ and leaves z unconstrained when $x = 0$. Converting the rule $x \Leftrightarrow z$ is almost trivial, as the logic is enforced by the constraint $x = z$. (Constraints of the form $x = z$ are often removed by the solver during pre-solve by substituting x for z in other constraints to improve solution efficiency.)

The rule x AND $y \Rightarrow z$ requires only a single constraint: $z \geq x + y - 1$. In this constraint, z is unconstrained unless both $x = y = 1$, in which case z is forced to also be one. The pattern can be generalized for rules of the form x_1 AND x_2 AND$\cdots$AND $x_k \Rightarrow z$ with the MILP constraint $z \geq x_1 + x_2 + \cdots + x_k - (k - 1)$.

There are two common methods to encode the disjunction x OR $y \Rightarrow z$. The first uses a single constraint: $z \geq (1/2)(x + y)$. Here, z is forced to be nonzero when either x or y is one. The coefficient of $1/2$ prevents the sum $x + y$ from ever forcing z to be greater than one, which is infeasible. Generalizing again, the rule x_1 OR x_2 OR$\cdots$OR $x_k \Rightarrow z$ can be expressed as $z \geq (1/k)(x_1 + x_2 + \cdots + x_k)$. As a second method, the rule x OR $y \Rightarrow z$ can be expressed as two constraints, $z \geq x$ and $z \geq y$. The rule x_1 OR x_2 OR$\cdots$OR $x_k \Rightarrow z$ requires k constraints $z \geq x_i$ for each $i \in \{1, \ldots, k\}$ (*see* **Note 2**).

Converting "if and only if" rules requires additional MILP constraints. For conjunctions, the rule x AND $y \Leftrightarrow z$ is equivalent to the two rules x AND $y \Rightarrow z$ and $z \Rightarrow x$ AND y. The first rule is converted to a constraint as described above. The latter can be enforced with the constraint $x + y \geq 2z$, which requires both $x = y = 1$ when $z = 1$. (By extension, the rule $z \Rightarrow x_1$ AND x_2 AND$\cdots$AND x_k is equivalent to $x_1 + x_2 + \cdots + x_k \geq kz$.) For disjunctions, the rule x OR $y \Leftrightarrow z$ requires the constraints x OR $y \Rightarrow z$ and $z \Rightarrow x$ OR y. The latter can be written as an MILP with the equation $x + y \geq z$ (or $x_1 + x_2 + \cdots + x_k \geq z$ for the general disjunction $z \Rightarrow x_1$ OR x_2 OR$\cdots$OR x_k).

Only two additional transformations are required to convert any Boolean rule into MILP constraints. First, complex rules can be reduced to a set of simpler rules via substitution. Consider the rule x AND (y OR w) $\Rightarrow z$. This rule does not match any of the above rules, so it must be simplified. Using a placeholder variable (I), we can split the original rule into two rules: x AND $I \Rightarrow z$, and y OR $w \Leftrightarrow I$. Both of these simple rules can be converted into MILP constraints. Substituting placeholder variables for parts of rules can be done recursively. Any rule, regardless of its complexity, can be reduced to a set of simple rules [10].

The second transformation required for Boolean rule conversion is the negation operator (NOT). While negations do not appear in GPR rules, they are common features of transcriptional regulatory models. When an expression NOT x appears in a rule, we substitute a new variable, $\bar{x}$, in its place. The relation that $\bar{x} =^{\text{def}}$ NOT x is enforced by adding the constraint $x + \bar{x} = 1$.

As a final example, we convert a rule from the iND750/iMH805 metabolic [17] and transcriptional regulatory [13] models for the yeast *Saccharomyces cerevisiae*:

$$\text{glucose AND hap1 AND (oxygen OR (NOT rox1))} \Leftrightarrow \text{erg11}$$

First, we substitute the variable $\overline{\text{rox1}}$ for the expression NOT rox1, giving two rules:

$$\text{glucose AND hap1 AND}\left(\text{oxygen OR }\overline{\text{rox1}}\right) \Leftrightarrow \text{erg11}$$

$$\overline{\text{rox1}} \Leftrightarrow \text{NOT rox1}$$

We create an indicator variable for the disjunction (oxygen OR $\overline{\text{ROX1}}$), yielding three simple rules:

$$\text{glucose AND hap1 AND } I \Leftrightarrow \text{erg11}$$

$$\overline{\text{rox1}} \Leftrightarrow \text{NOT rox1}$$

$$\text{oxygen OR }\overline{\text{rox1}} \Leftrightarrow I$$

Each of these rules can be directly converted into a MILP constraint using the relations from above [10]. The final five MILP constraints are

$$\text{erg11} \geq \text{glucose} + \text{hap1} + I - 2$$

$$\text{glucose} + \text{hap1} + I \geq 3\tilde{n}\text{erg11}$$

$$\text{rox1} + \overline{\text{rox1}} = 1$$

$$I \geq (1/2)\left(\text{oxygen} + \overline{\text{rox1}}\right)$$

$$\text{oxygen} + \overline{\text{rox1}} \geq I$$

Converting Boolean rules into MILP constraints can be a tedious and error-prone process. Fortunately, TIGER fully automates this process. We can process the same rules with TIGER by first creating an “empty” TIGER model, i.e., one without any reactions or constraints. In Matlab:

```
>> t = create_empty_tiger()t =
           A: []
           b: []
          lb: []
          ub: []
         obj: []
    varnames: {}
    rownames: {}
      ctypes: "
    vartypes: "
         ind: []
   indtypes: "
       param: [1x1 struct]
      bounds: [1x1 struct]
```

Now we use the add_rule function to convert a Boolean rule into MILP constraints. The function returns a modified TIGER model containing the new rule:

```
>> add_rule(t,'glucose and hap1 and (oxygen or not rox1) <=>
erg11')t =
           A: [7x8 double]
           b: [7x1 double]
          lb: [8x1 double]
          ub: [8x1 double]
         obj: [8x1 double]
    varnames: {8x1 cell}
    rownames: {7x1 cell}
      ctypes: [7x1 char]
    vartypes: [8x1 char]
         ind: [7x1 double]
    indtypes: [7x1 char]
       param: [1x1 struct]
      bounds: [1x1 struct]
```

The TIGER model now contains seven constraints and eight variables. Our manual conversion required five constraints in six variables. As we saw earlier, there are multiple ways to encode the same rule as MILP constraints, and TIGER uses its own heuristics to find a suitable transformation. All transformations, however, are functionally equivalent. To see the transformation TIGER chose for our example, we use the show_tiger function:

```
>> show_tiger(t)
Optimization sense: minimize
```

```
---- Objective ----
obj:---- Constraints ----
ROW1:  rox1 + NOT__rox1 = 1
ROW2:  -oxygen + 3*I0002 - NOT__rox1 >= 0
ROW3:  -oxygen + 3*I0002 - NOT__rox1 <= 2
ROW4:  2*hap1 - 4*I0001 + 2*I0002 <= 3
ROW5:  2*hap1 - 4*I0001 + 2*I0002 >= -1
ROW6:  -4*erg11 + 2*glucose + 2*I0001 <= 3
ROW7:  -4*erg11 + 2*glucose + 2*I0001 >= -1
```

3.2 Coupling Boolean Regulation

Converting Boolean rules into MILP constraints creates model variables for each entity in the rules. These binary variables need to be connected to other variables in the model, such as the reaction fluxes. Since fluxes are continuous variables, we cannot use the Boolean techniques described in Subheading 3.1. Instead, this section outlines methods for interfacing between binary and continuous variables in MILP models.

Consider the simplest GPR, where the reaction is catalyzed by the product of a single gene. In this case, the reaction flux (ν) can only be nonzero if the corresponding gene (g) is active. If the reaction is irreversible (only positive fluxes), we can enforce the dependence of the reaction flux on the gene's state with the constraint $\nu \leq Mg$, where M is a large constant [18]. When $g = 0$, the flux is forced to zero. Only when the gene is active ($g = 1$) can the flux can be nonzero. The constant M is necessary because ν is a continuous variable with an upper bound that could be greater than one. Without M, the flux would be restricted to always be less than one, even when the gene is active. A common choice for M is the upper bound for ν found in the ub field of the COBRA model structure.

This method for coupling the GPR to reaction fluxes can be extended in two ways. First, reversible reactions can be coupled to a gene with the constraint $M_l g \leq \nu \leq M_u g$, where M_l and M_u are the lower and upper bounds for ν, respectively. (For irreversible reactions, $M_l = 0$, and for reversible reactions $M_l < 0$.) For GPRs involving more than one gene, we create a "reaction coupling variable" that is coupled to the reaction flux. For example, if a reaction is catalyzed by the product of either gene g_1 or gene g_2, we create the rule g_1 OR $g_2 \Leftrightarrow R$ and couple R with the reaction flux ($M_l R \leq \nu \leq M_u R$).

GPR couplings restrict the range of continuous variables (reaction fluxes) using binary variables (genes or reaction coupling variables). The reverse is also possible, i.e., driving a binary variable with a continuous one. One common usage of continuous variables coupled to binary variables is "flux sensors" that activate transcription factors when a reaction's flux exceeds a threshold value. Consider the rule $\nu \geq k \Rightarrow I$, where ν is a reaction flux and k is a constant defining the threshold above which the binary variable I should be activated. To incorporate this rule into an MILP, two constraints are added:

$$v + s = k$$

$$MI + s \geq \epsilon$$

The first constraint introduces a "slack variable" s, which is a continuous variable equal to the difference (slack) between the flux v and the threshold k. If $v \geq k$, the slack variable s is non-positive. Thus requiring I to be active when the flux crosses the threshold is equivalent to activating I when $s \leq 0$. Some MILP solvers (e.g., Gurobi and CPLEX) implement "indicator constraints" that can couple a binary variable (I) to a slack variable (s), completing the conversion of the rule. For other solvers, the coupling between I and s must be done directly through the second constraint ($MI + s \geq \epsilon$). To understand this constraint, consider the related constraint $MI + s > 0$, where M is a large constant. When $s \leq 0$, the binary variable I must be one; otherwise, there is no way the left-hand side can be positive. The constraint requires that I be active whenever $s \leq 0$, which happens whenever v exceeds the threshold flux k. Unfortunately, MILP solvers cannot implement strict inequalities ($>$ or $<$), so we must use the approximation $MI + s \geq \epsilon$, where ϵ is a very small, positive constant (*see* **Note 3**).

We can convert "if and only if" rules ($v \geq k \Leftrightarrow I$) with additional constraints. We start by adding the same constraints for the forward implication ($v \geq k \Rightarrow I$). Next, we add constraints to enforce the reverse implication ($I \Rightarrow v \geq k$). This is equivalent to enforcing that $I \Rightarrow s \leq 0$, which can be added to an MILP as the inequality $s \leq M(1 - I)$.

It is possible to derive similar transformations for other inequalities. One approach is to transform the other inequalities into expressions containing the inequality $x \geq y$, which can be converted as described above. For example, the inequality $x \leq y$ is equivalent to the expression $-x \geq -y$, and $x > y$ can be rewritten as NOT($-x \geq -y$). A complete list of transformations is available in the supplementary material of the TIGER manuscript [10]. The same reference generalizes the above approach to inequalities involving multiple variables and provides methods for calculating the constants M from the variable bounds.

3.3 Example

To highlight the interplay between reaction fluxes, GPRs, and transcriptional regulation, consider the (vastly) simplified metabolic network for *Streptococcus mutans* shown in Fig. 1. Like many bacteria, *S. mutans* can catabolize a variety of sugars, including glucose and galactose [19]. However, glucose is the preferred carbon source for many Gram positive bacteria, most of whom use a global transcriptional regulator to suppress catabolism of all other carbohydrates in the presence of adequate glucose. In streptococci, this regulator is the protein ccpA [16], which is activated upon glucose uptake. ccpA suppresses galactose catabolic enzymes like galK while promoting glucose fermentation to lactate by the enzyme ldh [19].

In our model, we use flux through a glucose transport reaction to signal glucose uptake. The rule $\nu < 0 \Rightarrow$ ccpA requires ccpA expression whenever glucose enters the cell. (By convention in many COBRA models, transport fluxes are written with the positive flux flowing out of the cell; a negative flux implies uptake [17].) Once ccpA has been transcribed, it modifies the expression of two metabolic enzymes. Enzyme ldh is activated (ccpA $\Rightarrow$ ldh), and expression of galK is turned off (ccpA $\Rightarrow$ NOT galK) [20]. Notice that we chose only right implications ($\Rightarrow$) for these rules. An "if and only if" relationship would not be appropriate, since ldh can be expressed in the absence of ccpA (ldh is almost always transcribed in *S. mutans*, as the bacteria ferment nearly any sugar they find [19]), and galK can be turned off without ccpA. These modeling decisions are based entirely on our understanding of *S. mutans* metabolism and should not be influenced by the mathematical structure of the model.

The effects of regulating enzymes galK and ldh are enforced by the GPRs. Additional constraints shut down flux through the associated reactions. Although the system in Fig. 1 is a tiny subset of the metabolic network, it demonstrates a complete cycle through multiple cellular networks. The flux through the glucose transporter drives expression of a transcription factor (ccpA). The

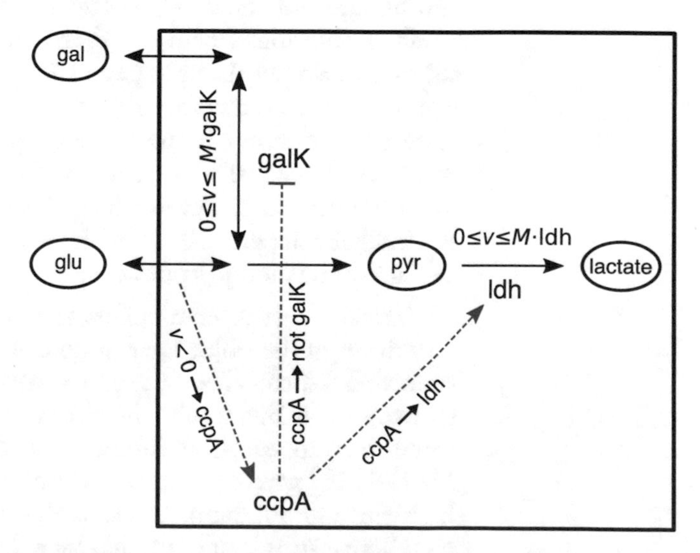

Fig. 1 Simplified schematic for fermentation in *Streptococcus mutans*. Circled metabolites include glucose (glu), galactose (gal), pyruvate (pyr), and lactate. Solid black lines are metabolic reactions. Dashed red lines are regulatory interactions

transcriptional factor modulates the state of metabolic enzymes, which in turn change the potential flux distributions. We have traveled from fluxes to logical constraints and back to fluxes. This trip required two changes from continuous to binary variables and two binary-binary interactions. When modeling both metabolic and transcriptional network simultaneously, interfaces between multiple types of variables are common.

3.4 Numerical Issues

3.4.1 Trickle Flows

The couplings between continuous and binary variables are often the source of numerical issues in MILPs. For COBRA models, a common problem is "trickle flows," or reactions that carry very small fluxes when they should be forced to zero by a binary variable. Trickle flows are not specific to COBRA models and are a result of the "Big-M" constraints used to block continuous variables based on the value of a binary variable. (The name "Big-M" refers to the large constants that appear before the binary variables in the coupling constraints. In the early literature, the symbol M was used, a convention we adopted in this chapter.) To understand the source of trickle flows, consider the coupling constraint introduced in Subheading 3.2 to bind flux v to the state of a gene g:

$$v \leq Mg$$

MILP solvers find solutions by solving a series of LP approximations to the original MILP. Although we think of g as an integer (zero or one) in the above equation, in practice g is solved as a finite-precision continuous variable that is restricted to be very close to either zero or one. Each solver typically limits this precision—called the "integer tolerance"—to between 10^{-5} and 10^{-10} [15, 16]. As a result, in solutions where $g = 0$, the numerical value of g used by the solver may actually be as large as 10^{-5}. This value seems trivial, but it is important to remember that the flux is not constrained by g, but instead by M times g in the above inequality. Since M is often chosen to be very large (10^3–10^5) to allow for very large fluxes, v could still carry a nontrivial flux even when g is zero. For example, if the solver has integer tolerance 10^{-5} and $M = 1000$, the flux could be as large as 0.01 even when g is not active.

Trickle flows often manifest as models that grow (have nonzero flux through the objective reaction) when growth should be infeasible. Small fluxes through some essential reactions may be enough to produce a metabolite necessary for growth. These situations sometimes appear after running algorithms like GIMME [21] or MADE [22] where gene states ("on" or "off") are calculated using high-throughput datasets. These algorithms produce tailored models where a subset of genes has been turned off to mimic reactions that are not active in a particular environment. Both GIMME and MADE require that the models remain functional as measured by a nonzero objective flux. Some modelers find that a gene state calculated by GIMME or MADE does not allow growth in subsequent simulations. In these cases, trickle flows that appeared during the GIMME or MADE solution process are often a culprit.

Three modifications to the model or solver can help remove trickle flows:

1. Reduce the solver's integer tolerance. Solver parameters like IntFeasTol for Gurobi and TolXInteger for CPLEX can be used to force the value of binary variables closer to the values 0 or 1.
2. Reduce M. Both TIGER and the COBRA toolbox choose default values for M based on the upper and lower bounds for each reaction. Often these bounds are set to a large, arbitrary value (± 1000, for example) to allow any physiologically reasonable flux. Usually, these bounds can be reduced without over-constraining the model, although this should be checked by running FBA after the bounds are adjusted.
3. Rescale the metabolic network. Metabolic network models with poor scaling have reactions with both very large and very small fluxes. Rescaling the fluxes to similar magnitudes reduces the chance that a small trickle flux will allow significant flux through the biomass reaction. Software techniques like reaction lifting [23] can detect and correct scaling issues in COBRA models.

Of the three options above, the first solution is easiest to implement and is a good first choice when faced with trickle flows. (TIGER's implementation of MADE automatically reduces the integer tolerance to prevent trickle flows.) All three can be applied simultaneously to handle especially difficult problems, which in our experience arise most often when solving large models with complex GPRs.

3.4.2 Infeasible Models

Genome-wide transcriptional regulatory models contain hundreds of interconnected rules. The resulting MILPs can easily become infeasible, or at least incapable of growth. Usually these problems arise from rules that include essential genes, where the rule set incorrectly disables an important gene. Finding the source of infeasibility in a large model is difficult, but software like TIGER can help. Given an infeasible model, the find_infeasible_rules function identifies the smallest set of rules that must be removed before the model is feasible again.

The find_infeasible_rules function rebuilds the infeasible model by replacing each regulatory rule with an "artificially satisfiable" rule. To understand this process, consider the rule $x \Rightarrow y$. If we modified the rule's right hand side to $x \Rightarrow y \text{ OR } r_i$, we could always satisfy the rule by setting $r_i = 1$, regardless of the values of x and y. Similarly, the "if and only if" rule $x \Leftrightarrow y$ can always be satisfied when transformed to x OR $l_i \Leftrightarrow y$ OR r_i and setting $l_i = r_i = 1$. By applying this transformation to every rule in the model (with separate variables l_i and r_i for each rule), the model is always feasible, i.e., artificially satisfiable. In an extreme case, the solver could set every l_i and r_i to one and make every rule true. On the other extreme, setting every l_i and r_i to zero makes the artificially satisfiable model equivalent to the original (infeasible) model. The problem of finding the rules that made the original problem infeasible can now be cast as an optimization problem—minimize the number of variables l_i and r_i that must be set to one before the model is feasible. The goal is not only to make the model feasible, but also to identify the smallest rule set that can explain the infeasibility. The find_infeasible_rules function returns a list of rules that required l_i or r_i to be activated. These rules should be checked manually to determine the root cause of the infeasibility.

Solutions like the find_infeasible_rules function require that the model be infeasible. This is different than a model that does not grow, i.e., a model that is solvable but has an objective flux of zero. Fortunately, the latter case can be made infeasible by adding a constraint that the objective flux be nonzero using TIGER's add_growth_constraint function. The resulting infeasible model can be passed to find_infeasible_rules to identify rules that prevent growth.

Like all MILPs, the optimization to find infeasible rule sets is degenerate. Other rule sets could be found that explain the infeasibility, although in practice we often find the true error in or near the rule set returned by find_infeasible_rules.

4 Notes

1. We have observed several MILPs that were declared "unfeasible" by GLPK only to be solved quickly by other commercial software packages.
2. Whether disjunctions are encoded using a single constraint or multiple simpler constraints is, for now, a matter of choice. We are unaware of any data comparing the efficiency of either approach.
3. Any solver that appears to accept strict inequalities automatically converts them to an inequality containing an ϵ, usually set to the numerical tolerance of the solver.

Acknowledgments

The author thanks Caroline Blassick for her assistance with Fig. 1.

References

1. Fell DA, Small JR (1986) Fat synthesis in adipose tissue. An examination of stoichiometric constraints. Biochem J 238:781–786
2. Edwards JS, Palsson BO (2000) Metabolic flux balance analysis and the in silico analysis of Escherichia coli K-12 gene deletions. BMC Bioinformatics 1:1. https://doi.org/10.1186/1471-2105-1-1
3. Kim J, Reed JL (2010) OptORF: optimal metabolic and regulatory perturbations for metabolic engineering of microbial strains. BMC Syst Biol 4:53. https://doi.org/10.1186/1752-0509-4-53
4. Suthers PF, Zomorrodi A, Maranas CD (2009) Genome-scale gene/reaction essentiality and synthetic lethality analysis. Mol Syst Biol 5:301. https://doi.org/10.1038/msb.2009.56
5. Shlomi T, Eisenberg Y, Sharan R, Ruppin E (2007) A genome-scale computational study of the interplay between transcriptional regulation and metabolism. Mol Syst Biol 3:101
6. Maranas CD, Zomorrodi AR (2016) Optimization methods in metabolic networks. Wiley, Hoboken, NJ. https://doi.org/10.1002/9781119188902
7. Becker SA, Feist AM, Mo ML et al (2007) Quantitative prediction of cellular metabolism with constraint-based models: the COBRA Toolbox. Nat Protoc 2:727–738. https://doi.org/10.1038/nprot.2007.99
8. Ebrahim A, Lerman JA, Palsson BØ, Hyduke DR (2013) COBRApy: COnstraints-Based Reconstruction and Analysis for Python. BMC Syst Biol 7:74. https://doi.org/10.1186/1752-0509-7-74
9. Bartell JA, Yen P, Varga JJ et al (2014) Comparative metabolic systems analysis of pathogenic Burkholderia. J Bacteriol 196:210–226. https://doi.org/10.1128/JB.00997-13
10. Jensen PA, Lutz KA, Papin JA (2011) TIGER: toolbox for integrating genome-scale metabolic models, expression data, and transcriptional regulatory networks. BMC Syst Biol 5:147. https://doi.org/10.1186/1752-0509-5-147
11. Covert MW, Palsson BØ (2003) Constraints-based models: regulation of gene expression reduces the steady-state solution space. J Theor Biol 221:309–325
12. Covert MW, Knight EM, Reed JL et al (2004) Integrating high-throughput and computational data elucidates bacterial networks. Nature 429:92–96. https://doi.org/10.1038/nature02456
13. Herrgård MJ, Lee B-S, Portnoy V, Palsson BØ (2006) Integrated analysis of regulatory and metabolic networks reveals novel regulatory mechanisms in Saccharomyces cerevisiae. Genome Res 16:627–635. https://doi.org/10.1101/gr.4083206
14. Oki E (2012) GLPK (GNU linear programming kit). In: Linear programming and algorithms for communication networks. CRC Press, Boca Raton, FL, pp 25–29
15. IBM ILOG CPLEX Optimizer (2010) http://www-01.ibm.com/software/integration/optimization/cplex-optimizer/
16. Gurobi Optimization, Inc. (2016) Gurobi Optimizer Reference Manual
17. Duarte NC, Herrgård MJ, Palsson BØ (2004) Reconstruction and validation of Saccharomyces cerevisiae iND750, a fully compartmentalized genome-scale metabolic model. Genome Res 14:1298–1309. https://doi.org/10.1101/gr.2250904
18. Burgard AP, Vaidyaraman S, Maranas CD (2001) Minimal reaction sets for Escherichia coli metabolism under different growth requirements and uptake environments. Biotechnol Prog 17:791–797. https://doi.org/10.1021/bp0100880
19. Zeng L, Das S, Burne RA (2010) Utilization of lactose and galactose by Streptococcus mutans: transport, toxicity, and carbon catabolite repression. J Bacteriol 192:2434–2444. https://doi.org/10.1128/JB.01624-09
20. Abranches J, Chen Y-YM, Burne RA (2004) Galactose metabolism by Streptococcus mutans. Appl Environ Microbiol 70:6047–6052. https://doi.org/10.1128/AEM.70.10.6047-6052.2004
21. Becker SA, Palsson BØ (2008) Context-specific metabolic networks are consistent with experiments. PLoS Comput Biol 4:e1000082. https://doi.org/10.1371/journal.pcbi.1000082
22. Jensen PA, Papin JA (2011) Functional integration of a metabolic network model and expression data without arbitrary thresholding. Bioinformatics 27:541–547. https://doi.org/10.1093/bioinformatics/btq702
23. Sun Y, Fleming RMT, Thiele I, Saunders MA (2013) Robust flux balance analysis of multiscale biochemical reaction networks. BMC Bioinformatics 14:240. https://doi.org/10.1186/1471-2105-14-240

Chapter 16

Dynamic Flux Balance Analysis Using DFBAlab

Jose Alberto Gomez and Paul I. Barton

Abstract

Bioprocesses are of critical importance in several industries such as the food and pharmaceutical industries. Despite their importance and widespread application, bioprocess models remain rather simplistic and based on unstructured models. These simple models have limitations, making it very difficult to model complex bioprocesses. With dynamic flux balance analysis (DFBA) more comprehensive bioprocess models can be obtained. DFBA simulations are difficult to carry out because they result in dynamic systems with linear programs embedded. Therefore, the use of DFBA as a modeling tool has been limited. With DFBAlab, a MATLAB code that performs efficient and reliable DFBA simulations, the use of DFBA as a modeling tool has become more accessible. Here, we illustrate with an example how to implement bioprocess models in DFBAlab.

Key words Dynamic flux balance analysis, Flux balance analysis, Bioprocess modeling, Linear programming, Dynamic systems

1 Introduction

Dynamic Flux Balance Analysis (DFBA) [1, 2] is a bioprocess modeling framework that relies on genome-scale metabolic network reconstructions (GENREs) of microorganisms. It is the dynamic extension of Flux Balance Analysis (FBA) [3], which has become popular with the advent of high-throughput genome sequencing. In fact, the number and level of detail of genome-scale metabolic network reconstructions has rapidly increased since 1999 (*see* Fig. 1 in [4]). Despite the ever-increasing availability of new and better metabolic network reconstructions, DFBA modeling remains challenging, and therefore its use has been limited.

Traditionally, bioprocess modeling relies on unstructured models to calculate the growth rates of microorganisms. This approach has significant limitations that make it impossible to simulate very complex bioprocesses. These limitations are countered by FBA by considering genome-scale metabolic networks of the microorganisms involved. FBA models the growth and metabolic fluxes rates of microorganisms as solutions of the following linear program (LP):

Marco Fondi (ed.), *Metabolic Network Reconstruction and Modeling: Methods and Protocols*, Methods in Molecular Biology, vol. 1716, https://doi.org/10.1007/978-1-4939-7528-0_16, © Springer Science+Business Media, LLC 2018

$$\begin{aligned} &\max v_{\text{growth}} \\ &\text{s.t.}\ \ \mathbf{Sv} = \mathbf{0}, \\ &\quad \mathbf{v}^{\text{LB}} \le \mathbf{v} \le \mathbf{v}^{\text{UB}}, \end{aligned}$$

where $\mathbf{S}$ is the stoichiometric matrix, $\mathbf{v}$ is the fluxes vector, $\mathbf{v}^{\text{UB}}$ and $\mathbf{v}^{\text{LB}}$ are the bounds on the metabolic fluxes given by thermodynamics, the extracellular environment and/or genetic modifications, and v_{growth} sums all growth associated fluxes. Given mass balance and thermodynamic constraints on a metabolic network, FBA finds a solution that satisfies these constraints and maximizes growth. If the LP becomes infeasible, it may indicate a lack of sufficient substrates and nutrients to provide the minimum maintenance energy for the respective microorganism to survive.

DFBA combines process models described by an ordinary differential equation (ODE) system, a differential-algebraic equation (DAE) system, or a partial differential-algebraic equation (PDAE) [5, 6] system with FBA to model bioprocesses. These models can be expressed as dynamic systems with LPs embedded [7, 8] which are challenging to simulate. The embedded LP poses difficulties in the form of non-unique solutions and premature LP infeasibilities. Fortunately, these complications have been addressed by efficient DFBA simulators, DSL48LPR in FORTRAN [7] and DFBAlab in MATLAB [9]. Both are free for academic research and can be found in the following webpage: http://yoric.mit.edu/software. The rest of this chapter will talk exclusively about how to use DFBAlab to perform DFBA simulations.

DFBAlab uses lexicographic optimization and the phase I of the simplex algorithm to deal with nonunique solutions and LP infeasibilities, respectively. Lexicographic optimization is a strategy that enables obtaining unique exchange fluxes. This strategy requires defining an objective function for each exchange flux of interest. More information can be found in [7, 9].

2 Materials

1. MATLAB: DFBAlab is compatible with MATLAB 7.12.0 and newer versions.
2. DFBAlab: DFBAlab is a MATLAB code that performs DFBA simulations efficiently and reliably. It is free for academic research and can be downloaded at http://yoric.mit.edu/software. DFBAlab contains three folders named "Functions," "Examples_Direct," and "Examples_DAE." The folder named "Functions" must be added to the MATLAB path. The Examples folders contain the examples described in [9]. It is recommended to use "Examples_DAE" since these examples run much faster than "Examples_Direct."

3. Relevant GENREs: To perform DFBA simulations, GENREs of the relevant microorganisms are needed. An extensive collection of GENREs can be found at the following webpage: http://systemsbiology.ucsd.edu/InSilicoOrganisms/OtherOrganisms. There are some protocols and methods that generate these reconstructions automatically. A list can be found at Table 1 of [10]. GENREs are usually published in the Systems Biology Markup Language (SBML) [11].
4. LP Solvers Gurobi or CPLEX: As mentioned before, DFBA calculates the growth rates and exchange fluxes rates of microorganisms by solving LPs. Therefore, LP solvers are needed. DFBAlab is currently compatible with CPLEX [12] and Gurobi [13]. Academic licenses are available for both LP solvers.
5. Cobra Toolbox: The Cobra Toolbox [14] is a package that runs in MATLAB that uses constraint-based modeling to predict the behavior or microorganisms. It can be obtained from https://opencobra.github.io/. The Cobra toolbox has a function that transforms a metabolic network in SBML format into a ".mat" format, which is the input used by DFBAlab. For this function to work, the SBML package must be obtained from http://sbml.org/Main_Page.

3 Methods

Here, we describe step by step how to model a DFBA problem using DFBAlab. The examples in "Examples_Direct" and "Examples_DAE" require the same modeling effort. Only the implementation in "Examples_DAE" will be discussed because the implementation in "Examples_Direct" is very similar. Examples in both folders contain three key files: main.m, DRHS.m, and RHS.m. In addition, the examples in "Examples_DAE" contain a file called evts.m. In addition, the Examples folders contain a folder called "Example-ModelImport." This folder contains a very simple script that converts an SBML file into a ".mat" file that can be used in MATLAB simulations. Next, we show the relevant parts of these files and the inputs required from the user. In addition, we use Example 3 in [9] to illustrate the use of DFBAlab. In this example, *Chlamydomonas reinhardtii* and *Saccharomyces cerevisiae* grow together in a pond open to the atmosphere. For simplicity, this pond is modeled as a single continuously stirred-tank reactor (CSTR). The model results in a DAE system as pH balances are considered. In Subheading 4, **Notes 1** and **2** describe some common problems regarding the use of DFBAlab.

3.1 Converting a GENRE in SBML Format into ".mat" Format

When published, GENREs are usually in SBML format. The file "ModelImport.m" in the "Example-ModelImport" folder in Examples calls the Cobra toolbox and converts GENREs in

SBML format into ".mat" format. This function is really easy to use. Just modify line 4 of the code,

```
NAME1=readCBModel('NAME2',DB,'SBML')
```

and replace NAME1 for the name you want to give your model, replace NAME2 with the name of the SBML file, and replace DB with the numerical value you want infinity to be replaced with. Usually, a value of DB = 1000 works fine. The output will be a model with the name "NAME1.mat." Every microorganism that needs to be modeled in DFBAlab must be transformed into a ".mat" file.

3.2 Inputs for the main.m File

The "main.m" file sets up the simulation. In this section, each part of the "main.m" file will be described and the required inputs will be specified. It is important to notice that the structure variable INFO contains important information that is then passed to other MATLAB files and functions. This structure can be used to pass other important parameter information to "DRHS.m," "RHS.m," and "evts.m." The first part of main.m clears the workspace and specifies the number of models that will be used in the simulation.

```
clear all
INFO.nmodel = NM;
```

Here, NM stands for the number of models in the simulation. For example, consider a dynamic model of a plug flow reactor (PFR) with two different species growing. The PFR can be discretized in N spatial slices to transform it from a partial differential equation (PDE) system to an ODE system. Then, $NM = 2 \times N$. The next part of the "main.m" file loads the models. Here, the models must already be in ".mat" format after going through the "ModelImport.m" code:

```
load NAMEMODEL.mat
model{i} = NAMEMODEL;
DB(i) = db;
INFO.DB = DB;
```

Each species must be loaded into the file using the load command and replacing NAMEMODEL for the actual name of the model. Then, for $i = 1,\ldots, nmodel$, the cell model stores each relevant metabolic network model. The array DB stores the default bound specified in "ModelImport.m" for each model. In this way, DFBAlab can change this bound information back to infinity. This array is stored in the INFO structure. The next part defines the *exID* cell, which carries the exchange fluxes information:

```
exID{i}=[ex1,ex2,...,exj];
INFO.exID = exID;
```

Again $i = 1,\ldots,$ *nmodel*, and *ex1, ex2, exj* correspond to the indices of the exchange fluxes that are needed in the "DRHS.m" file for model *i*. For example, let us assume that model 1 corresponds to an *E. coli* model. To model this bioprocess, we may care about the exchange fluxes of glucose, oxygen, and carbon dioxide. Then in *model*{1}*.rxns* the indices associated with the exchange fluxes reactions for glucose, oxygen, and carbon dioxide must be found. These are the indices that must be inserted in place of *ex1, ex2, exj*. The order of these indices is relevant for the "RHS.m" file as explained later.

Next, the cell *C* corresponding to the cost vectors is defined:

```
minim = 1;
maxim = -1;
% Maximize growth
C{i}(j).sense = maxim;
C{i}(j).rxns = [cin];
C{i}(j).wts = [1];
```

DFBAlab relies on lexicographic optimization to perform efficient DFBA simulations [9]. Each cost vector requires three entries: sense, reactions, and weights. The information in *C*{*i*}(*j*) corresponds to model *i* and cost vector *j*. Usually, the first cost vector for any model will correspond to maximization of biomass. Other relevant biological objectives can be listed as subsequent cost vectors. All the fluxes that appear in the right-hand side of the ODE system must be listed as cost vectors in order to guarantee a unique solution of the simulation. If you want the corresponding cost vector to be maximized, enter *maxim* in sense, otherwise enter *minim*. Then in *C*{*i*}(*j*)*.rxns* list the indices of the reactions corresponding to this cost vector. Finally in *C*{*i*}(*j*)*.wts* enter the coefficients corresponding to this cost vector. For example,

```
C{i}(j).sense = maxim;
C{i}(j).rxns = [1108,103];
C{i}(j).wts = [1,-1];
INFO.C = C;
```

is equivalent to defining cost vector *j* for model *i* as *maximize* $v_{1108} - v_{103}$. The cost vectors are stored in the INFO structure. Next the initial conditions, integration time and DFBAlab options are set.

```
Y0 = [ICVECTOR];
% Time of simulation
tspan = [ti,tf];
% LP Objects construction parameters
INFO.LPsolver = 0; % CPLEX = 0, Gurobi = 1.
INFO.tol = tol1; % Feasibility, optimality and convergence tolerance for Cplex (tol>=1E-9).
        % It is recommended it is at least 2 orders of magnitude
        % tighter than the integrator tolerance.
        % If problems with infeasibility messages, tighten this
        % tolerance.
INFO.tolPh1 = tol2; % Tolerance to determine if a solution to phaseI equals zero.
        % It is recommended to be the same as INFO.tol.
INFO.tolevt = tol3; % Tolerance for event detection. Has to be greater
        % than INFO.tol. Recommended to be 2 times the integration tolerance.
```

Y0 is a column vector containing the initial conditions. *tspan* contains the integration time interval: *ti* corresponds to the initial time and *tf* corresponds to the final time. Next, some DFBAlab parameters are established: *INFO.LPsolver* selects between Gurobi (enter 1) and CPLEX (enter 0). Next, the LP tolerance is set at *INFO.tol*. This tolerance has to be larger than 10^{-9}. Next, *INFO.tolPh1* corresponds to the threshold value under which the penalty function will be considered equal to zero. Here, we recommend using *INFO.tol* for this value. Finally, *INFO.tolevt* corresponds to the event detection tolerance that triggers when the FBA LP must be solved again. Here, we recommend using two times the absolute tolerance of the integration method. Next, the integration options are set:

```
M = [MASS];
options = odeset('AbsTol',tol4,'RelTol',tol5,'Mass',M,'Events',@evts);
```

If we are integrating a DAE system, a mass matrix must be defined. Using odeset, other MATLAB integration options can be set such as absolute and relative tolerances, nonnegativity constraints, and event detection. DFBAlab always uses event detection if you use the Examples in "Examples_DAE." The following parts of the "main.m" file do not require any further inputs.

```
%%%%%%%%%%%%%%%%%%%%%%%%%%%%%%%%%%%%%%%%%%%%%%%%%%%%%%%%%%%%%%%%%%%%%%%%%%%
[model,INFO] = ModelSetupM(model,Y0,INFO);
%%%%%%%%%%%%%%%%%%%%%%%%%%%%%%%%%%%%%%%%%%%%%%%%%%%%%%%%%%%%%%%%%%%%%%%%%%%
if INFO.LPsolver == 0
    [INFO] = LexicographicOpt(model,INFO);
elseif INFO.LPsolver == 1
    [INFO] = LexicographicOptG(model,INFO);
else
    display('Solver not currently supported.');
end
```

```
tic
tint = 0;
TF = [];
YF = [];
while tint<tspan(2)
% Look at MATLAB documentation if you want to change solver.
% ode15s is more or less accurate for stiff problems.
  [T,Y] = ode15s(@DRHS,tspan,Y0,options,INFO);
  TF = [TF;T];
  YF = [YF;Y];
  tint = T(end);
  tspan = [tint,tspan(2)];
  Y0 = Y(end,:);
  if tint == tspan(2)
        break;
    end
% Update b vector
[INFO] = bupdate(tint,Y0,INFO);
%Determine model with basis change
  value = evts(tint,Y0,INFO);
  ind = find(value<=0);
  fprintf('Basis change at time %d. ',tint);
  k = 0;
  ct = 0;
  while ~isempty(ind)
    k = k + 1;
    ct = ct + size(model{k}.A,1);
    ind2 = find(ind<=ct);
    if ~isempty(ind2)
       INFO.flagbasis = k;
       fprintf('Model %i. \n',k);
       % Perform lexicographic optimization
       if INFO.LPsolver == 0
          [INFO] = LexicographicOpt(model,INFO);
       elseif INFO.LPsolver == 1
          [INFO] = LexicographicOptG(model,INFO);
       else
          display('Solver not currently supported.');
       end
          ind(ind2)=[];
    end
  end
end

display(toc);
```

The function ModelSetupM takes the model cell and the *INFO* structure to transform the LPs into a standard form, which is the key for efficient integration with DFBAlab. The functions LexicographicOpt and LexicographicOptG solve the LPs with CPLEX and Gurobi, respectively. Then comes the integration loop using the numerical integrator ode15s, used for stiff systems which is the case for DFBA systems. The results of integration are stored in vectors *YF* for the states and *TF* for the times. After this, any plots can be generated using the information in *YF* and *TF*.

3.3 Example Inputs for main.m

The following MATLAB code is commented in the relevant sections.

```
clear all
INFO.nmodel = 2; % Number of models: one for algae and one for yeast.
load iND750.mat % This is the yeast model
model{1} = iND750;
DB(1) = 1000; % This is the default bound for the yeast model (1000 = Infinity)
load iRC1080.mat % This is the algae model
model{2} = iRC1080;
DB(2) = 1000; % This is the default bound for the algae model (1000 = Infinity)
INFO.DB = DB;
exID{1}=[428,458,407,420]; % These are the exchange fluxes with variable bounds for
the yeast model. The bounds are defined in RHS.m.
exID{2}=[26,28,24,25,1,2,3,4,5,6,7,8,9,10,11,12,13,62,63,64,65,27,81,47]; % These
are the exchange fluxes with variable bounds for the algae model. The bounds are
defined in RHS.m.
INFO.exID = exID;

% Next, we define the cost vectors for yeast and algae. For both cases, first, we
maximize growth, and then we maximize or minimize each one of the exchange fluxes that
appear in the ODE system.
minim = 1;
maxim = -1;
% Yeast
% Maximize growth
C{1}(1).sense = maxim;
C{1}(1).rxns = [1266];
C{1}(1).wts = [1];
% Glucose
C{1}(2).sense = maxim;
C{1}(2).rxns = [428];
C{1}(2).wts = [1];
% O2
C{1}(3).sense = maxim;
C{1}(3).rxns = [458];
C{1}(3).wts = [1];
% CO2
C{1}(4).sense = maxim;
C{1}(4).rxns = [407];
C{1}(4).wts = [1];
```

```
% Ethanol
C{1}(5).sense = maxim;
C{1}(5).rxns = [420];
C{1}(5).wts = [1];
% Algae
% Maximize growth
C{2}(1).sense = maxim;
C{2}(1).rxns = [63];
C{2}(1).wts = [1];
% Acetate
C{2}(2).sense = minim;
C{2}(2).rxns = [28];
C{2}(2).wts = [1];
% O2
C{2}(3).sense = maxim;
C{2}(3).rxns = [24,81];
C{2}(3).wts = [1];
% CO2
C{2}(4).sense = minim;
C{2}(4).rxns = [25];
C{2}(4).wts = [1];

INFO.C = C;

% Initial conditions
% Y1 = Volume (L)
% Y2 = Biomass Yeast (gDW/L)
% Y3 = Biomass Algae (gDW/L)
% Y4 = Glucose (mmol/L)
% Y5 = O2 (mmol/L)
% Y6 = Total Carbon (mmol/L)
% Y7 = Ethanol (mmol/L)
% Y8 = Acetate (mmol/L)
% Y9 = Total Nitrogen (mmol/L)
% Y10 = Penalty
% Y11 = NH4+
% Y12 = NH3
% Y13 = CO2
% Y14 = HCO3-
% Y15 = CO3 -2
% Y16 = H +

Y0 = [140 1.1048 1.8774 0.0140, 0.00065156 1.2211 8.2068 0.0237 0.1643 0 0.1643
2.4476E-5 1.0568 0.1643 2.5842E-6 2.6701E-6]';

% Time of simulation
tspan = [0,24];

% LP options and tolerances
```

```
INFO.LPsolver = 0; % CPLEX = 0, Gurobi = 1.
INFO.tol = 1E-9; % Feasibility, optimality and convergence tolerance for LP solver (tol>=1E-9). It is recommended it is at least 1 order of magnitude tighter than the integration tolerance.
INFO.tolPh1 = INFO.tol; % Tolerance to determine if a solution to phaseI equals zero. Usually, the best value here is to set it equal to INFO.tol.
INFO.tolevt = 2E-6; % Tolerance for event detection. Has to be greater than INFO.tol. Usually a value of two times the integration tolerance works fine.

% This is a DAE system because of the pH balances. The first 10 states are the differential states and the last 6 correspond to the algebraic states.
M = [eye(10) zeros(10,6); zeros(6,16)];
options = odeset('AbsTol',1E-6,'RelTol',1E-6,'Mass',M,'Events',@evts);

% This part of the code constructs the LP problem structures that will be solved during integration, and solves the LPs at the initial conditions.
[model,INFO] = ModelSetupM(model,Y0,INFO);
if INFO.LPsolver == 0
    [INFO] = LexicographicOpt(model,INFO);
elseif INFO.LPsolver == 1
    [INFO] = LexicographicOptG(model,INFO);
else
    display('Solver not currently supported.');
end

% This is the integration loop.
tint = 0;
% TF and YF will concatenate T and Y that are returned by the DAE numerical integrator.
TF = [];
YF = [];
while tint<tspan(2)
% Look at MATLAB documentation if you want to change solver. DFBA systems tend to be stiff systems and ode15s is more or less accurate for stiff problems.
    [T,Y] = ode15s(@DRHS,tspan,Y0,options,INFO);
    TF = [TF;T];
    YF = [YF;Y];
    tint = T(end);
    tspan = [tint,tspan(2)];
    Y0 = Y(end,:);
    if tint == tspan(2)
        break;
    end

% Update the right-hand sides of the LPs given the current time and states.
[INFO] = bupdate(tint,Y0,INFO);
% Determine which LPs had a basis change
    value = evts(tint,Y0,INFO);
    ind = find(value<=0);
```

```
    fprintf('Basis change at time %d. ',tint);
    k = 0;
    ct = 0;
    while ~isempty(ind)
        k = k + 1;
        ct = ct + size(model{k}.A,1);
        ind2 = find(ind<=ct);
        if ~isempty(ind2)
           INFO.flagbasis = k;
           fprintf('Model %i. \n',k);
           % Perform lexicographic optimization
           if INFO.LPsolver == 0
               [INFO] = LexicographicOpt(model,INFO);
           elseif INFO.LPsolver == 1
               [INFO] = LexicographicOptG(model,INFO);
           else
               display('Solver not currently supported.');
           end
           ind(ind2)=[];
        end
    end
end
% TF contains all the times and YF all the states at these times. Any plotting options
can be included here.
```

3.4 Inputs for the DRHS.m File

The DRHS function defined in "DRHS.m" takes time, states, and the INFO structure and returns the right-hand side vector of the ODE or DAE system:

```
function dy = DRHS(t, y, INFO).
```

This file is very flexible. A key command that takes place before the right-hand side values are set is:

```
[flux,penalty] = solveModel(t,y,INFO);
```

Here, the flux variable is a matrix with rows corresponding to each model and columns corresponding to each cost vector. Therefore, *flux(i,j)* corresponds to the optimal value of cost vector j and model i with the order defined in "main.m." The penalty vector contains the objective function values of the Phase I LPs. If *penalty (i)*>0, model i corresponds to an infeasible LP. Otherwise, model i is feasible. We recommend that a penalty state that integrates the sum of all penalty functions is set. If this penalty state is greater than zero at the end of the simulation, then, this DFBA simulation is infeasible.

3.5 Example Inputs for the DRHS.m File

The following MATLAB code is commented in the relevant sections.

```
function dy = DRHS(t, y, INFO)

% Y1 = Volume (L)
% Y2 = Biomass Yeast (gDW/L)
% Y3 = Biomass Algae (gDW/L)
% Y4 = Glucose (mmol/L)
% Y5 = O2 (mmol/L)
% Y6 = Total Carbon (mmol/L)
% Y7 = Ethanol (mmol/L)
% Y8 = Acetate (mmol/L)
% Y9 = Total Nitrogen (mmol/L)
% Y10 = Penalty
% Y11 = NH4+
% Y12 = NH3
% Y13 = CO2
% Y14 = HCO3-
% Y15 = CO3 -2
% Y16 = H +

% Assign values from states
Vol = y(1);
X(1) = y(2);
X(2) = y(3);
for i=1:13
   S(i) = y(3+i);
end

% Feed rates
Fin = 1;
Fout = 1;

% Biomass Feed concentrations
Xfeed(1) = 0;
Xfeed(2) = 0;

% Mass transfer coefficients
KhO2 = 0.0013;
KhCO2 = 0.035;

% Mass transfer expressions
MT(1) = 0;
MT(2) = 0.6*(KhO2*0.21*1000 - S(2));
MT(3) = 0.58*(KhCO2*0.00035*1000 - S(10));
MT(4) = 0;
MT(5) = 0;
MT(6) = 0;
```

```
% Substrate feed concentrations
Sfeed(1) = 15.01;
Sfeed(2) = KhO2*0.21*1000;
Sfeed(3) = KhCO2*0.00035*1000;
Sfeed(4) = 0;
Sfeed(5) = 40;
Sfeed(6) = 0;

% The elements of the flux matrix have the sign given to them by the
% coefficients in the Cost vector in main.
% Example, if:
% C{k}(i).rxns = [144, 832, 931];
% C{k}(i).wts = [3, 1, -1];
% Then the cost vector for this LP will be:
% flux(k,i) = 3*v_144 + v_832 - v_931

%% Update bounds and solve for fluxes
[flux,penalty] = solveModel(t,y,INFO);

% Yeast fluxes
for i=1:4
  v(1,i) = flux(1,i+1);
end
  v(1,5) = 0;
  v(1,6) = 0;

% Algae fluxes
  v(2,1) = 0;
  v(2,2) = flux(2,3);
  v(2,3) = flux(2,4);
  v(2,4) = 0;
  v(2,5) = flux(2,2);

%% Dynamics
dy(1) = Fin-Fout;  % Volume
dy(2) = flux(1,1)*y(2) + (Xfeed(1)*Fin - y(2)*Fout)/y(1);  % Biomass yeast
dy(3) = flux(2,1)*y(3) + (Xfeed(2)*Fin - y(3)*Fout)/y(1);  % Biomass algae
for i = 1:5
  dy(i+3) = v(1,i)*X(1) + v(2,i)*X(2) + MT(i) + (Sfeed(i)*Fin - S(i)*Fout)/y(1) ;
end
if (S(2)/1000 > KhO2 && dy(3+2)>0)
  dy(3+2) = 0;
end
if (S(3)/1000 > KhCO2 && dy(3+3)>0)
  dy(3+3) = 0;
end
dy(9) = 0; % Leave total nitrogen constant
dy(10) = penalty(1) + penalty(2); %Penalty function
```

```
% Algebraic Equations
Ka = 10^-9.4003;
K1c = 10^-6.3819;
K2c = 10^-10.3767;
Nt = S(6);
Ct = S(3);
x = y(11:16);
F = [-x(1) + Nt*x(6)/(x(6) + Ka) ;
   -x(4) + x(1)/(1 + 2*K2c/x(6));
   -x(3) + x(6)*x(4)/K1c;
   -x(5) + x(4)*K2c/x(6);
   -Nt + x(1) + x(2)
   -Ct + x(3) + x(4) + x(5)];
dy(11:16) = F;
end
```

3.6 Inputs for the RHS.m File

The RHS function defined in "RHS.m" takes time, states, and the INFO structure and returns two matrices containing the upper and lower bounds for the fluxes specified in the *exID* cell in "main.m."

```
function [lb,ub] = RHS( t,y,INFO )
```

Here, *lb* corresponds to the lower bounds and *ub* to the upper bounds. Element *lb(i,j)* contains the lower bound corresponding to flux *j* in *exID{i}*. The same indexing applies for the *ub* matrix containing the upper bounds. The lower and upper bound quantities are functions of time and states. These functions must be continuous functions. In addition, if any lower bound or upper bound is defined as infinity or minus infinity, the value of this bound must remain constant the entire simulation.

3.7 Example Inputs for the RHS.m File

```
function [lb,ub] = RHS( t,y,INFO )

% Y1 = Volume (L)
% Y2 = Biomass Yeast (gDW/L)
% Y3 = Biomass Algae (gDW/L)
% Y4 = Glucose (mmol/L)
% Y5 = O2 (mmol/L)
% Y6 = Total Carbon (mmol/L)
% Y7 = Ethanol (mmol/L)
% Y8 = Acetate (mmol/L)
% Y9 = Total Nitrogen (mmol/L)
% Y10 = Penalty
% Y11 = NH4+
% Y12 = NH3
```

```
% Y13 = CO2
% Y14 = HCO3-
% Y15 = CO3 -2
% Y16 = H +

% This subroutine updates the upper and lower bounds for the fluxes in the
% exID arrays in main. The output should be two matrices, lb and ub. The lb matrix
% contains the lower bounds for exID{i} in the ith row in the same order as
% exID. The same is true for the upper bounds in the ub matrix.
% Infinity can be used for unconstrained variables, however, it should be
% fixed for all time.
%%%%%%%%%%%%%%%%%%%%%%%%%%%%%%%%%%%%%%%%%%%%%%%%%%%%%%%%%%%%%%%%%%%%%%%%%%%
% Yeast bounds
% Glucose
if (y(4)<0)
    lb(1,1) = 0;
else
    lb(1,1) = -(20*y(4)/(0.5/0.18 + y(4)))*1/(1+y(7)/(10/0.046));
end
ub(1,1) = 0;
% Oxygen
lb(1,2) = -8*y(5)/(0.003/0.016 + y(5));
ub(1,2) = 0;
% CO2
lb(1,3) = 0;
ub(1,3) = Inf;
% Ethanol
lb(1,4) = 0;
ub(1,4) = Inf;
%%%%%%%%%%%%%%%%%%%%%%%%%%%%%%%%%%%%%%%%%%%%%%%%%%%%%%%%%%%%%%%%%%%%%%%%%%%
% Algae bounds
% HCO3
lb(2,1) = 0;
ub(2,1) = 0;
% Acetate
lb(2,2) = -14.9*y(8)/(2.2956+y(8)+y(8)^2/0.1557);
ub(2,2) = 0;
% Oxygen
lb(2,3) = -1.41750070911*y(5)/(0.009+y(5));
ub(2,3) = Inf;
% CO2
lb(2,4) = -2.64279793224*y(13)/(0.0009+y(13));
ub(2,4) = Inf;
% Light
Ke1 = 0.32;
Ke2 = 0.03;
L = 0.4; % meters depth of pond
Ke = Ke1 + Ke2*(y(3)+y(2)/2);
Io = 28*(max(sin(2*pi()*t/48)^2,sin(2*pi()*5/48)^2)-sin(2*pi()*5/48)^2)/(1-sin(2*pi
```

```
()*5/48)^2);
lb(2,5) = 0;
ub(2,5) = Io*(1-exp(-L*Ke))/(Ke*L);
% Other possible light sources set equal to zero
for i=6:16
   lb(2,i) = 0;
   ub(2,i) = 0;
end
% H+
lb(2,17) = -10;
ub(2,17) = Inf;
% Autotrophic growth
lb(2,18) = 0;
ub(2,18) = 0;
% Mixotrophic growth
lb(2,19) = 0;
ub(2,19) = Inf;
% Heterotrophic growth
lb(2,20) = 0;
ub(2,20) = 0;
% Non-growth associated ATP maintenance
lb(2,21) = 0.183;
ub(2,21) = 0.183;
% Starch
lb(2,22) = 0;
ub(2,22) = 0;
% Photoevolved oxygen
lb(2,23) = 0;
ub(2,23) = 8.28;
% Ethanol
lb(2,24) = 0;
ub(2,24) = 0;
end
```

3.8 Inputs for the evts.m File

This file is critical for DFBAlab to perform efficient DFBA simulations and will most likely not require any changes. This file needs to be changed only if event detection is needed in addition to the event detection associated with the LPs embedded. In this case, any event detection conditions can be added at the end of the vectors *value, isterminal*, and *direction*. The definition of these vectors can be found in the MATLAB documentation.

```
function [value,isterminal,direction] = evts(t,y,INFO)
eps = INFO.tolevt;
lexID = INFO.lexID;
nmodel = INFO.nmodel;
bmodel = INFO.b;
lbct = INFO.lbct;
```

```
indlb = INFO.indlb;
indub = INFO.indub;
U = INFO.U;
L = INFO.L;
P = INFO.P;
Q = INFO.Q;
%% Update solutions
[lbx,ubx] = RHS( t,y,INFO );
ct = 0;
total = 0;
for i=1:nmodel
   total = length(bmodel{i}) + total;
end
value = zeros(total,1);
isterminal = ones(total,1); % stop the integration
direction = -1*ones(total,1); % negative directionfor i=1:nmodel
    b = bmodel{i};
    lb = lbx(i,1:lexID(i));
    ub = ubx(i,1:lexID(i));
    lb(indlb{i}) = [];
    ub(indub{i})=[];
    b(1:length(lb)) = lb;
    b(length(lb)+lbct(i)+1:length(lb)+lbct(i)+length(ub)) = ub;
    x = (L{i}\(P{i}*b));
    x = U{i}\x;
    x = Q{i}*x;
%     x = U{i}\(L{i}\(P{i}*b));
%     x = INFO.B{i}\b;
% Detect when a basic variable crosses zero.
    value(1+ct:length(x)+ct) = x + eps;
    ct = ct + length(x);
end
ADD NEW CONDITIONS HERE
end
```

If the code is not modified, the length of the vectors *value*, *isterminal*, and *direction* is equal to the sum of all m_i where m_i corresponds to the number of rows of *model{i}.A*. In addition, the "main.m" file needs to be modified at the integration loop to distinguish events due to basis changes in the LPs from other types of events. In Example 3 in [9], this file is not modified.

4 Notes

1. In some instances, DFBAlab may fail due to infeasible or unbounded LPs. Although theoretically this should not happen, numerically it does happen some times. When

encountering this problem, change INFO.tol a bit and/or change the LP solver and try again.

2. DFBAlab is designed such that only the fluxes defined as cost vectors can be accessed at DRHS.m. This is to ensure uniqueness in the right-hand side of the ODE/DAE system. Future versions of DFBAlab may contain the option of extracting all other fluxes, although these other fluxes must be treated carefully as they can be nonunique.

References

1. Varma A, Palsson BØ (1994) Stoichiometric flux balance models quantitatively predict growth and metabolic by-product secretion in wild-type Escherichia coli W3110. Appl Environ Microbiol 60(10):3724–3731
2. Mahadevan R, Edwards J, Doyle FI (2002) Dynamic flux balance analysis of diauxic growth in Escherichia coli. Biophys J 83 (3):1331–1340
3. Orth JD, Thiele I, Palsson BØ (2010) What is flux balance analysis? Nat Biotechnol 28:245–248
4. Monk J, Nogales J, Palsson BØ (2014) Optimizing genome-scale network reconstructions. Nat Biotechnol 32:447–452
5. Chen J, Gomez J, Höffner K, Phalak P, Barton P, Henson M (2016) Spatiotemporal modeling of microbial metabolism. BMC Syst Biol 10:21
6. Chen J, Gomez J, Höffner K, Barton P, Henson M (2015) Metabolic modeling of synthesis gas fermentation in bubble column reactors. Biotechnol Biofuels 8:89
7. Höffner K, Harwood SM, Barton PI (2013) A reliable simulator for dynamic flux balance analysis. Biotechnol Bioeng 110(3):792–802
8. Harwood S, Höffner K, Barton P (2016) Efficient solution of ordinary differential equations with a parametric lexicographic linear program embedded. Numer Math 133(4):623–653
9. Gomez JA, Höffner K, Barton PI (2014) DFBAlab: a fast and reliable MATLAB code for Dynamic Flux Balance Analysis. BMC Bioinformatics 15:409
10. Fondi M, Liò P (2015) Genome-scale metabolic network reconstruction. In: Mengoni A, Galardini M, Fondi M (eds) Bacterial pangenomics: methods and protocols. Springer, New York, pp 233–256
11. Hucka M, Finney A, Sauro H et al (2003) The systems biology markup language (SBML): a medium for representation and exchange of biochemical network models. Bioinformatics 19(4):524–531
12. IBM ILOG. CPLEX Optimizer. http://www-01.ibm.com/software/commerce/optimization/cplex-optimizer/
13. Gurobi, Gurobi Optimization I: Gurobi Optimizer Reference Manual (2014) http://www.gurobi.com
14. Becker SA, Feist AM, Mo ML, Hannum G, Palsson BØ, Herrgard MJ (2007) Quantitative prediction of cellular metabolism with constraint-based models: the COBRA Toolbox. Nat Protoc 2:727–738

Chapter 17

Designing Optimized Production Hosts by Metabolic Modeling

Christian Jungreuthmayer, Matthias P. Gerstl, David A. Peña Navarro, Michael Hanscho, David E. Ruckerbauer, and Jürgen Zanghellini

Abstract

Many of the complex and expensive production steps in the chemical industry are readily available in living cells. In order to overcome the metabolic limits of these cells, the optimal genetic intervention strategies can be computed by the use of metabolic modeling. Elementary flux mode analysis (EFMA) is an ideal tool for this task, as it does not require defining a cellular objective function. We present two EFMA-based methods to optimize production hosts: (1) the standard approach that can only be used for small and medium scale metabolic networks and (2) the advanced dual system approach that can be utilized to directly compute intervention strategies in a genome-scale metabolic model.

Key words Genome-scale, Metabolic model, Metabolic pathway analysis, Elementary flux mode, Minimal cut set, Duality, Systems metabolic engineering, Optimized production hosts

1 Introduction

A main issue in the chemical industry is that the petrochemical production of (fine) chemical commodities, such as pharmaceuticals and fuels, is ecologically unsustainable, and involves numerous complex and expensive process steps. Many of these steps are readily implemented in living cells. Therefore, exploiting the natural capability of cells for the bio-based production of substances of interest might help to significantly decrease the costs of pharmaceuticals and reduce the use of fossil fuels. However, the production of chemicals by living cells is often inefficient, as wild-type cells do not typically evolve naturally to optimally produce a specific product of (human) interest. For instance, an organism might use some of the provided nutrition to produce not only the product of interest (POI) but also unwanted by-products. In recent years, much effort has been made to increase the efficiency of such cell factories, e.g., by selecting suitable host cells, by optimizing process and environmental conditions or by implementing targeted genetic

Marco Fondi (ed.), *Metabolic Network Reconstruction and Modeling: Methods and Protocols*, Methods in Molecular Biology, vol. 1716, https://doi.org/10.1007/978-1-4939-7528-0_17,

interventions in the host. Identifying an optimal intervention strategy that enforces the efficient production of the POI is a nontrivial task. Several methods for modeling and optimizing cells have been developed over the last decade. The most prominent and arguably also the most successful concept is based on constraint-based modeling (CBM) [1]. Central to CMB is the stoichiometric matrix. Two basic approaches for the analysis of the stoichiometric matrix have been developed which include on the one hand flux balance analysis (FBA) and related methods [2] and on the other hand metabolic pathway analysis, especially elementary flux mode analysis (EFMA). In this article the focus is on the latter. The main advantage of EFMA over FBA is that it is an unbiased modeling method. This means that EFMA aims to characterize the complete steady-state solution space. FBA on the other hand singles out optimal solutions with respect to physiologically relevant objectives. Such objectives are not required in an EFMA.

In this article, we discuss EFMA-based methods that unbiasedly identify rational engineering strategies to turn wild-type organisms into optimized production hosts. The goal of our numerical analysis is the computation of so-called constrained minimal cut sets (cMCSs) [3, 4]. cMCSs are minimal (genetic) intervention strategies, i.e., gene knockouts, that remove unwanted functionalities from the host organism (e.g., the production of unwanted by-products), and—at the same time—keep the desired functionalities (e.g., efficient production of a POI, such as ethanol).

We present and discuss two methods of cMCS computation: (a) the three-step standard approach, where first the network is reduced, followed then by the computation of all elementary flux modes (EFMs) which are finally used to calculate cMCS by a hitting set algorithm and (b) a single-step approach that uses the concept of dual systems and a mixed integer linear program (MILP) to determine cMCSs [5, 6]. Typically, a large number of potentially suitable cMCSs are computed by both methods. These cMCS represent strategies that can be implemented in the lab. One way to numerically evaluate the quality of these cMCSs—before verifying them experimentally—is by calculating the effect of the intervention on the host's structural robustness. A high robustness score of a cell factory indicates that it better performs a desired function despite of unfavorable external influences compared to a low robustness cell factory.

2 Basics and Background

2.1 Metabolic Model and Stoichiometric Matrix

Commonly, the entire set of metabolic reactions of an organism is stored and distributed by files formatted in the systems biology markup language (SBML). These files contain the essential information of all relevant chemical reactions that take place in the cell, such as the compartment where the reactions occur, the reactants and products of the reactions, their stoichiometry and the (ir-) reversibility of the reactions. SBML is based on XML. The main feature of XML is that it is human-readable as well as machine-readable.

In order to be able to numerically simulate a cell's metabolism, the SBML file needs to be converted to a stoichiometric matrix $\boldsymbol{S}$ with m rows (number of metabolites) and n columns (number of reactions). The stoichiometric matrix $\boldsymbol{S}$ is a linear transformation which relates a flux vector $\boldsymbol{f}$ with the time derivative of the metabolite concentration vector $\boldsymbol{c}$:

$$\frac{\mathrm{d}\boldsymbol{c}}{\mathrm{d}t} = \boldsymbol{Sf}$$

In the field of CMB we are typically interested in steady-state solutions which means that the time derivatives $\mathrm{d}/\mathrm{d}t$ are equal to zero:

$$0 = \boldsymbol{Sf}$$

Because of thermodynamic constraints some reactions might only be able to carry a flux in one direction, i.e.,

$$f_i \geq 0,$$

with $i\in$ of the set of irreversible reactions. Thus, the information about the (ir-) reversibility constraints of the model's reactions also needs to be extracted from the SBML file. Usually, this information is stored in a binary vector $\boldsymbol{v}$, where $v_i = 0$ if the reaction R_i is irreversible and $v_i = 1$ if the reaction R_i is reversible.

2.2 What Are EFMs?

An EFM $\boldsymbol{e}$ is a flux vector (1) which obeys $\boldsymbol{Se} = \boldsymbol{0}$, (2) where fluxes through irreversible reactions meet the irreversibility constraints, and (3) which consists of an indivisible set of flux-carrying reactions [7]. Hence, removing a single flux-carrying reaction from $\boldsymbol{e}$ will result in disobeying equation $\boldsymbol{Sf} = \boldsymbol{0}$ and, therefore, in violating the steady-state condition. The superposition (non-negative linear combination) of all EFMs defines the feasible solution space in which the organism can operate.

2.3 What Are Minimal Cut Sets (MCSs)?

Minimal cut sets (MCSs) are intervention strategies that define sets of reactions to be deactivated in order to suppress particular steady-state behaviors [3, 4].

In general, the design of an optimized production host requires the definition of undesired cellular functions, i.e., functions that should be eliminated. Such undesired functions could be, for example, all pathways with a product yield below a certain threshold. However, removing unwanted functions usually interferes with wanted functionalities too, as the eliminated reactions might also be participating in desired functions. If we want to keep some desired functionality we can exploit the concept of constrained minimal cut sets (cMCSs), which allows defining specific functions, e.g., production of ATP, that must be kept when undesired properties of the network are disabled.

3 The Standard cMCS Method

The typical workflow of the standard method to compute cMCSs is shown in Fig. 1. First, the metabolic model needs to be obtained. As mentioned above, these models are usually distributed in SBML format. It is essential that the obtained model is checked for correctness. Then, the syntactical correctness and validity of the SBML file should be tested, for instance by the online tool SBML Validator [8]. Next, the mass balance needs to be verified. Furthermore, for most applications it is essential that the model contains a working biomass function which guarantees that the organism is able to grow. Usually, these checks are performed with FBA and flux variability analysis (FVA).

As can be seen in Fig. 1, the standard cMCS method is mainly a three-step process. The first step concerns the reduction of the size of the metabolic network. The second step is the computation of all EFMs of the network. During the third step a so-called hitting set problem is solved in order to compute the cMCS from the EFMs. In the next sections these three steps are described in detail. As a running example, we aim to turn *E. coli* into an efficient ethanol producing cell factory. To represent *E. coli* in silico, we will use the genome-scale metabolic model *i*AF1260. It can be downloaded from the website of UC San Diego [9], and consists of 1668 metabolites and 2382 reactions. Growth is simulated in an anaerobic environment on a minimal medium with glucose as the sole carbon source. All other uptake reactions were removed from the model. Additionally, we removed all the reactions (identified by a FVA) that could not support any steady-state flux distribution under this condition [10]. Essentially, FVA checks for the existence of nonzero solutions by sequentially minimizing and maximizing the flux through each reaction in the network subject to the steady-state and the (ir)reversibility constraints. If both the minimum and the maximum flux through a reaction are zero, it cannot support a steady state, and can be removed without altering the total solution space. If FVA identified a reversible reaction to be irreversible under the used conditions, the reaction was appropriately redefined. This procedure resulted in a consistent network with 765 metabolites and 1413 reactions (1105 of which were irreversible), which represents the basis for the following analysis.

The standard cMCS method first requires the computation of all EFMs of the network. As the number of EFMs grows exponentially with the network size and as even medium scale metabolic models with only about 100 reactions can easily have several hundred

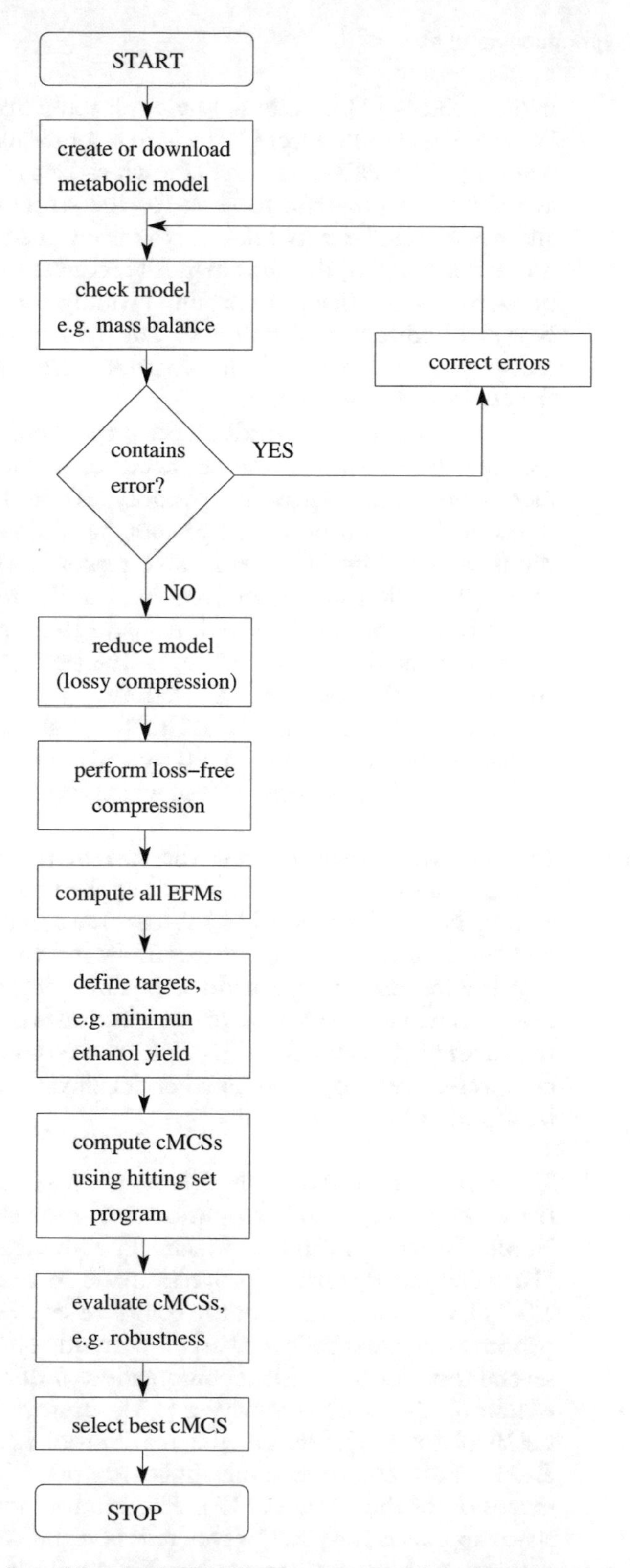

Fig. 1 The typical workflow of the standard method to compute cMCSs. The three main steps of this approach are (1) reduction of the model, (2) computation of all EFMs, and (3) solving a hitting set problem to determine the cMCSs

3.1 Network Reduction

3.1.1 Lossy Network Compression

million EFMs [11], an initial network reduction step is required. We use NetworkReducer [12] which is a module of the CellNetAnalyzer [13]. CellNetAnalyzer (for which free academic licenses are available) is a powerful toolbox for the structural and functional analysis of cellular networks. It is written in Matlab. NetworkReducer is a program that automatically reduces the size of metabolic networks by iteratively performing pruning and compression steps. NetworkReducer performs a so-called lossy compression which means that some metabolic information is irrecoverably lost during the compression procedure.

The input for NetworkReducer is the metabolic network to be reduced. Furthermore, the user needs to define a list of reactions, metabolites, and functions (phenotypes) to be protected from removal. For each protected phenotype, the algorithm calculates the flux variability of all removable reactions. Then, the reaction with the smallest flux range (considering the results of all phenotypes) is selected and removed. A final FBA is performed to guarantee that the flux constraints in all the protected phenotypes are still feasible after the reaction deletions.

This procedure was applied to the consistent network mentioned above and resulted in a (reduced) network with 431 metabolites and 448 reactions, 159 of which were reversible.

3.1.2 Loss-Free Network Compression

In order to further decrease the size of the reduced metabolic network, a loss-free compression is applied to the network generated by NetworkReducer [14]. A loss-free compression can be fully recovered without losing any metabolic information. For instance, a typical loss-free compression step is lumping together a chain of reactions where the product of one reaction is the educt of the next: the chain of reactions $R_1 \rightarrow R_2 \rightarrow R_3$ can be compressed to a single compressed reaction R_{123}. Further details on the compression may be found in [14, 15].

3.2 EFM Computation

The computation of the complete set of EFMs $\mathbf{E}$ of small metabolic networks is simple and straightforward. However, the number of EFMs of a network grows exponentially with the number of reactions [16]. Despite the major progress made in recent years [11, 14, 17–21], the computation of the complete set of EFMs for large and genome scale metabolic networks is currently out of reach. There are several tools for the efficient computation of EFMs available, such as Metatool [22], CellNetAnalyzer [13], Efmtool [17], and FluxModeCalculator [21]. We use the regEfmtool [23] to compute the EFMs of our compressed metabolic network. The regEfmtool is an extension of the Efmtool [17]. The Efmtool implements the null-space approach [14] and heavily relies on the concept of binary bit patterns to store and process the EFMs. Both Efmtool as well as regEfmtool are open source programs written in the multi-platform programming language Java and, hence, run on Windows, Linux, and Apple operating systems.

The computation of all EFMs for the reduced *E. coli* model with 431 metabolites and 448 reactions resulted in 596,865 EFMs and took approximately 5 min on a Linux server with 12 cores when using 10 parallel threads.

3.3 MCS Computation

After the computation of all EFMs of the network, we must decide which of the EFMs have to be removed (because they are responsible for unwanted functionalities) and which of the EFMs we want to keep (because they represent desired functions). Figure 2 shows all EFMs for our reduced *E. coli* model projected on a 2D plane spanned by the (molar) ethanol yield $\left(R_{\mathrm{ethanol}}^{\mathrm{secretion}}/R_{\mathrm{glucose}}^{\mathrm{uptake}}\right)$ and the (molar) biomass yield $\left(R_{\mathrm{biomass}}/R_{\mathrm{glucose}}^{\mathrm{uptake}}\right)$. As the non-negative linear combination of the EFMs defines the possible operating points, the organism can operate in any point that is within the envelope of all EFMs depicted in Fig. 2. In order to prevent the possibility that our host ends up in a state that does not produce any ethanol, we split the EFMs into two groups. One group contains all EFMs with a low yield. Here, we have chosen a limit of 1.4 as the minimum ethanol yield. The other group contains all EFMs with an ethanol yield higher than our limit (*see* the horizontal red line in Fig. 2).

In order to compute the cMCSs we convert all EFMs $\boldsymbol{e}$ from the numerical representation to their binary format $\boldsymbol{b}$:

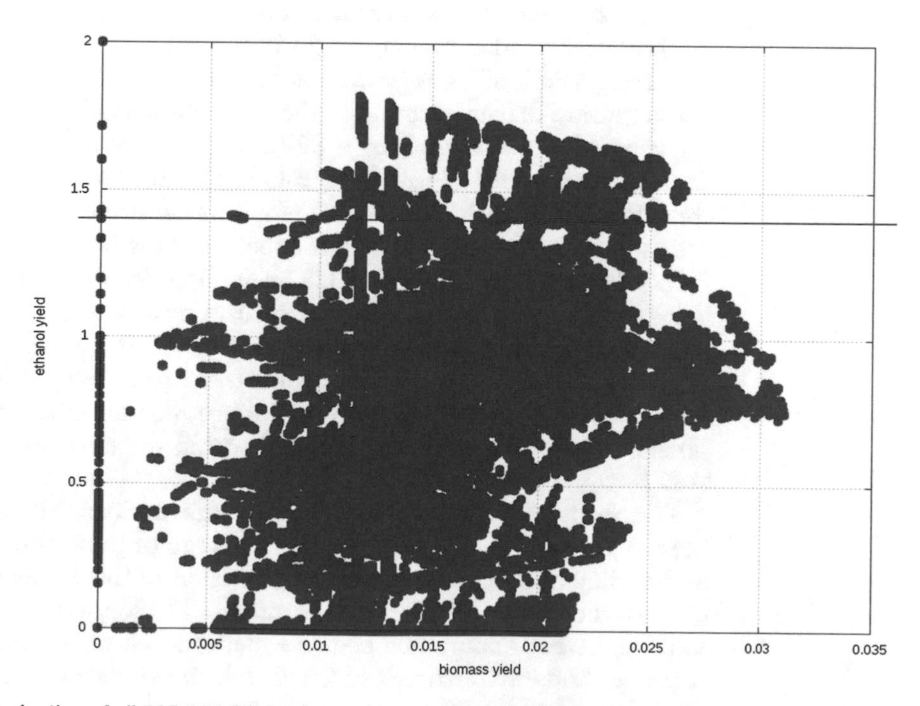

Fig. 2 Projection of all 596,865 EFMs of the reduced *E. coli* model on a 2D plane spanned by the molar ethanol yield and the molar biomass yield. The red line indicates the desired minimal ethanol yield of 1.4

$$b_i = \begin{cases} 0 \; if \; e_i = 0 \\ 1 \; if \; e_i \neq 0 \end{cases} \quad i = \{1..n\},$$

where the index i runs over all components of $\boldsymbol{e}$, i.e., over all of the n reactions.

$\boldsymbol{D}$ denotes the set of desired binary EFMs and $\boldsymbol{T}$ the set of binary EFMs we want to remove. Furthermore, we define the minimum number N_{min} of desired EFMs that have to survive the intervention. The computation of the cMCSs can now be solved by applying a hitting set algorithm [24]. A hitting set $\boldsymbol{h}$ representing a valid cMCS is characterized by

$$\boldsymbol{h} \cap \boldsymbol{t} \neq \emptyset \quad \forall \boldsymbol{t} \in \boldsymbol{T}$$

if for at least N_{min} of the desired EMFs $\boldsymbol{d}$ the following condition is met:

$$\boldsymbol{h} \cap \boldsymbol{d} = \emptyset.$$

Solving hitting set problems is numerically expensive and challenging, as hitting set problems are NP-hard [15]. We use the mhsCalculator that is a hitting set solver especially designed to efficiently compute cMCSs [25]. The mhsCalculator is a freely available open source software written in the programming language C. It supports two methods to compute the cMCS. The first method uses an integer linear program (ILP). This ILP is solved by utilizing a library of the commercial optimization toolbox CPLEX for which free academic licenses are available. The second method uses an adapted Berge algorithm [26].

Using a limit of 1.4 for the ethanol yield for the 596,865 EFMs and requiring that at least one of the most efficient EFMs survives, the mhsCalculator computed 8,837,604 potential cMCSs with a cardinality (number of reactions) ranging from 2 up to 18. Here, the efficiency of an EFM was defined as its substrate-specific productivity, i.e., product yield times specific growth rate (biomass flux/substrate uptake flux) [27]. The application of such a cMCS is depicted in Fig. 3. The diagram clearly shows that not a single EFM with an ethanol yield below 1.4 survived the genetic intervention. Consequently, it is impossible that our *E. coli* production host will end up in a state that does not produce any ethanol, as all possible operation states are confined within the envelope generated by the surviving EFMs.

Recently, further methods to compute optimal MCSs have been reported [28, 29]. The main advantage of these approaches is that they do not require a manual division of the EFMs in two groups, the desired EFMs and the undesired EFMs, as the optimal split is inherently computed by these methods. However, currently these methods are restricted to small scale metabolic networks.

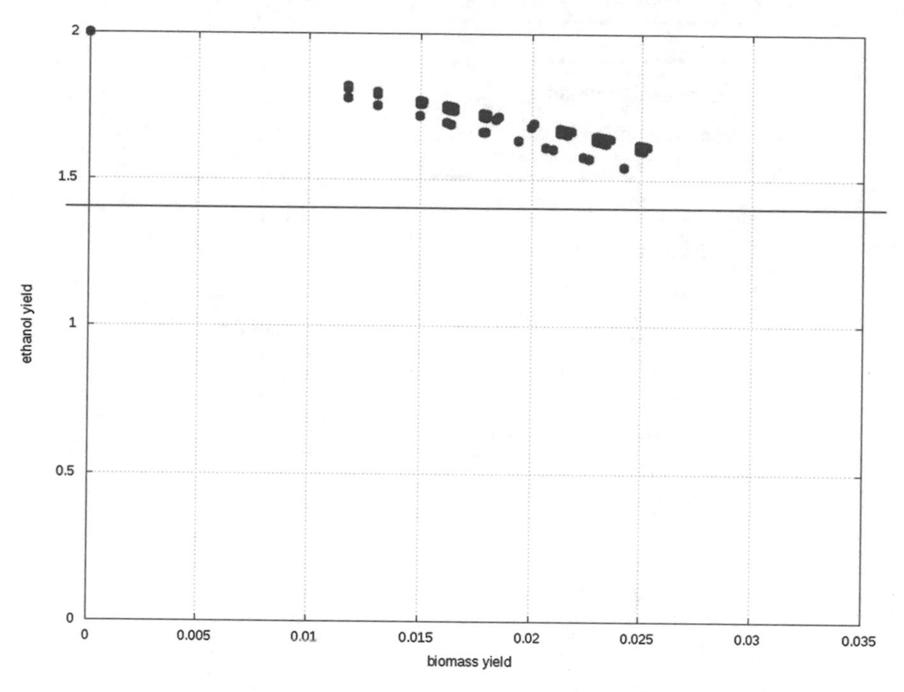

Fig. 3 The phenotypic solution space of the engineered *E. coli* strain when a cMCS with a cardinality of 8 is applied to the original network (*see* Fig. 2). In total, 863 EFMs survive the genetic intervention

4 Next-Generation cMCS Computation: The Dual System Approach

As the computation of all EFMs cannot be done for large and genome-scale metabolic networks, the previous approach to computing the MCSs for optimizing a production host is limited to small and medium scale networks. In this section we represent an approach that allows the calculation of cMCSs even for genome-scale metabolic networks by exploiting the fact that the EFMs of the "dual system" of the network are MCSs of the primal system [5]. The principal workflow of the next-generation dual approach is shown in Fig. 4. We start again with getting the metabolic model and checking the model for syntactical and biological correctness. A lossy network compression step is not required. However, in order to reduce the execution time a loss-free compression is usually performed (*see* Subheading 3.1.2). Next, the unwanted and desired functions are defined, and then, the system is transformed to its dual form that is a very fast and simple process. The dual system is translated to a MILP that is typically solved by a commercial optimization tool box, such as GUROBI or CPLEX. The output of the linear program is a set of potential cMCSs. Again, we need to analyze and evaluate the computed cMCS and select only proper candidates for the experimental realization. Note that the dual system approach only contains one numerically expensive step: the solution of the MILP.

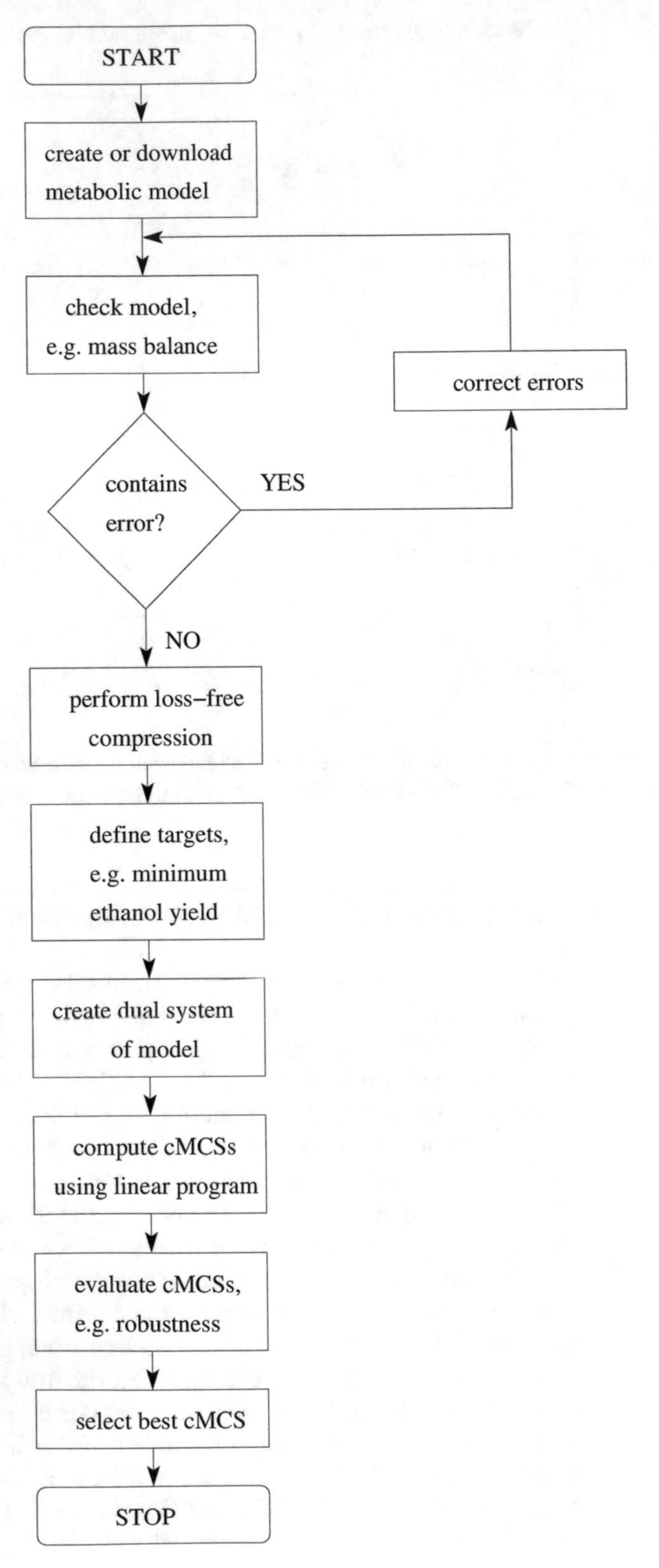

Fig. 4 The typical workflow of the advanced dual-system method to compute cMCSs. This method only involves one computationally expensive step: solving a linear program created of the dual system of the stoichiometric matrix

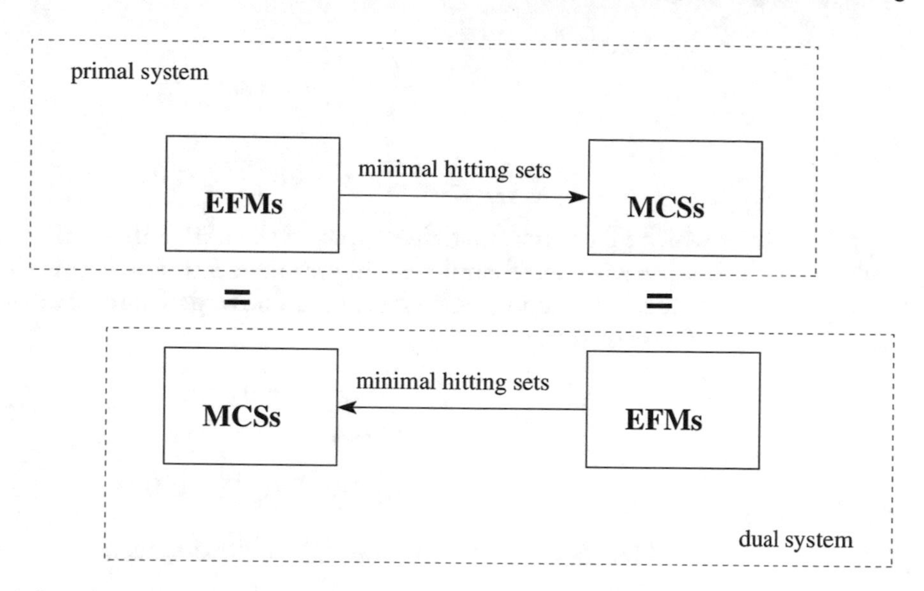

Fig. 5 Relation of primal and dual network properties

The dual system approach is based on Farkas's lemma [30] and is characterized by the fact that the MCSs of the primal system are equal to the EFMs of the dual system (*see* Fig. 5). As the creation of the dual system only involves very simple mathematical transformations, such as transposing and augmenting a matrix, we only need to compute the EFMs of the dual system to compute the wanted cMCSs. Furthermore, it is important to note that usually we are only interested in short cMCSs (cMCSs with a low cardinality), as these cMCSs require the fewest reaction deletions and thereby a minimum of wet lab work. This means that we can use the *k*-shortest EFM calculation approach introduced by de Figueiredo et al. [31] and we do not have to rely on computationally expensive algorithms based on the double description method [32].

We used the consistent genome-scale *E. coli* model from above which had been adapted to an anaerobic environment with glucose as the only carbon source. All EFMs that exhibit an ethanol yield of less than 1.0 are unwanted and, hence, the target constraint T was

$$\frac{R_{\text{ethanol}}^{\text{secretion}}}{R_{\text{glucose}}^{\text{uptake}}} \leq 1.0$$

$$R_{\text{ethanol}}^{\text{secretion}} - 1.0 R_{\text{glucose}}^{\text{uptake}} \leq 0$$

Without defining the desired functionality some of the computed MCSs might kill all other EFMs as well [37]. As we only wanted to calculate suitable MCSs, we defined the desired functionality D as follows

$$\frac{R_{\text{ethanol}}^{\text{secretion}}}{R_{\text{glucose}}^{\text{uptake}}} > 1.0$$

$$-R_{\text{ethanol}}^{\text{secretion}} + 1.0R_{\text{glucose}}^{\text{uptake}} < 0$$

which guarantees that after applying the cMCS there will always be EFMs with an ethanol yield higher than 1.0. Furthermore, we are only interested in solutions that are able to generate a biomass yield of at least 0.01:

$$\frac{R_{\text{biomass}}}{R_{\text{glucose}}^{\text{uptake}}} > 0.01$$

$$-R_{\text{biomass}} + 0.01R_{\text{glucose}}^{\text{uptake}} < 0$$

The system to be solved has the following form:

$$\begin{bmatrix} S_{\text{rev}}^T & I_{\text{rev}} & -I_{\text{rev}} & T_{\text{rev}}^T & 0 \\ S_{\text{irrev}}^T & I_{\text{irrev}} & -I_{\text{irrev}} & T_{\text{irrev}}^T & 0 \\ 0 & 0 & 0 & 0 & S \\ 0 & 0 & 0 & 0 & D \end{bmatrix} \begin{bmatrix} u \\ f_p \\ f_n \\ w \\ r \end{bmatrix} \begin{matrix} = \\ \geq \\ = \\ \leq \end{matrix} \begin{bmatrix} 0 \\ 0 \\ 0 \\ d \end{bmatrix}$$

$$t^T w \geq -c$$

$$u \in \Re^m, f_p, f_n, r \in \Re^n, w \in \Re^t, d \in \Re^d, f_p, f_n, r_{\text{irrev}}, w \geq 0, c > 0$$

MCSs of the primal system are EFMs of the above system of equations with a minimal support with respect to the reactions f_{pi} and f_{ni}. In order to simplify and speed up the linear program, the variables f_p and f_n were connected to the binary indicator variables z_p and z_n, respectively:

$$\begin{aligned} f_{pi} = 0 &\rightarrow z_{pi} = 0 \\ f_{pi} > 0 &\rightarrow z_{pi} = 1 \\ f_{ni} = 0 &\rightarrow z_{ni} = 0 \\ f_{ni} > 0 &\rightarrow z_{ni} = 1 \end{aligned}$$

To avoid futile two-cycles the following constraint was added:

$$z_{pi} + z_{ni} \leq 1$$

The objective function of the mixed integer linear program to be solved was

$$\min \sum_{i=1}^{n} \left(z_{pi} + z_{ni} \right)$$

which results in the minimization of the number of reactions participating in a cMCS. Each time when the solver finds a valid cMCS a simple constraint is added to the system and then the solver is restarted. This additional constraint prevents the solver from computing supersets of already computed cMCS [24].

The minimum number of reaction deletions for the above system is six. In total we computed 570 cMCS with a cardinality of six, which took approximately 12 h. Figure 6 shows the ethanol-biomass solution space for the original genome-scale network and the designed network. The red area depicts the potential operating states for the original network (anaerobic environment, minimal medium, and glucose as the only carbon source). The blue region depicts the possible ethanol-biomass space for the engineered genome-scale network which we obtained by applying one of the computed cMCS with six reaction deletions. As mentioned before, it is not possible to compute all EFMs of such a large network. Hence, unlike Figs. 2 and 3 that illustrate the discrete EFM values in the form of dots, we can only calculate the possible area of operating states by utilizing a flux variability analysis. Clearly, the blue region only allows states with an ethanol yield larger than 1.0. It can also be seen that the biomass yield of our designed strain can be 0.01.

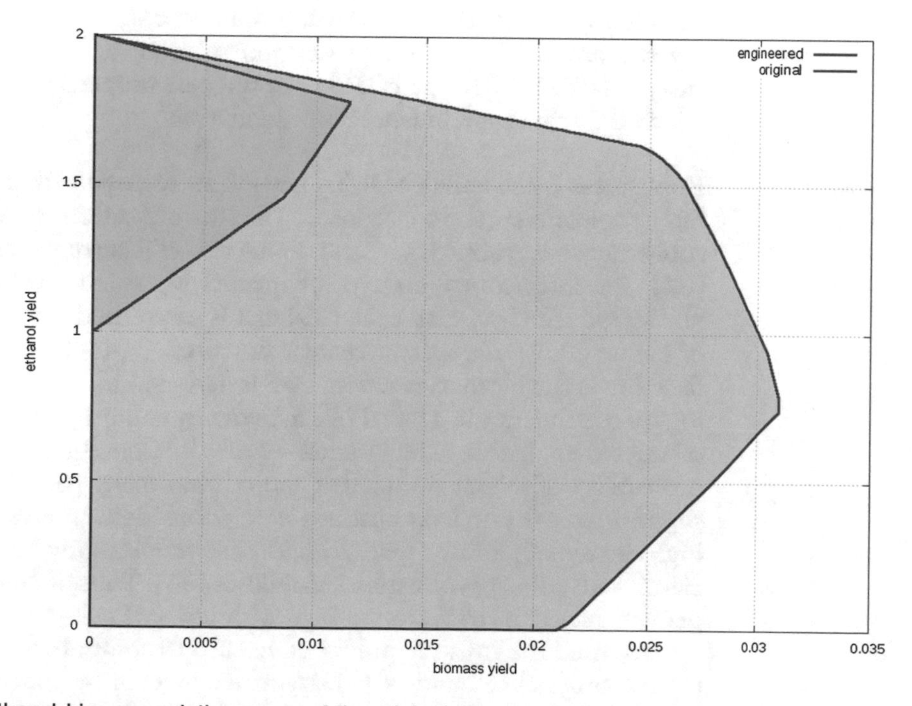

Fig. 6 Ethanol-biomass solution space of the original genome-scale network and a network engineered by applying a cMCS with six knockouts

5 MCS Evaluation and Selection

5.1 Feasibility of MCSs

There are different ways of deactivating reactions, such as knocking out a gene expressing an enzyme that is necessary for the reaction or disrupting the supply of a substrate. Some of the cell's reactions do not involve catalysts and simply rely on diffusion. Naturally, this type of reactions cannot be knocked out. Consequently, the set of obtained MCSs must be screened for any undeletable reaction. Preferably, these reactions are removed from the MCS computation process as early as possible, as this might have a beneficial influence on the runtime and hardware requirements.

5.2 Range of Solution Space

A specific cMCS removes all unwanted features from the host cell and, additionally, keeps wanted functionality. However, the deletion of a reaction—which is used to disable the undesired features—always also affects the desired properties of the production host. Some of the cMCSs might be better able to meet the desired requirements than others. For instance, if we want to design an *E. coli* strain that exhibits a minimum ethanol yield larger than 1.4, then some of the computed cMCSs might give a yield of just above 1.4 whereas other cMCSs might be able to produce a minimum yield of almost 1.6 (*see* Fig. 3).

Hence, one method of selecting the best cMCS is by computing the limits of the possible solution space by a flux variability analysis (FVA) [33]. The cCMS with the best properties might be selected for the experimental implementation.

5.3 Robustness

Robustness is the ability of living organisms to cope with perturbing (environmental) conditions. The quantification of cellular robustness is nontrivial and subject of current research [34]. Recently, a new concept of quantifying cellular robustness was introduced by Gerstl et al. [35] that is based on the probability of failure (PoF) of a specific cellular function. Typically, this target function is the biomass reaction. PoF is designed to measure topological redundancy in a metabolic network by counting the number of ways that disable—for instance—growth. Consequently, a low probability of failure means that the cell has many pathways that support growth and its robustness is therefore high. Interestingly, a high degree of pathway diversification has very recently been associated with robust production capabilities [36]. Thus, it is advantageous to select a cell factory design with low PoF.

We used the software published in [35] to compute the PoF of (a) the original cell and of (b) strains engineered by applying the calculated cMCSs to the original network. The PoF of the original *E. coli* strain was 9.9%, whereas the PoF of designed strains ranged from 11.2% to 11.4%. As expected, the PoF values of the designed strains are higher than in the wild type, because deleted reactions reduce the metabolic versatility of the cells. Furthermore, we see that the range of PoF values for the designed cells is rather small; hence, in this case, the evaluation of PoF values is only a weak indicator for selecting the best cMCS. Note, however, that we used the PoF with respect to growth as a proxy for assessing the robustness of ethanol production. This is possible as all our mutants are growth coupled and necessarily produce ethanol under all growth conditions and growth rates.

6 Conclusion

The computation of genetic intervention strategies to design and optimize production hosts is a nontrivial and computationally expensive task. The standard three-step method of computing cMCSs is not applicable to larger and genome-scale models, as all EFMs have to be calculated for this method. Hence, in order to run the standard cMCS computation, a lossy network-compression is required before the MCS computation. The main advantage of the advanced dual-system method is that a computation of the complete set of EFMs is not required. Thus, this method allows a direct computation of cMCSs from genome-scale metabolic models.

Acknowledgment

This work was supported by the Austrian BMWFW, BMVIT, SFG, Standortagentur Tirol, Government of Lower Austria and ZIT through the Austrian FFG-COMET-Funding Program. D.A.P.N. was funded by the Austrian Science Fund (FWF): Doctoral Program BioToP—Biomolecular Technology of Proteins (FWF W1224).

References

1. Lewis NE, Nagarajan H, Palsson BØ (2012) Constraining the metabolic genotype-phenotype relationship using a phylogeny of in silico methods. Nat Rev Microbiol 10:291–305. https://doi.org/10.1038/nrmicro2737
2. Orth JD, Thiele I, Palsson BØ (2010) What is flux balance analysis? Nat Biotechnol 28:245–248. https://doi.org/10.1038/nbt.1614
3. Klamt S, Gilles ED (2003) Minimal cut sets in biochemical reaction networks. Bioinformatics 20:226–234. https://doi.org/10.1093/bioinformatics/btg395
4. Hädicke O, Klamt S (2011) Computing complex metabolic intervention strategies using contrained minimal cut sets. Metab Eng 13:204–213. https://doi.org/10.1016/j.ymben.2010.12.004
5. Ballerstein K, von Kamp A, Klamt S, Haus UU (2012) Minimal cut sets in a metabolic network are elementary modes in a dual network. Bioinformatics 28(3):381–387. https://doi.org/10.1093/bioinformatics/btr674
6. von Kamp A, Klamt S (2014) Enumeration of smallest intervention strategies in genome-scale metabolic networks. PLoS Comput Biol 10:e1003378. https://doi.org/10.1371/journal.pcbi.1003378

7. Zanghellini J, Ruckerbauer DE, Hanscho M, Jungreuthmayer C (2013) Elementary flux modes in a nutshell: properties, calculation and applications. Biotechnol J 8:1009–1016. https://doi.org/10.1002/biot.201200269
8. Online SBML Validator (2016) http://sbml.org/validator/. Accessed 17 June 2016
9. BiGG Models (2016) http://bigg.ucsd.edu/models/iAF1260. Accessed 17 June 2016
10. Acuña V, Chierichetti F, Lacroix V, Marchetti-Spaccamela A, Sagot MF, Stougie L (2009) Modes and cuts in metabolic networks: complexity and algorithms. Biosystems 95:51–60. https://doi.org/10.1016/j.biosystems.2008.06.015
11. Jungreuthmayer C, Ruckerbauer DE, Gerstl MP, Hanscho M, Zanghellini J (2015) Avoiding the enumeration of infeasible elementary flux modes by including transcriptional regulatory rules in the enumeration process saves computational costs. PLoS One 10:e0129840. https://doi.org/10.1371/journal.pone.0129840
12. Erdrich P, Steuer R, Klamt S (2015) An algorithm for the reduction of genome-scale metabolic network models to meaningful core models. BMC Syst Biol 9:48. https://doi.org/10.1186/s12918-015-0191-x
13. Klamt S, Saez-Rodriguez J, Gilles ED (2007) Structural and functional analysis of cellular networks with CellNetAnalyzer. BMC Syst Biol 1:2. https://doi.org/10.1186/1752-0509-1-2
14. Gagneur J, Klamt S (2004) Computation of elementary modes: a unifying framework and the new binary approach. BMC Bioinformatics 5:175. https://doi.org/10.1186/1471-2105-5-175
15. Jungreuthmayer C, Nair G, Klamt S, Zanghellini J (2013) Comparison and improvement of algorithms for computing minimal cut sets. BMC Bioinformatics 14:318. https://doi.org/10.1186/1471-2105-14-318
16. Klamt S, Stelling J (2002) Combinatorial complexity of pathway analysis in metabolic networks. Mol Biol Rep 29:233–236. https://doi.org/10.1023/A:1020390132244
17. Terzer M, Stelling J (2008) Large-scale computation of elementary flux modes with bit pattern trees. Bioinformatics 24:2229–2235. https://doi.org/10.1093/bioinformatics/btn401
18. Hunt KA, Folsom JP, Taffs RL, Carlson RP (2014) Complete enumeration of elementary flux modes through scalable demand-based subnetwork definition. Bioinformatics 30:1569–1578. https://doi.org/10.1093/bioinformatics/btu021
19. Gerstl MP, Ruckerbauer DE, Mattanovich D, Jungreuthmayer C, Zanghellini J (2015) Metabolomics integrated elementary flux mode analysis in large metabolic networks. Sci Rep 5:8930. https://doi.org/10.1038/srep08930
20. Gerstl MP, Jungreuthmayer C, Zanghellini J (2015) tEFMA: computing thermodynamically feasible elementary flux modes in metabolic network. Bioinformatics 31:2232–2234. https://doi.org/10.1093/bioinformatics/btv111
21. van Klinken J, Willems van Dijk DK (2016) FluxModeCalculator: an efficient tool for large-scale flux mode computation. Bioinformatics 32:1265–1266. https://doi.org/10.1093/bioinformatics/btv742
22. von Kamp A, Schuster S (2006) Metatool 5.0: fast and flexible elementary modes analysis. Bioinformatics 22:1930–1931. https://doi.org/10.1093/bioinformatics/btl267
23. Jungreuthmayer C, Ruckerbauer DE, Zanghellini J (2013) regEfmtool: speeding up elementary flux mode calculation using transcriptional regulatory rules in the form of three-state logic. Biosystems 113:37–39. https://doi.org/10.1016/j.biosystems.2013.04.002
24. Jungreuthmayer C, Zanghellini J (2012) Designing optimal cell factories: integer programming couples elementary mode analysis with regulation. BMC Syst Biol 16:103. https://doi.org/10.1186/1752-0509-6-103.
25. Jungreuthmayer C, Beurton-Aimar M, Zanghellini J (2013) Fast computation of minimal cut sets in metabolic networks with a Berge algorithm that utilizes binary bit pattern trees. IEEE/ACM Trans Comput Biol Bioinform 10:1329–1333. https://doi.org/10.1109/TCBB.2013.116
26. Berge C (1989) Combinatorics of finite sets, Hypergraphs, vol 45. Elsevier, North Holland
27. Feist AM, Zielinski DC, Orth JD, Schellenberger J, Herrgard MJ, Palsson BØ (2010) Model-driven evaluation of the production potential for growth-coupled products of escherichia coli. Metab Eng 12:173–186. https://doi.org/10.1016/j.ymben.2009.10.003
28. Ruckerbauer DE, Jungreuthmayer C, Zanghellini J (2014) Design of optimally constructed metabolic networks of minimal functionality. PLoS One 9:e92583. https://doi.org/10.1371/journal.pone.0092583
29. Nair G, Jungreuthmayer C, Hanscho M, Zanghellini J (2015) Designing minimal microbial strains of desired functionality using a genetic

algorithm. Algorithms Mol Biol 10:29. https://doi.org/10.1186/s13015-015-0060-6

30. Farkas J (1902) Theorie der einfachen Ungleichungen. J Reine Angew Math 124:1–27
31. de Figueiredo LF, Podhorski A, Rubio A, Kaleta C, Beasley JE, Schuster S, Planes FJ (2009) Computing the shortest elementary flux modes in genome-scale metabolic networks. Bioinformatics 25:3158–3165. https://doi.org/10.1093/bioinformatics/btp564
32. Fukuda K, Prodon A (1996) Double description method revisited. Comb Comput Sci 1120:91–111
33. Mahadevan R, Schilling C (2003) The effects of alternate optimal solutions in constraint-based genome-scale metabolic model. Metab Eng 5:264–276. https://doi.org/10.1016/j.ymben.2003.09.002
34. Behre J, Wilhelm T, von Kamp A, Ruppin E, Schuster S (2008) Structural robustness of metabolic networks with respect to multiple knockouts. J Theor Biol 252:433–441. https://doi.org/10.1016/j.jtbi.2007.09.043
35. Gerstl MP, Klamt S, Jungreuthmayer C, Zanghellini J (2016) Exact quantification of cellular robustness in genome-scale metabolic networks. Bioinformatics 32:730–737. https://doi.org/10.1093/bioinformatics/btv649
36. Yang L, Srinivasan S, Mahadevan R, Cluett WR (2015) Characterizing metabolic pathway diversification in the context of perturbation size. Metab Eng 28:114–122. https://doi.org/10.1016/j.ymben.2014.11.013
37. Mahadevan R, von Kamp A, Klamt S (2015) Genome-scale strain designs based on regulatory minimal cut sets. Bioinformatics 31(17):2844–2851. https://doi.org/10.1093/bioinformatics/btv217

Chapter 18

Optimization of Multi-Omic Genome-Scale Models: Methodologies, Hands-on Tutorial, and Perspectives

Supreeta Vijayakumar, Max Conway, Pietro Lió, and Claudio Angione

Abstract

Genome-scale metabolic models are valuable tools for assessing the metabolic potential of living organisms. Being downstream of gene expression, metabolism is increasingly being used as an indicator of the phenotypic outcome for drugs and therapies. We here present a review of the principal methods used for constraint-based modelling in systems biology, and explore how the integration of multi-omic data can be used to improve phenotypic predictions of genome-scale metabolic models. We believe that the large-scale comparison of the metabolic response of an organism to different environmental conditions will be an important challenge for genome-scale models. Therefore, within the context of multi-omic methods, we describe a tutorial for multi-objective optimization using the metabolic and transcriptomics adaptation estimator (METRADE), implemented in MATLAB. METRADE uses microarray and codon usage data to model bacterial metabolic response to environmental conditions (e.g., antibiotics, temperatures, heat shock). Finally, we discuss key considerations for the integration of multi-omic networks into metabolic models, towards automatically extracting knowledge from such models.

Key words Multi-omics, metabolic models, Flux-balance analysis, Machine learning, Data integration, Multi-objective optimization

1 Introduction

Metabolism is the set of biochemical reactions in a cell which maintain its living state. As these reactions are indispensable, it is vital that metabolic networks in all living organisms are as well-characterized as possible. In the higher organization level of a microbial community, cells can act as either sinks or sources of metabolites in their environment, as they consistently produce or deplete a range of metabolites in the environmental metabolite pool [1]. Being downstream of gene expression, metabolism is increasingly being used as an indicator of the phenotypic outcome for drugs and therapies, as well as for cancer studies [2].

Supreeta Vijayakumar and Max Conway contributed equally to this work.

Marco Fondi (ed.), *Metabolic Network Reconstruction and Modeling: Methods and Protocols*, Methods in Molecular Biology, vol. 1716, https://doi.org/10.1007/978-1-4939-7528-0_18,

Constraint based reconstruction and analysis (COBRA) techniques are commonly used for modelling reconstructions of metabolic networks at the genome scale. The most widely used method is flux balance analysis (FBA), which has long been used to mathematically express the flow of metabolites through a network of biochemical pathways. FBA uses the assignment of stoichiometric coefficients to represent each of the metabolites involved in any given reaction [3]. Through these coefficients, mass-balance constraints can be imposed on the system to identify a range of points representing all possible flux distributions, which correspond to the set of feasible phenotypic states. In this solution space, there exists a global optimal value which satisfies a given objective function (usually the maximisation of biomass). For purposes of mass conservation, all fluxes within this system are calculated under the steady state assumption where the total amount of any metabolite being produced must be equal to the total amount of that metabolite consumed [4], and the cell can utilize resources optimally in time-invariant and spatially homogeneous extracellular conditions [5, 6]. Linear programming can be used to maximize an objective function indicating the extent to which each reaction contributes to a certain phenotype, under constraints which can be defined by a cell's metabolic potential, stoichiometry, and limits of reaction and transport rates [1].

The main advantage of using FBA is that it does not invariably require the definition of kinetic parameters. In fact, fluxes are calculated in a pseudo-steady state using stoichiometric coefficients and mass balances; this affirms its suitability for building mechanistic predictive models from genome-scale metabolic networks [7]. Using the optimal value obtained through FBA, flux variability analysis (FVA) [8] returns the maximum and minimum values for fluxes through each reaction whilst keeping the formation of biomass to a minimum, which can help in calculating the rate of metabolite consumption or production [9].

More detailed analyses can be carried out to provide a deeper insight into certain aspects of metabolic processes. To overcome the limitations of the steady state assumption, dynamic FBA can be carried out by monitoring time dependent changes in the concentration of metabolites and reaction fluxes over time [5]. This involves calculating the conservation of mass for each of the metabolites consumed and produced in reactions and imposing additional constraints on the rates of flux changes, non-negative metabolite and flux levels and transport fluxes [10].

Several genome-scale metabolic models are readily available in online repositories such as KEGG [11], BIGG [12], BioCyc [13], and SEED [14]. These are prepared by building a genome-scale reconstruction of all metabolic reactions taking place in the organism followed by manual curation, gap-filling and annotation of specific genes, metabolites and pathways with descriptive metadata.

Recently, an increasing number of genome-scale signalling and regulatory networks are also being compiled in order to garner a better understanding of the underlying mechanisms of metabolic pathways [5], and approaches to extract pathway cross-talks have been proposed [15].

Parsimonious enzyme usage FBA (pFBA) is a variant of FBA which aims to maximize the stoichiometric efficiency of a metabolic network by identifying a subset of genes which contribute to maximizing the growth rate in silico. These include both essential and non-essential genes, as well as those which are enzymatically and metabolically less efficient and those which are completely unable to carry flux in experimental conditions [16].

For a more detailed introduction to constraint-based metabolic models, the interested reader is referred to the following texts: [17, 18]. After reviewing the available methods for optimization of metabolic networks, we also provide a tutorial for multi-objective optimization using METRADE. The tutorial illustrates how to predict bacterial multi-response under varying environmental conditions, by computing the trade-off between contrasting metabolic objectives.

Finally, we recognize that systematic fusion of multiple data types into a single, cohesive network is a challenge faced by many modellers, particularly when measuring bacterial response at multiple omic levels. In view of this, we include a critical perspective describing key considerations for the integration of multi-omic networks into metabolic models, towards automatically extracting knowledge from such models.

2 Materials

2.1 Multi-Target Optimization of Multi-Omic Metabolic Networks

Available methods for analysis of metabolic networks and metabolic engineering usually define gene lethality in terms of effect on the growth rate only. In fact, organisms often have multiple objectives to satisfy in addition to the maximization of biomass. To this end, a number of approaches have been recently proposed to take into account multi-target optimization of cellular tasks. Unlike single-objective approaches, these allow for simultaneous maximization or minimization of two or more properties of interest.

Gene knockout simulation is one of the most consistently used methods for determining the essentiality of genes, and has been successfully applied to the design and optimization of strains for metabolic engineering. However, it has been contended that single gene perturbations can often fail to capture the essentiality of genes or localize gene function owing to genetic redundancy. As a result, when a metabolic function is encoded by two or more genes, the removal of any one of these genes will not result in an altered phenotype, and it may therefore be falsely concluded that they are superfluous [19]. The regulatory on/off minimization (ROOM) algorithm uses mixed integer linear programming to predict the metabolic state of an organism following knockout [20]. This is achieved by searching for the flux distribution of the perturbed strain that minimizes the number of significant flux changes (which may allude to underlying regulatory changes after knockout) whilst satisfying all stoichiometric, thermodynamic, and flux capacity constraints applied during FBA. On the other hand, multiple genetic perturbations carried out concurrently may lead to issues relating to technical and conceptual scaling. Hence, pairwise gene knockouts may be considered better for identifying which deletions have a damaging effect. For instance, a computational approach has been presented for identifying dosage lethality effects (IDLE) in genome-scale models of cancer metabolism [21] using synthetic dosage lethality to simulate the pairwise knockout of non-essential enzymes by overexpressing the first enzyme-coding gene but underexpressing the second.

On the whole, performing complete gene knockouts is still likely to present a number of complications such as: (1) the lack of information regarding the effect of removing essential reactions; (2) increased compression of the flux distribution following the removal of flux values during knockout; (3) difficulty in optimizing fluxes if they are limited to their Boolean definition of having either a lethal or neutral phenotypic effect [22].

To address the problem of the state-space explosion when considering all possible combinations of multiple gene knockouts, evolutionary algorithms have been proposed, both searching in the discrete space of gene knockouts [23] and in the continuous space of gene partial overexpression/underexpression [24]. This enables the consideration of more than one objective function and expands the phenotypic solution space as there are a greater number of feasible optimal points. Multi-objective optimization can help to resolve trade-offs between conflicting metabolic objectives through simulating a series of optimal, non-dominated vectors in the multi-dimensional objective space. In metabolic engineering, each vector may represent a Boolean gene knockout strategy, or a real-valued partial knockdown/overexpression strategy. For such vectors, there is no better solution which exists for a given objective without sacrificing the performance of another [25]. This is known as a Pareto front and enables the consideration of multiple conditions and constraints affecting each objective in a multi-objective optimization problem.

The key advantage of multi-objective optimization is that it seeks a trade-off between multiple cellular objectives, without the need to define individual weights and combine them into a single objective [26] or hierarchically order objectives [27]. This eliminates difficulties associated with choosing the most suitable objective function or selecting weights which uniformly represent the

Pareto front. The use of multi-objective evolutionary algorithms (MOEAs) such as NSGA-II [28], SPEA2 [29] and MOEA/D [30] quickly renders all Pareto-optimal solutions when objectives are simultaneously optimized. Linear physical programming-based flux balance analysis (LPPFBA) orders objectives by their Pareto-optimal solutions to identify those which are in conflict [31]. This helps to select regions of the solution space which contain feasible fluxes. Optimal flux vectors can also be found using comprehensive polyhedra enumeration flux balance analysis (CoPE-FBA) through finding the topology of sub-networks corresponding to these vectors [32]. In this method, dividing reversible reactions into separate forward and backward reactions further simplifies the solution space for finding non-decomposable flux routes [33].

Multi-objective optimization can be implemented into FBA using the noninferior set estimation (NISE) method to approximate Pareto curves for conflicting objectives and examine flux at all Pareto-optimal solutions [34]. More recently, variations of MOFBA and MOFVA have been used to compute metabolic trade-offs for multiple species within microbial communities in terms of growth rates and associated reactions [35]. Thermodynamic states have also been incorporated in such analyses to inform responses to environmental conditions. Estimations of maximum yields using single objective optimization can be extended for multiple objectives to find the area for which one factor cannot be increased without sacrificing another (i.e., a Pareto surface of yield versus productivity), through which it is possible to devise strategies for improving performance by increasing metabolic flexibility [36].

As a pre-processing step, sensitivity analysis can be carried out to discover the most influential inputs for the multi-objective optimization problem by interrogating the pathway, reaction or species spaces of the model. In particular, pathway-oriented sensitivity analysis [23] has proved to be useful in metabolic engineering for improving the robustness of strains by determining the most sensitive metabolic pathways; this is achieved by identifying which knockouts or genetic manipulations contribute the most towards a certain output.

2.2 Integration of Multi-Omic Data Types into Genome-Scale Metabolic Models

Several methods for integration of gene expression data into metabolic models have been proposed; for a comprehensive review, the reader is referred to Machado and Herrgård [37]. However, it has readily been established that multi-omic integration of data allows for a more comprehensive evaluation of model predictions, rather than solely relying on gene expression profiling for the observation of metabolic responses over a range of different environmental conditions. The optimization of transcriptomic and proteomic layers with respect to different growth conditions serves to refine predictions of metabolic phenotypes (Fig. 1).

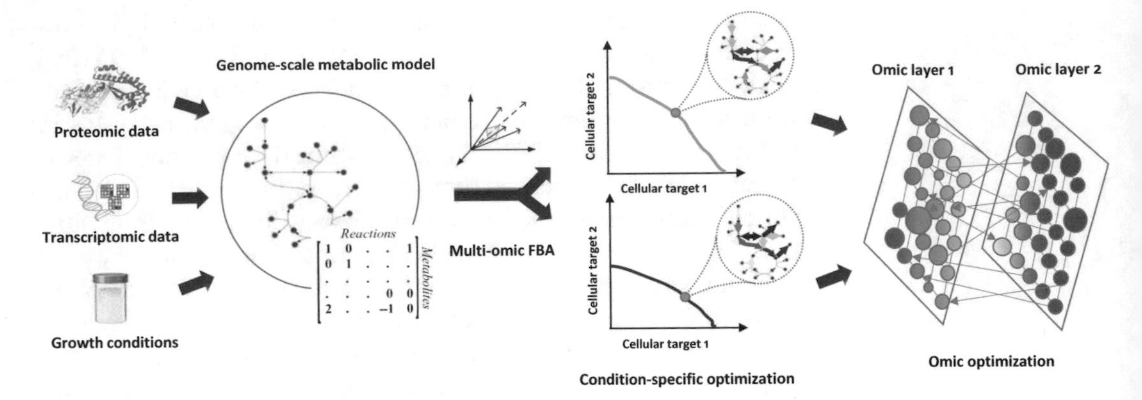

Fig. 1 Through the collection of transcriptomic, proteomic, and other omic data across various growth conditions from in-vivo experiments and existing literature, a genome-scale metabolic model can be constructed and FBA carried out at multiple levels. The simulation of growth under different conditions allows for condition-specific optimization of each of the omic layers, which can then be combined to form a multi-omic network

Regulatory FBA (rFBA) is an extension of FBA which adds the dimension of transcriptional regulation to improve flux predictions for dynamic models by recording transcriptional events and protein activity as well as simulating the uptake of metabolites, biomass production, and the secretion of by-products [38]. Alternatively, the probabilistic regulation of metabolism (PROM) method combines gene expression data with transcriptional regulatory networks by quantifying the interactions from high-throughput data in an automated fashion [39] to overcome limitations associated with Boolean logic. This is achieved through the use of conditional probabilities to represent gene states and gene–transcription factor interactions [40]. Therefore, a greater number of interactions can be modelled, consequently improving the prediction of phenotypic states for various transcriptional perturbations.

Conditional FBA applies conditional dependencies present in the metabolic model as constraints for each flux. In other words, each flux is constrained by the activity of the compound that facilitates it. For example, temporal variations in response to varying light intensity and associated conditional dependencies were included in a constrained genome-scale metabolic model, in order to simulate the phototrophic growth of the cyanobacterium *Synechocystis* sp. PCC 6803 over a diurnal cycle [41]. More recently, a system was devised using *Synechococcus elongatus* PCC 7942 as a model to study issues concerning resource allocation encountered during phototrophic growth [42].

A unified measure of bacterial responses computed by a condition-specific model allows for the detection of coordinated responses shared between different data types as well as the variation in responses across differing growth conditions. In this regard, a method for the concatenation of disparate omics data types (layers) has been proposed over varying growth conditions (nodes) into an aggregated model [43]. Using multilayer network models, the omics were weighted for the reliability of the flux rate predictions. Additionally, calculating flux distributions with multiple levels allowed for exploration of the total metabolic potential of the organism and the use of a non-binary measure of gene expression. By coupling fluxomic and proteomic data, a novel biological relationship was uncovered between protein structure and translational pausing, as well as an improved in vivo estimation of genome-wide enzyme turnover rates [44]. This approach helped to develop a parameterized model to predict responses to conditions, and consequently inform metabolic cost-benefit ratios at the cellular level.

The minimization of metabolic adjustments (MOMA) uses quadratic programming to solve its optimization problem. The objective function is calculated as the distance between two different flux distributions: the flux distribution for optimal growth rate and the flux distribution following the generation of a knockout mutant through genetic perturbation [45]. This accounts for the fact that knockout mutants are likely to display a lower growth rate than the wild type, therefore their flux distribution is better predicted by the minimal flux response to the knockout rather than by an optimal growth rate [46]. The inactivation of genes imposes additional constraints on the system, arguably leading to a shift towards a more valid and biologically meaningful representation of flux distribution as close as possible to that of the wild type [47, 48]. Similarly, integrative omics metabolic analysis (IOMA) uses a mechanistic model to determine reaction rates by incorporating quantitative proteomic and metabolomic data into the model to deliver more accurate predictions of flux alterations following genetic perturbation [49].

In the context of metabolic engineering, multi-omic integration has been used to find strain-specific differences for the improved selection and design of optimal strains. The goal is threefold: (1) maximizing the theoretical yield of a particular metabolic product by comparing high flux reactions between strains using physiological data added to the model; (2) quantifying differential gene expression using transcriptomic profiles; (3) analyzing gene expression across different conditions, thus characterizing the specific metabolic capabilities of individual strains [50]. Gene expression measurements can be obtained from microarray and/or RNA sequencing data from public repositories for integration with metabolic networks. Gene inactivity moderated by metabolism and expression (GIMME) is a switch-based algorithm which can be used to perform discretization (i.e., binary classification) of gene expression data to reduce the amount of experimental noise, by finding inactive genes in the dataset and re-enabling flux associated with false negative values [51]. Chiefly, the algorithm scores the consistency of gene expression data for a given metabolic objective [52].

Conversely, there are a number of valve-based algorithms such as E-flux [53] and METRADE [24], which treat gene expression data as continuous rather than discrete. Lower and upper bounds are set so that the maximum allowable flux for a reaction is a function of the normalized expression of genes controlling that reaction. The idea is to tightly constrain the maximum and minimum flux when the expression for a gene is low, but relaxing these constraints when the expression is high. Due to the addition of these constraints, performing FBA returns an altered flux distribution, which may consequently alter the corresponding metabolic state or optimal metabolic capacity identified. There is another branch of methods which employ "pruning" so that only a core set of reactions are retained in the metabolic model. Methods using this approach to integrate models with tissue-specific data include MBA [54], FASTCORE [55], and mCADRE [56].

Since an increasing number of genome-scale transcriptional regulatory networks are now available, methods like PROM [39] should be preferred to examine cellular transcriptional activity, as they do not rely on assigning a Boolean on/off state to each gene. Regulatory elements may also be incorporated into models by performing enrichment analysis of transcription factors for differential control of genes [50], or by merging transcriptional regulatory networks with constraint-based metabolic models [57]. A multilayer model was constructed for *Escherichia coli* [58] which merged sub-models of transcriptional regulatory networks, signal transduction pathways and metabolic networks; trained parameters were fed into the model to return information for an objective function and set of constraints with subsequent model predictions improved through supplementation with experimental data. To bridge the gap (and the still debated assumption of strong correlation) between gene expression levels and protein abundance, a method was recently proposed to account for the synonymous codon usage bias [59].

We believe that the large-scale comparison of the metabolic responses between different environmental conditions will be an important challenge for genome-scale modelling. In the following section, a tutorial is presented for METRADE [24], which gives a step-by-step guide to perform optimization of metabolic models. This is achieved by mapping gene expression values to the objective space of a genome-scale metabolic model and performing multi-objective optimization for identifying optimal phenotypes through the comparison of predicted flux rates for multiple objectives. METRADE develops a multi-omic model of *Escherichia coli* that includes a multi-objective optimization algorithm to find the allowable and optimal metabolic phenotypes through concurrent maximization or minimization of multiple metabolic markers. A number of experimental conditions are mapped to the model through transcriptomic data, and then mapped to a phenotypic multidimensional objective space.

3 Methods

The framework for the metabolic and transcriptomics adaptation estimator (METRADE) incorporates multi-objective optimization by constructing a Pareto front which displays gene expression profiles and codon usage arrays in a condition-phase space, where each profile is associated with a growth condition [24]. This allows for comparison of objectives to identify the best trade-off, where the maximal number of cellular objectives are simultaneously optimized. Sets of Pareto-optimal solutions in the front may be represented using a hypervolume indicator [60], enabling comparison between mapped conditions and examination of Pareto set evolution towards an optimal configuration over time.

In the context of metabolic engineering, strains may be compared for their ability to simultaneously fulfil multiple objectives and optimize production of multiple metabolites at the same time. It is also possible to establish the optimal growth conditions necessary to achieve this output and devise strategies for further optimization through performing gene knockouts or changing flux rates in-vitro. Additional insights into bacterial adaptability can be obtained through principal component analysis (PCA) [61], pseudospectra [62], and community detection [63]. PCA aids investigation of components (i.e., expression profiles) with the greatest variance for multiple objectives, whereas the pseudospectra and community detection methods elucidate the community structure of bacteria in the condition phase-space.

METRADE can be run (1) as a standalone program to find the optimal gene expression values for maximization of given cellular objectives, and (2) on a dataset of growth conditions to find the predicted flux rates in any given condition.

3.1 Initial Settings

METRADE is fully compatible with the COBRA 2.0 toolbox [64]. The full code needed for METRADE can be downloaded from http://www.nature.com/articles/srep15147. The user can download COBRA toolbox for MATLAB from http://opencobra.github.io/ and set the local COBRA folder in the MATLAB path with the instruction

```
addpath ( genpath ( ' local_path_to_COBRA_toolbox ' ) );
```

Load the model, e.g., the one included in the folder, *Escherichia coli* iJO1366 [65] with acetate-biomass set as objectives:

```
load ( ' iJO1366_Ecoli_ac. mat ' )
```

The variable *fbamodel.f* selects the first objective (default: biomass). The variable *fbamodel.g* selects the second objective (default:

acetate). To find the indices of the reactions for oxygen, succinate and acetate import/export, type

```
ix_o2 =  find (ismember (fbamodel. rxns,  'EX_o2( e ) ')==1);
ix_succ = find (ismember ( fbamodel. rxns, 'EX_succ( e ) ')==1);
ix_ac = find (ismember ( fbamodel. rxns, 'EX_ac( e ) ')==1);
```

The pair of objective functions can be changed. For instance, to change the second objective to succinate, use:

```
fbamodel. g( ix_ac ) =  0;
fbamodel. g( ix_succ ) =  1;
```

To change between aerobic and anaerobic conditions, we have to set a new lower bound for the reaction importing oxygen. For instance, to simulate an anaerobic condition, set the lower bound to zero (no import allowed).

```
fbamodel. lb ( ix_o2 ) =  0;
```

Note that default aerobic conditions are considered with a lower bound of − 10 mmol/h/gDW. A negative lower bound represents the maximum rate available for the import of that metabolite. Anaerobic conditions are with a null lower bound.

3.2 Mapping Gene Expression Compendia to Multidimensional Objective Spaces

Run *pareto_microarray_fluxes.m*. This will generate flux rates for 466 given growth conditions [66], and will save them in a variable called *fluxes*. The two fluxes chosen as objectives (default: biomass and acetate) will be saved in a file called *points*. These values represent the coordinates of the points in the bi-dimensional objective space shown in the paper.

Listing 18.1 Mapping growth conditions on multidimensional phenotypic spaces:

```
format long
% starts the p a r a l l e l toolbox to use four cores
i f ( matlabpool ( ' size ' ) == 0) % opens only i f i t i s closed
  matlabpool ( ' open ', ' local ',4)
end
% i n i t i a l i s e s the Cobra toolbox
initCobraToolbox
% loads variables
load ( ' genes. mat ' );
load ( ' reaction_expression. mat ' );
load ( ' probe_genes. mat ' );
load ( ' glucose. mat ' );
load ( ' oxygen. mat ' );
load ( ' name_conditions_with_replicates. mat ' );
load ( ' name_conditions. mat ' );
```

```
if evalin('base','exist(''data_only'',''var'')')==0
  load('data_only.mat');
end
load('gene_variances.mat')
max_gene_importance = 10000;
% The following instructions find the locations of the gene 'bXXXX' (genes in the fbamodel
) in the array probe_genes (sequence of genes appearing in the microarray data available)
position_gene = cell(length(genes),1);
for i=1:length(genes)
  matches = strfind(probe_genes, genes{i});
  position_gene{i} = find(~cellfun('isempty', matches));
end
points = zeros(size(data_only,2),2);  %table of points coordinates. The number of points
is equal to the number of conditions in the microarray data
gene_importance = zeros(length(genes),1);  % array of the coefficients indicating
gene importance. The size is equal to the number of genes in the model
min_var = min(gene_variances);
probe_gene_importance = max_gene_importance * 1./(gene_variances/min_var);  %this
way the importance of a gene can range from 0 to max_gene_importance
for i=1:length(genes)
  if isempty(position_gene{i})
    gene_importance(i) = max_gene_importance/2;
  else
    gene_importance(i) = probe_gene_importance(position_gene{i});
  end
end
number_conditions = size(data_only,2);
points = zeros(number_conditions,2);
fluxes = zeros(length(reaction_expression), number_conditions);
for index_cond = 1 : number_conditions
  microarray_data = data_only(:, index_cond);
  [v1, out] = process_conditions(microarray_data, index_cond, genes, position_gene,
fbamodel, oxygen, glucose, name_conditions, name_conditions_with_replicates, reactio-
n_expression, gene_importance); %it is necessary to pass the original fbamodel that will
be changed in the subfunction (oxygen will be put to zero or not according to the anaerobic or aero-
bic condition, the default condition in fbamodel is aerobic)
  points(index_cond,:) = out;
  fluxes(:, index_cond) = v1;
  disp(index_cond);
end
string = ['points_gene_importance_' num2str(max_gene_importance)];
save(string, 'points');
string = ['fluxes_gene_importance_' num2str(max_gene_importance)];
save(string, 'fluxes');
```

3.3 Multi-Objective Optimization of Gene Expression

We will now solve the inverse problem, namely finding the best genome-wide expression values that allow for the maximization of two given cellular objectives. This part implements a multi-objective optimization algorithm using a genetic algorithm based on NSGA-II [28] (the comments in the genetic algorithm code below are adapted from the original NSGA-II implementation). The trade-off between multiple metabolic objectives is found as a result. Such methods can guide genetic engineering to find the best gene expression values for specific goals. Furthermore, they can elucidate the metabolic capability of an organism and the relationship between contrasting cellular objectives.

To start the optimization, launch *RUN.m* (by editing the file, it is possible to set the number of cores and select the model). The number of populations of the optimization algorithm is set to 150 by default, and the number of individuals per population is set to 100. We suggest keeping this proportion. The results in the METRADE paper have been obtained with 1500 populations of 1000 individuals each.

Listing 18.2 Multi-objective optimization of metabolic models. The code has been parallelized to work on all the available cores when executed on a multi-core processor using the parfor function.

```
load ' genes. mat ';
load ' reaction_expression. mat ';
M = 2; % number of objective functions
V = length ( genes ); %length of the input individuals with-
out ranking, crowding distance and outputs
N = pop; %population size
min_range= zeros (1,V);    % the expression of each gene is >= 0
max_range= 100 * ones (1,V); % the expression of each gene is <= 100%
% Initialize the population
 if (( last_gen==0))
  chromosome =ones (pop,V+M); % gene expressions initialized
as1, i.e. all the genes are normally expressed ( reference state ). A
chromosome means, in our case, an array of gene expression values%
chromosome= chromosome+ 0.1.* ( rand (pop,V+M) -0.5.* ones (pop,V
+M) ); % adds some initial random noise
  chromosome= chromosome+ 2.* ( rand (pop,V+M) -0.5.* ones (pop,V+
M) ); % adds some initial random noise
  [ v1, fmax ] = flux_balance ( fbamodel, true );
  for i =1:pop
    chromosome( i,V +1)= - fmax;% acetate
    chromosome( i,V +2)= - fbamodel. f ' * v1;%biomass
  end
  %% Sort the Initialized population
  % Sort the population using non - domination-sort. This returns
two columns for each individual which are the rank and the crowding dis-
tance corresponding to their position in the front they belong.
```

```
At this stage the rank and the crowding distance for each chromosome i
s added to the chromosome vector for easy of computation.
  chromosome = non_domination_sort_mod (chromosome, M, V);
 else
  sol =[ ' solution ' num2str ( last_gen ) '. mat ' ];
  load ( sol );
end
%% Start the evolution process
% The following are performed in each generation
% * Select the parents which are fit for reproduction
% * Perform crossover and mutation operators on the selected parents
% * Perform selection from the parents and the offspring
% * Replace the unfit individuals with the fit individuals
to maintain a
% constant population size.
for i = last_gen+1 : gen
  % Select the parents
  % Parents are selected for reproduction to generate offspring. The o
riginal NSGA -II uses a binary tournament selection
based on the crowded-comparison operator. The arguments are
  % pool - size of the mating pool. It is common to have this to be half
the population size.
  % tour - Tournament size. Original NSGA -II uses a binary tourna-
ment selection, but to see the effect of tournament size this is
kept arbitrary, to be chosen by the user. pool = round (pop /2);
  tour = 2;
  % Selection process
   % A binary tournament selection is employed in NSGA -II.
In a binary tournament selection process two individuals
are selected at random and their fitness is compared.
The individual with better fitness is selected as a parent.
Tournament selection is carried out until the pool size is fille
d. Basically a pool size is the number of parents to be selected.
The input arguments to the function tournament_selection are chromo-
some, pool, tour. The function uses only the information from last
two elements in the chromosome vector.
    % The last element has the crowding distance information
while the penultimate element has the rank information. Selection is
based on rank and if individuals with same rank are encountered,
crowding distance is compared. A lower rank and higher crowding dis-
tance is the selection criteria.
 parent_chromosome = tournament_selection (chromosome, pool, tour
); % We now apply crossover and mutation operators
 mu = 20;
 mum = 20;
 if (num_cores >1)
     offspring_chromosome = genetic_operator_parallel ( parent_-
chromosome, M, V, mu, mum, min_range, max_range, fbamodel, genes,
```

```
reaction_expression );
  else
   offspring_chromosome= genetic_operator (parent_chromosome, M,
V, mu, mum, min_range, max_range, fbamodel, genes, reaction_ex-
pression );
  end
   % We now create the intermediate population, namely the com-
bined population of parents and offspring of the current generation.
The population size is two times the initial population.  [
main_pop, temp ] = size (chromosome );
  [ offspring_pop, temp ] = size ( offspring_chromosome );
  clear temp
  % intermediate_chromosome is a concatenation of current popula-
tion and the offspring population.
  intermediate_chromosome (1: main_pop,: ) = chromosome;
    intermediate_chromosome (main_pop +  1 :  main_pop +  off-
spring_pop,1 : M + V) = offspring_chromosome;
  % Non – domination–sort of intermediate population
   % The intermediate population is sorted again based on non
–domination sort  before  the replacement operator is  per-
formed on the intermediate population.
   intermediate_chromosome = non_domination_sort_mod ( interme-
diate_chromosome, M, V);
  % Perform Selection
 % Once the intermediate population is sorted only the best solution
is selected based on it rank and crowding distance. Each front is fi
lled in ascending order until the addition of population size is
reached. The last front is included in the population based on the in-
dividuals with least crowding distance
  chromosome= replace_chromosome ( intermediate_chromosome, M, V,
pop );
%  chromosome= delete_redundant (chromosome, fbamodel );
  solution=[ ' solution ' num2str ( i ) ];
  save ( solution, 'chromosome ' );
end
```

After the optimization, *append_and_plot_solutions.m* computes the Pareto front. The file *non_dominated.mat* contains all the Pareto optimal points, while *others.mat* contains the dominated points. The first two columns of both output files contain the predicted values for the two objective functions. The 4th column is the number of population in which that solution has been found, while the 5th column is the position of that solution in that population.

Finally, *plot_and_export_color.m* plots the final version of the Pareto front. An example of a Pareto front obtained for 1,2-propanediol and biomass is shown in Fig. 2.

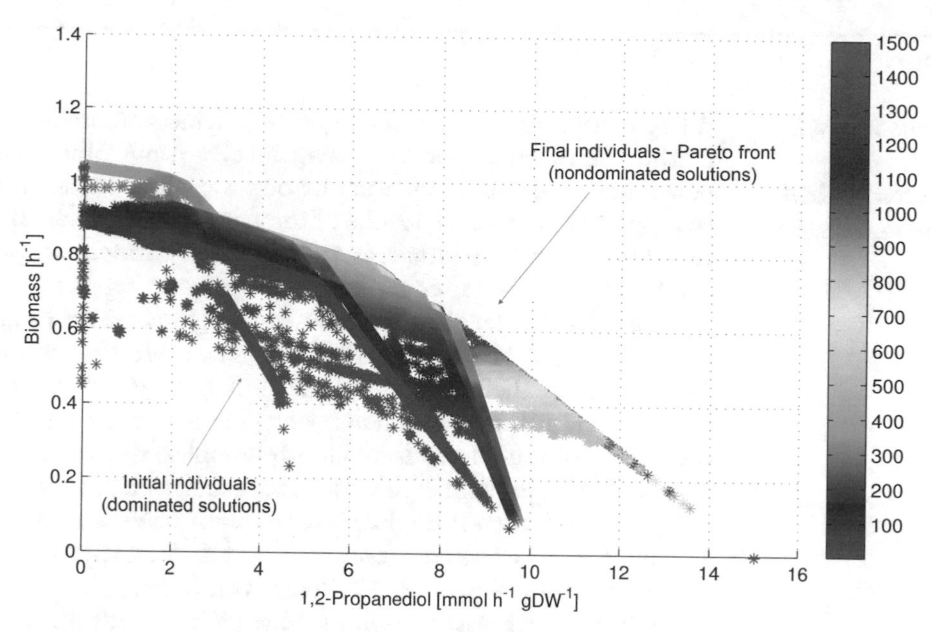

Fig. 2 Pareto front produced by METRADE when maximizing for 1,2-propanediol and biomass in *E. coli* (adapted from [24]). The trade-off sheds light on the regions where the bacterium operates. Solutions are asterisks denoted by progressively warmer colors according to the time step of the genetic algorithm in which they have been generated. Although discrete, the Pareto front can be approximated by a piecewise linear function

To validate with the proteomic dataset by Hui et al. [67] included in METRADE, load *iJO1366_Ecoli_ac_lactoseMedium.mat* and run *pareto_proteomic.m*. The dataset is composed of 14 expression profiles in different growth conditions with: (1) titrated catabolic flux through controlled inducible expression of the lacY gene; (2) titrated anabolic flux through controlled expression of GOGAT; (3) inhibition of protein synthesis with chloramphenicol. To run the pseudospectrum analysis on the growth conditions as detailed in METRADE, run *plot_eigenvector.m*. The code requires an updated *eigtoollib* toolbox.

There are numerous factors to consider when integrating such multi-omic datasets into metabolic models, many of which are discussed in the following perspective. In order to extract the most meaning from multi-omic models, systematic fusion of the multiple data types into a single, cohesive network is essential for measuring bacterial response at multiple omic levels. Whilst considering the structure of multi-omic data to be used for integration, the techniques used to integrate these data into the model are of equal importance.

4 Notes

4.1 Omic Network Integration in Metabolic Models: A (Critical) Perspective

A large proportion of the techniques which incorporate multi-omic methods into metabolic modelling involve using other omics to constrain the metabolome: they are one-way procedures. However, to properly interpret the results of these procedures, techniques are required which can integrate the different datasets to produce something that is easier to interpret than the separate datasets, and then provide feedback on how those separate datasets affected the integrated dataset. For example, in gene expression constrained FBA methods, the genome and metabolome are integrated into a combined model that enumerates feasible metabolic states. However, the resulting model is inherently complex: the actual relationship between a particular gene and a particular outcome can be hard to understand, even though it is deterministic in the model. Additionally, there is a lack of consensus about the best approach to take when estimating flux rates in different conditions.

Any approach based upon FBA has an inherently linear character: the outputs (fluxes of interest) are linearly dependent on some subset of the inputs (the bounds and objective function). The complexity comes from the fact that, while the output is only linearly dependent on a small fraction of the inputs in any given configuration, all of the other inputs affect which subset this is. This relationship is a piecewise linear equation with a large number of terms, but where most of the coefficients are zero in any given piece. This means that the challenge in understanding these models in an intuitive way is not so much in understanding how each variable affects the model, as when.

When looking to understand the effects of genetic or proteomic data on simulated phenotypes, naturally, the first place to start is with techniques used for understanding the effects of genetic or proteomic data on real phenotypes. With regression style techniques, it becomes clear that the reaction rates induced by FBA are often multimodal, since the best values are likely to be at either the maximum or minimum of the possible range. This multimodality violates normality assumptions, and it is therefore difficult to sensibly normalize such distributions. This issue has been demonstrated in a correlation analysis between expression levels and Pareto front position [22]. More specifically, even if there are several layers of normalization and the Pareto front acts to smooth flux values, there are two clear peaks in the distribution of flux rates. Figure 3 shows how this pattern occurs across a number of reactions in a knockout simulation.

The obvious choice when faced with distributions with several narrow peaks is to regard these values as fully categorical. However, this approach eventually ends up mired in overfitting; a good approach to combat this is to incorporate structure from the

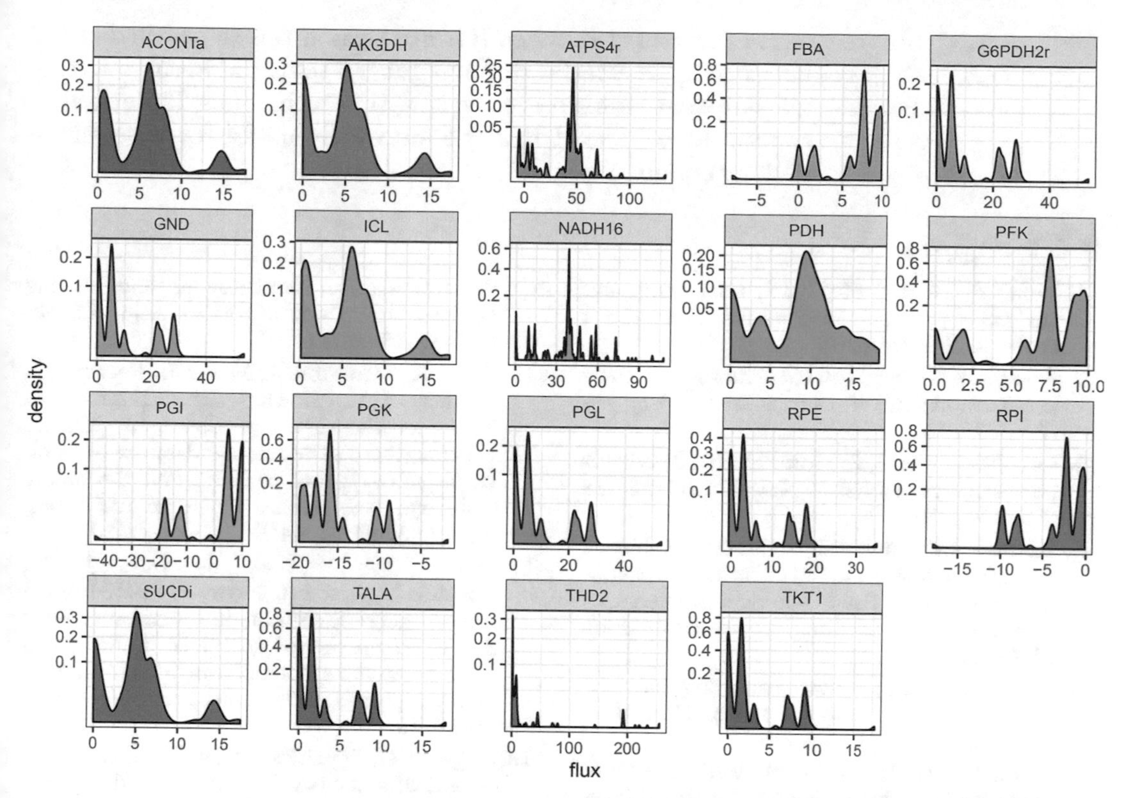

Fig. 3 Density plots of reaction fluxes for 19 reactions across 4560 simulations of one and two reaction knockouts on a model of *E. coli* core metabolism. Data was filtered to remove fluxes for reactions when they were knocked out, to remove simulations with low biomass flux, and to remove reactions with low variation. These reactions all show unsurprising peaks at a flux of 0, but more interestingly show a multimodal distribution, with a small number of other preferred values

network, e.g. by using a network regularized regression [68] technique to tie the values at nodes to those at nearby nodes.

Using multi-omic data, it is possible to go a step further than network regularised regression, and merge multiple omic layers together to form a single network where the value at each node incorporates information both from equivalent nodes in multiple layers, and also neighbors at each level. For instance, Similarity Network Fusion has been proposed to integrate information from a large number of simulations in genotype, metabolome, and phenotype domains [43]. This step was an unsupervised precursor to a supervised decision tree algorithm, which was used to explore the information that various reactions supply about phenotypes.

Ultimately, however, these techniques can only go so far. At their best, they identify under what circumstances certain variables are important, what their effects are, and how they can be clustered. This is a good start, but in order to understand why variables have the effects they do, a view on the network is required that is simple enough to understand but contains the detail necessary to elucidate a given type of regulation. It is not clear at this stage whether it is better to approach this through general statistical learning techniques or more domain-specific analytical techniques. Either way, it appears to be a goal that will be widely useful for the systems biology community.

References

1. Louca S, Doebeli M (2015) Calibration and analysis of genome-based models for microbial ecology. Elife 4:e08208
2. Nilsson A, Nielsen J (2016) Genome scale metabolic modeling of cancer. Metab Eng 43 (B):103–112
3. Orth JD, Thiele I, Palsson BØ (2010) What is flux balance analysis? Nat Biotechnol 28 (3):245–248
4. Zieliński ŁP, Smith AC, Smith AG, Robinson AJ (2016) Metabolic flexibility of mitochondrial respiratory chain disorders predicted by computer modelling. Mitochondrion 31:45–55
5. Palsson BØ (2011) Systems biology: simulation of dynamic network states. Cambridge University Press, Cambridge
6. Jayaraman A, Hahn J (2009) Methods in Bioengineering: systems analysis of biological networks. Artech House methods in bioengineering series. Artech House, Boston. https://books.google.co.uk/books?id=Haod3KR-tR8C
7. Bordbar A, Monk JM, King ZA, Palsson BO (2014) Constraint-based models predict metabolic and associated cellular functions. Nat Rev Genet 15(2):107–120
8. Burgard AP, Vaidyaraman S, Maranas CD (2001) Minimal reaction sets for Escherichia coli metabolism under different growth requirements and uptake environments. Biotechnol Prog 17(5):791–797
9. Mahadevan R, Schilling C (2003) The effects of alternate optimal solutions in constraint-based genome-scale metabolic models. Metab Eng 5(4):264–276
10. Mahadevan R, Edwards JS, Doyle FJ (2002) Dynamic flux balance analysis of diauxic growth in Escherichia coli. Biophys J 83 (3):1331–1340
11. Kanehisa M, Furumichi M, Tanabe M, Sato Y, Morishima K (2016) Kegg: new perspectives on genomes, pathways, diseases and drugs. Nucleic Acids Res. http://dx.doi.org/10.1093/nar/gkw1092
12. King ZA, Lu J, Dräger A, Miller P, Federowicz S, Lerman JA, Ebrahim A, Palsson BO, Lewis NE (2016) Bigg models: a platform for integrating, standardizing and sharing genome-scale models. Nucleic Acids Res 44 (D1):D515–D522
13. Caspi R, Billington R, Ferrer L, Foerster H, Fulcher CA, Keseler IM, Kothari A, Krummenacker M, Latendresse M, Mueller LA et al (2016) The metacyc database of metabolic pathways and enzymes and the biocyc collection of pathway/genome databases. Nucleic Acids Res 44(D1):D471–D480
14. Devoid S, Overbeek R, DeJongh M, Vonstein V, Best AA, Henry C (2013) Automated genome annotation and metabolic model reconstruction in the seed and model seed. In: Systems metabolic engineering: methods and protocols. Humana Press, New York, pp 17–45
15. Angione C, Pratanwanich N, Lió P (2015) A hybrid of metabolic flux analysis and bayesian factor modeling for multiomic temporal pathway activation. ACS Synth Biol 4(8):880–889
16. Lewis NE, Hixson KK, Conrad TM, Lerman JA, Charusanti P, Polpitiya AD, Adkins JN, Schramm G, Purvine SO, Lopez-Ferrer D et al (2010) Omic data from evolved E. coli are consistent with computed optimal growth from genome-scale models. Mol Syst Biol 6 (1):390
17. Palsson B (2015) Systems biology: constraint-based reconstruction and analysis. Cambridge University Press, Cambridge. https://books.google.co.uk/books?id=QNBpBgAAQBAJ
18. Voigt C (2011) Synthetic biology, part b: computer aided design and DNA assembly. Methods in enzymology. Elsevier Science, Amsterdam. https://books.google.co.uk/books?id=9uPvZWiabr4C
19. Deutscher D, Meilijson I, Schuster S, Ruppin E (2008) Can single knockouts accurately single out gene functions? BMC Syst Biol 2(1):50
20. Shlomi T, Berkman O, Ruppin E (2005) Regulatory on/off minimization of metabolic flux changes after genetic perturbations. Proc Natl Acad Sci USA 102(21):7695–7700
21. Megchelenbrink W, Katzir R, Lu X, Ruppin E, Notebaart RA (2015) Synthetic dosage lethality in the human metabolic network is highly predictive of tumor growth and cancer patient

survival. Proc Natl Acad Sci 112 (39):12217–12222

22. Conway M, Angione C, Liò P (2016) Iterative multi level calibration of metabolic networks. Curr Bioinforma 11(1):93–105
23. Costanza J, Carapezza G, Angione C, Lió P, Nicosia G (2012) Robust design of microbial strains. Bioinformatics 28(23):3097–3104
24. Angione C, Lió P (2015) Predictive analytics of environmental adaptability in multi-omic network models. Sci Rep 5:15147
25. Angione C, Costanza J, Carapezza G, Lió P, Nicosia G (2015) Multi-target analysis and design of mitochondrial metabolism. PloS One 10(9):e0133825
26. Xu G (2011) An iterative strategy for bi-objective optimization of metabolic pathways. In: 2011 fourth international joint conference on computational sciences and optimization
27. Sendin J, Exler O, Banga JR (2010) Multi-objective mixed integer strategy for the optimisation of biological networks. IET Syst Biol 4 (3):236–248
28. Deb K, Pratap A, Agarwal S, Meyarivan T (2002) A fast and elitist multiobjective genetic algorithm: NSGA-II. IEEE Trans Evol Comput 6(2):182–197. http://dx.doi.org/10.1109/4235.996017
29. Zitzler E, Thiele L (1999) Multiobjective evolutionary algorithms: a comparative case study and the strength Pareto approach. IEEE Trans Evol 3(4):257–271
30. Zhang Q, Li H (2007) Moea/d: a multiobjective evolutionary algorithm based on decomposition. IEEE Trans Evol Comput 11 (6):712–731
31. Nagrath D, Avila-Elchiver M, Berthiaume F, Tilles AW, Messac A, Yarmush ML (2010) Soft constraints-based multiobjective framework for flux balance analysis. Metab Eng 12 (5):429–445
32. Kelk SM, Olivier BG, Stougie L, Bruggeman FJ (2012) Optimal flux spaces of genome-scale stoichiometric models are determined by a few subnetworks. Sci Rep 2:580
33. Maarleveld TR, Wortel MT, Olivier BG, Teusink B, Bruggeman FJ (2015) Interplay between constraints, objectives, and optimality for genome-scale stoichiometric models. PLoS Comput Biol 11(4):e1004166
34. Oh YG, Lee DY, Lee SY, Park S (2009) Multiobjective flux balancing using the NISE method for metabolic network analysis. Biotechnol Prog 25(4):999–1008
35. Budinich M, Bourdon J, Larhlimi A, Eveillard D (2017) A multi-objective constraint-based approach for modeling genome-scale microbial ecosystems. PloS One 12(2):e0171744
36. John PCS, Crowley MF, Bomble YJ (2016) Efficient estimation of the maximum metabolic productivity of batch systems. arXiv preprint. arXiv:161001114
37. Machado D, Herrgård M (2014) Systematic evaluation of methods for integration of transcriptomic data into constraint-based models of metabolism. PLoS Comput Biol 10(4): e1003580
38. Covert MW, Schilling CH, Palsson B (2001) Regulation of gene expression in flux balance models of metabolism. J Theor Biol 213 (1):73–88
39. Chandrasekaran S, Price ND (2010) Probabilistic integrative modeling of genome-scale metabolic and regulatory networks in Escherichia coli and Mycobacterium tuberculosis. Proc Natl Acad Sci 107(41):17845–17850
40. Chandrasekaran S, Price ND (2013) Metabolic constraint-based refinement of transcriptional regulatory networks. PLoS Comput Biol 9 (12):e1003370
41. Rügen M, Bockmayr A, Steuer R (2015) Elucidating temporal resource allocation and diurnal dynamics in phototrophic metabolism using conditional FBA. Sci Rep 5, 15247
42. Reimers AM, Knoop H, Bockmayr A, Steuer R (2016) Evaluating the stoichiometric and energetic constraints of cyanobacterial diurnal growth. arXiv preprint. arXiv:161006859
43. Angione C, Conway M, Lió P (2016) Multiplex methods provide effective integration of multi-omic data in genome-scale models. BMC Bioinf 17(4):83
44. Ebrahim A, Brunk E, Tan J, O'Brien EJ, Kim D, Szubin R, Lerman JA, Lechner A, Sastry A, Bordbar A et al (2016) Multi-omic data integration enables discovery of hidden biological regularities. Nat Commun 71:13091
45. Segre D, Vitkup D, Church GM (2002) Analysis of optimality in natural and perturbed metabolic networks. Proc Natl Acad Sci 99 (23):15112–15117
46. Raval A, Ray A (2013) Introduction to biological networks. CRC Press, Boca Raton
47. Machado D, Costa RS, Ferreira EC, Rocha I, Tidor B (2012) Exploring the gap between dynamic and constraint-based models of metabolism. Metab Eng 14(2):112–119
48. Brochado AR, Andrejev S, Maranas CD, Patil KR (2012) Impact of stoichiometry representation on simulation of genotype-phenotype relationships in metabolic networks. PLoS Comput Biol 8(11):e1002758

49. Yizhak K, Benyamini T, Liebermeister W, Ruppin E, Shlomi T (2010) Integrating quantitative proteomics and metabolomics with a genome-scale metabolic network model. Bioinformatics 26(12):i255–i260
50. Monk JM, Koza A, Campodonico MA, Machado D, Seoane JM, Palsson BO, Herrgård MJ, Feist AM (2016) Multi-omics quantification of species variation of Escherichia coli links molecular features with strain phenotypes. Cell Syst 3(3):238–251
51. Vivek-Ananth R, Samal A (2016) Advances in the integration of transcriptional regulatory information into genome-scale metabolic models. Biosystems 147:1–10
52. Becker SA, Palsson BØ (2008) Context-specific metabolic networks are consistent with experiments. PLoS Comput Biol 4(5): e1000082
53. Colijn C, Brandes A, Zucker J, Lun DS, Weiner B, Farhat MR, Cheng TY, Moody DB, Murray M, Galagan JE (2009) Interpreting expression data with metabolic flux models: predicting Mycobacterium tuberculosis mycolic acid production. PLoS Comput Biol 5(8): e1000489
54. Jerby L, Shlomi T, Ruppin E (2010) Computational reconstruction of tissue-specific metabolic models: application to human liver metabolism. Mol Syst Biol 6(1):401
55. Vlassis N, Pacheco MP, Sauter T (2014) Fast reconstruction of compact context-specific metabolic network models. PLoS Comput Biol 10(1):e1003424
56. Wang Y, Eddy JA, Price ND (2012) Reconstruction of genome-scale metabolic models for 126 human tissues using mCADRE. BMC Syst Biol 6(1):153
57. Imam S, Schäuble S, Brooks AN, Baliga NS, Price ND (2015) Data-driven integration of genome-scale regulatory and metabolic network models. Front Microbiol 6:409
58. Carrera J, Estrela R, Luo J, Rai N, Tsoukalas A, Tagkopoulos I (2014) An integrative, multiscale, genome-wide model reveals the phenotypic landscape of Escherichia coli. Mol Syst Biol 10(7):735
59. Kashaf SS, Angione C, Lió P (2017) Making life difficult for clostridium difficile: augmenting the pathogen's metabolic model with transcriptomic and codon usage data for better therapeutic target characterization. BMC Syst Biol 11(1):25
60. Zitzler E, Brockhoff D, Thiele L (2007) The hypervolume indicator revisited: on the design of pareto-compliant indicators via weighted integration. In: International conference on evolutionary multi-criterion optimization. Springer, Berlin, pp 862–876
61. Ringnér M (2008) What is principal component analysis? Nat Biotechnol 26(3):303
62. Trefethen LN, Embree M (2005) Spectra and pseudospectra: the behavior of nonnormal matrices and operators. Princeton University Press, Princeton
63. Newman M (2013) Spectral community detection in sparse networks. arXiv preprint. arXiv:13086494
64. Schellenberger J, Que R, Fleming RM, Thiele I, Orth JD, Feist AM, Zielinski DC, Bordbar A, Lewis NE, Rahmanian S et al (2011) Quantitative prediction of cellular metabolism with constraint-based models: the cobra toolbox v2. 0. Nat Protoc 6 (9):1290–1307
65. Orth J, Conrad T, Na J, Lerman J, Nam H, Feist A, Palsson B (2011) A comprehensive genome-scale reconstruction of Escherichia coli metabolism. Mol Syst Biol 7(1):535
66. Faith JJ, Hayete B, Thaden JT, Mogno I, Wierzbowski J, Cottarel G, Kasif S, Collins JJ, Gardner TS (2007) Large-scale mapping and validation of Escherichia coli transcriptional regulation from a compendium of expression profiles. PLoS Biol 5(1):e8
67. Hui S, Silverman JM, Chen SS, Erickson DW, Basan M, Wang J, Hwa T, Williamson JR (2015) Quantitative proteomic analysis reveals a simple strategy of global resource allocation in bacteria. Mol Syst Biol 11(2):784
68. Li C, Li H (2008) Network-constrained regularization and variable selection for analysis of genomic data. Bioinformatics 24 (9):1175–1182

Index

Marco Fondi (ed.), *Metabolic Network Reconstruction and Modeling: Methods and Protocols*, Methods in Molecular Biology, vol. 1716, https://doi.org/10.1007/978-1-4939-7528-0, © Springer Science+Business Media, LLC 2018